COMBINATORICS AND GRAPH THEORY

Third Edition

Dr. S. B. SINGH

Deptt. of Maths Statistic & Computer Science,
G.B. Pant University of Agriculture & Technology,
Pantnagar

KHANNA BOOK PUBLISHING CO. (P) LTD.

Publisher of Engineering and Computer Books

4C/4344, Ansari Road, Darya Ganj, New Delhi-110002
Phone : 011-23244447-48 **Mobile:** +91-9910909320
E-mail : contact@khannabooks.com
Website : www.khannabooks.com

Dear Readers,

To prevent the piracy, this book is secured with HIGH SECURITY HOLOGRAM on the front title cover. In case you don't find the hologram on the front cover title, please write us to at contact@khannabooks.com or whatsapp us at +91-99109 09320 and avail special gift voucher for yourself.

Specimen of Hologram on front Cover title:

Moreover, there is a SPECIAL DISCOUNT COUPON for you with EVERY HOLOGRAM.

How to avail this SPECIAL DISCOUNT:

Step 1: Scratch the hologram
Step 2: Under the scratch area, your "coupon code" is available
Step 3: Logon to www.khannabooks.com
Step 4: Use your "coupon code" in the shopping cart and get your copy at a special discount
Step 5: Enjoy your reading!

Price: ₹ 425.00

ISBN: 978-81-90645-10-2

Combinatorics and Graph Theory
by S.B. Singh

First Edition: 2002
Third Edition: 2016
Revised & Reprint: 2022

To view complete list of books,
Please scan the QR Code:

Published by:
Khanna Book Publishing Co. (P) Ltd.
CIN: U22110DL1998PTC095547

Visit us at: **www.khannabooks.com**
Write us at: **contact@khannabooks.com**

Printed in India by: City Press, Delhi.

Dedicated To My Father

Shri Bhairo Nath Singh

PREFACE TO THIRD EDITION

The second edition of this book has been well received by teachers and students. The author thanks all of them for their appreciation and suggestions. The present edition been thoroughly revised, enlarged and various suggestions of the teachers have incorporated. A number of sections have been written, rewritten and various additions, deletions and refinements have been made.

I hope the book in the present form will be found more useful.

The authors take the opportunity to thank Dr. Jai Kishore, Dr. Ekata, Mrs. Geetam Sharma and many other faculty members from all over the country who have suggested in many ways to improve the old edition of the book for their suggestions to improve the book. I want to express my gratitude and appreciation to my wife Mrs. Neelam Singh and my sons Pratham and Saksham for their constant support, patience and encouragement throughout the time of revising of this book. The author is thankful to Mr. Puneet Khanna of Ms. Khanna Publishing Company (Pvt.) Ltd. and printer for publishing this work in present nice form. Thanks are also due to Mrs. Shikha for her help in preparing the book.

Suggestions for improvement of the book will be gratefully acknowledged.

S.B. Singh

PREFACE TO FIRST EDITION

With the advancement of computer age, the study of Combinatorics and Graph Theory becomes increasingly important. The book has been written to fulfill the growing needs of students of M.C.A., B.Tech., M.Sc. of various Indian universities. The book is primarily written on the basis of the syllabus of U.P. Tech. Univ., Lucknow. A special care has been taken to present the book with an approach to explain the concepts in lucid and communicable language solution that it should also be helpful to the students of M.C.A. from arts and commerce background. The book is divided into nine chapters. The chapter 2 on graph theory and chapter 7 on directed graphs are related that's whey many topics which are interrelated are taken in chapter 2. Numerous worked out examples have been given which will help the readers in understanding various concepts easily. A problem set is given at the end of each chapter so that students may try the problems and access their confidence of answering the problems after studying whole topics of the book. A miscellaneous examples set which contains examples related to all chapters of the book has been taken at last.

The author is indebted and obliged to Dr. P.P. Gupta Director, Dr. K.N. Modi Inst. of Engg. & Tech. Modinagar, the faculty members, friends, staff and students for the suggestion and help in preparing this book. Finally, I want to express my love and appreciation to my wife Mrs. Neelam Singh for her constant support patience and encouragement throughout the time of writing of this book. The author is thankful to Ms. Khanna Book Publishing Company (PVT) Ltd. and printer for publishing this work in present nice form.

Suggestion for improvement of the book will be gratefully acknowledge.

Dr. S.B. Singh

CONTENTS

1 COMBINATORICS

1.0 INTRODUCTION

In a department of a certain engineering college, there are 60 students. Director of the college wants to nominate three students to the department post of president, vice president and secretary. The Director first selects a group of three students, say Pratham, Mansi and Saksham. Now, he wants to nominate them for the three department posts. He can do this in different ways as under:

President	Vice-President	Secretary
Pratham	Mansi	Saksham
Pratham	Saksham	Mansi
Mansi	Pratham	Saksham
Mansi	Saksham	Pratham
Saksham	Pratham	Mansi
Saksham	Mansi	Pratham

Here we can see that there are six possible ways in which Director can nominate after selecting a group of three students out of 60 students. We realize that the Director can face following two problems:

(a) Selecting a group of three students in all possible ways.

(b) Naming the selected students for different posts in all possible ways.

In Mathematics both the above problems are discussed under the heading Combinatorial Mathematics or Combinatorics, which deals with the problem of counting or enumeration. In particular number of ways of **combinations** deals with problems of selection of object and **permutations** deal with problems of number of ways of arrangements of objects.

1.1 FUNDAMENTAL PRINCIPLE OF COUNTING

In the study of combinatorial mathematics, two types of problems related to addition and multiplication are of great importance. The problems are respectively based on addition and multiplication principles, which are stated below:

(a) ADDITION PRINCIPLE

If an event can happen in m possible ways and another event can occur in n possible ways and both the events are mutually exclusive (i.e. only one event can occur at a time either first or second), then either of the two events can be occurred in $(m + n)$ ways. For example, suppose we want to find two digits even number that can be formed with the help of digits 1, 2, 3, 4 and 5. The possibility that a number of two digits will be even if the number formed has either 2 or 4 in the unit's place. The possibilities that the number ending with 2 are 12, 32, 42 and 52. Similarly the possibilities that the number ending with 4 are 14, 24, 34 and 54. Now the total possibilities that the two digits number will be even are 4 + 4 = 8 (4 numbers end with 2 and 4 numbers end with 4).

Example 1.1

There are four books on Discrete Mathematics and six books on Numerical Analysis in a certain show room. In how many ways can Ramesh buy a book either of Discrete Mathematics or of Numerical Analysis?

Solution

Book on Discrete Mathematics can be purchased in four ways and Numerical Analysis can be purchased in six ways. Then Ramesh can purchase either a Discrete Mathematics or a Numerical Analysis book in 4 + 6 = 10 ways.

(a) MULTIPLICATION PRINCIPLE

If an event can occur in m ways and corresponding to each way of occurring this event another event can occur in n ways independent of the first. Then the number of ways of happening both the events simultaneously is $m \times n$.

Example 1.2

Suppose there are 8 different courses offered in the morning and six different courses offered in the afternoon. Then there will be 7×5 choices for students who want to enroll one course in the morning and one in the afternoon.

Example 1.3

A carpenter has twelve patterns of chairs and five patterns of tables. In how many ways can he make a pair of chair and table?

Solution

For each ways of making a chair, a table can be made in five patterns, so for twelve ways of making chairs, the total number of ways of making a pair of chair and table by the multiplication principle are $12 \times 5 = 60$.

Example 1.4

In how many ways can we get a sum of 4 or of 8 when two distinguishable dice (say one is brown and the other is blue) are rolled? How many ways can we get an even sum?

Solution

Let us label the outcome of a 1 in brown dice and a 3 on the blue dice as the ordered pair (1, 3). The other possibilities of getting sum 4 are (2, 2) and (3, 1). Hence there are 3 ways to get sum 4.

Likewise, we get a sum 8 by the outcomes (2, 6), (3, 5), (4, 4), (5, 3) and (6, 2). Therefore, total number of possibilities in which sum is either 4 or 8 is 3 + 5 = 8.

Now, to get an even sum, the possibilities are 2, 4, 6, 8, 10 or 12. The possible ways to get the sum 2 is one, to get the sum 4 is three, to get the sum 6 is five, to get the sum 8 is five, to get the sum 10 is three and to get the sum 12 is one. Therefore, there are 1 + 3 + 5 + 5 + 3 + 1 = 18 ways to get an even sum.

NOTE

When the dice are indistinguishable, then to get an even sum is possible in $1 + 2 + 3 + 3 + 2 + 1 = 12$ ways (In this case outcomes of the type (6, 2) and (2, 6) will be considered same).

Example 1.5

In how many ways 100 voters can vote for three candidates standing for the election of the post of president of their association?

Solution

Every member of the association can vote to either of the three candidates, thus by addition principle the number of ways that a member can vote is 3, hence every member can vote in 3 ways. Thus by multiplication principle 100 members can vote in $3 \times 3 \times 3 \times \ldots \times 3 \times 3$ (100 times) i.e. 3^{100} ways.

Example 1.6

Suppose that license plates of a certain state require 3 English letters followed by 4 digits.

(a) How many different plates can be manufactured if repetition of letters and digits are allowed?

(b) How many plates are possible if only the letters can be repeated?

(c) How many plates are possible if only the digits can be repeated?

(d) How many plates are possible if no repetitions are allowed at all?

Solution

(a) Since there are 26 possibilities for each of the 3 letters, thus it is possible in 26^3 ways and each of the 4 digits are possible in 10 ways. Thus the total possibilities for this case is 10^4. Therefore $26^3.10^4$ plates can be manufactured if repetition of letters and digits are allowed.

(b) $26^3 . 10.9.8.7$ ways.

(c) $26.25.24. 10^4$ ways.

(d) 26.25.24.10.9.8.7 ways.

Example 1.7

(i) How many numbers are there between 100 and 1000 such that 7 is in the unit's place?

(ii) How many numbers are there between 100 and 1000 such that at least one of their digits is 7?

(iii) How many of them have exactly one of their digits as 7?

Solution

(i) As all these numbers have three digits. We want 7 in the unit's place. The middle digit can be any of the 10 integers between 0 and 9. The digit in hundred's place can be any one of the 9 numbers between 1 and 9. Therefore, by the multiplication principle, there are $10 \times 9 = 90$ numbers.

(ii) The numbers between 100 and 1000 such that at least one of their digits is 7 can be classified into three types. If the 7 is at the unit's place then possible ways as seen in (i) is 90. If the 7 is at middle place, but not in the unit's place, then possible ways are $9 \times 9 = 81$. If the 7 is at the hundred's place only, there are $9 \times 9 = 81$ possible ways.

Totally, therefore, there are $90 + 81 + 81 = 252$ numbers are required.

(iii) Here again in this case there are three kinds:

(a) The numbers that have 7 in the unit's place but not in any other place.

(b) The numbers that have 7 in the ten's place but not in any other place

(c) The numbers that have 7 in the hundred's place but not in any other place.
The required number is 72 + 72 + 81 = 225.

1.2 FACTORIALS

Consider the product 1, 1.2, 1.2.3, 1.2.3.4, so on, we denote them respectively by 1!, 2!, 3!, 4! so, on.

For each positive integer we define

$n! = n(n-1)(n-2) \ldots 3.2.1 =$ the product of all integers from 1 to n.

$n!$ is read as "***n* factorial**". It is also denoted by the notation $\lfloor n$.

Also define $0! = 1$. Note that $1! = 1$

In general $\quad n! = n[(n-1)!]$

In particular, $\quad 4! = 4\,(3!)$

$3! = 3\,(2!)$.

NOTE

The factorial of proper fractions and negative integers are not defined. The factorial is defined only for whole numbers.

Example 1.8

Write each of the following in expanded form and compute its value

(i) $\frac{9!}{5!}$ (ii) $\frac{10}{5!.5!}$.

Solution

(i) $\frac{9!}{5!} = \frac{9.8.7.6.5.4.3.2.1}{5.4.3.2.1} = 9.8.7.6 = 3024$

(ii) $\frac{10!}{5!5!} = \frac{10.9.8.7.6.5.4.3.2.1}{(5.4.3.2.1)(5.4.3.2.1)} = \frac{10.9.8.7.6}{5.4.3.2.1} = 252$

Example 1.9

Prove that $n!\,(n+2) = n! + (n+1)!$.

Solution

$$
\begin{aligned}
\text{R.H.S.} &= n! + (n+1)! \\
&= n! + n!(n+1) \\
&= n!\,[1 + (n+1)] \\
&= n!(n+2) \\
&= \text{L.H.S.}
\end{aligned}
$$

Example 1.10

If $(n+2)! = 2550.\,(n!)$, find n.

Solution

We know

$$
\begin{aligned}
(n+2)! &= (n+2)\,(n+1).n \ldots 3.2.1 \\
&= (n+2)\,(n+1)n!
\end{aligned}
$$

Its given $\quad (n+2)! = 2550\,n!$

i.e. $\quad (n+2)\,(n+1)\,n! = 2550\,n!$

or $\quad (n+2)(n+1) = 2550$

or $\quad n^2 + 3n - 2548 = 0.$

Which is a quadratic equation. Its roots are

$$n = \frac{-3 \pm \sqrt{9 + 4 \times 2548}}{2}$$

$$= 49 \text{ or } -52.$$

Out of these two roots, the second root is ruled out because factorials of negative integers are not defined.

EXERCISE 1.1

1. How many two digit number is there in the set of natural numbers?
2. A lady wants to select one cotton saree and one polyester saree from a textile shop. If there are 10 cotton varieties and 12 polyester varieties, in how many ways can she choose the two sarees?
3. How many possible telephone numbers are there when there are seven digits, the first two of which are between 2 and 9 inclusive, the third digit between 1 and 9 inclusive, and each of the remaining may be between 0 and 9 inclusive?
4. (i) How many ways are there to select 2 cards (without replacement) from a deck of 52?

   How many ways are there to select the two cards such that:

   (ii) The first card is an ace and the second card is a king?
   (iii) The first card is an ace and the second is not a king?
   (iv) The first card is a heart and the second is a club?
   (v) The first card is a heart and the second is a king?
   (vi) The first card is a heart and the second is not a king?
   (vii) Neither card is an ace?
   (viii) At least one of the cards drawn is an ace?
5. In a crossword puzzle there are two solutions to each of the three given places and three solutions for four other places. How many different solutions could be sent in to win a prize?
6. There are five different roads from city A to city B, three different roads from city B to city C, and three different roads that go directly from A to C.

   (i) How many different ways are there to go from A to C via B?
   (ii) How many different ways are there from A to C altogether?
   (iii) How many different ways are there from A to C and then back to A?
   (iv) How many different trips are there from A to C and back again to A that visit B both going and coming?
   (v) How many different trips are there that go from A to C via B and return directly from C to A?
   (vi) How many different trips are there that go directly from A to C and return to A via B?
   (vii) How many different trips are there from A to C and back to A that visit B at least once?
   (viii) Suppose that once a road is used it is closed and cannot be used again. Then how many different trips are there from A to C via B and back to A again via B?

(ix) Using the assumption in (viii) how many different trips are there from A to C and back to A again?

7. The students in a class are seated according to their marks in the previous examination. Once it so happens that four of these students get equal marks and therefore the same rank. To decide their seating arrangement, the teacher wants to write down all possible arrangements, one in each of separate bits of paper, in order to choose one of these by lots. How many bits of paper are required?

8. How many 5-letter words are there where the first and last letters
   (i) are consonants ?
   (ii) are vowels ?
   (iii) are vowels and middle letters are consonants?
   (iv) How many 5-letter words are there if vowels can only appear (if at all) as the first or last letter?

9. How many different four letter words can be formed out of the word 'LOGARITHMS' if repetition of letters is not allowed. (The word may or may not have a meaning)

10. In a certain programming language, an identifier is a sequence of a certain number of characters where the first character must be a letter of the English alphabet and remaining characters may be either a letter or a digit.
   (i) How many identifiers are there of length 5?
   (ii) In particular, in some implementations of Pascal an identifier is a sequence of from 1 upto 8 characters with the above restrictions. How many Pascal identifiers are there?

11. Each section in the first year of M.C.A. course has exactly 40 students. If there are 4 sections, in how many ways can a set of 4 students representatives be selected, one from each section?

12. A chain letter is sent to 10 people in the first week of the year. The next week each person who received a letter sends letter to 10 new people, and so on.
   (i) How many people have received letters after 10 weeks?
   (ii) At the end of the year?

13. A mother distributes 5 different apples among 8 children.
   (i) How many ways can this be done if each child receives at most one apple?
   (ii) How many ways can this be done if there is no restriction on the number of apples a child can receive?

14. If $(n+1)! = 12\ [(n-1)!]$, find n.

15. Prove that $\dfrac{2n!}{n!} = 2^n\left(1.3.5.....\overline{2n-1}\right)$

16. Find the total number of positive integers that can be formed from the digits 1, 2, 3, 4 and 5 if no digit is repeated in any one integer.

17. Find the sum of all 4-digit numbers that can be obtained by using (without repetition) the digits 2, 3, 5 and 7.

18. Prove that

$$(n+1)\left[n.n! + (2n-1)\overline{n-1}! + (n-1)\left(\overline{n-2}\right)!\right] = \overline{n+2}!.$$

19. Compute (i) $\dfrac{15!}{9!.6!}$ (ii) $\dfrac{17!}{12!}$.

ANSWERS

1. 90   2. 120   3. $8 \cdot 8 \cdot 9 \cdot 10^4$

4. (i) $52 \cdot 51$ (ii) $4 \cdot 4$ (iii) $4 \cdot 47$
   (v) $1 \cdot 48 + 12 \cdot 47$ (vi) $48 \cdot 47$ (vii) $52 \cdot 51 - 48 \cdot 47$

5. 648   6. (i) $5 \cdot 3$ (ii) $15 + 3$
   (iii) 18^2 (iv) 15^2 (v) $15 \cdot 3$
   (vi) $3 \cdot 15$ (vii) $15 \cdot 3 + 3 \cdot 15 + 15^2 = 18^2 - 3^2 = 15 \cdot 18 + 3 \cdot 15$
   (viii) 15.8 (ix) $15 \cdot 8 + 15 \cdot 3 + 3 \cdot 15 + 3 \cdot 2$ 7. 24

8. (i) $21^2 \cdot 26^3$ (ii) $5^2 \cdot 26^3$
   (iii) $5^2 \cdot 21^3$
   (iv) $26^2 \cdot 21^3 = 21^5 + 2 \cdot 5 \cdot 21^4 + 5^2 \cdot 21^3$ 9. 5040.

10. (i) In (i) identifiers are of length 5 and the first character can be any letter out of 26 letters, while the remaining 4 characters may be any of the 26 letters *A* through *Z*, or 10 digits 0 through 9. Thus, each of the remaining 4 characters can be filled in 36 ways.
Therefore, the length of identifiers of length 5 are $26 \cdot 36^4$.
   (ii) $26 + 26 \cdot 36 + 26 \cdot 36^2 + \ldots + 26 \cdot 36^7$, Pascal identifiers.

11. 2560000

12. (i) $10 + 10^2 + 10^3 + \ldots + 10^{10}$
    (ii) $10 + 10^2 + 10^3 + \ldots + 10^{52}$

14. 3

16. $5 + 5 \cdot 4 + 5 \cdot 4 \cdot 3 + 5 \cdot 4 \cdot 3 \cdot 2 + 5 \cdot 4 \cdot 3 \cdot 2 \cdot 1$

17. There are 4! such numbers. Every one of the 4 positions each digit occurs 6 times. The sum of the digits is 17. Hence the sum of these 24 numbers is $(6) \cdot (17) \cdot (1111)$.

19. (i) 5005   (ii) 742560

1.3 PERMUTATIONS

The different number of ways in which three books (i) Hindi (ii) English and (iii) Bengali can be arranged on a shelf are

(i)	Hindi	English	Bengali
(ii)	Hindi	Bengali	English
(iii)	Bengali	Hindi	English
(iv)	Bengali	English	Hindi
(v)	English	Bengali	Hindi
(vi)	English	Hindi	Bengali

The above six ways are total number of possible ways of arranging the books on shelf in which no book is repeated, such types of arrangements are called **permutations**. We define the word permutation as under:

The ways in which a set of objects can be arranged without duplication of any of the object which can be made by taking some or all of the given number of distinct objects is called **permutation**. The total number of such arrangements is called the **number of permutation**.

The permutation of n objects taken r at a time is denoted by $P(n, r)$ or ${}^nP_r$.

THEOREM 1.1

The number of possible permutations of r distinct elements taken from a set of n elements is given by

$$P(n,r) = n(n-1)(n-2)...(n-r+1) = \frac{n!}{(n-r)!}$$

PROOF

While arranging n objects taken r at a time, the position can be filled by any one of the n objects. This can be done in n ways. Any one of the remaining n -1 objects can fill the second position. This can be done in n -1 ways and so on. Finally, the rth position can be filled in $n - r + 1$ ways, because after filling up the previous r - 1 positions, exactly $n - (r-1) = n - r + 1$ objects remain. After filling up the rth position, we are through in arranging r objects. By the fundamental principle of multiplication, the total number of ways of arranging r objects from n objects is $n(n-1)... (n-r+1)$. Thus

$$P(n,r) = n(n-1)\ .\ .\ .\ (n-r+1)$$

$$= \frac{n(n-1)(n-2)....(n-r+1)(n-r)!}{(n-r)!}$$

$$P(n,r) = \frac{n!}{(n-r)!}$$

[Here it is to be noted that $1 \le r \le n$].

COROLLARY

Out of n objects, if we take all at a time then the number of permutations will be

$$P(n, n) = n(n-1)(n-2) \ldots 3.2.1 = n!$$

NOTE

In Theorem 1.1 we have got the number of permutations i.e. the total number of arrangements without repetition. In case an object is repeated certain number of times, then we call the total number of ways as arrangements and not permutations. In the following theorem we will discuss the total number of arrangements when certain object is repeated any number of times.

THEOREM 1.2

Prove that the number of arrangements of n different objects taken r at a time, when each object may be repeated any number of times in one arrangement, is n^r (here r can be greater than n also).

PROOF

Here we want to fill up r places with n different ways. Since any one of n objects can be placed at first place, hence first place can be filled up with n different ways. Now to fill up the second place, again we have n objects as the object placed at first place can be repeated. Thus there are altogether $n.\ n = n^2$ ways of filling first two places. The third place can again be filled in n ways and each of fillings in the third place may be associated with each of the n^2 ways of filling in the first two places. Therefore, first three places can be filled in n^3 ways. Generalizing the result we find that the filling in r places is n^r. Hence the required number of arrangements is n^r.

Example 1.11

Find the value of $P(4, 3)$.

Solution

We know $P(n, r) = n(n-1) \ldots (n-r+1)$

Here $n = 4,\ r = 3,$

$\therefore n - r + 1 = 4 - 3 + 1 = 2.$

Hence $P(n, r) = P(4, 3) = 4 \cdot 3 \cdot 2 = 24$

Aliter: We know

$$P(n,r) = \frac{n!}{(n-r)!}$$

Here $n = 4,\ r = 3$

$$\therefore P(4,3) = \frac{4!}{(4-3)!} = \frac{4!}{1!} = 4 \times 3 \times 2 = 24$$

Example 1.12

How many five different letter words can be formed out of the word " LOGARITHMS"?

Solution

The total letters in the word "LOGARITHMS" is 10 i.e. $n = 10$ and we have to form five letter words thus $r = 5$.

The required number of words is

$$P(10,5) = \frac{10!}{(10-5)!} = \frac{10!}{5!}$$

$$= \frac{10 \times 9 \times 8 \times 7 \times 6 \times 5!}{5!}$$

$$= 10 \times 9 \times 8 \times 7 \times 6 = 30240$$

Example 1.13

How many different license plates are there that involve 1, 2 or 3 letters followed by 4 digits?

Solution

Out of total 26 letters, 3 letter words can be formed with no repeated letters in $P(26, 3)$ ways. Complete yourself.

Example 1.14

Find n if $P(n, 4) = 42 . P(n, 2)$.

Solution

$$P(n, 4) = n(n-1)(n-2)(n-3) \quad \ldots (1)$$

And $$P(n, 2) = n(n-1) \quad \ldots (2)$$

Given $$P(n, 4) = 42 \cdot P(n, 2) \quad \ldots (3)$$

substituting values (i) and (ii) in (iii), we get

$$n(n-1)(n-2)(n-3) = 42\, n(n-1)$$

or $$(n-2)(n-3) = 42$$

or $n^2 - 5n + 6 = 42$

or $n^2 - 5n - 36 = 0$

or $(n - 9)(n + 4) = 0$

or $n = 9$ or $n = -4$,

Out of these two answers, $n = -4$ is not possible because for negative numbers permutation is not defined, hence $n = 9$ is the required answer.

Example 1.15

It is required to sit 5 men and 4 women in a row so that the women occupy the even places. How many such arrangements are possible?

Solution

Here, there are exactly four even places and exactly four women to occupy them. Hence, these even positions can be filled in $P(4, 4) = 4!$ ways. Now we are left with 5 positions and we have exactly 5 men. These 5 positions can be filled in $P(5, 5) = 5!$ ways. Therefore, the required number of ways by the fundamental principle of multiplication is

$$P(4, 4) \cdot P(5, 5) = 4!.5! = 24.120 = 2880.$$

Example 1.16

In how many ways can the letters of English alphabet be arranged so that there are exactly 5 letters between the letters *a* and *b*?

Solution

5 letters between *a* and *b* can be arranged in $P(24, 5)$ ways. *a* and *b* can be arranged in two ways and then 20! ways to arrange any 7 letter word treated as one unit along with the remaining 19 letters. Hence the total number of possible ways are $P(24, 5) \cdot (20!) \cdot 2$.

Example 1.17

There are four prizes to be distributed among 6 students. In how many ways can it be done when (i) no student gets more than one prize; (ii) there is no restriction as to the number of prizes any student gets; (iii) no student gets all the prizes.

Solution

(i) Number of students = 6

Number of prizes = 4

When no student to get more than one prize. This can be done in

$$P(6,4) = \frac{6!}{(6-4)!} = \frac{6!}{2!} = 6 \times 5 \times 4 \times 3 = 360 \text{ ways}$$

(ii) As there is no restriction to the number of prizes to be won by the students, the first prize can be distributed in 6 ways. Corresponding to each way of distribution of the prize there are again 6 ways of distributing the second prize since it can be awarded to the same student who has won the first prize. Therefore, the first two prizes can be distributed in $6 \times 6 = 6^2$ ways. Similarly each of the third and fourth prizes can be awarded in 6 different ways.

$\therefore$ The required number $= 6 \times 6 \times 6 \times 6 = 6^4 = 1296$

(iii) Total number of ways a prize can be awarded = 6 different ways. Therefore, the total number of ways in which no student gets all the prizes $6^4 - 6 = 1290$.

Example 1.18

How many three-digit numbers are there with no digit repeated?

Solution

Since total number of digits are 10. The number of possible ways that a three-digit number can be formed is *P* (10, 3). But to form the three digits number 0 should not be in the left-most position. For such position total number of permutation is *P* (9, 2). Hence, *P* (10, 3) – *P* (9, 2) is the required number of ways a three-digit number can be formed with no digit repeated.

THEOREM 1.3

Prove that $P(n, r) = P(n-1, r) + r \cdot P(n-1, r-1)$.

PROOF

$$\text{R.H.S} = P(n\text{-}1, r) + r \cdot P(n\text{-}1, r\text{-}1)$$

$$= \frac{(n-1)!}{(n-1-r)!} + \frac{r \cdot (n-1)!}{(n-1-r-1)!}$$

$$= (n-1)!\left\{\frac{1}{(n-r-1)!} + \frac{r}{(n-r)!}\right\}$$

$$= (n-1)!\left\{\frac{n-r+r}{(n-r)!}\right\} = \frac{(n-1)!\,n}{(n-r)!}$$

$$= \frac{n!}{(n-r)!} = P(n,r) = \text{L.H.S.}$$

1.3-1 CIRCULAR PERMUTATIONS

The permutations we have considered so far are more properly called **linear permutations** for the objects are being arranged in a line. If instead of arranging objects in a line, we arrange them in a cycle, then the permutation is said to be **circular permutations**.

Let us arrange 5 integers 1, 2, 3, 4, and 5 around a circular table. The five possible arrangements of the integers are 12345, 23451, 34512, 45123, and 51234.

Here we can see that all above five arrangements are different if they are to be placed in a straight line. But if the arrangements are made in the cyclic order then they are not different as seen from the above Fig.1.1. We can note that in a circular arrangement, there is neither a beginning nor an

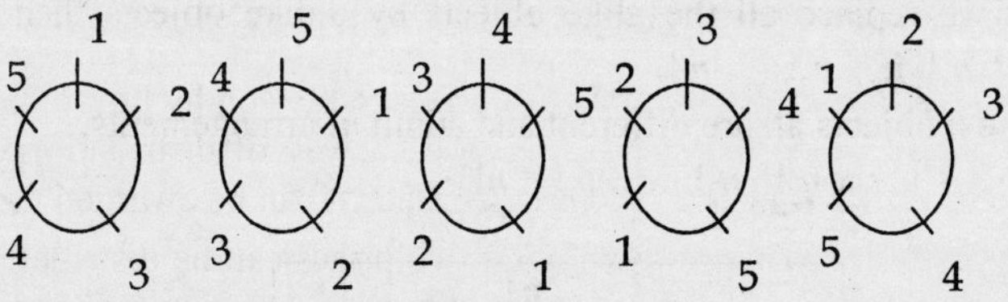

Fig. 1.1

end. In this situation, we have to check the relative position of the given object. The circular permutations are different only when the relative order of the object is changed, otherwise these are the same.

As the circular position depends upon the relative position of the objects, we fix the position of one object and then arrange the remaining $(n - 1)$ objects in all possible ways. Thus the number of circular arrangements of five objects 1, 2, 3, 4, 5 will be $(5 - 1)! = 4! = 24$.

Example 1.19

In how many ways can a party of seven gentlemen and six ladies be seated at a round table so that no two ladies are seated together?

Solution

As there are 7 gentlemen and they can be arranged around the table in (7-1)! ways. Now when the gentlemen have been seated, there are seven places for ladies one each between two gentlemen. Hence the ladies can be seated in $P(7, 6)$ ways. Therefore, the required number of ways this arrangements can be made by the fundamental principle of multiplication is $P(7,6) \times 6! = 362880$.

1.3-2 Permutations with Repetitions

THEOREM 1.4

The number of arrangements of a multi-set objects some of which are alike say n_1 are alike of the first kind, n_2 are alike of the second kind, ... n_r are alike the rth kind, is given by

$$P(n; n_1, n_2, \ldots\ldots, n_r) = \frac{n!}{n_1! n_2! \ldots\ldots n_r!}$$

Where $n_1 + n_2 + n_3 + \ldots + n_r \leq n.$

PROOF

Let n objects be denoted by n letters, n_1 of them being alike each be denoted by p, n_2 being alike each be denoted by q and so on.

Let t be the required number of arrangements. Let us replace one of these t arrangements by n_1 alike letters 'p' by unlike letters, say $p_1, p_2, p_3, \ldots\ldots, P_{n_1}$ different from any of the rest and arrange these n_1 new letters amongst themselves without changing the positions of the remaining letters. Then, we have $n_1!$ new arrangements and if we change each of t arrangements, the total number of arrangements will be $t \cdot n_1!$.

Similarly for n_2 alike objects, 'q' be replaced by n_2 new letters different from each other and different from the rest, the total number of arrangements would be $t \cdot n_1! \cdot n_2$.

Similarly, if we replace all the alike objects by unlike objects then the total number of arrangements will be $t \cdot n_1! \cdot n_2! \cdot n_3! \ldots. n_r!$.

But here these n objects all are different and admit $n!$ arrangements.

Hence $x \cdot n_1! \cdot n_2! \ldots.. n_r! = n!$.

or

$$\boxed{x = \frac{n!}{n_1! \cdot n_2! \cdot n_3! \ldots. n_r!}.}$$

Example 1.20

How many seven-letter words can be formed using the letters of the word "BENZENE"?

Solution

Here total number of letters are seven of which three are alike (the three Es) and two are alike (the two *N*s). Thus the total number of such words is

$$P(7,\ 3,2) = \frac{7!}{3!.2!} = 420$$

Example 1.21

How many number greater than one million can be formed with the digits 4, 6, 6, 0, 3, 6, 3?

Solution

For numbers to be greater than one million, it must consist of 7 or more digits. Here, we are given with 7 digits in all. Out of these 7 digits, there are 3 sixes, 2 threes and the rest all digits are different. Therefore,

$$\text{Total number of possible numbers} = \frac{7!}{2!\cdot 3!} = 420$$

Out of these numbers, some begin with zero and are thus less than one million, which must be rejected. The members beginning with zero

$$= \frac{6!}{2!\cdot 3!} = 60$$

$\therefore$ The required numbers greater than one million

$$= 420 - 60 = 360.$$

Example 1.22

How many different signals, each consisting of eight flags hung in a vertical line, can be formed from a set of four indistinguishable red flags, three indistinguishable white flags, and a blue flag?

Solution

As there are four indistinguishable red flags, three indistinguishable white flags. Therefore, the required numbers of permutations of eight objects are

$$P(8;4,3) = \frac{8!}{4!\cdot 3!} = 280.$$

Example 1.23

In how many ways can four mathematics books, three history books, three chemistry books, and two sociology books be arranged on a shelf so that all books of the same subject are together?

Solution

There are four mathematics books that can be arranged in 4! ways, the history books can be arranged in 3! ways, the chemistry books in 3! ways, and sociology books in 2! ways. Thus, altogether, there are $4!\cdot 4!\cdot 3!\cdot 3!\cdot 2! = 41472$ ways.

1.3-3 Restricted Permutation

Theorem 1.5

The numbers of permutations of *n* different objects take *r* at a time in which *m* particular objects are:

(a) never included will be $P(n - m, r)$

(b) always included will be $P(n - m, r - m)\, P(r, m)$, where $(n \geq m, r \geq m)$.

PROOF

(a) When we are not interested in including m particular objects, then arrange r objects out of remaining $n - m$ objects. This can be done in $P(n - m, r)$ ways.

(b) When we are interested in including m particular objects always, we keep them aside and form the permutation of the remaining $(n - m)$ objects taking $(r - m)$ at a time. This can be done in $P(n - m, r - m)$ ways.

Now we take the case of m particular objects kept aside one by one. Out of total m objects kept aside, the first object can be occupied by $(r - m + 1)$ places as $(r - m)$ objects are already arranged. Thus the first can be introduced in $(r - m + 1)$ ways. Similarly second object can be introduced in $(r - m + 2)$ ways and proceeding in the same ways the mth object can be introduced in $(r - m + m) = r$ ways. Therefore, m objects can be arranged in

$$(r - m + 1)(r - m + 2) \cdot \cdot \cdot \cdot (r) = P(r, m) \text{ ways.}$$

$\therefore$ The required number of permutations are

$$P(n - m, r - m) \cdot P(r, m).$$

Example 1.24

How many words of four letters can be formed with the letters a, b, c, d, e, f, g, h and i when

(i) e and f are not to be included and

(ii) e and f are to be included.

Solution

(i) We are given with eight letters out of which we are to take four letters after keeping aside 2. After leaving 2 letters we are left with 6 letters, now the 4 letters out of 6 can be arranged in $P(6, 4) = 360$ ways i.e. there are 360 four lettered word which do not contain e and f.

(ii) When two letters e and f are included 2 letters e and f for 2 places out of 4 places of 4 lettered words and the remaining 2 letters are to be taken out of the remaining 6 letters a, b, c, d, e, g, h, i. Thus the number of words of 4 letters in this case will be

$$P(6, 2) \cdot P(4,2) = (6 \cdot 5) \cdot (4 \cdot 3) = 360.$$

Example 1.25

There are 15 points in a plane no three of, which are collinear. Find the number of straight lines formed by joining them.

Solution

A straight line can be formed whenever two points are joined. Now selection of two points out of 15 points can be made. i.e. $C(15, 2) = 105$.

EXERCISE 1.2

1. Find n if
   (i) $P(n, 2) = 72$
   (ii) $P(n, 3) + 3 \cdot P(n, 2) + P(n, 1) = n^3$
   (iii) $2P(n, 2) + 50 = P(2n, 2)$.
2. In how many ways the letters of word (i) LIBERTY (ii) RATION (iii) REPUBLIC can be arranged ?

3. In how many ways can the letter of word 'PENCIL' be arranged so that N is always next to E?
4. In how many ways 10 boys can be arranged in a line if two particular boys are (i) always together (ii) never together?
5. The Director wants to arrange 5 students on the platform such that the boy SALIM occupies the second position and such that the girl GITA is always adjacent to the girl RITA. How many such arrangements are possible?
6. An electric network contains 14 switches such that each switch may have three possible positions. How many different switching are there?
7. (a) How many automobile license plates can be made if each plate contains two different letters followed by three different digits?
   (b) Solve the problem if first digit cannot be 0.
8. Ten students are participating in a race. In how many ways the first three prizes are won?
9. In how many ways can we arrange 15 books on a shelf; 7 on mathematics, 6 on Numerical Analysis, and 2 on Operations Research, when no distinction is to be made in the books of the same subject?
10. If there are six periods in each working day of a institute, in how many ways can one arrange 5 subjects such that each subject is allowed at least one period?
11. (i) Find the number of ways in which five persons can sit in a row.
    (ii) How many ways are there if two of the persons insist on sitting next to one another?
    (iii) Solve part (i) assuming they sit around a circular table.
    (iv) Solve part (ii) assuming they sit around a circular table.
12. In how many ways can 7 persons address a meeting so that out of these, three persons *A, B, C* are to address in such a way that *A* will speak before *B* and *B* before *C* ?
13. Four alphabets *E, K, S* and *V*, one each, were purchased from a plastic warehouse. How many ordered pair of alphabets, to be used as initials, can be formed from them?
14. In how many ways can 7 boys and 5 girls be seated together in a row, so that no two girls are together?
15. In an examination hall, there are four rows of chairs. Each row has 8 chairs one behind the other. There are two classes sitting for the examination, with 16 students in each class. It is desired that in each row all students belong to the same class. In how many ways can these 32 students be seated?
16. Find the number of arrangements, which can be made out of the letters of the word "ALGEBRA" without altering the relative positions of vowels and consonants.
17. From among the 36 teachers in a college, one principal and one vice-principal are to be appointed. In how many ways can this be done?
18. Five boys and five girls are to be seated together. In how many ways can they be seated if :
    (i) All boys must be seated in the five leftmost seats?
    (ii) No two boys can be seated together?
    (iii) Mansi and Mary must be seated together?
19. (i) In how many ways can three boys and two girls sit in a row?
    (ii) In how many ways can they sit in a row if the boys and girls are to sit together?
    (iii) In how many ways can they sit in a row if just the girls are to sit together?
20. How many words of four words can be formed with the letters of the words "RATION" if :
    (i) R and O are not to be included .

(ii) R and O are to be included.

21. When a group photograph is taken, all the seven teachers should be in the first row and all the twenty students should be in the second row. If the two corners of the second row are reserved for the two tallest students, interchangeable only between them, and if the middle seat of the front row is reserved for the principal, how many arrangements are possible?
22. Find the number of natural number less than 1000 and divisible by 5 which can be formed with the digits 0, 1, 2, 3, 4, 5, 6, 7, 8, 9 each digit not occurring more than once in each number.
23. Find the number of ways that a party of seven persons can arrange themselves: (i) in a row of seven-chair (ii) around a circular table.
24. In how many ways can 9 cabinet ministers occupy their seats at a round table if one of the seats is to be reserved for the Prime Minister and there are eight other seats?
25. Find the number of ways of forming a ring with the help of 10 girls.
26. Three boys and three girls are to be seated around a table in a circle. Among them, the boy X doesn't want any girl neighbour and the girl Y doesn't want any boy neighbour. How many such arrangements are possible?
27. If the different permutations of the word EXAMINATION are listed as in a dictionary, how many items are there in this list before the first word starting with E?
28. How many numbers lying between 3000 and 4000 and divisible by 5 can be formed with the digits 3, 4, 5, 6, 7 and 8 repetition of the digits are not allowed?

ANSWERS

1. (i) 9	(ii) all natural numbers	(iii) 5	2.(i) 5040
(ii) 720	(iii) 40320	3. 120 4.(i) $2\cdot 9!$	(ii) $8\cdot 9!$
5. 8	6. 3^{14}	7. (i) $26\cdot 25\cdot 10\cdot 9\cdot 8=468000$	
8. 720	9. 180180	10. 3600	11. (i) 120
(ii) 48	(iii) 24	(iv) 12	12. 840
13. 12	14. 3386880	15. $2\cdot(16!)^2$	16. 72
17. 1260	19. (i) 120	(ii) 24	(iii) 48
20. (i) 24	(ii) 144	21. $(1440)\cdot 18!$	22. 154
23. (i) 7!	(ii) 6!	24. 40320	25. 9!
26. 4	27. 907200	28. 12	

1.4 COMBINATIONS

On many occasion we are not interested in the ordering or arrangements but only in selecting r objects from n objects. For example if a company wants to recruit 6 of the applicants for a position; a student may want to choose 4 books from a library at a time.

Suppose we want to select two balls out of four balls marked *A, B, C* and *D*. We may choose either *A, B* or *A, C* or *A, D* or *B, C* or *B, D* or *C, D*. Note that we do not list *B, A* separately, because it is the same as the choice *A, B*. Thus there are six ways of selecting 2 persons out of 4 persons. Here, we can see that in this selection, we didn't care the order but only two different balls. We denote the above fact by writing $C(4, 2) = 6$.

Consider a finite collection of *n* objects. A combination of these *n* objects taken *r* at a time is, any selection of *r* objects where order does not count. The number of such selection is called the **number of combination.**

The combination of *r* objects out of *n* given objects is denoted by ^nC_r or $C(n, r)$ or $C_{n,r}$ or C_r^n or $\binom{n}{r}$.

1.4-1 Difference Between a Permutation and a Combination

In a combination only selection is made whereas in permutation not only a selection is made but also an arrangement in a definite order is also taken into consideration.

In a combination, the ordering of the selected objects is immaterial. In a permutation, this ordering is essential. For example *A, B* and *B, A* are same as combinations but different as permutations.

Example 1.26

Find the number of combinations of the four objects *A, B, C,* and *D* taken three at a time.

Solution

Here total number of given objects are 4 and we are interested in finding combination consisting of three objects which is 3! = 6 permutations of the objects in the combinations as shown in the table below.

Here we can see that number of combinations multiplied by 3! equals the number of permutations, i.e.

$$C(4, 3).\ 3! = P(4, 3)$$

$$\text{or} \quad C(4,3) = \frac{P(4,3)}{3!}$$

Combination	Permutations
ABC	ABC, ACB, BAC, BCA, CAB, CBA
ABD	ABD, ADB, BAD, BDA, DAB, DBA
ACD	ACD, ADC, CAD, CDA, DAC, DCA
BCD	BCD, BDC, CBD, CDB, DBC, DCB
Total no. 4	**24**

Since $\quad P(4,3) = 4.3.2 = 24$ and $3! = 6$.

Therefore $\quad C(4,3) = 4$

We can observe from the above example that the number of permutations exceeds the number of combinations which is usually true except for some trivial cases.

In higher mathematics, the combinations $C(n, r)$ occur more frequently than the permutations.

1.4-2 COMBINATION OF *n* DIFFERENT OBJECTS

THEOREM 1.6

The number of combination of *n* distinct objects taken *r* at a time is given by

$$C(n,r) = \frac{n!}{r!(n-r)!} = \frac{P(n,r)}{r!}$$

PROOF

As we have seen in the above example the selection of *r* objects from *n* objects, the selected *r* objects can be arranged in $P(r, r) = r!$ ways.

Thus for each combination counted in $C(n, r)$, there are $r!$ permutations counted in $P(n, r)$. Therefore, we conclude that

$$P(n, r) = r!\, C(n, r)$$

$$\text{or} \quad C(n,r) = \frac{P(n,r)}{r!} = \frac{n!}{r!(n-r)!}$$

COROLLARY 1 $C(n, 0) = 1$

$$C(n,0) = \frac{n!}{0!(n-0)!} = \frac{n!}{1.n!} = 1$$

COROLLARY 2 $C(n, n) = 1$

$$C(n,n) = \frac{n!}{n!(n-n)!} = \frac{n!}{n!\cdot 0!} = 1$$

THEOREM 1.7

Prove that $C(n, r) = C(n, n - r)$.

PROOF

$$\text{L.H.S.} = C(n,r) = \frac{n!}{r!(n-r)!}$$

$$\text{R.H.S.} = C(n,n-r) = \frac{n!}{(n-r)!(n-n+r)!}$$

Hence $C(n, r) = C(n, n - r)$

NOTE

$C(n, r)$ and $C(n, n-r)$ are called **complementary combinations.**

THEOREM 1.8

Prove that

$$C(n + 1, r) = C(n, r - 1) + C(n, r).$$

PROOF

$$\text{R.H.S.} = C(n, r - 1) + C(n, r)$$

$$= \frac{n!}{(r-1)!(n-r+1)!} + \frac{n!}{r!(n-r)!}$$

$$= n!\left[\frac{1}{(n-r+1)(n-r)!(r-1)!} + \frac{1}{r(r-1)!(n-r)!}\right]$$

$$= \frac{n!}{(n-r)!(r-1)!}\left(\frac{1}{(n-r)!(r-1)!} + \frac{1}{r}\right)$$

$$= \frac{n!}{(n-r)!(r-1)!}\left(\frac{n+1}{r(n-r+1)}\right)$$

$$= \frac{(n+1)!}{r!(n-r+1)!} = C(n+1,r) = \text{L.H.S.}$$

NOTE

Above identities are combinatorial identities. They can alternatively be proved by combinatorial arguments.

Example 1.27

Compute $C(13, 5)$.

Solution

$$C(13,5) = \frac{13!}{5!(13-5)!} = \frac{13!}{5!.8!}$$

$$= \frac{13.12.11.10.9.8!}{5.4.6.1.8!}$$

$$= \frac{13.12.11.10.9}{5.4.3.2.1} = 13.11.9$$

$$= 1287$$

Example 1.28

Find n, if $C(n, 7) = C(n, 5)$.

Solution

Given $C(n, 7) = C(n, 5)$

$$\frac{n!}{7!(n-7)!} = \frac{n!}{5!(n-5)!}$$

or

$$(n!)(5!)(n-5)! = (n!)(7!)(n-7)!$$

or $(n-6)(n-5) = 6 \times 7$

or $n^2 - 11n - 12 = 0$

or $n = 12, -1$

In this problem n should be non-negative Therefore, $n = 12$ is the only solution.

Example 1.29

(i) In how many ways can a committee of 5 are chosen from 9 people?

(ii) How many committees of 5 or more can be chosen from 9 people?

(iii) In how many ways can a committee of 5 teachers and 4 students be chosen from 9 teachers and 15 students?

(iv) In how many ways can the committee in (iii) be formed if teacher A refuses to serve if student B is on the committee?

Solution

(i) $C(9, 5)$ ways.

(ii) $C(9, 5) + C(9, 6) + C(9, 7) + C(9, 8) + C(9, 9)$.

(iii) The 4 students can be chosen out of 15 students in $C(15, 4)$ ways and the teachers can be selected in $C(9, 5)$ ways. Hence the committee can be formed is $C(9, 5)\cdot C(15, 4)$ ways.

(iv) First we count the number of committees where both A and B are on the committee. Now, we are left with 8 teachers from whom 4 teachers are to be chosen. Similarly we are left with 14 students from whom 3 more students are to be chosen. There are $C(8,4)\cdot C(14,3)$ committees containing both A and B, hence there are

$$C(9, 5)\cdot C(15, 4) - C(8, 4)\cdot C(14, 3)$$

Committees that do not have both A and B on the committee.

Example 1.30

In how many ways can a student choose a programme of 5 courses if 9 courses are available and two courses are compulsory for all the students?

Solution

Here two courses are given to be compulsory, the students are required to choose 3 out of 7 courses and this can be done in $C(7,3) = \dfrac{7!}{3!\cdot 4!} = 35$.

Example 1.31

The question paper of mathematics contains two questions divided into two groups of five questions each. In how many ways can an examinee answer six questions taking at least two questions from each group?

Solution

The examinee can answer questions from two groups in following ways:

(i) 2 from first group and 4 from the second group.

(ii) 3 from first group and 3 from the second group.

(iii) 4 from the first group and 2 from the second group.

Now, the number of ways of selecting the questions in case (i)

$$= C(5,2)\cdot C(5,4) = 10\times 5 = 50$$

Likewise the number of ways of selecting the question in case (ii)

$$= C(5,3)\cdot C(5,3) = 10\times 10 = 100$$

and the number of ways of selecting the question in case (iii)

$$= C(5,4)\cdot C(5,2) = 5\times 10 = 50$$

Hence the examinee can answer the questions in either of the three ways.

Therefore, the required number of ways

$$= 50 + 100 + 50$$
$$= 200.$$

Example 1.32

(i) How many 5-card hands consist only of hearts?

Solution

Out of total 52 cards we have 13 hearts. Since there are 13 hearts to choose from, each such hand is a 5 combination of 13 objects. Therefore,

$$C(13,5)=\frac{13!}{5!.8!}=\frac{13.12.11.10.9}{5.4.3.2.1}=13.11.9=1287$$

(ii) How many 5-card hands consist of cards from a single suit?

Solution

Out of each 4 suits, spades, hearts, diamonds or clubs there are 5·*C* (13, 5) cards in hands. Hence, there are a total of 4·*C* (13, 5) such hands.

Example 1.33

A farmer buys 3 hens, 2 pigs and 4 cows from a man who has 6 hens, 5 pigs and 8 cows. How many choices does the farmer have?

Solution

The farmer can choose the hens in *C* (6, 3) ways, the pigs in *C* (5, 2) ways, and hens in *C* (8,4) ways. Hence by the fundamental principle of multiplication, required number of ways is

$$C(6,3)\cdot C(5,2)\cdot C(8,4)=\frac{6\cdot5\cdot4\cdot5\cdot4\cdot8\cdot7\cdot6\cdot5}{1\cdot2\cdot3\cdot1\cdot2\cdot1\cdot2\cdot3\cdot4}$$

$$=20\cdot10\cdot70$$

$$=14{,}000 \text{ ways.}$$

Example 1.34

If there are 12 persons in a party, and if each two of them shake hands with each other, how many handshakes happen in the party?

Solution

When two persons shake hands, it is counted as one handshake, not two. Therefore, the total number of handshakes is the same as number of ways of selecting 2 persons from among 12 persons. This is given by

$$C(12,2)=\frac{12\cdot11}{1\cdot2}=66$$

Example 1.35

A bag contains six white marbles and five red marbles. Find the number of ways four marbles can be drawn from the bag if (i) they can be any colour, (ii) two must be white and two red (iii) they must all be of same colour.

1.4-3 Restricted Combinations

Theorem 1.9

The number of combinations of *n* object taken *r* at a time in which *m* particular objects are

(i) always included is *C* (*n* - *m*, *r* - *m*) ways and

(ii) never included is *C* (*n* - *m*, *r*) ways, where ($n \geq m$).

Proof

(i) When we are to include *m* particular objects compulsorily then for our selection we keep them aside and selection of (*r* - *m*) out of remaining (*n* - *m*) objects is made. This can be done in *C* (*n* - *m*, *r* - *m*) ways.

(ii) When we are to exclude m particular objects compulsorily then for selections we should ignore them. Thus we shall be selecting r out of remaining $(n - m)$ objects, which can be done in $C(n - m, r)$ ways.

Example 1.36

In how many ways a football eleven can be chosen out of 17 players when (i) five particular players are to be always included (ii) two particular players are to be always excluded?

Solution

Here, first we include five particular players. Then, we are left with 12 players. Out of which we are to select remaining 6 players and this can be done in $C(12, 6)$ ways, i.e.

$$\frac{12}{6!.6!} = \frac{12 \times 11 \times 10 \times 9 \times 8 \times 7!}{6!.6!}$$

$$= \frac{12 \times 11 \times 10 \times 9 \times 8 \times 7!}{6 \times 5 \times 4 \times 3 \times 2 \times 1} = 924 \text{ ways}$$

(ii) Here, we wish to exclude two particular players. Hence, we are left with 15 players. Out of which we are to select football eleven, which can be done in $C(15,11)$ ways.

1.4-4 Combinations with Repetitions

Theorem 1.10

The total number of combinations of $(n_1 + n_2 + n_3 \ldots)$ objects where n_1 are alike, n_2 are alike, and so on taken any number at a time is

$$(n_1 + 1)(n_2 + 1)(n_3 + 1) \ldots -1.$$

Proof

First of all consider n_1 like objects. These n_1 like objects can be dealt in $n_1 + 1$ ways for we may take 1, 2, 3, ... , n_1 or none in any selection. Similarly n_2 like objects can be dealt in $(n_2 + 1)$ ways and so on. By the principle of mathematical induction, the total number of selection is $(n_1 + 1)(n_2 + 1)(n_3 + 1) \ldots$ ways. But these selections include one case where in all objects has been left out. Therefore, the total number of ways in which at least one object has been selected $(n_1 + 1)(n_2 + 1)(n_3 + 1) \ldots -1$.

Example 1.37

How many different choices can be made of selections out of 15 Maths, 10 Numerical Analysis and 12 Operations Research books when at least one book is to be selected?

Solution

Total number of ways mathematics books can be dealt is (15 + 1) ways including one of not selecting. Similarly books on Numerical Analysis and Operations Research can be dealt in (10 + 1) and (12 + 1) ways respectively. Then the total numbers of ways of dealing with these books are $(15 + 1). (10 + 1)\cdot(12 + 1) = 16\cdot11\cdot13 = 2288$. But there is only one way in which none of these books are selected. Therefore, the required number of ways of selecting at least one of them is $2288 - 1 = 2287$.

Example 1.38

Prove that the total number of ways in which it is possible to make a selection by taking some or all of the given n different objects, is $2^n - 1$.

PROOF

As each object can be disposed of in two different ways, i.e. it may be accepted or it may be rejected. And each way of disposing of one object may be associated with each way of disposing of any other object. Therefore, the total number of ways in which all objects may be disposal of is $2\times2\times2\times\ldots n$ times $= 2^n$ ways.

But there is one way in which nothing is accepted.

Hence the required number of ways = $2^n - 1$.

Example 1.39

In how many ways Gaurav can invite eight of his friends to a party by inviting at least one of them?

Solution

There are two ways in which Gaurav can deal with his friends i.e. he may invite him or he may not invite him. Therefore, he can deal with eight friends 2^8 ways, but this includes the case when he does not invite any one of them which can be possible only in one way. Therefore, the required number of ways he can invites his friends $2^8 - 1 = 255$ ways.

1.4-5 COMBINATORIAL IDENTITIES AND AGREEMENTS

Here are the list of certain identities or equations involving the symbols $P(n, r)$ and $C(n, r)$ including the ones that we have already seen:

1. $P(n, n) = P(n, n-1)$
2. $P(n, n) = n\,P(n-1, n-1)$
3. $(n+1)\,P(n, r) = (n-r+1)\,P(n+1, r)$
4. $P(n, n) = r!\,P(n, n-r)$
5. $P(n, r) = P(n-1, r) + r\,P(n-1, r-1)$
6. $P(n, r) = r!\,C(n, r)$
7. $C(n, r) + C(n, r-1) = C(n+1, r)$
8. $r\,C(n, r) = n\,C(n-1, r-1)$
9. $C(n, r) = C(n, n-r)$
10. $C(n, 0) + C(n, 1) + C(n, 2) \ldots + C(n, n) = 2^n$

All these can be proved using the formula $C(n, r)$ and $P(n, r)$ or other agreements. The readers are advised to try each of the above identities or equations.

EXERCISE 1.3

1. Evaluate (i) $C(14, 2)$ (ii) $C(20, 17)$ (iii) $C(52, 4)$ (iv) $C(10, 2)$.

2. If $C(n, 8) = C(n, 6)$. Find $C(n, 2)$.

3. Find n, if $C(n, 6) : C(n-3, 3) = 91 : 4$.

4. Prove that $C(2, 1) + C(3, 1) + C(5, 1) = C(3, 2) + C(4, 2)$.

5. Show that

   (i) $C(n, 0) + C(n, 1) + C(n, 2) + \ldots + C(n, n) = 2^n$.

   (ii) $C(n, 0) - C(n, 1) + C(n, 2) - C(n, 3) + \ldots + C(n, n) = 0$.

6. Prove that relation

$C(n, r) = C(n-1, r) + C(n-1, r-1$.

7. How many committees can be formed consisting of 2 professors, 3 Asst. professors and 5 lecturers in a college having 9 professors, 12 Asst. professors and 20 lecturers?

8. A woman has 11 close friends
   (i) In how many ways can she invite five of them to lunch?
   (ii) In how many ways if two of the friends are married and will not attend separately?
   (iii) In how many ways if two of them are not on speaking terms and will not attend together?

9. A collection of 100 light bulbs contains 8 defective ones.
   (i) In how many ways can a sample of 10 bulbs be selected?
   (ii) In how many ways can a sample of 10 bulbs be selected which contain 6 good bulbs and 4 defective ones?
   (iii) In how many ways can the sample contains either the sample contains 6 good ones and 4 defective ones or 5 good ones and 5 defective ones?

10. A question paper has two parts, part A and part B, each containing 10 questions. If the student has to choose 8 from part A and 5 from part B, in how many ways can he choose the questions?

11. A student is to answer 10 out of 13 question on an examination.
   (i) How many choices have he?
   (ii) How many if he must answer the first two questions?
   (iii) How many if he must answer the first or second question but not both?
   (iv) How many if he must answer exactly three out of the first five questions?
   (v) How many if he must answer at least three of the first five questions?

12. A man has 15 close friends, 6 of them are women.
   (i) In how many ways can he invite 3 or more of his friends to a party?
   (ii) In how many ways can he invite 3 or more of his friends if he wants the same number of men (including him) as woman?

13. A woman has 11 colleagues in her office, of which 8 are men. She could like to invite some of her colleagues to dinner. Find the number of her choices if she decides to invite:
   (i) at least 9 of them,
   (ii) All her women colleagues and sufficient men colleagues to make the number of women and men equal.

14. In how many ways can we select a cricket eleven from 17 players in which 5 players can bowl? Each cricket team must have at least two bowlers.

15. In how many ways can 5 days are chosen from each of the 12 months of an ordinary year of 365 days.

16. Suppose there are 15 red balls and 5 white balls. Assume that the balls are distinguishable and that a sample of 5 balls is to be selected.
   (i) How many samples of 5 balls are there?
   (ii) How many samples contain all red balls?
   (iii) How many samples contain 3 red balls and 2-white balls?
   (iv) How many samples contain at least four red balls?

17. In how many ways can a committee consisting of 5 members be formed each containing 3 consonants and 2 vowels?

18. From a class of 25 students, 10 are to be chosen for an excursion party. There are three students who decide that either all of them will join or none of them will join. In how many ways can they be chosen?

19. A committee of 8 is to be selected out of 6 males and 8 females. How many ways can it be done so that male may not be out numbered?

20. A group of 8 scientists is composed of 5 physicists and 3 mathematicians:
    (i) In how many ways a committee of 5 be formed.
    (ii) In how many ways can a committee of 5 be formed that has 3 physicist and 2 mathematicians?

21. Shyam invited a party of 10 students to a dinner and places half of them on one side of a long table and half on the other side. If 3 particular men desires to sit one side and two on the other. Find the number of ways in which he can arrange the guests?

22. There are 16 points in a plane, no three of which are collinear.
    (i) Find the number of straight lines.
    (ii) Number of triangles formed by joining them.

23. How many triangles can be formed by joining 10 points, 5 of which are in the same straight line.

24. In a town, there are 87 families, of which 52 families have almost 2 children. In a rural development programme, 20 families are to be helped chosen for assistance, of which at least 18 families must have at most 2 children. In how many ways can the choice are made?

25. A committee of eight is to be selected out consisting of five gentlemen and three ladies out of eleven gentlemen and nine ladies with the condition that Miss X refuses to serve on the committee when Mr. Y is selected. In how many ways can the committees are formed.

26. Compute the number of 6-letter combination of the letters of the English alphabet if no letter is to appear in a combination more than two times.

27. For the posts of 5 teachers, there are 23 applicants. 2 posts are reserved for SC candidates and there are 7 SC candidates among the applicants. In how many ways can the selection are made?

28. A candidates is required to answer 6 out of 10 questions which are divided into two groups each containing five questions and he is not permitted to attempt more than four from any group. In how many ways can he make up his choice?

29. A committee of 5 is to be formed out of 6 gentlemen and 4 ladies. In how many ways can this be done if
    (i) At least two ladies are to be included?
    (ii) At most two ladies are to be included?

30. A farmer buys 3 cows, 8 pigs, and 12 chickens from a man who has 9 cows, 25 pigs, and 100 chickens. How many choices does the farmer have?

31. Find the number of diagonals of a polygon of 80 sides.

32. A teacher intends to take eight students to a factory taking three of them at a time. How many times will the teacher go and how many times each student will go?

33. A woman has 11 close friends of whom six are also women
    (i) In how many ways can she invite three or more to a party?
    (ii) In how many ways can she invite three or more of them if she wants the same number of men as women (including her)?

ANSWERS

1. (i) 91 (ii) 1140 (iii) 270725 (iv) 45
2. 91 3. 15 7. 122791680
8.(i) 462 (ii) 210 (iii) 252
9.(i) C(100,10) (ii) C(92, 6)·C(8, 4) (iii) $C(92,6)\cdot C(8,4)+C(92,5)\cdot C(8,5)$
10. 11340 11.(i) 286 (ii) 165 (iii) 110 (iv) 80
(v) 276 12.(i) $2^{15}-1-C(15,1)-(15,2)=C(15,3)+C(15,4)+\ldots+C(15,4)$
(ii) $C(9,1)\cdot C(6,2)+C(9,2)\cdot C(6,3)+C(9,3)\cdot C(6,4)+C(9,4)\cdot C(6,5)+C(9,5)\cdot C(6,6)$
14. 50050 15. $\{C(30,5)\}^{4}\{C(31,5)\}^{7}\cdot C(28,5)$
16. (i) $C(20,5)$ (ii) $C(15,5)$ (iii) $C(15,3)\cdot C(5,2)$
17.(i) 56 (ii) 56 18. $C(12,4)\cdot C(8,1)\cdot C(14,2)$
19. 141420.(i) $C(8,5)$ (ii) $C(5,3)\cdot C(3,2)$ 21. 288000 22. (i) 120
(ii) 56 24. $C(52,18)\cdot C(35,2)+C(52,19)\cdot C(35,1)+C(52,20)$
25. 32928
26. There are $C(26,6)$ combinations with all 6 letters distinct; $C(26,1)\ C(25,4)$ combinations with 1 pair of letters and 4 other letters etc.. Total: $C(26,6)+C(26,1)\cdot C(25,4)+C(26,3)+C(26,2)\cdot C(24,2)$.
27. There are $C(26,6)$ combinations with all 6 letters distinct; $C(26,1)\ C(25,4)$ combinations with 1 pair of letters and 4 other letters etc. Total: $C(26,6)+C(26,1)\cdot C(25,4)+C(26,3)+C(26,2)\cdot C(24,2)$.
27. $C(7,2)\cdot C(16,3)=11760$ 28. 200
29 .(i) 186 (iii) 126 30. $C(9,3)\cdot C(25,8)\cdot C(100,12)$
31. 3080 32.(i) 56 (ii) 21
33 .(i) $2^{11}-1-C(11,2)-C(11,2)=1981$ or $C(11,3)+C(11,4)+\ldots+C(11,11)=1981$
(ii) $C(5,5)\cdot C(6,4)+C(5,4)\cdot C(6,3)+C(5,3)\cdot C(6,2)+C(5,2)\cdot C(6,1)=325$.

1.5 RECURRENCE RELATIONS AND GENERATING FUNCTIONS

1.5-1 Polynomial Expression

An nth degree polynomial ($n \geq 0$) is an expression of the form

$$a_0x^n + a_1x^{n-1} + a_2x^{n-2} + \ldots a_n, \quad a_0 \neq 0$$

where $a_0, a_1, \ldots\ldots, a_n$ are called set of coefficients.

This is a **non-recursive** definition of polynomial.

1.5-2 Zeroth Degree Polynomial

Polynomials of zeroth degree are called constant polynomials and are simply elements of set of coefficients.

1.5-3 Evaluation of Polynomials (Non-Recursively)

Let $$f(x) = 5x^4 + 12x^3 - 6x^2 + x + 6 \qquad \ldots(1)$$

be a polynomial of degree four. To evaluate it for $x=2$, put it in the polynomial (1).

$$f(2)=5(2)^4+12(2)^3-6(2)^2+2+6=160$$

Here we can see that to evaluate this polynomial we have to perform seven multiplication and four additions/subtractions operations.

1.5-4 Polynomial Expression (Recursive)

Let A be a set of coefficients, then

(i) A polynomial expression of degree zero (zeroth degree) is an element of set A.

(ii) An nth degree polynomial expression for $n \geq 1$ is an expression of the form $[\varphi(x)]\,x + a$, where $\varphi(x)$ is an $(n-1)^{th}$ degree polynomial expression and $a \in A$.

1.5-5 Evaluation of Polynomials (Recursive)

The polynomial expression (1) can be expressed in recursive form as

$$\begin{aligned} f(x) &= 5x^4 + 12x^3 - 6x^2 + x + 6 \\ &= (5x^3 + 12x^2 - 6x + 1)x + 6 \\ &= ((5x^2 + 12x - 6)\,x + 1)x + 6 \\ &= (((5x + 12)x - 6)\,x + 1)x + 6 \\ &= ((((5)x + 12)x - 6)x + 1) + 6 \qquad \text{... (2)} \end{aligned}$$

Now, we can see in the above polynomial (5) is a zeroth degree polynomial $(5)x + 12$ is a first degree polynomial, $((5)x + 12)x - 6$ is a second degree polynomial $(((5)x + 12)\,x - 6)x + 1$ is a third degree polynomial, $((((5)x + 12)x - 6)x + 1)x + 6$ is a fourth degree polynomial.

Putting $x = 2$ in (2), we have

$$f(2) = ((((5)\,2 + 12)2 - 6)\,2 + 1)\,2 + 6 = 160.$$

In this form we can see that only four multiplication and four additions/subtractions are used. We can see that operations required to evaluate (1) with the help of recursive form are less than that of non-recursive form polynomial.

1.5-6 Telescopic Form

The final expression (2) of the polynomial (1) is called **telescoping form** of (1).

Method of evaluating a polynomial, after converting into telescoping form is called **Horners Method.**

Example 1.40

(i) The telescoping form of $f(x) = 3x^4 - 5x^3 - 2x^2 + 7x + 8$ is

$$((((3)\,x - 5)x - 2)x + 7)x + 8.$$

(ii) The telescoping from of $f(x) = x^5 + 3x^4 + 3x^3 - 2x^2 + x$ is

$$f(x) = (((((-1)\,x - 5)\,x + 3)\,x + 3)\,x + 2)\,x + 1)\,x + 0.$$

1.6 RECURSION

We give the name recursion to the technique of defining a function, a set, or an algorithm in terms of itself, where it is generally understood that the definition will be in terms of previous values or simpler objects. Describing an object in a simple definition form is usually called the **basis** for recursion.

Example 1.41

Out of several recursively-defined functions, let us take one very familiar recursively defined function $f_n = n!$ (factorial function).

Solution

(i) $f_0 = 1$ (ii) $f_{n+1} = (n+1)f_n$ for all $n \geq 0$.

Here $f(0) = f_0$ is called the initial condition or basis. When $n = 0, 1, 2$.

(ii) becomes

$$f_1 = 1.f_0 = 1.1 = 1,\ f_2 = (1+1)f_1 = 2.1 = 2,\ f_3 = (2+1)f_2 = 3\cdot 2 = 6 \text{ and so on.}$$

1.6-1 Recursion Theorem

Let A be a given function from a set A into A. Let $a_0 \in A$ and N denote the set of positive integers. Then there exist a unique function $f : N \rightarrow A$ such that

(i) $f(0) = a_0$ and

(ii) $f(n+1) = F(f(n))\ \forall\ n \in N$.

i.e $f(n+1)$ in a function of $f(n)$.

The condition (i) is called the initial condition and condition (ii) is called **recurrence relation** or generating rule, which will be defined later on in this chapter. The proof of the theorem is beyond the scope of this book.

Example 1.42

Consider the famous Fibonacci sequence, which is defined as

$f_{n+1} = f_n + f_{n-1}$ for all integers $n \geq 1$. $f_0 = 1 = f_1$, here f_0 and f_1 are called basis or initial condition.

Now, if we want to compute f_3 then,

$$f_3 = f_2 + f_1 = (f_1 + f_0) + f_1 = (1+1) + 1 = 3$$

One can see here that we have moved backward from the third number to second and first which is closer to the basis. Similarly, we can compute f_5,

$$f_5 = f_4 + f_3 = (f_2 + f_3) + (f_1 + f_2)$$
$$= f_1 + 2f_2 + f_3 = f_1 + 2f_2 + f_1 + f_2$$
$$= 2f_1 + 3f_2 = 2f_1 + 3(f_0 + f_1)$$
$$= (3f_0 + 5f_1) = 3.1 + 5.1 = 8.$$

1.7 ITERATION

Iteration is a process by which with the help of initial value(s) we can to find the values of higher order terms or we can say forward values.

For example, if we want to compute f_5 of Fibonacci with the help of iteration, then

$$f_2 = f_1 + f_0 = 1 + 1 = 2$$
$$f_3 = f_2 + f_1 = 2 + 1 = 2$$
$$f_4 = f_3 + f_2 = 3 + 2 = 5$$

$$f_5 = f_4 + f_3 = 5 + 3 = 8$$

Here, we can see that iterative computation is usually faster than the computations that apply recursion. Therefore, if it is possible better to transform recursive function into an equivalent iterative process for computation or a faster method if possible.

NOTE

The drawback of recursion is that usually the execution of a recursive program is slower than that of its iterative counterparts, and a recursive program is more difficult to debug than a corresponding iterative one.

1.8 MATHEMATICAL INDUCTION

Induction is a process in which we guess and conclude from careful observations of a number of special examples. There may emerge a pattern or a rule, which may found to be common to all the observations. We then formulate a statement in general terms, which applies to all the individual cases under consideration. Consider the following cases of the sums of natural numbers.

$$1+2+3+4=10=\frac{4}{2}(4+1);$$

$$1+2+3+4+5=15=\frac{5}{2}(5+1);$$

$$1+2+3+4+5+6=21=\frac{6}{2}(6+1)$$

....

We may conclude that the sum of first n natural numbers is $n(n+1)/2$. But is this statement true for all n? As seen above it is true for all cases from $n=1$ to $n=6$. To verify that this statement is always true, we take it to be true for any arbitrary n and then show it to be true for the successor.

Sum of first n natural number $n(n+1)/2$ to be true, so that

$$1+2+3+4+\ldots+n = n(n+1)/2$$

Add its successor i.e. $(n + 1)$ to both the sides, we get

$$\therefore \quad 1+2+3+\ldots+n+(n+1) = \frac{n}{2}(n+1)+(n+1)$$

$$= \left(\frac{n}{2}+1\right)(n+1) = \left(\frac{1}{2}(n+2)\right)(n+1)$$

$$= \frac{n+1}{2}\left(\overline{n+1}+1\right).$$

Hence that the statement $1+2+3+4+\;.\;.\;.\;+n=\frac{n}{2}(n+1)$ is also true statement when n is replaced by $n + 1$. The above is an illustration of principle of mathematical induction. We state the principle as follows:

Let $P(n)$ be a statement such that

(a) $P(1)$ is true

(b) $P(r+1)$ is true whenever $P(r)$ is true where $n = r$.

Then $P(n)$ is true for all natural numbers n.

Example 1.43

Show that $1^2+2^2+\ldots+n^2=\dfrac{n(n+1)(2n+1)}{6}$ by principle of mathematical induction.

Proof

Let $P(n)$ be the statement such that

$$P(n)=1^2+2^2+\ldots+n^2=\frac{n(n+1)(2n+1)}{6}$$

Then $$P(1)=\frac{1.(1+1)(2.1+1)}{6}=1$$

Hence, $P(1)$ is true

If $P(n)$ is true for an arbitrary r, then

$$1^2+2^2+3^3+\ldots+r^2=\frac{r(r+1)(2r+1)}{6}$$

Add $(r+1)^2$ to both the sides, we have

$$1^2+2^2+3^3+\ldots+r^2+(r+1)^2=\frac{r(r+1)(2r+1)}{6}+(r+1)^2$$

$$=\left((r+1)\left[\frac{r(2r+1)}{6}+(r+1)\right]\right)$$

$$=(r+1)\left[\frac{(2r^2+7r+6)}{6}\right]$$

$$P(r+1)=\frac{(r+1)(r+2)(2r+3)}{6}$$

$\therefore$ $P(r+1)$ is true. Hence the given statement is true for all $n \in N$.

The formula that we have just prove for *P(n)* is called a **closed form expression**. It involves no recursion.

1.9 CLOSED FORM EXPRESSION

Let $F=F(x_1,x_2,\ldots,x_n)$ be an algebraic expression in which variables x_i $(i = 1, 2, \ldots n)$ can take any values from some fixed set. F is said to be in closed form expression if there exist a number D such that evaluation of F will take no more than D operations for any allowed values of x_i $(i = 1, 2, \ldots, n)$.

Example 1.44

Closed form expression for sum of first n natural number, i.e.

$$\sum_{m=1}^{n} m = 1+2+3+\ldots+n=\frac{n(n+1)}{2}$$

Some more examples of closed form expressions are

Expression	Closed form
$\sum_{m=1}^{n} 2m = 2 + 4 + 6 + \ldots + 2n$	$n(n+1)$
$\sum_{m=1}^{n} 2m - 1 = 1 + 3 + 5 + \ldots + (2n-1)$	n^2

EXERCISE 1.4

(1) Write the telescoping form of

(a) $f(x) = 5x^4 + 12x^3 - 6x^2 + x + 6$

(b) $f(x) = x^5 + 3x^4 - 15x^3 + x - 10$ and calculate the value of $f(3)$.

(2) Prove that function $f(n) = 2(3^n) - 5$ is the unique function defined by $f(n) = 4f(n-1) - 3f(n-2)$ for $n \geq 2$, where $f(0) = -3$, $f(1) = 1$.

(3) Use mathematical induction to prove that each of the following statements is true for all positive integers n.

(i) $1^3 + 3^3 + 5^3 + \ldots + (2n-1)^3 = n^2(2n^2 - 1)$

(ii) $1 \cdot 2 + 2 \cdot 3 + 3 \cdot 4 + \ldots\ldots + n \cdot (n+1) = \dfrac{n(n+1)(n+2)}{3}$.

(4) Prove by induction that $g(k) = 3k + 2$, $k \geq 0$ is a closed form expression for the g defined recursively by

$$g(0) = 2 \text{ and } g(k) = g(k-1) + 3 \text{ for } k \geq 1.$$

(5) Prove by induction that $g(n) = (3)\,5^n + (7)\,2^n$ is a closed form expression for the sequence g defined recursively by $g(0) = 10$, $g(1) = 29$ and $g(n+1) = 7g(n) - 10g(n-1)$ for $n \geq 1$.

(6) Prove by induction that $g_n = 2^n - 1$ is a closed form expression for the sequence g defined recursively by

$$g_0 = 0, g_1 = 1 \text{ and } g_n = 3g_{n-1} - 2g_{n-2} \text{ for } n \geq 2.$$

(7) Using Recursion theorem verify that the definition $g(n+1) = 3g(n)^2 + 7$ for $n \geq 0$, $g(0) = 1$ do in fact define function.

(8) What is wrong in the definition $f: R \to R$?

$$f(0) = 1 \text{ and } f(x) = f(x/2)/2\,?$$

ANSWERS

1. (a) $((((5)x + 12)x - 6)x + 1)x + 6$

   (b) $(((((1)x + 3)x - 15)x)\,x + 1)x - 10$.

8. The basis is not reached in a finite number of steps if you try to compute $f(x)$ for x not zero.

1.10 SEQUENCES OR DISCRETE FUNCTIONS

Let us consider the ordered collection of numbers 2, 4, 6, 8, 10, . . . and assign the order amongst its members. We observe that 2 is the first member, 4 is the second member, 6 is the third member and so on. These respective numbers are called the terms of the sequence, i.e the first member is called the Ist **term**, the second member the 2nd term and so on. We in general denote the terms of the sequence by $u_1, u_2, u_3, \ldots, u_n$ or $a_1, a_2, a_3 \ldots, a_k$ or $t_1, t_2, t_3, \ldots, t_n$ the subscript denoting the place (or position) of the term in the sequence.

We define a sequence in a nonempty set X to be a map $f: N \to X$. If $X = R$ (real number), we call the sequence f a **real sequence** and if $X = C$ (complex no.) we call f a **complex sequence.** If domain of f is finite N (set of Natural numbers), the sequence f is said to be **finite sequence** and if domain of $f = N$, it is called an **infinite sequence**.

A **sequence of integers** is a function from the set of natural numbers into the set of integers, i.e. if f is a sequence then $f: N \to Z$. The image of any $k \in N$ can be written as $f(k)$ or f_k.

A sequence is often called a **discrete function.**

Example 1.45

(i) The sequence $f(n) = 2n^2 + 3n$, $n \geq 0$ is a sequence of integers.

(ii) The sequence $f(n) = 2n^3 + 4n^2 - 3n$, $n \geq 0$ is a sequence of integers.

(iii) $f(n) = n^2 - 1$, $n \in N$.

Example 1.46

The sequence t defined recursively by $t_1 = 1, t_2 = 1$ and $t_n = t_{n-1} + 2t_{n-2}, n \geq 3$ is a sequence of integers. Various terms of t can be computed either by applying recursion formula or by iteration.

By recursion

$$\begin{aligned} t_5 &= t_4 + 2t_3 \\ &= (t_3 + 2t_1) + 2t_3 \\ &= ((t_2 + 2t_1 + 2t_1) + 2(t_2 + 2t_1)) \\ &= t_2 + 4t_1 + 2(t_2 + 2t_1) \\ &= 1 + 4 \cdot 1 + 2 \cdot (1 + 2) = 5 + 6 = 11 \end{aligned}$$

By iteration

$$\begin{aligned} t_3 &= t_2 + 2t_1 = 1 + 2 \cdot 1 = 3 \\ t_4 &= t_3 + 2t_2 = 3 + 2 \cdot 1 = 5 \\ t_5 &= t_4 + 2t_3 = 5 + 2 \cdot 3 = 11 \end{aligned}$$

1.11 RECURRENCE RELATION

For a numeric function $(a_0, a_1, a_2, \ldots, a_r \ldots)$ or for sequence of numbers an equation relating a_r, for any r, to one or more of the a_i's, $i < r$, is called **Recurrence Relation**. Recurrence relations are also called **difference equations**, because they can be written in terms of the difference between the consecutive terms of a sequence.

We will see that there is a striking resemblance between recurrence relations of Discrete Mathematics and differential equation of continuous mathematics. Although there is some difference between the two, but many concepts and results about Recurrence relations and their solutions are similar to the concepts and solutions of differential equations.

EXAMPLES OF RECURRENCE RELATIONS

(1) If a_n denotes the sum of first n terms then $a_n = n + a_{n-1}$.

(2) If t_n denotes the nth term of a geometric progression with common ratio r, then $t_n = r\, t_{n-1}$.

(3) $a_n = 3a_{n-1} - 2a_{n-2},\ n \geq 2.$

(4) $a_n = (n-1)\,a_{n-1} - (n-1)\,a_{n-2},\ n \geq 2.$

(5) $a_n = 9a_{n-1} + 26\,a_{n-2} - 24\,a_{n-3} + 5n,\ n \geq 3.$

(6) $a_n = 2\,(a_{n-1})^2 - n\,a_{n-3},\ n \geq 3.$

(7) $a_n^2 = -(a_n - 1)^2 - 1,\ n \geq 1.$

In above examples, relations (1) and (2) are first order recurrence relations because a_n and t_n depend upon only one previous term of a and t respectively. Examples (3) and (4) are second order recurrence relation as they depend upon previous two terms. Examples (5) and (6) are third order recurrence relation as they depend upon previous three terms a_0, a_1 and a_2.

Example 1.47

The Fibonacci sequence is defined by

$$F_k = F_{k-2} + F_{k-1},\ F_0 = F_1 = 1,\ k \geq 2.$$

This is a second order recurrence relation. This sequence is first given by Fibonacci, that's why it is called Fibonacci sequence.

1.11-1 Solutions of Recurrence Relations

In elementary algebra solving a quadratic equation means finding the values of x which satisfy the quadratic equation. Here we see that $x = 2, 3$ are two values which satisfies the $x^2 - 6x + 5 = 0$ and hence they are the solutions.

Sequences are often most easily defined with a recurrence relation. It is clear that according to the recurrence relation, we can carry out step-by-step computation to determine a_r from $a_{r-1}, a_{r-2}, \ldots$. The value of the function at one or more points is given so that computation can be initiated, however the calculation of terms by directly applying a recurrence relation can be time consuming. The process of finding a closed form expression for the terms of sequence from its recurrence relation is called **solving the relation.**

Just as in algebra, recurrence relation may have no solution. Recurrence relation $a_n^2 + (a_{n-1})^2 = -1$ have no solution since there are no real valued function which satisfy the relation, as the squares of real numbers are always nonnegative. Unfortunately, no general method or algorithm of solution for handling all recurrence relations is known. In the following section, we shall study a class of recurrence relations known as linear recurrence relations with constant coefficients.

1.12 n^{th} ORDER RECURRENCE RELATIONS WITH CONSTANT COEFFICIENTS

Let S be a sequence of numbers with domain $k \geq 0$. Then a recurrence relation of the form

$$S(k) + c_1 S(k-1) + \ldots + c_n S(k-n) = f(k) \qquad \ldots (1)$$

where c_i's are constant and f is a numeric function defined for $k \geq n$ is called a **linear recurrence** relation of order n with constant coefficients.

Example 1.48

(i) $2\,S(k) + 3\,S(k-1) = 2^k$ is a first order recurrence relation with constant coefficients. And

$3a_r - 5a_{r-1} + 2a_{r-2} = r^2 + 5$ and $a_r + 2a_{r-3} = r^2$ are 2nd are 3rd order recurrence relation with constant coefficients respectively.

If b_1 and b_2 are arbitrary constants then $a_n = b_1 2^n + b_2 5^n$ satisfies the recurrence relation

$$a_n - 7a_{n-1} + 10a_{n-2} = 0,\ n \geq 2. \qquad \ldots (1)$$

Substituting the value of a_n in (1), we have

$$b_1 2^n + b_2 5^n - 7\left(b_1 2^{n-1} + b_2 5^{n-1}\right) + 10\left(b_1 2^{n-2} + b_2 5^{n-2}\right)$$

$$= b_1 2^n + b_2 5^n - 7b_1 2^{n-1} - 7b_2 5^{n-1} + 10b_1 2^{n-2} + 10b_2 5^{n-2}$$

$$= b_1 2^n + b_2 5^n - 7b_1 2^{n-1} - 7\ b_2 5^{n-1} + 5.2\ b_1 2^{n-2} + 5.2b_1 5^{n-2}$$

$$= b_1 2^n + b_2 5^n - 7b_1 2^{n-1} - 7b_2 5^{n-1} + 7b_1 2^{n-1} + 5b_1 5^{n-1}$$

For a nth order linear recurrence relation with constant coefficients, fewer than n values of the numeric function will not be sufficient to determine the numeric function uniquely. As we know for a given set of boundary conditions the solution to a linear recurrence relation with constant coefficients is unique, as long as we are able to find a numeric function that satisfies the recurrence relation as well as initial conditions or boundary conditions. Before giving an algorithm for solving finite order linear relations, we will examine some recurrence relations that arise from some closed form expressions.

Example 1.49

Solve the recurrence relation $T(k) = 2T(k - 1),\ \ T(0) = 1$.

Solution

Given $T(k) = 2T(k - 1),\ \ T(0) = 1$

$= 2 \cdot 2\ T(k - 2)$

$= 2^2\ T(k - 2) = 2^2 \cdot 2\ T(k - 3)$

$- - - - - - - - - - - - - - - - - - - -$

$- - - - - - - - - - - - - - - - - - - -$

$= 2^k\ T(k - k)$ [By Mathematical induction]

$= 2^k\ \mathrm{T}(0) = 2^k.1$ [Using initial condition]

$= 2^k\ \ \forall \mathrm{k} \geq 1$

which is the required solution.

ALITER

The same problem we can solve with the help of mathematical induction

$T(k) = 2\ T(k - 1)$.

$k = 1,\ \ T(1) = 2.T(0) = 2.1 = 2$

$k = 2,\ \ T(2) = 2.T(1) = 2.2 = 2^2$

$k = 3,\ \ T(3) = 2.T(2) = 2.\ 2^2 = 2^3$

$k = 4,\ \ T(4) = 2^4$

We can see that $T(k) = 2^k,\ k \geq 1$.

Example 1.50

Solve the recurrence relation $a_k = (1.08)\ a_{k-1}$ for $k \geq 1$, given $a_0 = 100$.

Solution

Given $a_k = (1.08)\, a_{k-1}$

When $k = 1,\ a_1 = (1.08)\, a_0 = 1.08\,(100)$

$k = 2,\ a_2 = (1.08)\, a_1 = (1.08)^2\,(100)$

$k = 3,\ a_3 = (1.08)^3\,(100)$

By mathematical induction, we have

$$a_k = (100)\ (1.08)^k$$

Which is the required solution.

ALITER

Given recurrence relation is

$$a_k = (1.08)\, a_{k-1} = (1.08)\,(1.08)\, a_{k-2}$$
$$= (1.08)^2\, a_{k-2} = (1.08)^3\, a_{k-3}$$
$$\text{-------------------------}$$
$$= (1.08)^k\, a_0 = 100\,(1.08)^k$$

Example 1.51

Obtain recurrence relation for the solution $D(k) = k^2 - k$.

Solution

Given $D(k) = k^2 - k.$ ...(1)

$\therefore\quad D(k-1) = (k-1)^2 - (k-1)$

$= k^2 - 2k + 1 - k + 1$

$= k^2 - 3k + 2$ ... (2)

Similarly $D(k-2) = (k-2)^2 - (k-2)$

$= k^2 - 5k + 6$ ... (3)

Multiplying (2) by 2 and subtracting from (1) and then adding (3), we get

$$D(k) - 2D(k-1) + D(k-2) = k^2 - k - 2k^2 + 6k - 4 + k^2 - 5k + 6 = 2$$

Putting $k = 0$, $k = 1$ in (1), we get

$$D(0) = 0,\ D(1) = 0.$$

$\therefore$ The recurrence relation is

$$D(k) - 2D(k-1) + D(k-2) = 2$$

With initial conditions $D(0) = 0$, $D(1) = 0$.

Example 1.52

If $S(k) = 3^{k-1} + 2^{k+1} + k$, $k \geq 0$. Find the recurrence relation.

Solution

Given $S(k) = 3^{k-1} + 2^{k+1} + k$ ...(1)

$S(k-1) = 3^{k-2} + 2^k + (k-1)$ ...(2)

Multiplying (ii) by 3 and subtracting from (i), we get

$$S(k) - 3S(k-1) = 3^{k-1} - 3\cdot 3^{k-2} + 2^{k+1} - 3.\ 2^k + k - 3(k-1)$$

$$S(k) - 3S(k-1) = -2^k - 2k + 3 \quad \text{(First order recurrence relation)}$$

Similarly, we can get the second order recurrence relation

$$S(k)-3S(k-1)+6S(k-2) = 2k-7$$

With initial conditions

$$C(6) = 7/3 \text{ and } C(1) = 5 \text{ for } k \geq 2.$$

Example 1.53

Find the recurrence relation of $a_n = A3^n + B(-4)^n$.

Solution

Given $a_n = A\cdot 3^n + B(-4)^n$ ... (1)

$\therefore\quad a_{n+1} = A\cdot 3^{n+1} + B\cdot(-4)^{n+1}$ ... (2)

and $a_{n+2} = A\cdot 3^{n+2} + B(-4)^{n+2}$ ... (3)

Now adding (2) and (3) and subtracting (1) from it, we get

$$a_{n+2} + a_{n+1} - a_n = 9A3^n + 3A3^n - A3^n + 16B(-4)^n + 4B(-4)^n - B(-4)^n$$

$$a_{n+2} + a_{n+1} - 12a_n = 0$$

which is the required recurrence relation.

EXERCISE 1.5

Find the recurrence relations for the closed form expressions:

(1) $B(k) = 5\cdot 2^k$

(2) $C(k) = 2k^2 + 1$

(3) $a_n = \frac{5}{2}.2^n + \frac{3}{2}n.2^n$

(4) $a_n = \frac{2}{5}.4^n + \frac{3}{5}(-1)^n$

(5) $D(k) = 2.4^k - 5(-3)^k$

(6) $E(k) = (3+k)2^k$

ANSWERS

(1) $B(k) - 2B(k-1) = 0$

(2) $C(k) - 2C(k-1) + C(k-2) = 4$

(3) $a_n - 4a_{n-1} + 4a_{n-2} = 0$ for $n \geq 2$ $a_0 = 5/2$, $q = 8$.

(4) $a_n - 3a_{n-1} - 4a_{n-2} = 0$ for $n \geq 2$ and $a_0 = a_1 = 1$.

(5) $D(k) - D(k-1) + 12D(k-2)$ for $k \geq 2$.
$D(0) = -3$, $D(1) = 23$.

(6) $E(k) - 4E(k-1) + 4E(k-2) = 0$ $k \geq 2$, $E(0) = 3$, $E(1) = 8$.

1.13 HOMOGENEOUS RECURRENCE RELATION

Suppose n and k are nonnegative integers. A linear recurrence relation of the form

$$S(k) + C_1 S(k-1) + \ldots\ldots\ldots + C_n S(k-n) = f(k)$$

or

$$S_k + C_1 S_{k-1} + \ldots\ldots\ldots + C_n S_{k-n} = f(k)$$

said to be **homogeneous** if $f(k)$ is identically zero i.e.

$$S(k) + C_1 S(k-1) + \ldots\ldots\ldots + C_n S(k-n) = 0$$

or

$$S_k + C_1 S_{k-1} + \ldots\ldots + C_n S_{k-n} = 0$$

otherwise it is said to be **inhomogeneous.**

Example 1.54

(i) $D(k) - 3D(k-1) + 2D(k-2) = 0$ is a second order homogeneous recurrence relation.

(ii) $a_n - 2a_{n-1} = 0$ is a first order homogeneous recurrence relation with $a_0 = 5$.

1.13 -1 HOMOGENEOUS SOLUTIONS

The total solution of a linear recurrence relation (or linear difference equation) with constant coefficient consist of two parts, the **homogeneous solution** and **particular solution.** The homogeneous solution satisfies the recurrence relation when R.H.S. is set to zero. We will discuss the particular solution later.

Let $a^{(h)} = (a_0, a_1, \ldots\ldots, a_r, \ldots.)$ denote the homogeneous solution of

$$S_k + c_1 S_{k-1} + \ldots\ldots\ldots + c_n S_{k-n} = 0 \qquad \ldots(1)$$

then

$$a_r + c_1 a_{r-1} + \ldots\ldots\ldots + c_n a_{r-n} = 0 \qquad \ldots(2)$$

A homogeneous solution of a linear recurrence relation with constant coefficients is of the form $A\alpha^r$, where α is called a **characteristic root** and A is a constant to be determined by the initial / boundary conditions. Putting $S_k = A\alpha^r$ in (1), we get

$$A\alpha^r + c_1 A\alpha^{r-1} + c_2 A\alpha^{r-2} + \ldots + c_n A\alpha^{r-n} = 0$$

$$A\alpha^{r-n} \times \left[\alpha^n + c_1\alpha^{n-1} + c_2\alpha^{n-2} + \ldots + c_n\right] = 0$$

$$\alpha^n + c_1\alpha^{n-1} + c_2\alpha^{n-2} + \ldots + c_n = 0 \qquad \ldots(3)$$

(3) is called the **characteristic equation** of the recurrence relation (difference equation). If α_1 is one of the roots (characteristic roots) of the characteristic equation then $A\alpha_1^r$ will be a homogeneous solution to the recurrence relation.

Example 1.55

Find the characteristic equation of

$$a_n - 7a_{n-1} + 12a_{n-2} = 0,\ n \geq 2 \qquad \ldots(1)$$

Putting $a_n = A \cdot \alpha^n$ in (1), we get

$$A(\alpha^n - 7\alpha^{n-1} + 12\alpha^{n-2}) = 0$$

or

$$A\alpha^{n-2}[\alpha^2 - 7\alpha + 12] = 0$$

or

$\alpha^2 - 7\alpha + 12 = 0$ is the characteristic equation.

Example 1.56

Characteristic equation of $S(k) + 2S(k-1) - 3S(k-2) - 6S(k-4) = 0$ is given by

$$\alpha^4 + 2\alpha^3 - 3\alpha^2 - 6 = 0.$$

1.13-2 ALGORITHM FOR SOLVING n^{th} ORDER HOMOGENEOUS LINEAR RECURRENCE RELATION

(1) First write down the characteristic equation of the relation

$$S_k + c_1 S_{k-1} + c_2 S_{k-2} + \dots\dots + c_n S_{k-n} = 0$$

which is $\alpha^n + c_1\alpha^{n-1} + c_2\alpha^{n-2} + \dots\dots + c_n = 0$.

(2) Find all the characteristic roots of characteristic equations. Let it be $\alpha_1, \alpha_2, \dots \alpha_n$.

(3) If all the roots are distinct and real, i.e. $\alpha_1 \neq \alpha_2 \neq \alpha_3 \neq \dots\dots \neq \alpha_n$, then the general solution is $A_1\alpha_1^k + A_2\alpha_2^k + A_3\alpha_3^k + \dots\dots + A_n\alpha_n^k = 0$, where A_1, A_2 ......, A_n are constants which are to be determined by initial conditions.

(4) If there is a multiple roots, say $\alpha_1 = \alpha_2, \alpha_3, \alpha_4, \dots, \alpha_n$,

then $S_k = (A_1 + A_2 k)\alpha_1^k + A_3\alpha_3^k + \dots\dots + A_n\alpha_n^k$ is a general solution.

Here α_1 is a double root.

(5) Find the values of the constants $A_1, A_2, \dots\dots, A_n$ by substituting it into initial conditions.

(6) To get the required solution, substitute the values of $A_1, A_2, \dots\dots, A_n$ in the general solution.

(7) When the characteristic equation has imaginary roots.

Let $\alpha + i\beta$ is a root of the characteristic equation. Since imaginary roots always occur in pair, hence $\alpha - i\beta$ will also be the root of the characteristic equation.

The solution corresponding to these roots will be given by

$$S_k = A_1(\alpha + i\beta)^k + A_2(\alpha - i\beta)^k \quad \dots (1)$$

Let $$\alpha + i\beta = r\cos\theta + ir\sin\theta \quad \dots (2)$$

then $$\alpha - i\beta = r\cos\theta - ir\sin\theta \quad \dots (3)$$

Now putting (2) and (3) in (1) and using Demoivre's theorem, we have

$$S_k = A_1(r\cos\theta + ir\sin\theta)^k + A_2(r\cos\theta - ir\sin\theta)^k$$

$$= r^k\left[A_1\cos k\theta + iA_1\sin k\theta + A_2\cos k\theta - iA_1\sin k\theta\right]$$

Let $B = A_1 + A_2$ and $C = A_1 - A_2$, then

$$S_k = r^k\left[B\cos k\theta + C\sin k\theta\right]$$

Where $r = \sqrt{\alpha^2 + \beta^2}$ and $\varphi = \tan^{-1}\dfrac{\beta}{\alpha}$

If the pair of complex roots $\alpha \pm i\beta$ occur twice, then the solution is given by

$$S_k = r^k\left[(B_1 + B_2 k)\cos k\theta + (C_1 + C_2 k)\sin k\theta\right]$$

REMARK

If the coefficients of characteristic equation are all integers and constant term equal to m, then the only possible rational characteristic roots are divisors of m, both positive and negative.

Example 1.57

Solve $a_n - 5a_{n-1} + 6a_{n-2} = 0$, where $a_0 = 2$ and $a_1 = 5$.

Solution

Given $$a_n - 5a_{n-1} + 6a_{n-2} = 0 \quad ...(1)$$

The characteristic equation is

$$\alpha^2 - 5\alpha + 6 = 0$$

or $$(\alpha - 3)(\alpha - 2) = 0$$

or $\alpha = 3, 2$ are the roots, hence the general solution

$$a_n = B_1 3^n + B_2 2^n \quad ...(2)$$

The given initial conditions are $a_0 = 2$ and $a_1 = 5$. Substituting these values in (2), we get

$$a_0 = B_1 3^0 + B_2 2^0$$

or $$2 = B_1 + B_2 \quad ...(3)$$

and $$5 = 3B_1 + 2B_2 \quad ...(4)$$

Solving (3) and (4) for B_1 and B_2, we get $B_1 = 1$ and $B_2 = 1$.

Therefore, the required solution is $a_n = 3^n + 2^n \; \forall \, n \geq 0$.

Example 1.58

Solve $T(k) - 7T(k-2) + 6T(k-3) = 0$,
where $T(0) = 8$, $T(1) = 6$ and $T(2) = 22$.

Solution

Given relation is

$$T(k) - 7T(k-2) + 6T(k-3) = 0 \quad ...(1)$$

The characteristic equation is

$$\alpha^3 - 7\alpha + 6 = 0$$

The roots of the characteristic equation are $\alpha = 1, 2, -3$,
Therefore, the general solution is

$$A_1 . 1^k + A_2 . 2^k + A_3 (-3)^k = 0 \quad ... (2)$$

The given initial conditions are

$$T(0) = 8, \; T(1) = 6 \text{ and } T(2) = 22$$

Putting the initial conditions in (2), we get

$$T(0) = A_1 + A_2 + A_3 = 8 \quad ... (3)$$

$$T(1) = A_1 + 2A_2 - 3A_3 = 6 \quad ... (4)$$

$$T(2) = A_1 + 4A_2 + 9A_3 = 22 \quad ...(5)$$

Solving (3), (4) and (5) for the values of A_1, A_2 and A_3, we get

$$A_1 = 5, \; A_2 = 2, \; A_3 = 1$$

Putting the values in (2), one can get the required solution

$$T(k) = 5 + 2^{k+1} + (-3)^k .$$

Example 1.59

Solve $a_r + 6a_{r-1} + 12a_{r-2} + 8a_{r-3} = 0$.

Solution

Given $a_r + 6a_{r-1} + 12a_{r-2} + 8a_{r-3} = 0$.

The characteristic equation is

$$\alpha^3 + 6\alpha^2 + 12\alpha + 8 = 0$$

The roots of the equation is α = -2, -2, -2 (triple characteristic root)

Therefore, the general solution is

$$a_r = \left(A_1 + A_2 r + A_3 r^2\right)(-2)^r.$$

Example 1.60

Solve $4t_r - 20t_{r-1} + 17t_{r-2} - 4t_{r-3} = 0$.

Solution

The characteristic root of the given recurrence relation is

$$4\alpha^3 - 20\alpha^2 + 17\alpha - 4 = 0$$

The characteristic roots are $\frac{1}{2}, \frac{1}{2}, 4$.

The general solution is given by

$$t_r = \left(A_1 + A_2 r\right)\left(\frac{1}{2}\right)^r + A_3\left(4^r\right).$$

Example 1.61

Solve the recurrence relation $a_r + a_{r-1} + a_{r-2} = 0$.

Solution

Given $a_r + a_{r-1} + a_{r-2} = 0$

The characteristic equation is

$$\alpha^2 + \alpha + 1 = 0$$

or

$$\alpha = \frac{-1 \pm i\sqrt{3}}{2}$$

The solution is given by

$$a_r = A_1\left(\frac{-1 + i\sqrt{3}}{2}\right)^r + A_2\left(\frac{-1 - i\sqrt{3}}{2}\right)^r.$$

Example 1.62

Solve the recurrence relation $a_{r+4} + 8a_{r+2} + 16a_r = 0$.

Solution

Given $a_{r+4} + 8a_{r+2} + 16a_r = 0$.

The characteristic equation is

$$\alpha^4 + 8\alpha^2 + 16 = 0$$

or

$$(\alpha^2 + 4)^2 = 0$$

The roots of the equation is

$$\alpha = \pm 2i, \pm 2i$$

Now comparing the roots with polar form, we have

$$\alpha = 0 \pm 2i = r\cos\theta + ir\sin\theta$$

$\Rightarrow$ $\quad r\cos\theta = 0$ and $r\sin\theta = 2$

Therefore $\quad r = 2$ and $\theta = \pi/2$

Hence the complementary function is given by

$$a_r = 2^r\left[(B_1 + B_2 r)\cos\left(r\frac{\pi}{2}\right) + (C_1 + C_2 r)\sin\left(r\frac{\pi}{2}\right)\right]$$

EXERCISE 1.6

Solve the following recurrence relations:

1. $a_r - 3a_{r-1} - 4a_{r-2} = 0$ for $r \geq 2$; $a_0 = a_1 = 1$.
2. $a_r - 4a_{r-1} + 4a_{r-2} = 0$ for $r \geq 2$; $a_0 = 5/2$, $a_1 = 8$.
3. $a_r - 7a_{r-1} + 10a_{r-2} = 0$; $a_0 = 0, a_1 = 3$.
4. $a_r + 5a_{r-1} + 5a_{r-2} = 0$; $a_0 = 0, a_1 = 2\sqrt{5}$.
5. $a_r - 2a_{r-1} + 2a_{r-2} - a_{r-3} = 0$; $a_0 = 2, a_1 = 1$ and $a_2 = 1$.
6. $a_r - 20a_{r-1} + 100a_{r-2} = 0$; $a_0 = 2, a_1 = 30$.
7. $a_r - a_{r-1} - 6a_{r-2} = 0$; $a_0 = 12, a_1 = -1$.
8. $a_r + a_{r-1} - 5a_{r-2} + 3a_{r-3} = 0$; $a_0 = 0, a_1 = 1$ and $a_2 = 2$.
9. $a_r - 4a_{r-1} - 11a_{r-2} + 30a_{r-3} = 0$, $a_0 = 0, a_1 = -35, a_2 = -85$.
10. $a_r - 9a_{r-1} + 18a_{r-2} = 0$, $a_0 = 0, a_1 = 4$.
11. $a_r - 7a_{r-2} + 10a_{r-4} = 0$; $a_0 = 7, a_1 = 8\sqrt{2} + 5\sqrt{5}, a_2 = 41, a_3 = 16\sqrt{2} + 25\sqrt{5}$.
12. $a_r - 6a_{r-1} + 12a_{r-2} - 8a_{r-3} = 0$.
13. $a_r - 7a_{r-1} + 16a_{r-2} - 12a_{r-3} = 0$, $a_0 = 1, a_1 = 4$ and $a_2 = 8$.
14. $2a_r - 7a_{r-1} + 3a_{r-2} = 0$; $a_0 = 1$ and $a_1 = 1$.
15. $6a_r - 19a_{r-1} + 15a_{r-2} = 0$; $a_0 = 0, a_1 = -\frac{1}{6}$.
16. $a_n = 2a_{n-1} - a_{n-2}$; $a_1 = 1.5$ and $a_2 = 3$.

ANSWERS

1. $a_r = \frac{2}{5}4^r + \frac{3}{5}(-1)^r$.
2. $a_r = \frac{5}{2}2^n + \frac{3}{2}r2^n$.
3. $a_r = 5^r - 2^r$.
4. $a_r = 2\left(\frac{-5+\sqrt{5}}{2}\right)^r - 2\left(\frac{-5-\sqrt{5}}{2}\right)^r$.

6. $(2 + r)10^r$.

9. $a_r = 4(-3)^r + 2^r - 5^{r+1}$.

10. $a_r = \left(\frac{2}{3}\right)3^r + \left(\frac{1}{3}\right)6^r$

11. $a_r = 3(\sqrt{2})^r - 5(-\sqrt{2})^r + 7(\sqrt{5})^r + 2(-\sqrt{5})^r$.

12. $a_r = \left(A_1 + A_2 r + A_3 r^2\right) 2^r$. 13. $a_r = 5.2^r + 3.n2^r - 4.3^r$.

1.13-3 Solution of Inhomogeneous Recurrence Relation

As discussed earlier the total solution of a difference equation consist of two parts i.e. homogeneous solution and particular solution. **Particular solution** is a solution which satisfies the difference equation with $f(k)$ on the R.H.S. i.e.

$$S_k + c_1 S_{k-1} + c_2 S_{k-2} + \dots\dots\dots + c_n S_{k-n} = f(k)$$

Let $a^{(h)} = \left(a_0^{(h)}, a_1^{(h)}, \dots\dots a_r^{(h)}\right)$ denote the homogeneous solution and $a^{(p)} = \left(a_0^{(p)}, a_1^{(p)}, \dots\dots a_r^{(p)}\right)$ denote the particular solution, then

$$a_r^{(h)} + c_1 a_{r-1}^{(h)} + c_2 a_{r-2}^{(h)} + \dots\dots\dots + c_n a_{r-n}^{(h)} = 0$$

and $$a_r^{(p)} + c_1 a_{r-1}^{(p)} + c_2 a_{r-2}^{(p)} + \dots\dots\dots + c_n a_{r-n}^{(p)} = f(k)$$

and complete solution will be

$$a_r^{(h)} + c_1 a_{r-1}^{(h)} + c_2 a_{r-2}^{(h)} + \dots\dots\dots + c_n a_{r-n}^{(h)} + a_r^{(p)} + c_1 a_{r-1}^{(p)} + c_2 a_{r-2}^{(p)} + \dots\dots\dots + c_n a_{r-n}^{(p)} = f(k)$$

i.e. $$S = a^{(h)} + a^{(p)}$$

Here one may recall that many discrete numeric function satisfy the given difference equation but only one of them satisfies the boundary conditions at the same time. Particular solution alone in general does not satisfy the boundary conditions but we can adjust the homogeneous solution so that the total solutions will satisfy the difference equation as well as boundary conditions.

There is no general method for determining the particular solution of a difference equation. In simple cases, however it can be determined by the method of inspection or by taking an educational guess. **Following is the algorithm for solving inhomogeneous finite order linear recurrence relation.**

Let the recurrence solution is $S_k + c_1 S_{k-1} + \dots\dots + c_n S_{k-n} = f(k)$

(i) Find the homogeneous solution by putting $f(k) = 0$.

(ii) To find the particular solution of the relation, take an educated guess in the form of a particular solution.

(iii) Substitute your educated guess from step (ii) in the recurrence relation. If your guess is correct you will be able to determine the unknown coefficient of your guess. If your guess is wrong, it should be apparent from the result of this substitution, so go back to step (ii).

(iv) The general solution of the recurrence relation is the sum of homogeneous and particular solutions.

(v) The solution obtained in the step (iv) is over if no initial conditions are given. If n initial conditions are given, obtain n linear equations in n unknowns and solve the system, and get a complete solution.

Form of Particular Solutions for Given R.H.S. of the Recurrence Relation

Table 1.1

S.No.	R.H.S. of Recurrence relation	Form of particular Solution
1	D, a constant	A, a constant
2	$D\,a^n$ and if a is not a root of characteristic polynomial	$A\,a^n$

3	Da^n and if a is a root of multiplicity m	$An^m a^n$
4	D_0+D_1k, a linear function	$A_0+A_1 k$,a linear function
5	$D_0+D_1k + D_2k^2+\ldots\ldots + D_mk^m$ an mth degree polynomial an mth degree polynomial	$A_0+A_1k+A_2k^2+\ldots.+A_mk^m$
6	$Dn^s a^n$, and if a is not a root of characteristic polynomial	$\left(A_0 + A_1n+\ldots..A_sn^s\right)a^n$
7	$Dn^s a^n$, if a is a root of multiplicity m	$n^m\left(A_0 + A_1n+\ldots..+ A_sn^s\right)a^n$
	Dn^s and if 1 is not a root	$A_0 + A_1n+\ldots..A_sn^s$
	Dn^s and if 1 is a root of multiplicity m	$n^m\left(A_0 +\ldots\ldots..A_sn^s\right)$

Example 1.63

Solve $T(k) + 5T(k-1) = 9$, $T(0) = 6$.

Solution

Associated homogeneous relation is

$$T(k) + 5T(k-1) = 0$$

And its characteristic equation is $\alpha+5=0$ or $\alpha=-5$.

$\therefore$ The homogeneous solution is

$$T^{(h)}(k)=A_1(-5)^k$$

To find the particular solution, we assume

$$T^{(p)}(k) = A \text{ as R.H.S.} = 9\text{, a constant}$$

Substituting it into the given relation we get

$$A + 5A = 9 \Rightarrow A = 1\cdot 5 \Rightarrow T^{(p)}(k) = 1.5$$

$\therefore$ The general solution is

$$T(k)=T^{(h)}(k)+T^{(p)}(k)$$

or $$T(k) = A_1(-5)^k + 1.5 \quad \ldots(1)$$

Initial condition is $T(0) = 6$, putting it into (1), we get

$$T(0) = A_1 + 1.5 = 6$$

or $$A_1 = 4.5$$

$\therefore$ The required solution is

$$T(k) = 4.5\,(-5)^k + 1.5.$$

Example 1.64

Solve $a_n - 6a_{n-1} + 8a_{n-2} = 3^n$, where $a_0 = 3$ and $a_1 = 7$.

Solution

The associated homogeneous relation of the given relation is

$$a_n - 6a_{n-1} + 8a_{n-2} = 0$$

and its characteristic equation is

$$\alpha^2 - 6\alpha + 8 = 0 \quad \Rightarrow \quad \alpha = 2, 4$$

$$a_n^{(h)} = A_1(2)^n + A_2(4)^n$$

Now, to find the particular solution, as R.H.S. = 3^n , let us assume $a_n^{(p)} = A\cdot 3^n$.
Putting it into the given recurrence relation, we get

$$A\cdot 3^n - 6\cdot 3^{n-1}A + 8\cdot 3^{n-2}A = 3^n$$

or $$3^{n-2}\left(3^2 - (6)^3 + 8\right)A = 3^n$$

or $$[9 - 18 + 8]\,A = 3^2$$

or $$-A = 9 \text{ or } A = -9$$

∴ $$a_n^{(p)} = (-9)3^n$$

The complete solution is

$$a_n = (-9)3^n + A_1(2)^n + A_2(4)^n \qquad \text{...(1)}$$

Given initial conditions are $a_0 = 3$ and $a_1 = 7$
Putting these in (1), we get

$$a_0 = (-9)\,3^0 + A_1(2)^0 + A_2(4)^0 = 3$$

or $$-9 + A_1 + A_2 = 3$$

or $$A_1 + A_2 = 12 \qquad \text{...(2)}$$

and $$a_1 = (-9)\,3^1 + A_1(2)^1 + A_2(4)^1 = 7$$

or $$-27 + 2A_1 + 4A_2 = 7$$

or $$2A_1 + 4A_2 = 34$$

or $$A_1 + 2A_2 = 17 \qquad \text{...(3)}$$

Solving for the values of A_1 and A_2 with the help of equation (2) and (3), we get

$$A_1 = 7,\ A_2 = 5$$

∴ The complete solution is $a_n = (-9)3^n + 5(4)^n + 7(2^n)$.

Example 1.65

Solve $S(k) - 7S(k-1) + 10S(k-2) = 6 + 8k$.

Solution

The associated homogeneous relation is

$$S(k) - 7S(k-1) + 10S(k-2) = 0$$

Its characteristic equation is $\alpha^2 - 7\alpha + 10 = 0 \Rightarrow \alpha = 2,5$

∴ The homogeneous solution is $S^{(p)}(k) = B_1 2^k + B_2 5^k$

Now to find the particular solution, we assume

$$S^{(p)}(k) = A_0 + A_1 k$$

Substituting this in given recurrence solution, we get

$$(A_0 + A_1k) - 7\,[A_0 + A_1(k-1)] + 10\,[A_0 + A_1(k-2)] = 6 + 8k$$

or $$4A_0 + 4A_1k - 13A_1 = 6 + 8k$$

or $$(4A_0 - 13A_1) + 4A_1k = 6 + 8k$$

Two polynomials are equal only if their coefficients are equal

$$4A_0 - 13A_1 = 6 \text{ and } 4A_1 = 8$$

⇒ $$A_1 = 2 \text{ and } A_0 = 8$$

$\therefore \qquad S^{(p)}(k) = 8 + 2k$

$\therefore$ The complete solution is

$$S(k) = B_1 2^k + B_2 5^k + 8 + 2k \qquad \ldots (1)$$

Given initial conditions are $S(0) = 1$ and $S(1) = 2$,
Substituting these in (1), we get

$$S(0) = B_1 + B_2 + 8 = 1 \Rightarrow B_1 + B_2 = -7$$

and $\qquad S(1) = 2B_1 + 5B_2 + 8 + 2 = 2$

$\Rightarrow \qquad 2B_1 + 5B_2 = -8$

Solving these, we get

$$B_1 = -9 \text{ and } B_2 = 2$$

$\therefore$ The complete solution is

$$S(k) = -9.2^k + 2.5^k + 8 + 2k.$$

Example 1.66

Solve $a_n - 6a_{n-1} + 8a_{n-2} = n4^n$ with initial conditions $a_0 = 8$ and $a_1 = 22$.

Solution

Its characteristic equation is

$$\alpha^2 - 6\alpha + 8 = 0 \Rightarrow \alpha = 2, 4$$

$\therefore \qquad a_n^{(h)} = B_1(2)^n + B_2(4)^n$

To find particular solution let us assume $a_n^{(p)} = n(A_0 + A_1 n)4^n$ as R.H.S. involves an exponential function (i.e. $n4^n$) with a base that equals characteristic equation [consult Table 1.1, S.No.7].

Substituting $a_n^{(p)}$ in the given recurrence relations, we have, we get

$$n(A_0 + A_1 n)4^n - 6(n-1)(A_0 + A_1(n-1))4^{n-1} + 8(n-2)(A_0 + A_1(n-2)4^{n-2} = n4^n$$

or $\qquad 4^{n-2}[16n(A_0 + A_1 n) - 24(n,-1)((A_0 + A_1)n - 1) + 8(n-2)(A_0 + A_1)(n-2) = n4^{n-2} \cdot 4^2$

or $\qquad 16n(A_0 + A_1 n) - 24(n-1)(A_0 + A_1(n-1)) + 8(n-2)(A_0 + A_1(n-2) = 16n$

Simplifying, we get $a_0 = 1$, $a_1 = -7$,

$\therefore \qquad a_n^{(p)} = n(n-7)4^n$

Hence, general solution is $a_n = B_1 2^n + B_2 4^n + n(n-7)4^n$.

Example 1.67

Solve the difference equation $a_r + 5a_{r-0} + 6a_{r-2} = 3r^2 - 2r + 1$.

Solution

Given difference equation is

$$a_r + 5a_{r-1} + 6a_{r-2} = 3r^2 - 2r + 1 \qquad \ldots (1)$$

The associated homogeneous difference equation is

$$a_r + 5a_{r-1} + 6a_{r-2} = 0$$

and its characteristic equation is

$$\alpha^2 + 5\alpha + 6 = 0 \Rightarrow \alpha = -2, -3,$$

hence the homogeneous solution is

$$a_r^{(h)} = B_1(-2)^r + B_2(-3)^r$$

For finding particular solution, let us assume

$$a_r^{(p)} = A_1 r^2 + A_2 r + A_3$$

Substituting this in (1), we get

$$A_1r^2 + A_2r + A_3 + 5A_1(r-1)^2 + 5A_2(r-1) + 5A_3 + 6A_1(r\text{-}2)^2 + 6A_2(r\text{-}2) + 6A_3 = 3r^2 - 2r + 1$$

Simplifying, we get

$$12A_1r^2 - (34A_1 - 12A_2)r + (29A_1 - 17A_2 + 12A_3) = 3r^2 - 2r + 1$$

Equating the coefficient of same degree variables, we get

$$12A_1 = 3 \quad \Rightarrow \quad A_1 = 4$$

$$34A_1 - 12A_2 = 2 \quad \Rightarrow \quad A_2 = \frac{13}{24}$$

$$29A_1 - 17A_2 + 12A_3 = 1 \quad \Rightarrow \quad A_3 = \frac{71}{288}$$

∴ The particular solution is

$$a_r^{(p)} = \frac{1}{4}r^2 + \frac{13}{24}r + \frac{71}{288}$$

And the complete solution is

$$a_r = B_1(-2)^r + B_2(-3)^r + \frac{1}{4}r^2 + \frac{13}{24}r + \frac{71}{288}.$$

Example 1.68

Find the general solution of $S_r - 4S_{r-1} + 4S_{r-2} = 2^r + r2^r$.

Solution

Given difference equation is

$$S_r - 4S_{r-1} + 4S_{r-2} = 2^r + r2^r \qquad ...(1)$$

The associated homogeneous difference equation is

$$S_r - 4S_{r-1} + 4S_{r-2} = 0$$

Its characteristic equation is

$$\alpha^2 - 4\alpha + 4 = 0 \quad \Rightarrow (\alpha - 2)^2 = 0$$

or $\alpha = 2, 2$

∴ The homogeneous solution is

$$S_r^{(h)} = (B_1 + B_2 r)2^r$$

Now to find the particulars solution let us assume

$$S_r^{(p)} = r^2(A_1 r + A_2)2^r \qquad ...(2)$$

(Since 2 is a double characteristic root).

Substituting (2) in (1), we get

$$r^2(A_1r + A_2)\, 2^r - 4(r-1)^2\, \{A_1(r-1) + A_2\}\, 2^{r-1}$$

$$+ 4(r-1)^2 \{A_1 (r-2) + A_2\} 2^{r-2} = (r+1) 2^r$$

After simplification, we obtain

$$6A_1 2^r r = 2^r r$$

and $$(-6A_1 + 2A_2) 2^r = 2^r$$

which gives $A_1 = \frac{1}{6}$ and $A_2 = 1$

$\therefore$ The particular solution is

$$S_r^{(p)} = r^2\left(\frac{r}{6}+1\right)2^r$$

The complete solution is

$$S_r = (B_1 + B_2 r)\, 2^r + r^2\left(\frac{r}{6}+1\right)2^r.$$

Example 1.69

Find the general solution is $a_r - a_{r-1} - 12a_{r-2} = (-3)^r + 6\cdot 4^r$.

Solution

The characteristic roots are –3 and 4, therefore homogeneous solution is

$$a_r^{(h)} = B_1(3)^r + B_2(4)^r$$

And the particular solution will be of the form

$$a_r^{(p)} = A_1 r(-3)^r + A_2 r 4^r$$

Try general solution yourself.

Example1.70

Solve the recurrence relation $a_{r+2} + a_r = r^2 + r + 1$.

Solution

Given recurrence relation is

$$a_{r+2} + a_r = r^2 + r + 1 \qquad \text{... (1)}$$

The associated homogeneous recurrence relation is

$$a_{r+2} + a_r = 0$$

and its characteristic equation is given by

$$\alpha^2 + 1 = 0 \quad \Rightarrow \quad \alpha = \pm i$$

Now, comparing the values with the polar form, we get

$$0 \pm i = r\cos\theta \pm ir\sin\theta$$

$$r\cos\theta = 0, \quad r\sin\theta = 1$$

$\Rightarrow$ $$r = \sqrt{0+1} = 1, \theta = \tan^{-1}\frac{1}{0} = \frac{\pi}{2}$$

hence the homogeneous solution is

$$a_r^{(h)} = 1^r\left[B_1 \cos r\frac{\pi}{2} + B_2 \cos r\frac{\pi}{2}\right]$$

For finding particular solution, let us assume

$$a_r^{(p)} = A_1 r^2 + A_2 r + A_3$$

Substituting this in (1), we get

$$\left(A_1(r+2)^2 + A_2(r+2) + A_3\right) + A_1 r^2 + A_2 r + A_3 = r^2 + r + 1$$

or

$$A_1\left(r^2 + 4r + 4\right) + A_2 r + 2A_2 + A_3 + A_1 r^2 + A_2 r + A_3 = r^2 + r + 1$$

or

$$2A_1 r^2 + 4A_1 r + 2A_2 r + 4A_1 + 2A_2 + 2A_3 = r^2 + r + 1$$

Equating the coefficients of same degree variables, we get

$$2A_1 = 0 \Rightarrow A_1 = \frac{1}{2}$$

$$4A_1 + 2A_2 = 1 \Rightarrow A_2 = -\frac{1}{2}$$

$$4A_1 + 2A_2 + 2A_3 = 1 \Rightarrow A_3 = 0$$

Therefore the particular solution is given by

$$a_r^{(p)} = \frac{1}{2}r^2 - \frac{1}{2}r + 0 = \frac{1}{2}r(r-1)$$

And the complete solution is given by

$$a_r = B_1 \cos r\frac{\pi}{2} + B_2 \sin r\frac{\pi}{2} + \frac{1}{2}r(r-1).$$

Example1.71

Solve the difference equation $a_{r+1} - 3a_r = 2^r \cos r\frac{\pi}{2}$.

Solution

Given difference equation is

$$a_{r+1} - 3a_r = 2^r \cos r\frac{\pi}{2} \qquad \ldots (1)$$

The associated homogeneous difference equation is

$$a_{r+1} - 3a_r = 0$$

and its characteristic equation is

$$\alpha - 3 = 0 \quad \text{or} \quad \alpha = 3$$

hence the homogeneous solution is given by

$$a_r^{(h)} = B3^r$$

For finding particular solution, let us assume

$$a_r^{(h)} = 2^r\left[A_1 \cos r\frac{\pi}{2} + A_2 \sin r\frac{\pi}{2}\right]$$

Substituting this in (1), we get

$$2^{r+1}\left(A_1 \cos(r+1)\frac{\pi}{2} + A_2 \sin(r+1)\frac{\pi}{2}\right) - 3\,.2^r\left[A_1 \cos r\frac{\pi}{2} + A_2 \sin r\frac{\pi}{2}\right] = 2^r \cos r\frac{\pi}{2}$$

or $$2^r\left(2A_1\cos\left(r\frac{\pi}{2}+\frac{\pi}{2}\right)+2A_2\sin\left(r\frac{\pi}{2}+\frac{\pi}{2}\right)\right)-3A_1\cos r\frac{\pi}{2}-3A_2\sin r\frac{\pi}{2}=2^r\cos r\frac{\pi}{2}$$

or $$2A_1\sin r\frac{\pi}{2}+2A_2\cos r\frac{\pi}{2}-3A_1\cos r\frac{\pi}{2}-3A_2\sin r\frac{\pi}{2}=\cos r\frac{\pi}{2}$$

or $$(2A_2-3A_1)\cos r\frac{\pi}{2}-(2A_1+3A_2)\sin r\frac{\pi}{2}=\cos r\frac{\pi}{2}$$

Equating the coefficients of $\cos r\frac{\pi}{2}$ and $\sin r\frac{\pi}{2}$ on both sides, we get

$$2A_2-3A_1=1 \qquad \text{... (2)}$$

$$-2A_1-3A_2=0 \qquad \text{... (3)}$$

From (2) $$A_1=-\frac{3}{2}A_2$$

Substituting this in (3), we have

$$2A_2-3\left(-\frac{3}{2}A_2\right)=1$$

or $$4A_2+9A_2=2 \Rightarrow A_2=\frac{2}{3}$$

Therefore $$A_1=-\frac{3}{2}\cdot\frac{2}{3}=-\frac{3}{13}$$

Hence the particular solution is

$$a_r^{(p)}=2^r\left(-\frac{3}{13}\cos r\frac{\pi}{2}+\frac{2}{13}\sin r\frac{\pi}{2}\right)$$

And the complete solution is given by

$$a_r=B3^r+2^r\left(-\frac{3}{13}\cos r\frac{\pi}{2}+2\sin r\frac{\pi}{2}\right).$$

1.13-4 METHOD OF OPERATOR

Consider the difference equation

$$S_{k+n}+C_1S_{k+n-1}+C_2S_{k+n-2}+\ldots+C_nS_n=f(k)$$

Which can be written in symbolic form as

$$\phi(E)S_k=f(k)$$

where $$\phi(E)=E^n+C_1E^{n-1}+\ldots+C_n$$

and E is the shifting operator which is defined as

$$S_{k+n}=E^nS_k$$

Then the particular solution is given by

$$S^{(p)}=\frac{1}{\phi(E)}f(k)$$

Now, let us take the following particular cases:

(1) When right hand side of difference equation is a^k, i.e.

$$f(k) = a^k$$

Then particular solution

$$S^{(p)} = \frac{1}{\phi(E)} a^k$$

Putting $E = a$, we have

$$S^{(p)} = \frac{1}{\phi(a)} a^k \quad (\text{provided } \phi(a) \neq 0)$$

If $\phi(a) = 0$, then

(a) $(E - a)S_k = a^k$

$$S^{(p)} = \frac{1}{(E-a)} a^k = k\, a^{k-1}$$

(b) $(E - a)^2 S_k = a^k$

$$S^{(p)} = \frac{1}{(E-a)^2} a^k = \frac{k(k-1)}{2!} a^{k-2}$$

(c) $(E - a)^3 S_k = a^k$

$$S^{(p)} = \frac{1}{(E-a)^3} a^k = \frac{k(k-1)(k-2)}{3!} a^{k-3}$$

and in general

(d) $(E - a)^k S_k = a^k$

$$S^{(p)} = \frac{1}{(E-a)^n} a^k = \frac{k(k-1)(k-2)\ldots(k-\overline{n-1})}{n!} a^{k-n}$$

When $f(k) = \sin ck$, where c is a constant then particular integral

$$S^{(p)} = \frac{1}{\phi(E)} \sin ck$$

$$= \frac{1}{\phi(E)} \left[\frac{e^{ick} - e^{-ick}}{2i} \right]$$

$$= \frac{1}{2i} \left[\frac{1}{\phi(E)} e^{ick} - \frac{1}{\phi(E)} e^{-ick} \right]$$

Now apply $\dfrac{a^k}{\phi(E)}$

(2) when $f(k) = \cos ck$, where c is a constant, then particular integral is given by

$$S^{(p)} = \frac{1}{\phi(E)}.\cos ck = \frac{1}{\phi(E)}\left[\frac{e^{ick} + e^{-ick}}{2}\right]$$

$$= \frac{1}{2}\left[\frac{1}{\phi(E)}e^{ick} + \frac{1}{\phi(E)}e^{-ick}\right]$$

(3) When $f(k) = \psi(k)$
then particular integral

$$S^{(p)} = \frac{1}{\phi(E)}.\psi(k)$$

$$= \frac{1}{\phi(1+\Delta)}\psi(k) \qquad \text{[since } E = 1 + \Delta\text{]}$$

Expanding by Binomial theorem, we have

$$S^{(p)} = \left(C_0 + C_1\Delta + ... + C_n\Delta^n\right)\psi(k)$$

It is to be noted here that the difference of order higher than that of polynomial will always be zero, i.e. $\Delta^m\psi(k) = 0$, if $m >$ degree of $\psi(k)$.

(4) When $f(k) = C^k f(k)$, where $f(k)$ is a polynomial.
Then the particular integral

$$S^{(p)} = \frac{1}{\phi(E)}.C^k f(k)$$

or

$$S^{(p)} = C^k \frac{1}{\phi(CE)}.f(k)$$

Now use case (3).

Example1.72

Solve the following difference equation $a_{k+2} - 7a_{k+1} + 6a_k = 2\cdot 3^k$.

Solution

Given difference equation is

$$a_{k+2} - 7a_{k+1} + 6a_k = 2\cdot 3^k$$

The associated homogeneous difference equation can be written as

$$E^2 a_k - 7Ea_k + 6a_k = 2\cdot 3^k$$

or

$$\left(E^2 - 7E + 6\right)a_k = 2\cdot 3^k$$

And its characteristic equation is given by

$$\alpha^2 - 7\alpha + 6 = 0$$

or

$$\alpha = 1, 6$$

Therefore, the homogeneous solution is

$$a_k^{(h)} = B_1 1^k + B_2 6^k$$

Now, particular solution is obtained by

$$a_k^{(p)} = \frac{2\cdot 3^k}{E^2 - 7E + 6} = 2\cdot\frac{3^k}{E^2 - 7E + 6}$$

$$= 2\cdot\frac{3^k}{3^2 - 7\cdot 3 + 6} = 2\cdot\frac{3^k}{-6} = -\frac{1\cdot 3^k}{3}$$

Hence the complete solution is

$$a_k = B_1 + B_2 6^k - \frac{1}{3}\cdot 3^k .$$

Example 1.73

Solve the recurrence relation $t_{r+2} + p^2 t_r = \sin pr$.

Solution

Associated homogeneous difference equation is

$$t_{r+2} + p^2 t_r = 0$$

Which can be written with the help of shifting operator as

$$\left(E^2 + p^2\right) t_r = 0$$

The characteristic equation is given by

$$\alpha^2 + p^2 = 0 \quad \text{or} \quad \alpha = 0 \pm ip$$

This can be written in polar form as

$$r\cos\theta \pm ir\sin\theta = 0 \pm ip$$

$$\Rightarrow \qquad r\cos\theta = 0, \quad r\sin\theta = p$$

i.e. $$\theta = \tan^{-1}\frac{p}{0} = \tan^{-1}\infty = \frac{\pi}{2}$$

$$r^2 = p^2 \Rightarrow r = p$$

Hence the complementary function (homogeneous solution) is given by

$$t^{(h)} = p^r\left[B_1 \cos r\frac{\pi}{2} + B_2 \sin r\frac{\pi}{2}\right]$$

Now to find the particular solution

$$t^{(p)} = \frac{\sin pr}{E^2 + p^2}$$

Here, we use Demoiver's theorem to find the particular integral, as we know

$$e^{i\theta} = \cos\theta + i\sin\theta$$

since here we have to find the particular integral corresponding to sine function, hence after writing with exponential function, we find particular integral for imaginary parts (I.P.).

Now, $$t_r^{(p)} = \text{I.P. of}\ \frac{e^{ipr}}{E^2 + p^2}$$

$$= \text{I.P. of}\ \frac{e^{ipr}}{(e^{ip})^2 + p^2}$$

$$= \text{I.P. of } \frac{e^{ipr}(e^{-2ip}+p^2)}{(e^{2ip}+p^2)(e^{-2ip}+p^2)}$$

$$= \text{I.P. of } \frac{e^{ip(r-2)}+p^2e^{ipr}}{1+p^2(e^{2ip}+e^{-2ip})+p^4}$$

$$= \text{I.P. of } \frac{\lfloor \cos p(r-2)+i\sin p(r-2)+p^2(\cos pr+i\sin pr)\rfloor}{1+2p^2\cos 2p+p^4}$$

hence
$$t_r^{(p)} = \frac{\sin p(r-2)+p^2\sin pr}{1+2p^2\cos 2p+p^4}$$

and the complete solution is given by

$$t_r = p^r\left[B_1\cos r\frac{\pi}{2}+B_2\sin r\frac{\pi}{2}\right]+\frac{p^2\sin pr+\sin p(r-2)}{1+2p^2\cos 2p+p^4}$$

Example1.74

Solve the difference equation $t_r+4t_{r-1}+4t_{r-2}=r^2-3r+5$.

Solution

The associated homogeneous equation of the given equation is

$$t_r+4t_{r-1}+4t_{r-2}=0$$

which can be written with the help of shifting operator as

$$(1+4E^{-1}+4E^{-2})t_r=0$$

The corresponding characteristic equation is

$$1+4\alpha^{-1}+4\alpha^{-2}=0$$

or
$$\alpha^2+4\alpha+4=0$$

or
$$\alpha=-2,-2$$

Hence the homogenous solution is

$$t_r^{(h)}=(B_1+B_2r)(-2)^r$$

Now to find the particular solution

$$t_r^{(p)}=\frac{1}{(E^2+4E+4)}(r^2-3r+5)$$

$$=\frac{1}{(E+2)^2}(r^2-3r+5)$$

$$=\frac{1}{(1+\Delta+2)^2}(r^2-3r+5) \qquad \text{[Since } E=1+\Delta]$$

$$=\frac{1}{(3+\Delta)^2}(r^2-3r+5)$$

$$= \frac{1}{9\left(1+\frac{\Delta}{3}\right)^2}(r^2-2r+5)$$

Let us express r in factorial notation $r^{(n)} = r(r-1)(r-2)...(r-n-1)$, and $r^{/} = r$.

$$t_r^{(p)} = \frac{1}{9}\left(1+\frac{\Delta}{3}\right)^{-2}\left(r^2-2r+5\right)$$

$$= \frac{1}{9}\left(1-\frac{2\Delta}{3}+\frac{3\Delta^2}{9}+...\right)\left(r^2-2r+5\right)$$

$$= \frac{1}{9}\left[r^2-2r+5-\frac{2\Delta}{3}\left(r^2-2r+5\right)+\frac{3\Delta^2}{9}\left(r^2-2r+5\right)+...\right]$$

$$= \frac{1}{9}\left[r^2-2r+5-\frac{2}{3}(2r-2)+\frac{3}{9}(2)\right]$$

$$= \frac{1}{9}\left[r^2-2r-\frac{4r}{3}+5+\frac{4}{3}+\frac{2}{3}\right]$$

$$= \frac{1}{9}\left[r^2-\frac{10r}{3}+\frac{21}{3}\right]$$

$$= \frac{1}{9}\left[r^2-\frac{10r}{3}+7\right]$$

Hence the complete solution is

$$t_r = (B_1+B_2r)(-2)^r + \frac{1}{9}\left[r^2-\frac{10r}{3}+7\right].$$

Example1.75

Solve the recurrence relation $t_{r+2}+4t_{r+1}-12t_r = 2^r(r-1)$.

Solution

The given equation with the help of characteristic equation can be written as

$$(E^2+4E-12)t_r = 2^r(r-1)$$

The characteristic equation is given by

$$\alpha^2+4\alpha-12=0$$

$$\Rightarrow \qquad \alpha=2,\ \alpha=-6$$

Hence the homogenous solution is

$$t_r^{(h)} = B_1 2^r + B_2(-6)^r$$

Now, particular solution is obtained by

$$t_r^{(p)} = \frac{2^r(r-1)}{E^2+4E-12} = \frac{2^r(r-1)}{(E-2)(E+6)}$$

$$= \frac{1}{(E-2)}\left[\frac{1}{(E+6)}2^r\right](r-1)$$

$$= \frac{1}{(E-2)}\left[\frac{2^r}{(2+6)}\right](r-1)$$

$$= \frac{1}{8}\cdot\frac{2^r\cdot(r-1)}{(E-2)}$$

$$= \frac{1}{8}\cdot\frac{2^r}{(E-2)}(r-1)$$

$$= \frac{1.2^r}{16}\cdot\frac{1}{(E-1)}(r-1)$$

$$= \frac{2^r}{16}\cdot\frac{1}{\Delta}(r-1) \qquad \text{[since } E = 1 + \Delta]$$

$$= \frac{2^r}{16}\cdot\frac{1}{\Delta}(r-1)$$

$$= \frac{2^r}{16}\cdot\Delta^{-1}(r^{(1)} - r^{(0)}) \qquad \text{[since } r^{(0)} = 1]$$

$$= \frac{2^6}{16}\left[r^{(2)} - r^{(1)}\right]$$

$$= \frac{2^r}{16}\left[\frac{r(r-1)}{2} - r\right] \qquad \text{[Operator } \Delta^{-1} \text{ is analogous with integral]}$$

$$= \frac{2^r}{16}\frac{r}{2}(r-1-2) = \frac{2^r}{32}r(r-3)$$

Hence the complete solution is given by

$$t_r = B_1 2^r + B_2(-6)^r + \frac{2^r}{32}r(r-3)$$

EXERCISE 1.7

Find the solution of following recurrence relations:

1. $a_r - 2a_{r-1} + a_{r-2} = 2$, $a_0 = 25$ and $a_1 = 16$.
2. $a_r + 6a_{r-1} + 9a_{r-2} = 3$, $a_0 = 0$ and $a_1 = 1$.
3. $a_r - 5a_{r-1} + 6a_{r-2} = 2$, $a_0 = 1$ and $a_1 = -1$.
4. $a_r = a_{r-1} + 7$.
5. $a_r - a_{r-1} - 6a_{r-2} = 30$, $a_0 = 20$ and $a_1 = -5$.

6. $a_r - 5a_{r-1} + 6a_{r-2} = 1$.

7. $a_r - 4a_{r-1} + 4a_{r-2} = 2^r$.

8. $a_r - 5a_{r-1} = 5^r$, $a_0 = 3$.

9. $a_r - 6a_{r-1} + 8a_{r-2} = 4^r$ $a_0 = 8$ and $a_1 = 22$.

10. $a_r - 9a_{r-1} + 20a_{r-2} = 2.5^r$.

11. $a_r - 7a_{r-1} + 10a_{r-2} = 3^r$, $a_0 = 1$ and $a_1 = 1$.

12. $a_r - 10a_{r-1} + 25a_{r-2} = 2^r$, $a_0 = 2/3$ and $a_1 = 1$.

13. $a_r + 2a_{r-1} = r + 3$ for $r \geq 1$ and $a_0 = 3$.

14. $a_r - 4a_{r-1} + 4a_{r-2} = 3r + 2^r$ $a_0 = 1$ and $a_1 = 1$.

15. $a_r = 1.08a_{r-1} + 2^r$ $a_0 = 1$.

16. $a_r - 2a_{r-1} = 2^r$.

17. $a_r - 7a_{r-1} + 12a_{r-2} = r\,4^r$.

18. $a_r - 7a_{r-1} + 12a_{r-2} = 3^r$.

19. $a_r - 2a_{r-1} + a_{r-2} = 4$.

20. $a_r + 3a_{r-1} - 10a_{r-2} = r^2 + r + 1$.

21. $a_r - 3a_{r-1} + 2a_{r-2} = 5r + 3$.

22. Given that $a_0 = 0$, $a_1 = 1$, $a_2 = 4$ and $a_3 = 12$ satisfy the recurrence relation $a_r + c_1 a_{r-1} + c_2 a_{r-2} = 0$, determine a_r.

23. Solve $a_r - a_{r-1} = 7r$ find also the solution when $f(r) = 7r^2$.

24. Solve the recurrence relation $a_r + 3a_{r-1} + 2a_{r-2} = f(r)$, given

$$f(r) = \begin{cases} 1 & r = 2 \\ 0 & \text{otherwise} \end{cases}$$ with boundary condition $a_0 = a_1 = 0$

25. $a_{r+2} - 2a_{r+1} + a_r = r^2 2^r$.

26. $a_{r+1} - 3a_r = 2^r \cos r\dfrac{\pi}{2}$.

27. $a_{r+2} - 7a_{r+1} + 12a_r = \cos r$.

28. $a_{r+2} - 6a_{r+1} + 8a_r = r^2 + 3^r$.

29. $a_{r+2} + a_r = \sin r\dfrac{\pi}{2}$.

30. $a_{r+2} + k^2 a_r = \cos kr$.

31. $a_{r+2} - 3a_{r+1} + 2a_r = \cos 2r$.

32. $a_{r+2} + a_{r+1} + a_r = r^2 + r + 1$.

33. $a_{r+3} + 2a_{r+2} + a_{r+1} = 2^r + r^2 + r$.

34. $t_{r+1} - at_r = a^2(2r+1)$ with $t_0 = \frac{1}{a}$.

ANSWERS

1. $a_r = r^2 - 10r + 25$

3. $a_r = -2 \cdot 3^r + 2^{r+1} + 1$

4. $a_r = B \cdot (1)^r + 7r$ (since 1 is a characteristic root $a_r^{(p)} = pr$)

5. $a_r = 11(-2)^r + 4 \cdot 3^r + 5$

6. $B_1 2^r + B_2 3^r + \frac{1}{2}$

7. $a_r = \frac{r^2}{2} \cdot 2^r + A_1 2^r + A_2 r 2^n$

8. $a_r = (3+r)5^r$

9. $a_r = r(r-1)4^r + 3(4^r) + 5 \cdot 2^r$

12. $a_r = \frac{2}{9} \cdot 5^r + \frac{1}{5} r 5^r + \frac{2^{r+2}}{9}$

13. $a_r = \frac{16}{9} \cdot (-2)^r + \frac{r}{3} + \frac{11}{9}$

14. $a_r = 12 + 3r + (r^2 + 7r + 22) \cdot 2^{r-1}$

15. $a_r = (-1.174)(1.08)^r + (2.174) \cdot 2^r$

16. $a_r = B2^r + Ar2^r$ (since 2 is a charateristic root)

17. $a_r = B_1 4^r + B_2 3^r + r(A_1 + A_2 r)4^r$

18. $a_r = B_1 4^r + B_2 3^r + Ar3^r$ (since 3 is a characteistic root)

19. $a_r = (B_1 + B_2 r) \cdot 1^r + A r^2$ (since 1 is a charateristic root(find the value of A))

25. $B_1 + B_2 r + 2^r (r^2 - 8r + 20)$

26. $B\, 3^r + \frac{2^r}{13}\left(-3\cos r\frac{\pi}{2} + 2\sin r\frac{\pi}{2}\right)$

27. $B_1 3^r + B_2 4^r + \frac{\cos(r-2) - 7\cos(r-1) + 12\cos r}{194 + 12\cos 2 - 91\cos 1}$

28. $B_1 2^r + B_2 4^r + \frac{1}{3}\left(r^2 + \frac{8}{3}r + \frac{38}{9}\right) - 3^r$

29. $B_1 \cos r\frac{\pi}{2} + B_2 \sin r\frac{\pi}{2} \frac{1}{2} \sin r\frac{\pi}{2}$

30. $k^r\left(B_1 \cos r\frac{\pi}{2} + B_2 \sin r\frac{\pi}{2}\right) + \frac{k^2 \cos k^r \cos(k(r-2))}{1 + 2k^2 \cos 2k + k^4}$

31. $B_1 + B_2 2^r + \frac{\cos\{2(r-2)\} - 3\cos\{2(r-1)\} + 2\cos 2r}{14 - 18\cos 2k + 4\cos 4}$

32. $B_1 \cos\frac{2\pi r}{3} + B_2 \sin\frac{2\pi r}{3} + \frac{r^2}{3} - \frac{r}{3} + \frac{1}{9}$

25. $(-B_1 + B_2 - B_3 r)(-1)^r + \frac{2^r}{18} + \frac{1}{9}\left(r^2 - 3r - \frac{1}{2}\right)$

26. $\frac{1}{a} \cdot a^r + r^2 a^{r-1}$.

1.14 DISCRETE NUMERIC FUNCTIONS

As we know that a function is a binary relation which associates to each element in the domain a unique element in the co-domain. In this section we shall study those functions whose domain is the set N of natural numbers and whose range is the set R of real numbers. These types of functions are known as **discrete numeric functions** or, in short, **numeric functions.** These functions are useful in digital computations to solve the counting problems involving non-negative integers.

Definition

A function whose domain is the set N of natural numbers and whose range is the set R of real numbers is said to be discrete numeric functions or in brief, numeric functions.

From the definition, it is clear that a numeric function associates to each non-negative integer r, a unique real number. We shall use bold face lower case letters to denote numeric functions. For a numeric function $\boldsymbol{a}$ we use $a_0, a_1, a_2, \ldots, a_r, \ldots$ to denote the values of the function at 0, 1, 2, ..., r instead of $\boldsymbol{a}(0), \boldsymbol{a}(1), \boldsymbol{a}(2), \ldots, \boldsymbol{a}(r), \ldots$

1.14-1 Representation of Numeric Functions

We can represent a numeric function is several different ways:

(i) By listing in order $(a_0, a_1, a_2, \ldots, a_r, \ldots)$ the values of a numeric function. For example (2, 3, 10, 29, ..., $r^3 + 2$, ...) is the numeric function a whose r^{th} term a_r is $r^3 + 2$.

(ii) By using a formula for its rth term. For example, the numeric function (2, 3, 10, 29, ..., $r^3 +$ 2, ...) can be represented as $a_r = r^3 + 2, \quad r \geq 0$.

The numeric function (0, 1, 2, 3, 17, 26, 37, ...) can be represented as

$$a_r = \begin{cases} r, & 0 \leq r \leq 3 \\ r^2 + 1, & r \geq 4 \end{cases}$$

(iii) A numeric function $(a_0, a_1, a_2, \ldots, a_r, \ldots)$ can also be represented by recurrence relation. For example, $a_0 = 2$ and $a_r = 2r\, a_{r-1}$, $r \geq 1$ describe the numeric function (2, 4, 16, 96, ...)

Example 1.76

A ball is dropped to the floor from a height 100 *m*. Suppose that the ball always rebounds to reach three fourth of the height from it falls. Find the numeric function $\boldsymbol{a}$, where a_r, denote the height it reaches in the rth rebound.

Solution

The numeric function so formed is given by the formula. $a_0 = 100$ and $a_r = \frac{3}{4} a_{r-1}$, $r \geq 1$

or
$$a_r = 100\left(\frac{3}{4}\right)^r, \quad r \geq 1$$

Example 1.77

In a process control system, a monitoring device measures the temperature inside a chemical reaction chamber once every 30 S. Let a_r denote the rth reading in degrees in centigrade. Find an expression for a_r if it is known that the temperature rises from 100 to 120° at a constant rate in the first 300 S and stays at 120° from then on.

Solution

We have

$$a_r = \begin{cases} 100 + \dfrac{r}{15}, & 0 \le r \le 300 \\ 120, & r \ge 330 \end{cases}$$

1.14-2 Manipulation of Numeric Functions

Sum of Numeric Functions

Let $\boldsymbol{a}$ and $\boldsymbol{b}$ be any two numeric functions. The sum $\boldsymbol{a} + \boldsymbol{b}$ is another numeric function $\boldsymbol{c}$ whose value at r is equal to the sum of the values of $\boldsymbol{a}$ and $\boldsymbol{b}$ at r, i.e. $c_r = a_r + b_r$.

Example 1.78

Consider two numeric functions $\boldsymbol{a}$ and $\boldsymbol{b}$,

where $$a_r = \begin{cases} 2r, & 0 \le r \le 2 \\ 3^r - 1, & r \ge 3 \end{cases}$$

and $$b_r = \begin{cases} 0, & 0 \le r \le 1 \\ r + 5, & r \ge 2 \end{cases}$$

Let $\boldsymbol{c}$ be the sum of $\boldsymbol{a}$ and $\boldsymbol{b}$ denoted by $\boldsymbol{a} + \boldsymbol{b}$. Then

$$c_r = a_r + b_r$$

$$= \begin{cases} a_0 + a_1 + b_0 + b_1\,, & 0 \le r \le 1 \\ a_2 + b_2, & r = 2 \\ a_r + b_r, & r \ge 3 \end{cases}$$

$$= \begin{cases} 2\,, & 0 \le r \le 1 \\ 11, & r = 2 \\ 3^r + r + 4, & r \ge 3 \end{cases}$$

1.14-3 Scalar Multiplication of a Numeric Function

Let $\boldsymbol{a}$ be any numeric function and λ be a real number (scalar). Then $\lambda \boldsymbol{a}$ be a numeric function whose value at r is λ times the value of a_r. This numeric function $\lambda \boldsymbol{a}$ is known as a scaled version of $\boldsymbol{a}$ with scaling factor λ.

Example1.79

Consider a numeric function $\boldsymbol{a}$, where

$$a_r = \begin{cases} r, & 0 \le r \le 3 \\ 3^r - 5, & r \ge 4 \end{cases}$$

then 5*a* is a numeric function whose value at *r* is $5a_r$, i.e.

$$5a_r = \begin{cases} 5r, & 0 \le r \le 3 \\ 5(3^r - 5), & r \ge 4 \end{cases}$$

1.14-4 PRODUCT OF NUMERIC FUNCTIONS

Let *P* be the product of two numeric functions *a* and *b*, denoted by *ab*. The value of *P* at is equal to the product of the values of two numeric function *a* and *b* at *r*, i.e

$$P_r = a_r b_r$$

Example1.80

Let
$$a_r = \begin{cases} r, & 0 \le r \le 3 \\ 3^r - 5, & r \ge 4 \end{cases}$$

And
$$b_r = \begin{cases} 2r, & 0 \le r \le 2 \\ r + 3, & r \ge 3 \end{cases}$$

Then
$$P_r = a_r b_r = \begin{cases} r \cdot 2r, & 0 \le r \le 2 \\ 3 \cdot (3+3), & r = 3 \\ (3^r - 5)(r+3), & r \ge 4 \end{cases}$$

$$= \begin{cases} 2r^2, & 0 \le r \le 2 \\ 18, & r = 3 \\ r3^r - 5r + 3^{r+1} - 15, & r \ge 4 \end{cases}$$

Verification

(i) Here $a_0 = 0,\ b_0 = 0 \Rightarrow P_0 = a_0 \cdot b_0 = 0.0 = 0$

(ii) $a_1 = 1,\ b_1 = 2 \Rightarrow p_1 = a_1 \cdot b_1 = 1{\cdot}2 = 2$

(iii) $a_2 = 2,\ b_2 = 4 \Rightarrow p_2 = a_2 \cdot b_2 = 2{\cdot}4 = 8$

(iv) $a_3 = 3,\ b_3 = 6 \Rightarrow p_3 = a_3 \cdot b_3 = 3{\cdot}6 = 18$

(v) $a_4 = 76,\ b_4 = 7 \Rightarrow p_4 = a_4 \cdot b_4 = 76{\cdot}7 = 532$

All the values are verified.

1.14-5 NUMERIC FUNCTIONS (SPECIAL)

(1) $S^n a$: Let *a* be a numeric function and *n* be a positive integer. A numeric function denoted by $S^n a$ is a function whose value at *r* is equal to 0 for *r* = 0, 1, 2, ..., *n* - 1 and is a_{r-n}, for $r \ge n$. In other words,

$$S^n a = \begin{cases} 0, & 0 \le r \le n-1 \\ a_{r-n}, & r \ge n \end{cases}$$

Example 1.81

Let $$a_r = \begin{cases} 0, & 0 \le r \le 7 \\ 2^{-r} + 5, & r \ge 8 \end{cases}$$

Find $S^7 a$.

Solution

By definition of $S^n a$.

$$S^7 a = \begin{cases} 0, & 0 \le r \le 7-1 \\ a_{r-7}, & r \ge 7 \end{cases}$$

$$= \begin{cases} 0, & 0 \le r \le 6 \\ a_{r-7}, & r \ge 7 \end{cases}$$

Here $$a_r = \begin{cases} 0, & 0 \le r \le 7 \\ 2^{-r} + 5, & r \ge 8 \end{cases}$$

Replacing *r* by *r* - 7, we get

$$a_{r-7} = \begin{cases} 0, & 0 \le r - 7 \le 7 \\ 2^{-(r-7)} + 5, & r - 7 \ge 8 \end{cases}$$

$$= \begin{cases} 0, & 7 \le r \le 14 \\ 2^{-(r+7)} + 5, & r \ge 15 \end{cases}$$

Therefore,

$$S^7 a = \begin{cases} 0, & 0 \le r \le 6 \\ 0, & 7 \le r \le 14 \\ 2^{-r+7} + 5, & r \ge 15 \end{cases}$$

$$= \begin{cases} 0, & 0 \le r \le 14 \\ 2^{-r+7} + 5, & r \ge 15 \end{cases}$$

Example1.82

Let *a* be *b* numeric function such that

$$a_r = \begin{cases} 2r, & 0 \le r \le 3 \\ 3^r \cdot r, & r \ge 4 \end{cases}$$

Find $S^2 a$.

Solution

By definition,

$$S^2 a = \begin{cases} 0, & 0 \le r \le 2-1 \\ a_{r-2}, & r \ge 2 \end{cases}$$

$$= \begin{cases} 0, & 0 \le r \le 1 \\ a_{r-2}, & r \ge 2 \end{cases}$$

Here $$a_r = \begin{cases} 2r, & 0 \le r \le 3 \\ 3^r \cdot r, & r \ge 4 \end{cases}$$

Replace r by r - 2, we get

$$a_{r-2} = \begin{cases} 2(r-2), & 0 \le r-2 \le 3 \\ 3^{r-2}(r-2), & r-2 \ge 4 \end{cases}$$

$$= \begin{cases} 2r-4, & 2 \le r \le 5 \\ 3^{r-2}(r-2), & r \ge 6 \end{cases}$$

Therefore

$$S^2 a = \begin{cases} 0, & 0 \le r \le 1 \\ 2r-4, & 2 \le r \le 5 \\ 3^{r-2}(r-2), & r \ge 6 \end{cases}$$

Let a be a numeric function and n be a positive integer. A numeric function denoted by $S^{-n} a$ is a function whose value at r is equal to a_{r+n} for $r \ge 0$. In other words,

$$S^{-n} a = a_{r+n}, \; r \ge 0.$$

Example1.83

Let $$a_r = \begin{cases} 1, & 0 \le r \le 10 \\ 2, & r \ge 11 \end{cases}$$

Find $S^{-5} a$.

Solution

By definition, $$S^{-5} a = a_{r+5}, \; r \ge 0$$

Here $$a_r = \begin{cases} 1, & 0 \le r \le 10 \\ 2, & r \ge 11 \end{cases}$$

Replacing r by $r + 5$ in a_r, we get

$$a_{r+5} = \begin{cases} 1, & 0 \le r+5 \le 10 \\ 2, & r+5 \ge 11 \end{cases}$$

$$= \begin{cases} 1, & -5 \le r \le 5 \\ 2, & r \ge 6 \end{cases}$$

But we know that r is a non-negative integer. Thus

$$a_{r+5} = \begin{cases} 1, & 0 \le r \le 5 \\ 2, & r \ge 6 \end{cases}$$

Hence $$S^{-5} a = \begin{cases} 1, & 0 \le r \le 5 \\ 2, & r \ge 6 \end{cases}.$$

1.14-6 ACCUMULATED SUM OF NUMERIC FUNCTIONS

The accumulated sum of a numeric function a is defined as $a_0 + a_1 + a_2 + \ldots + a_r$ then

$$b_r = \sum_{n=0}^{r} a_n, \quad r \geq 0.$$

Example1.84

A person deposits Rs. 200 each year in a saving account at an yearly compound interest rate of 6 percent. What is his total amount after r years.

Solution

$$\text{The total amount} = \text{sum}\left(1+\frac{\text{rate}}{100}\right)^{\text{time}}$$

Let $\boldsymbol{a}$ be a numeric function which gives the amount in the saving account at the end of rth year, then

$$a_r = 200\left(1+\frac{6}{100}\right)^r$$

$$= 200(1.06)^r, \quad r \geq 0$$

Clearly, the total amount in the account after r years is equal to the accumulated sum of the numeric function $\boldsymbol{a}$ i.e.

$$b_r = \sum_{n=0}^{r} a_n = \sum_{n=0}^{r} 200(1.06)^n$$

$$= 200\left(1+(1.06)+(1.06)^2+\ldots+(1.06)^r\right)$$

$$= \frac{20000}{6}\left[(1.06)^{r+1}-1\right], \; r \geq 0.$$

1.14-7 Forward Difference

Let $\boldsymbol{a}$ be a numeric function. The forward difference of a is a numeric function, whose value at r is equal to $a_{r+1} - a_r$ and denoted by $\Delta\boldsymbol{a}$. If $\boldsymbol{b}$ be the forward difference of $\boldsymbol{a}$ i.e. $\boldsymbol{b} = \Delta\boldsymbol{a}$; then

$$b_r = a_{r+1} - a_r.$$

Example1.85

Let $$a_r = \begin{cases} 2, & 0 \leq r \leq 3 \\ 2^{-r}+5, & r \geq 4 \end{cases}$$

Find $\Delta\boldsymbol{a}$.

Solution

Here $$a_r = \begin{cases} 2, & 0 \leq r \leq 3 \\ 2^{-r}+5, & r \geq 4 \end{cases}$$

$$a_{r+1} = \begin{cases} 2, & 0 \leq r+1 \leq 3 \\ 2^{-(r+1)}+5, & r+1 \geq 4 \end{cases}$$

$$= \begin{cases} 2, & 0 \leq r \leq 2 \\ 2^{-r-1}+5, & r \geq 3 \end{cases}$$

Let $\boldsymbol{b} = \Delta\boldsymbol{a}$, then

$$b_r = a_{r+1} - a_r$$
$$= \begin{cases} 0, & 0 \le r \le 2 \\ \dfrac{49}{16}, & r = 3 \\ -2^{-r-1}, & r \ge 4 \end{cases}$$

Verification

(i) $b_0 = a_1 - a_0 = 2 - 2 = 0$

(ii) $b_1 = a_2 - a_1 = 2 - 2 = 0$

(iii) $b_2 = a_3 - a_2 = 2 - 2 = 0$

(iv) $b_3 = a_4 - a_3 = \dfrac{81}{16} - 2 = \dfrac{49}{16}$

(v) $b_4 = a_5 - a_4 = (2^{-5} + 5) - (2^{-4} + 5)$
$= 2^{-5} - 2^{-4} = -2^{-4-1}$

(vi) $b_r = a_{r+1} - a_r = -2^{-r-1}, \quad r \ge 4$

Example 1.86

Show that $S^{-1}(\nabla \boldsymbol{a}) = \Delta \boldsymbol{a}$.

Solution

Let $\boldsymbol{a}$ be any numeric function such that $\boldsymbol{b} = \nabla \boldsymbol{a}$ and $\boldsymbol{c} = \Delta \boldsymbol{a}$, then, $b_r = a_r - a_{r-1}$, and $c_r = a_{r+1} - a_r$.

Let $\boldsymbol{d} = S^{-1}(\nabla a)$, then

$$d_r = b_{r+1}, \quad r \ge 0$$
$$= a_{r+1} - a_r = c_r$$

Hence, $S^{-1}(\nabla \boldsymbol{a}) = \Delta \boldsymbol{a}$.

1.14-8 Backward Difference

Let $\boldsymbol{a}$ be a numeric function. The backward difference of $\boldsymbol{a}$ is a numeric function, whose value at r is equal to $a_r - a_{r-1}$ at $r \ge 1$ and denoted by $\nabla \boldsymbol{a}$.

If $\boldsymbol{c}$ is the forward difference of $\boldsymbol{a}$ i.e. $\boldsymbol{c} = \Delta \boldsymbol{a}$; then

$$c_r = a_r - a_{r-1}.$$

Example1.87

Let $$a_r = \begin{cases} 2, & 0 \le r \le 3 \\ 2^{-r} + 5, & r \ge 4 \end{cases}$$

Find $\nabla \boldsymbol{a}$.

Solution

Here $$a_r = \begin{cases} 2, & 0 \le r \le 3 \\ 2^{-r} + 5, & r \ge 4 \end{cases}$$

$$a_{r-1} = \begin{cases} 2, & 0 \le r-1 \le 3 \\ 2^{-r+1}+5, & r-1 \ge 4 \end{cases}$$

$$= \begin{cases} 2, & 1 \le r \le 4 \\ 2^{-r+1}+5, & r \ge 5 \end{cases}$$

Let $c = \nabla a$, then

$$c_r = a_r - a_{r-1}$$

$$= \begin{cases} 0, & 0 \le r \le 3 \\ \dfrac{49}{16}, & r = 4 \\ -2^{-r}, & r \ge 5 \end{cases}$$

Example1.88

Determine $a * b$ if

$$a_r = 2^r \text{ for all } r \text{ and } b_r = \begin{cases} 0, & 0 \le r \le 2 \\ 2^r, & r \ge 3 \end{cases}$$

Solution

Let $c = a * b$. Then

$$c_0 = a_0 b_0 = 2^0 \cdot 0 = 0$$

$$c_1 = a_0 b_1 + a_1 b_0 = 2^0 \cdot 0 + 2^1 \cdot 0 = 0$$

$$c_2 = a_0 b_2 + a_1 b_1 + a_2 b_0 = 2^0 \cdot 0 + 2^1 \cdot 0 + 2^2 \cdot 0 = 0$$

$$c_3 = a_0 b_3 + a_1 b_2 + a_2 b_1 + a_3 b_0 = 2^0 \cdot 2^3 + 2^1 \cdot 0 + 2^2 \cdot 0 + 2^3 \cdot 0$$

$$= 1.2^3$$

$$c_4 = a_0 b_4 + a_1 b_3 + a_2 b_2 + a_3 b_1 + a_4 b_0$$

$$= 2^0 \cdot 2^4 + 2^1 \cdot 2^3 + 2^2 \cdot 0 + 2^3 \cdot 0 + 2^4 \cdot 0$$

$$= 2^4 + 2^4$$

$$= 2.2^4$$

$$c_5 = a_0 b_5 + a_1 b_4 + a_2 b_3 + a_3 b_2 + a_4 b_1 + a_5 b_0$$

$$= 2^0 \cdot 2^5 + 2^1 \cdot 2^4 + 2^2 \cdot 2^3 + 2^3 \cdot 0 + 2^4 \cdot 0 + 2^5 \cdot 0$$

$$= 1 \cdot 2^5 + 2^5 + 2^5$$

$$= 3.2^5$$

$$\ldots\ldots\ldots$$

$$c_r = a_0 b_r + a_1 b_{r-1} + \ldots + a_r b_0$$

$$= (r-2)2^r$$

Thus $c = a * b$ is given by $c_r = \begin{cases} 0, & 0 \le r \le 2 \\ (r-2)2^r, & r \ge 3 \end{cases}$

Example 1.89

Show that $S^{-1}(\nabla a) = \Delta a$.

Solution

Let $\boldsymbol{a}$ be any numeric function such that $\boldsymbol{b} = \nabla \boldsymbol{a}$ and $\boldsymbol{c} = \Delta a$, then, $b_r = a_r - a_{r-1}$, and $c_r = a_{r+1} - a_r$. Let $\boldsymbol{d} = S(\nabla \boldsymbol{a})$, then

$$\begin{aligned} d_r &= b_{r+1},\ r \geq 0 \\ &= a_{r+1} - a_r \\ &= c_r \end{aligned}$$

Hence, $S^{-1}(\nabla \boldsymbol{a}) = \Delta \boldsymbol{a}$.

1.14-9 CONVOLUTION OF NUMERIC FUNCTION

The convolution of two numeric functions $\boldsymbol{a}$ and $\boldsymbol{b}$, denoted by $\boldsymbol{a} * \boldsymbol{b}$, is a numeric function $\boldsymbol{c}$ such that

$$\begin{aligned} c_r &= \sum_{i=0}^{r} a_i\, b_{r-i} \\ &= a_0\, b_r + a_1 b_{r-1} + a_2\, b_{r-2} + \ldots + a_{r-1}\, b_1 + a_r\, b_0 \end{aligned}$$

Example1.90

Let $a_r = 2^r,\ r \geq 0$ and $b_r = 5^r,\ r \geq 0$

then, for $\boldsymbol{c} = \boldsymbol{a} * \boldsymbol{b}$, $c_r = \sum_{i=0}^{r} 2^i \cdot 5^{r-1},\ \ r \geq 0$.

EXERCISE 1.8

1. A ball is dropped to the floor from a height 40m. Suppose that the ball always rebounds to reach half of the height from it falls. Find the numeric function $\boldsymbol{a}$, where a_r denotes the height it reaches in the rth rebound.
2. Let a_r denotes the altitude of an aircraft, in thousands of feet, at the rth minute. Suppose the aircraft takes off after spending 10 minutes on the ground, climbs up at a uniform speed to a cruising altitude of 30,000 feet in 10 minutes, starts to descend uniformly after 110 minutes of flying time, and lands 10 minutes later. Find the numeric function $\boldsymbol{a}$.
3. Let $\boldsymbol{a}$ and $\boldsymbol{b}$ be any two numeric functions where:

$$a_r = \begin{cases} 0, & 0 \leq r \leq 2 \\ 2^{-r} + 5, & r \geq 3 \end{cases} \quad \text{and} \quad b_r = \begin{cases} 3 - 2^r, & 0 \leq r \leq 1 \\ r + 2, & r \geq 2 \end{cases}$$

   Find (i) Sum of $\boldsymbol{a}$ and $\boldsymbol{b}$.
   (ii) Product of $\boldsymbol{a}$ and $\boldsymbol{b}$.
   (iii) $\Delta \boldsymbol{a}$ and $\nabla \boldsymbol{a}$.
4. Let $\boldsymbol{a}$ be a numeric function such that

$$a_r = \begin{cases} 1, & 0 \leq r \leq 10 \\ 2, & r \geq 11 \end{cases}$$

   Find (i) $S^5(a)$ (ii) $S^{-7}(a)$.

5. Determine the convolution of ***a*** and ***b***, where

$$a_r = 3^r \ , \quad r \geq 0 \quad \text{and} \quad b_r = 2^r, \quad r \geq 0.$$

6. Let ***a, b*** and ***c*** be numeric functions such that ***a**b* = c.** Given that

$$a_r = \begin{cases} 1, & r = 0 \\ 2, & r = 1 \\ 0, & r \geq 2 \end{cases} \quad \text{and} \quad c_r = \begin{cases} 1, & r = 0 \\ 0, & r \geq 1 \end{cases}$$

determine ***b***.

7. Let $a_r = \begin{cases} 1, & r = 0 \\ 3, & r = 1 \\ 2, & r = 2 \\ 0, & r \geq 3 \end{cases}$ and $c_r = 5^r \ \forall r$.

Determine ***b*** such that ***a * b*** = *c*.

8. A person deposit Rs. 200 in a saving account at an interest rate 6% per year compounded annually. Obtain the amount in this account at the end of *r* year. Find the numeric function.

9. Let **a** be a numeric function such that

$$a_r = \begin{cases} 1 & 0 \leq r \leq 20 \\ 2 & r \geq 21 \end{cases}$$

Find (i) $S^{11}(\boldsymbol{a})$ (ii) $S^{-11}(\boldsymbol{a})$.

ANSWERS

1. $a_r = 40\left(\dfrac{1}{2}\right)^r, \ r \geq 0$

2. $a_r = \begin{cases} 0 & 0 \leq r \leq 10 \\ 3(r-10) & 11 \leq r \leq 20 \\ 30 & 21 \leq r \leq 120 \\ 3(130-r) & 121 \leq r \leq 130 \\ 0 & r \leq 131 \end{cases}$

3. (i) $c_r = \begin{cases} 3 - 2^r & 0 \leq r \leq 1 \\ 4 & r = 2 \\ 2^{-r} + r + 7 & r \geq 3 \end{cases}$

(ii) Product $= \begin{cases} 0 & 0 \leq r \leq 2 \\ r \cdot 2^{-r} + 2^{-r+1} + 5r + 10 & r \geq 3 \end{cases}$

(iii) $$\Delta a = \begin{cases} 0 & 0 \le r \le 1 \\ \dfrac{41}{8} & r = 2 \\ -2^{-r-1} & r \ge 3 \end{cases}$$

(iv) $$\Delta a = \begin{cases} 0 & 0 \le r \le 2 \\ \dfrac{41}{8} & r = 3 \\ -2^{-r} & r \ge 4 \end{cases}$$

4. (i) $$S^5(a) = \begin{cases} 0 & 0 \le r \le 4 \\ 1 & 5 \le r \le 15 \\ 2 & r \ge 16 \end{cases}$$

(ii) $$S^{-7}(a) = \begin{cases} 1 & 0 \le r \le 3 \\ 2 & r \ge 4 \end{cases}$$

5. $$a * b = \sum_{i=0}^{r} 3^i \, 2^{r-i}, \quad r \ge 0$$

6. $$b_r = (-2)^r, \quad r \ge 0$$

8. $$a_r = 200(1.06)^r, \quad r \ge 0$$

9. (i) $$S^{11}(a) = \begin{cases} 0 & 0 \le r \le 10 \\ 1 & 11 \le r \le 31 \\ 2 & r \ge 32 \end{cases}$$

(ii) $$S^{-11}(a) = \begin{cases} 1 & 0 \le r \le 9 \\ 2 & r \ge 10 \end{cases}.$$

1.15 GENERATING FUNCTIONS

As we know for a Computer science and Engineering student notion of alternative representation is very important. As decimal numbers can alternatively be represented with the help of binary numbers. And if we want to add, subtract, multiply or divide decimal numbers, we represent them as binary numbers, use a computer to carry out the arithmetic operations on binary numbers, and then obtain results of our computation by converting the results in binary numbers into decimal numbers. It can be noted here that suitably chosen alternatively form leads to the easiness and efficiency of operations.

We introduce here an alternative way to represent a sequence of numbers called **Generating function**, and it is one of very useful tools for solving the recurrence relation.

For a sequence S with terms $\langle S_0, S_1, S_2 \ldots\ldots, S_n \rangle$ we define an infinite series.

$$G(S,z) = S_0 + S_1 z + S_2 z^2 + S_3 z^3 + \ldots\ldots$$

or $$G(S,z) = \sum_{n=0}^{\infty} S_n z^n \text{, where } z \text{ is a variable.}$$

Readers are advised to see examples (11-18), chapter 10 also.

Example 1.91

Generating function of the sequence ⟨1, 3, 5, 7....⟩ is $G(S,z) = 1 + 3z + 5z^2 + 7z^3 + ...$.

Example 1.92

Generating function of the sequence $\langle 2^0, 2^1, 2^2, 2^3, 2^4, \ldots \rangle$ is

$$G(S,z) = 1 + 2z + 2^2 z^2 + 2^3 z^3 + \ldots$$
$$= 1 + 2z + (2z)^2 + (2z)^3 + \ldots$$
$$= \sum_{n=0}^{\infty} 2^n z^n$$

1.15-1 Closed form Expression for Generating Functions

As we know closed form expression is a compact way to represent a sequence, we see that from the closed form expression of a generating function a sequence can be manipulated conveniently.

Example 1.93

Find the closed form expression of the generating function for the sequence $\langle 3^0, 3^1, 3^2, 3^3, 3^4, 3^5, \ldots \rangle$.

Solution

The generating function is given by

$$G(S,z) = 1 + 3z + 3^2 z^2 + 3^3 z^3 + ...$$
$$= 1 + 3z + (3z)^2 + (3z)^3 + ...$$
$$= (1 - 3z)^{-1} = \frac{1}{(1 - 3z)}$$

is the required closed form expression.

Example 1.94

Obtain partial fraction composition and identify the sequence having the expression $\dfrac{5 + 2z}{1 - 4z^2}$ as a generating function.

Solution Given $$G(S,z) = \frac{5 + 2z}{1 - 4z^2}$$

$$= \frac{5 + 2z}{(1 + 2z)(1 + 2z)} = \frac{2}{1 + 2z} + \frac{3}{1 - 2z}$$

$$= 2(1+2z)^{-1} + 3(1-2z)^{-1}$$

Expanding with the help of binomial theorem, we get

$$= 2\left(1 - 2z + (2z)^2 - (2z)^3 + ...\right) + 3\left(1 + 2z + (2z)^2 - (2z)^3 + ...\right)$$

$$= 2\sum_{n=0}^{\infty} (-1)^n 2^n z^n + 3\sum_{n=0}^{\infty} 2^n Z^n$$

∴ The required sequence is

$$S(r) = 2(-2)^r + 3.2^r.$$

Example 1.95

Find the closed form expression for the Fibonacci sequence defined by $F_r = F_{r-1} + F_{r-2}$, $r \geq 2$.

Solution

Given Fibonacci sequence is

$$F_r = F_{r-1} + F_{r-2} \quad ... (1)$$

$$r \geq 2,\ F_0 = 0 \text{ and } F_1 = 1$$

And the generating function is given by

$$F(S,z) = F_0 + F_1 z + F_2 z^2 + F_3 z^3 + ...$$

$$= \sum_{n=0}^{\infty} F_n z^n \quad ...(2)$$

Multiplying both sides of (1) by z^r and sum over $r \geq 2$, we get

$$\sum_{r\geq 2} F_r z^r = \sum_{r\geq 2} F_{r-1} z^r + \sum_{r\geq 2} F_{r-2} z^r$$

$$F(S, z) - F_0 - F_1 z = z\sum_{r\geq 2} F_{r-1} z^{r-1} + z^2 \sum_{r\geq 2} F_{r-2} z^{r-2}$$

$$= z\,(F(S, z) - F_0) + z^2\,(F(S,z)) \quad \text{[by (2)]}$$

$$\therefore \quad F(S, z) - z\,F(S, z) - z^2 F(S,z) = F_0 + F_1 z - z\,F_0$$

Using initial condition, $F_0 = 0$ and $F_1 = 1$, we get

$$\left(1 - z - z^2\right) F(S, z) = z$$

$$\therefore \quad F(S, z) = \frac{z}{1 - z - z^2}$$

This is the closed form expression for the Fibonacci sequence as a rational function.

NOTE

1. Finite Geometric Series:

$$a + ar^2 + ... + ar^n = a\frac{(1 - r^{n-1})}{(1 - r)}.$$

2. Infinite Geometric Series

$$a + ar + ar^2 + \ldots\ldots = \frac{a}{1-r}.$$

3. Arithmetic Series

$$a+(a+d)+(a+2d)+\ldots+(a+(n-1)d)=\frac{n}{2}\left\{2a+\overline{(n-1)}d\right\}.$$

CLOSED FORM EXPRESSIONS OF THE GENERATING FUNCTION FOR SOME SEQUENCES

	Sequence	Generating function
1.	1	$\frac{1}{1-z}$
2.	$(-1)^n$	$\frac{1}{1+z}$
3.	a^n	$\frac{1}{1-az}$
4.	$(-a)^n$	$\frac{1}{1+az}$
5.	$C(k, n)$	$(1+z)^n$
6	$(C(k-1+n,n))\,a^n$	$\frac{1}{(1-x)^k}$
7.	n	$\frac{z}{(1-z)^2}$
8.	$n+1$	$\frac{1}{(1-z)^2}$
9.	n^2	$\frac{z(1+z)}{(1-z)^3}$
10.	n^3	$\frac{z(1+4z+z^2)}{(1-z)^4}$
11.	na^n	$\frac{az}{(1-az)^2}$
12.	$(n+1)a^n$	$\frac{1}{(1-az)^2}$
13.	ba^n	$\frac{b}{(1-az)}$
14.	n^2a^n	$\frac{az(1+az)}{(1-az)^3}$

15.	$(n+1)n$	$\frac{2z}{(1-z)^3}$
16.	$\frac{1}{n!}$	e^z
17.	bna^n	$\frac{abz}{(1-az)^2}$

EXERCISE 1.9

Find the closed form expression for the generating function of the sequence given below

1. $T(k) = 4,\ k \geq 0.$
2. $T(n) = S(n) + P(n) + U(n)$

   $= 9 \cdot 5^n + 4 + 3(-1)^n.$

   [Hint $G(T, z) = G(S, z) + G(P, z) + G(U, z)$].
3. $a_n = 3^n.$
4. $a_n = 3^n + 5^n.$
5. $a_n = 7(n+3)(n+2)(n+1).$
6. $a_n = n^2 3^n.$
7. $a_n = 9.5^n,\ n \geq 0.$
8. $a_n - 6a_{n-1} + 5a_{n-2} = 0, a_0 = 2, a_1 = 2.$
9. Obtain fractional decomposition and identify the sequence having the expression as a generating function :

   (i) $\frac{32-22z}{2-3z+z^2}$ (ii) $\frac{1}{5-6z+z^2}$

   (iii) $\frac{6-29z}{1-11z+30z^2}$ (iv) $\frac{7z^2}{(1-2z)(1+3z)}$

   (v) $\frac{(1+z)^2}{(1-z)^4}$ (vi) $\frac{3-5z}{1-2z-3z^2}$

   (vii) $\frac{z^5}{5-6z+z^2}$ (viii) $(1+z)^n + (1+z)^n.$

ANSWERS

1. $G(T, z) = \frac{4}{1-z}$ 2. $\frac{16-34z-11z^2}{(1-5z-z^2+5z^3)}$

3. $\dfrac{1}{1-3z}$ 7. $\dfrac{9}{1-5z}$

8. $a_n = 2, n \geq 0;\quad G(a,z) = \dfrac{2}{1-z}$

9. (i) $G(S,z) = \dfrac{10}{1-z} + \dfrac{12}{2-z},\ S(k) = 10 + 6\left(\dfrac{1}{2}\right)^k$

(iii) $G(S,Z) = \dfrac{-1}{1-5z} + \dfrac{7}{1-6z},\ S(k) = 7.6^k - 5^k$

(vi) $G(S,z) = \dfrac{1}{1-3z} + \dfrac{2}{(1+z)},\ S(k) = 3^k + 2.(-1)^k.$

1.15-2 Properties of Generating Function

Let S and T be two sequences and their generating functions are $G(S, z)$ and $G(T,z)$ respectively, then $G(S, z) = \sum_{n=0}^{\infty} S_n z^n$ and $G(S, z) = \sum_{n=0}^{\infty} T_n z^n$

1. Uniqueness $G(S, z) = G(T, z)$ if and only if $S_n = T_n \forall n$.
2. Sum of two generating functions is again a generating function.

$$G(S, z) + B(S, z) = \sum_{n=0}^{\infty} S_n z^n + \sum_{n=0}^{\infty} T_n z^n$$

$$= \sum_{n=0}^{\infty} (S_n + T_n) z^n$$

$$= \sum_{n=0}^{\infty} C_n z^n = G(C, z) \qquad \text{(Here } C_n \text{ is sum of series } S_n \text{ and } T_n\text{)}$$

3. Let C be a number, then scalar product $C.G(S, z)$ is again a generating function.

$$C.G(S, z) = C. \sum_{n=0}^{\infty} S_n z^n = \sum_{n=0}^{\infty} (C.S_n) z^n$$

4. The product of two generating functions is again a generating function.
Let $G(S, z)$ and $G(T, z)$ be two sequences, then

$$G(S, z) \cdot G(T, z) = \sum_{n=0}^{\infty} S_n z^n \sum_{n=0}^{\infty} T_n z^n$$

$$= \sum_{n=0}^{\infty} C_n z^n \text{, where } C_n = \sum_{n=0}^{\infty} S_i T_{n-i}$$

5. The monomial product $z^n\ G(S,z),\ n \geq 1$ is again a generating function.

$$z^n G(S,z) = z^n \sum_{m-0}^{\infty} S_m z^m = \sum_{m=0}^{\infty} S_m z^{m+n} = \sum_{m=n}^{\infty} S_{m-n} z^m$$

6. **Differentiation**

Let $G(S, z)$ be an infinitely differentiable function, then

$$G(S, z) = S_o + S_1 z + S_2 z^2 + ... = \sum_{n=0}^{\infty} S_n z^n$$

Its first differential coefficient is given by

$$G'(S, z) = S_1 + 2S_2 z + 3S_3 z^2 + ... = \sum_{n=1}^{\infty} nS_n z^{n-1}$$

Similarly, second differential coefficient is given by

$$G''(S, z) = 2S_2 + 6S_3 z + ... = \sum_{n=2}^{\infty} n(n-1)S_n z^{n-2}$$

By mathematical induction,

$$G^{m}(S,z) = \sum_{n=1}^{\infty} n(n-1)(n-2)..(n-m+1)\, S_n z^{n-m} .$$

7. If $G(A\ Z)$ is generating function of $A(r)$.

(a) Then $Z^i G(A, Z)$ is generating function of $S^i A$. and

(b) $z^{-i}[G(A,Z) - A_0 - A_1 Z - ... - A_{i-1} Z^{i-1}]$ is generating function of $S^{-i} A$,

where $S^i A = \begin{cases} 0 & 0 \le r \le i-1 \\ A_{r+i} & r \ge i \end{cases}$ and $S^{-i} A = A_{r+i} \quad r \ge 0$

8. If $S(r) = \alpha^r T(r)$ then $G(S, Z) = G(T, \alpha Z)$, where α is any constant

1.15-3 Solution of a Recurrence Relation using Generating Function

Outline of the method of generating Functions.

To solve the recurrence relation with constant coefficients of degree n;

$$S_k + C_1 S_{k-1} + ... + C_n S_{k-n} = 0 ,$$

where $C_1, C_2, C_3, ... , C_n$ are constants $C_n \neq 0$ and $k \ge n$.

Let $S(z) = \sum_{k=0}^{\infty} a_k z^k$,

(i) Multiply each term of the recurrence relation by z^n, sum from n to ∞, and replace all infinite sums by equivalent expressions, then transform the recurrence relation into an algebraic equation.

$$S(z) = \frac{P(z)}{Q(z)},$$

where $P(z) = S_0 + (S_1 + C_1 S_0)z + (S_2 + C_1 S_1 + C_2 S_0)z^2 + . . . + (S_{n-1} + C_1 S_{n-2} + . . . + C_{n-1} S_0)z^{n-1}$

and $Q(z) = 1 + C_1 z + C_2 z^2 + . . . + C_n z^n$

(ii) After getting the values of $P(z)$ and $Q(z)$, transform $S(z)$ back to obtain the coefficients of S_n. This can be done in following two ways :

(a) Factorize $Q(z)$ and then use partial fractions to get $S(z)$ as a sum of familiar series and hence that S_k is the sum of the coefficients of known series.

(b) If the initial conditions are given then, we can find as many coefficients of S(z) as we desire even if we cannot factorize $Q(z)$ by long division of $Q(z)$ into $P(z)$ [or by finding the multiplicative inverse of $Q(z)$ since the constant term of $Q(z)$ is non zero].

Example 1.96

Solve $a_n - 2a_{n-1} - 3a_{n-2} = 0,\ \ n \geq 2$ with $a_0 = 3,$ and $a_1 = 1$.

Solution

Given recurrence relation is

$$a_n - 2a_{n-1} - 3a_{n-2} = 0 \qquad \text{... (1)}$$

The generating function is given by

$$G(a, z) = \sum_{n=0}^{\infty} a_n z^n = a_0 + a_1 z + a_2 z^2 + \ldots \qquad \text{.. (2)}$$

Multiplying (1) by z^n and sum over all $n \geq 2$, we get

$$\sum_{n\geq 2} a_n z^n - 2\sum_{n\geq 2} a_{n-1} z^n - 3\sum_{n\geq 2} a_{n-2} z^n = 0$$

$$(G(a,z))a_0 - a_1 z) - 2z \sum_{n\geq 2} a_{n-1} z^{n-1} - 3z^2 \sum_{n\geq 2} a_{n-2} z^{n-2} = 0 \qquad \text{[Using (2)]}$$

Now using initial conditions $a_0 = 3$ and $a_1 = 1$, we get

$$\{G(a,z) - 3 - z\} - 2z\{G(a,z) - a_0\} - 3z^2 G(a,z) = 0$$

or $$G(a,z) - 3 - z - 2z\{G(a,z) - 3\} - 3z^2 G(a,z) = 0$$

or $$(1 - 2z - 3z^2)\, G(a,z) = 3 - 5z$$

$\therefore$ $$G(a,z) = \frac{3-5z}{1-2z-3z^2}$$

$$= \frac{3-5z}{(1-3z)(1+z)}$$

$$= \frac{A}{1-3z} + \frac{B}{1+z}$$

$\therefore$ The required solution is

$$a_n = 1.3^n + 2(-1)^n = 3^n + 2(-1)^n.$$

Example 1.97

Solve the following recurrence relation $a_n - 9a_{n-1} + 26a_{n-2} - 24a_{n-3} = 0$ for $n \geq 3$ with $a_0 = 0, a_1 = 1$ and $a_2 = 10$.

Solution

Given recurrence relation is

$$a_n - 9a_{n-1} + 26a_{n-2} - 24a_{n-3} = 0 \qquad \ldots(1)$$

The generating function is given by

$$G(a,z) = \sum_{n=0}^{\infty} a_n z^n = a_0 + a_1 z + a_2 z^2 + \ldots. \qquad \ldots(2)$$

Multiplying throughout (1) by z^n and sum from 3 to ∞. Since $n \geq 3$, we get

$$\sum_{n=3}^{\infty} a_n z^n - 9\sum_{n=3}^{\infty} a_{n-1} z^n + 26\sum_{n=3}^{\infty} a_{n-2} z^n - 24\sum_{n=3}^{\infty} a_{n-3} z^n = 0$$

Replacing the infinite sum by equivalent expressions, we get

$$\left(G(a,z) - a_0 - a_1 z - a_2 z^2\right) - 9z(G(a,z) - a_0 - a_1 z) + 26z^2(G(a,z) - a_0) - 24z^3 G(a,z) = 0.$$

Simplifying, we get

$$G(a,z)(1 - 9z + 26)z^2 - 24z^3) = a_0 + a_1 z + a_2 z^2 - 9a_0 z - 9a_1 z^2 + 26a_0 z^2$$

or $$G(a,z) = \frac{a_0 + (a_1 - 9a_0)z + (a_2 - 9a_1 + 26a_0)z^2}{1 - 9z + 26z^2 - 24z^3}$$

Now, using initial conditions $a_0 = 0, a_1 = 1$ and $a_2 = 10$, we get

$$G(a,z) = \frac{z + z^2}{(1-2z)(1-3z)(1-4z)}$$

$$= \frac{A_1}{(1-2z)} + \frac{A_2}{(1-3z)} + \frac{A_3}{(1-4z)}$$

$$= \frac{3}{2(1-2z)} - \frac{4}{1-3z} + \frac{5}{2(1-4z)}$$

$\therefore$ $$G(a,z) = \frac{3}{2}\left(2^n\right) - 4\left(3^n\right) + \frac{5}{2}\left(4^n\right)$$

Example 1.98

Solve $a_n - 8a_{n-1} + 21a_{n-2} - 18a_{n-3} = 0$ for $n \geq 3$.

Solution

The given recurrence relation is $a_n - 8a_{n-1} + 21a_{n-2} - 18a_{n-3} = 0$ ... (1)

The generating function is given by

$$G(a,z) = \sum_{n=0}^{\infty} a_n z^n = a_0 + a_1 z + a_2 z^2 + \ldots. \qquad \ldots (2)$$

Multiplying (1) by z^n and sum over all $n \geq 3$, we get

$$\sum_{n\geq 3}^{\infty} a_n z^n - 6\sum_{n\geq 3}^{\infty} a_{n-1} z^n + 12\sum_{n\geq 3}^{\infty} a_{n-2} z^n - 8\sum_{n\geq 3}^{\infty} a_{n-3} z^n = 0$$

Replacing the infinite sum by equivalent expression and simplifying, we get

$$G(a,z) = \frac{a_0 + (a_1 - 6a_0)z + (a_2 - 6a_1 + 12a_0)z^2}{1 - 6z + 12z^2 - 8z^3}$$

$$= \frac{a_0 + (a_1 - 6a_0)z + (a_2 - 6a_1 + 12a_0)z^2}{(1-2z)^3}$$

$$G(a,z) = \frac{A_1}{1-2z} + \frac{A_2}{(1-2z)^2} + \frac{A_3}{(1-2z)^3}$$

$\Rightarrow$ $$G(a,z) = A_1 2^n + A_2 (n+1) 2^n + A_3 \frac{(n+2)(n+1)}{2} 2^n$$

Example1.99

Solve the recurrence relation $a_{n+2} - 5a_{n+1} + 6a_n = 2$, $a_0 = 1$, $a_0 = 2$, by the method of generating function.

Solution

Given recurrence relation is

$$a_{n+2} - 5a_{n+1} + 6a_n = 2, \qquad \text{... (1)}$$

The generating function is given by

$$G(a,z) = \sum_{n=0}^{\infty} a_n z^n = a_0 + a_1 z + a_2 z^2 + \ldots \qquad \text{... (2)}$$

Multiplying (1) by z^n and sum from $r = 0$ to ∞, we have

$$\sum_{n=0}^{\infty} a_{n+2} z^n - 5 \sum_{n=0}^{\infty} a_{n+1} z^n + 6 \sum_{n=0}^{\infty} a_n z^n = \sum_{n=0}^{\infty} 2z^n$$

or $$(a_2 + a_3 z + a_4 z^2 + \ldots) - 5(a_1 + a_2 z + a_3 z^2 + \ldots) + 6G(a,z) = 2(1 + z + z^2 + \ldots)$$

or $$\frac{(a_2 z^2 + a_3 z^3 + \ldots)}{z^2} - 5\frac{(a_1 z + a_2 z^2 + \ldots)}{z} + 6G(a,z) = 2(1-z)^{-1}$$

$$\frac{(a_0 + a_1 z + a_2 z^2 + \ldots) - a_0 - a_1 z}{z^2} - \frac{5(a_0 + a_1 z + a_2 z^2 + \ldots) - a_0}{z^2} + 6G(a,z) = \frac{2}{1-z}$$

or $$\frac{G(a,z) - a_a - a_1 z}{z^2} - 5\frac{G(a,z) - a_0}{z} + 6G(a,z) = \frac{2}{1-z}$$

Now using initial condition $a_0 = 1$ and $a_1 = 2$, we have

$$\frac{G(a,z) - 1 - 3z}{z^2} - 5\frac{G(a,z) - 1}{z} + 6G(a,z) = \frac{2}{1-z}$$

or $$(1 - 5z + 6z^2)G(a,z) = \frac{2z^2}{1-z} + 1 - 3z$$

or $$G(a,z) = \frac{1 - 4z + 5z^2}{(1 - 5z + 6z^2)(1-z)}$$

$$G(a,z) = \frac{A}{1-z} + \frac{B}{1-2z} + \frac{C}{1-3z}$$
$$= \frac{1}{1-z} - \frac{1}{1-2z} + \frac{1}{1-3z}$$

Hence the solution after inverse transformation is given by

$$a_n = 1 - 2^n + 3^n$$

Example1.100

Solve $4a_{n+2} - 4a_{n+1} + a_n = 2^{-n}$, $a_0 = 0$, $a_1 = 1$, by the method of generating function.

Solution

Given recurrence relation is

$$4a_{n+2} - 4a_{n+1} + a_n = 2^{-n} \quad \dots (1)$$

The generating function is given by

$$G(a,z) = \sum_{n=0}^{\infty} a_n z^n = a_0 + a_{1z} + a_2 z^2 + \dots \quad \dots (2)$$

Multiplying (1) by z^n and summing from $n = 0$ to $n = \infty$, we have

$$4\sum_{n=0}^{\infty} a_{n+2} z^n - 4\sum_{n=0}^{\infty} a_n z^n + \sum_{n=0}^{\infty} a_n z^n = \sum_{n=0}^{\infty} 2^{-n} z^n$$

or $$4(a_2 + a_3 z + a_4 z^2 + \dots) - 4(a_1 + a_2 z + \dots) + G(a,z) = 1 + \frac{z}{2} + \left(\frac{z}{2}\right)^2 + \dots$$

or $$\frac{4(a_2 z^2 + a_3 z^3 + \dots)}{z^2} - 4\frac{(a_1 z + a_2 z^2 + \dots)}{z} + G(a,z) = \frac{1}{1 - \frac{z}{2}}$$

or $$\frac{4(a_0 + a_1 z + a_2 z^2 + a_3 z^3 \dots) - a_0 - a_1 z}{z^2} - \frac{4(a_0 + a_1 z + a_2 z^2 + \dots) - a_0}{z} + G(a,z) = \frac{2}{2-z}$$

or $$4[G(a,z) - a_0 - a_1 z] - 4z[G(a,z) - a_0] + z^2 G(a,z) = \frac{2z^2}{2-z}$$

Using initial condition $a_0 = 1$ and $a_1 = 1$, we have

$$4[G(a,z) - 0 - 1.z] - 4z[G(a,z) - 0] + z^2 G(a,z) = \frac{2z^2}{1-z}$$

or $$4G(a,z) - 4z - 4zG(a,z) + z^2 G(a,z) = \frac{2z^2}{1-z}$$

or $$(4 - 4z + z^2) G(a,z) = 4z + \frac{2z^2}{1-z}$$

or $$(2-z)^2 G(a,z) = \frac{8z-2z^2}{2-z}$$

or $$G(a,z) = \frac{8z-2z^2}{(2-z)^3} = \frac{2z(4-z^2)}{(2-z)^3}$$

To resolve into partial fraction, let us put $2 - z = t$, or $z = 2 - t$, then

$$\frac{8z-2z^2}{(2-z)^3} = \frac{8(2-t)-(2-t)^2}{t^3} = \frac{8-2t^2}{t^3}$$

$$= \frac{8}{t^3} - \frac{2}{t} = \frac{8}{(2-z)^3} - \frac{2}{2-z}$$

or $$G(a,z) = \frac{8}{(2-z)^3} - \frac{2}{2-z}$$

$$= \frac{8}{8(1-z)^3} - \frac{2}{2\left(1-\frac{z}{2}\right)}$$

$$= (1-z)^{-3} - \left(1-\frac{z}{2}\right)^{-1}$$

or $$\sum_{r=0}^{\infty} a_n z^n = \sum_{n=0}^{\infty} \frac{(n+1)(n+2)}{2}\left(\frac{z}{2}\right)^n - \sum_{n=0}^{\infty}\left(\frac{z}{2}\right)^n$$

Equating the coefficient of z^n on both sides, we have

$$a_n = \frac{(n+1)(n+2)}{2} \cdot \frac{1}{2^n} - \frac{1}{2^n}$$

$$= \frac{n^2+2n+2-2}{2^{n+1}}$$

or $$a_n = (n^2+3n) \cdot 2^{-n-1}$$

Example 1.101

Solve the recurrence relation $t_{r+1} - t_r = r$ with $t_0 = 1$ by using method of generating functions.

Solution

The given recurrence relation is

$$t_{r+1} - t_r = r \qquad \text{... (1)}$$

The generating function is given by

$$G(a,z) = \sum_{r=0}^{\infty} t_r z^r = t_0 + t_1 z + t_2 z^2 + \ldots \qquad \text{... (2)}$$

Multiplying (1) by z^r and sum over $r = 0$ to $r = \infty$, we have

$$\sum_{r=0}^{\infty} t_{r+1}z^r - \sum_{r=0}^{\infty} t_r z^n = \sum_{r=0}^{\infty} r\cdot z^r$$

$$\sum_{r=0}^{\infty} t_{r+1}z^r - \sum_{r=0}^{\infty} t_r z^r = \sum_{r=0}^{\infty} r\cdot z^r$$

$$\left(t_1 + t_2 z + t_3 z^2 + ...\right) - G(a,z) = z + 2z^2 + 3z^2 + ...$$

$$\frac{\left(t_1 z + t_2 z + t_3 z^2 + ...\right)}{z} - -G(a,z) = z(1 + 2z + 3z^2 + ...)$$

or $$\frac{\left(t_0 + t_1 z + t_2 z^2 + ...\right) - t_0}{z} - G(a,z) = z(1-z)^{-1}$$

Since $t_0 = 1$, we have $$G(a,z) - 1 - zG(a,z) = z^2(1-z)^{-1}$$

or $$(1-z)\,G(a,z) = 1 + z^2(1-z)^{-2}$$

or $$G(a,z) = 1 + z^2(1-z)^{-2}$$

$$G(a,z) = (1-z)^{-1} + z^2(1-z)^{-3}$$

or $$\sum_{r=0}^{\infty} t_r z^r = \sum_{r=0}^{\infty} z^r + z^2 \sum_{r=0}^{\infty} \frac{(r+1)(r+2)}{2} z^r$$

or $$\sum_{r=0}^{\infty} t_r z^r = \sum_{r=0}^{\infty} z^r + z^2 \sum_{r=0}^{\infty} \frac{(r+1)(r+2)}{2} z^{r+2}$$

Equating coef. of z^r on both sides, we get $\quad t_r = 1 + \frac{1}{2}(r-1)r$.

Note: *Readers can see solved examples (11-18) in Chapter 10 related to these articles.*

1.15-4 Some More Applications of Generating Functions

1. Generating function is one of the very important tools to solve not only the recurrence relations but any other problems in discrete mathematics.
2. With the help of generating function, one can find the coefficient of n^r from the expression of the type $\left(1 + x^2 + x^3\right)\left(x^2 + x^3\right)\left(x + 4\right)$.
3. It can be used to find the number of committees having the particular size that can be formed from a number of delegates from different countries subject to some restrictions.

EXERCISE 1.10

Solve the following recurrence relation using generating functions:

1. $a_n - 9a_{n-1} + 20a_{n-2} = 0$ for $n \geq 2$, $a_0 = -3$ and $a_1 = -10$.
2. $a_n - a_{n-1} - 9a_{n-2} + 9a_{n-3} = 0$ for $n \geq 3$; $a_0 = 0$, $a_1 = 1$ and $a_2 = 2$.

3. $a_n - 7a_{n-1} + 10a_{n-2} = 0$, $a_0 = 0$ and $a_1 = 3$.

4. $a_n - 5a_{n-1} + 6a_{n-2} = 0$ *for* $n \geq 2$, $a_0 = 1$ and $a_1 = -2$.

5. $a_n - 3a_{n-2} + 2a_{n-3} = 0$ for $n \geq 3$ and $a_0 = 1$, $a_1 = 0$, $a_2 = 0$.

6. $a_n = a_{n-1} + a_{n-2}$ for $n \geq 3$, $a_1 = 2$ and $a_2 = 3$.

7. $a_n - 4a_{n-2} = 0$ for $n \geq 2$, $a_0 = 0$ and $a_1 = 1$.

8. $a_n - 10a_{n-1} + 33a_{n-2} - 36a_{n-3} = 0$ for $n \geq 3$, $a_0 = 1$, $a_1 = 1$ and $a_2 = -23$

9. $a_{n+2} + a_{n+1} - 6a_n = 0$, $a_0 = 3$, $a_1 = 1$.

10. $a_n - 5a_{n-1} + 6a_{n-2} = 2^n + n$, $n \geq 2$ $a_0 = 1$ and $a_1 = 1$.

11. Show that generating function for $a_n = \binom{2n}{n}$ is $(1-4z)^{-1/2}$.

12. Solve the recurrence relation using generating function $a_{n+2} - 4a_{n+1} + 3a_n = 0$, given $a_0 = 2, a_1 = 4$.

13. Find a simple expression for the generating function of the numeric function 1,-2, 3, -4, 5, -6.

14. Solve the recurrence relation $2a_{r+1} - a_r = 4$.

ANSWERS

1. $a_n = 2.5^n - 5.4^n$

2. $a_n = \{-3 + 4.3^n - (-3)^n\}/12$

3. $a_n = 2^{n+1} + (-3)^n$

4. $a_n = 5.2^n - 4.3^n$

5. $a_n = \frac{8}{9} - \frac{6}{9}n + \frac{1}{9}(-2)^n$

6. $a_n = \left(\frac{5+3\sqrt{5}}{10}\right)\left(\frac{1+\sqrt{5}}{2}\right)^n + \left(\frac{5-3\sqrt{5}}{10}\right)\left(\frac{1-\sqrt{5}}{2}\right)^n$

7. $a_n = 0$ if n is even, $a_n = 2^{n-1}$ if n is odd

8. $a_n = 15.3^n + 6(n+1)3^n - 20.4^n$

9. $a_n = 2^{n+1} + (-3)^n$

11. $1 + 3^n$

14. $a_r = 4 - \frac{1}{2^r}$.

1.16 PIGEONHOLE PRINCIPLE

The **pigeonhole principle** is also known as **Dirichlet Drawer Principle** or the **Shoe Box Principle.** This principle states that if a flock of pigeons flies into a set of pigeonholes to roost and the number of pigeons are more than the number of pigeonholes then there must be at least one pigeonhole with at least two pigeons in it. In practical life situation this principle applies to other objects also. The pigeonhole principle tells us nothing about how to locate the pigeons that contain two or more pigeons.

THEOREM 1.11 PIGEONHOLE PRINCIPLE

If k pigeons fly into n pigeonholes and $(k > n)$ then at least one pigeonhole contains two or more pigeons.

PROOF

If none of n pigeonholes contains more than one pigeon. Then the total number of pigeons would be at most n. But this is a contradiction, since there are $(k > n)$ pigeons.

Example 1.102

If there is a group of 27 English words, there must exist at least two that begin with the same letter, since there are 26 letters in the English alphabet.

Example 1. 103

Determine the minimum number of elements that one needs to take from the set $A = \{1, 2, 3, 4, 5, \ldots 9\}$ to be sure that two of the numbers add up to 10.

Solution

We have five pigeonholes here namely $\{1, 9\}$, $\{2, 8\}$, $\{3, 7\}$, $\{4, 6\}$, $\{5, 5\}$. One can observe from here that any choice of six elements (pigeons) of A will guarantee that two of the numbers add upto ten.

Example 1. 104

Let there are five men and three women at a function. Show that if these people are lined up in a row, at least two men will be next to each other.

Solution

Do yourself.

Example 1. 105

Determine the number of students must be in class X to guarantee that at least two students receive the same score in the final exam, if the exam is graded on a scale from 0 to 100 points.

Solution

There are 101 possible scores on the final exam. From the pigeonhole principle it is clear that among any 102 students there must be at least 2 students with the same score.

GENERALIZED PIGEONHOLE PRINCIPLE

If there exist n pigeonholes and more than $2n$ pigeons, than three or more pigeons will have to be assigned to at least one of the pigeonholes. In general, if the number of pigeons is much larger than the number of pigeonholes. For this situation theorem 1 can be restated as follows.

THEOREM 1.12

If n pigeons are to occupy m pigeonholes, then there is at least one pigeonhole containing at least $\lceil n/m \rceil$* objects.

---

* The ceiling function at x is denoted by $\lceil x \rceil$ and assigns to the real number x the smallest integer that is greater than or equal to x. For example $\lceil 1/2 \rceil = 1, \lceil 4{\cdot}1 \rceil = 5, \lceil 9 \rceil = 9$.

PROOF

Let none of the pigeonholes are occupied by more than $\lceil n/m \rceil$-1 pigeons. Then total number of pigeons is at most

$$m(\lceil n/m \rceil - 1 < m(((n/m) + 1) - 1) = n$$

where the inequality

$$\lceil n/m \rceil < (n/m) + 1$$

But this is a contradiction as there are only n objects.

Example 1.106

If 30 books in a school contain a total of 61,327 pages, then show that one of the books must have at least 2045 pages.

Solution

From the extended pigeonhole principle we see that if we assign each page (pigeon) to the books (pigeonholes), then one book must contain at least

$$\left\lceil \frac{61,327}{30} \right\rceil = \lceil 2044.2 \rceil = 2045$$

Example 1.107

Find the minimum number of persons in a joint family to be sure that 4 of them born in the same month.

Solution

Total number of months (pigeonholes) is 12. Complete yourself.

Example 1.108

What is the minimum number of students required in a particular class to be sure that at least six students would receive the same division if there are five possible divisions.

Solution

The least number of students required to make sure that at least 6 student get the same grade in the smallest integer n such that $\left\lceil \frac{4}{5} \right\rceil = 6$. The smallest such integer is $n = 5.5 + 1 = 26$.

$\Rightarrow$ the minimum number of students required to be sure that at least 6 students will get the same division.

EXERCISE 1.11

1. Show that among any group of five (not necessarily consecutive) integers, there are two with the same remainder when divided by 4.
2. A box contains 10 black balls and 10 green balls. Shyam selects balls at random without looking at them.
   (i) How many balls must Shyam select to be sure of having at least three balls of the same colour.
   (ii) How many balls must Shyam select to be sure of having at least three green balls.
3. If thirteen people are assembled in a get together party. Show that at least two of them must have their birthday in the same month.

4. Consider a tournament in which each of n players plays against every other player and each player wins at least once. Show that at least two players having the same number of wins.
5. How may students each of whom comes from one of the 50 districts must be enrolled in an engineering college to guarantee that there are at least 100 who come from the same state?
6. There are 51 electricity poles on a street. Each pole has a number between 1000 and 1099 inclusive. Show that at least two poles have numbers that are consecutive integers.
7. There are seven colours, which are to be used to paint 50 cars, show that at least eight bicycle will be of same colour.
8. Prove that if any 14 numbers from 1 to 25 are chosen, then one of them is multiple of another.
9. Show that if five integers are selected from the first eight positive integers, there must be a pair of these integers with a sum equal to 9.
10. If there are 38 different time periods during which classes at an engineering college can be scheduled. If there are 677 different classes, how many different rooms will be needed.
11. Show that there are at least four people in Modinagar (population: 25 million) with the same three initials who were born on the same days of the year).
12. How many friends you must have to guarantee at least five of them will have birthdays in the same month?

ANSWERS

1. Total number of possible remainder when an integer is divided by 4 is four, hence from the pigeonhole principle there are at least two have the same remainder.
3. Let the birth month is like pigeons and the calendar months, the pigeonhole i.e. there are 13 pigeons and 12 pigeonholes. Hence from the pigeonhole principle at least two people were born in the same month.
4. One can observe that the number of wins for a player is at least 1 and at most n -1. Here, this n -1 numbers correspond to n -1 pigeonholes which can not accommodate n player (pigeons). Thus, at least two players will have the same number as wins.
5. 4951.
7. Use extended pigeonhole principle.
8. By labeling the pigeonholes with 1, 3, 5, .., 25 (inclusive both). If we assign each of the selected 14 numbers to the pigeonhole labelled with its odd part. But there exist only 13 pigeonholes, so two numbers must have the same odd part. One is a multiple of other.
9. Grouping the first eight positive integers of each pair which are four in number 9 : {1, 8}, {2, 7}, {3, 6} and {4, 5}. If we select first five integers from the first eight positive integers. By the pigeonhole principle at least two of them must come from the same subset. Two such integers have a sum of 9, as required.
10. 18.
11. There are 6,432,816 possibilities for the three initials and a birthday. Hence from the generalized pigeonhole principle, there are at least [25,000,000/6,432,816] = 4 people who share the same initials and birthday.
12. At least 49 friends.

2 GRAPH THEORY

2.0 INTRODUCTION

Consider a number of towns connected by roads. Given a road map and we want to know whether the two towns are connected. Here, let $S = \{a, b, c, d, e, f, \ldots\}$ be the set of towns and R be a Binary Relation on S such that (a, b) is in R if town a is linked to town b by road. Furthermore, we want to know whether for a given a and b in S, there exist $c, d, e, \ldots, h$ in S such that $\{(a, c), (c, d), (d, e) \ldots, (h, b)\} \subseteq R$.

We can see many situations and problems of this type in real life that can be dealt with discrete objects and binary relations. These objects and binary relations can be represented conveniently with the help of graphs. This leads us naturally to a study of the theory of graphs. In this chapter, we will identify and discuss the basic components of a graph and some optional features that many graphs have, the most commonly asked questions related to graphs is how graphs are used to model different situations?

2.1 GRAPH

A graph $G = (V, E)$ consists of :

(i) A set of objects $V = \{v_1, v_2, \ldots\}$ whose elements are called **vertices, points, or nodes** of G.

(ii) A set $E = \{e_1, e_2, e_3, \ldots\}$ of unordered pairs of distinct vertices whose elements are called **edges** of G such that each edge e_r is identified with an unordered pair (v_i, v_j) of vertices. Fig.2.1 represents the graph $G(V, E)$.

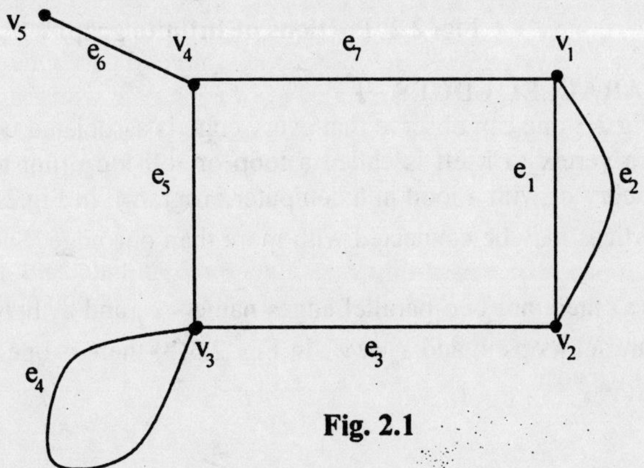

Fig. 2.1

The vertices v_i and v_j are said to be **adjacent** if there is an edge $v_iv_j = e_k$ to which it is associated. In such a case v_i and v_j are called the **end points** of e_k and e_k is said to **connect** or **join** of the vertices v_i and v_j.

The graphs are commonly represented by diagram. Every diagram consists of a set of vertices, which are represented by dots or circles, and in every diagram lines or arcs connect certain pairs of such points. The other details, such as geometry of the edges, their lengths and the positions of the points are of no importance at present. In the graph shown in the Fig.2.1 we can see it has five vertices and seven edges.

2.2 FINITE GRAPH, INFINITE GRAPH AND TRIVIAL GRAPH

A graph is said to be **finite** if it has finite number of vertices as well as finite number of edges, otherwise it is an **infinite graph**. It is to be noted here that if a graph has finite number of vertices it must automatically have a finite number of edges and hence must be finite. The finite graph with one vertex and no edge, i.e. a single point is called the **trivial graph**. Portions of two infinite graphs are shown in Fig. 2.2. The graphs shown in Fig.2.3 (a) and (b) are examples of finite graphs

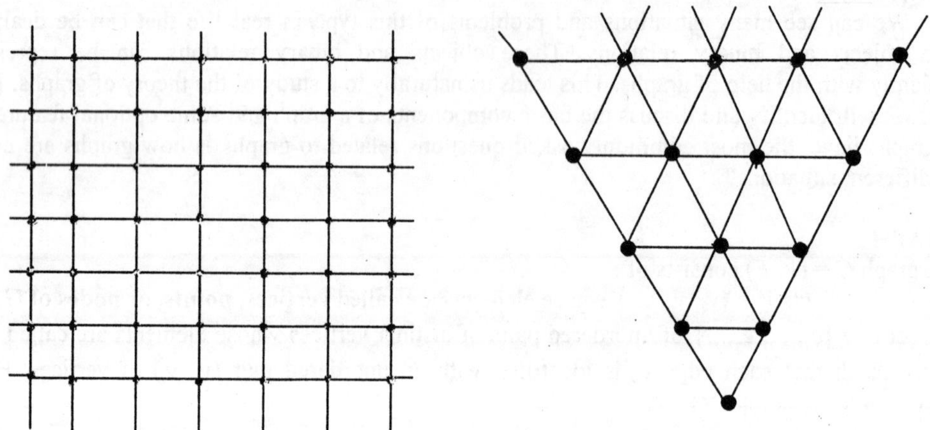

Fig. 2.2 Portions of Infinite graphs

2.3 LOOP AND PARALLEL EDGES

From the Fig.2.3 one can observe that every edge is associated with two vertices. An edge of a graph which join a vertex to itself is called a **loop** or **self-loop** (not to be confused with loop of electrical network theory or with a loop in a computer program). In Fig.2.3 (b) e_7 is a loop. It is quite possible that two vertices may be connected with more than one edge. Such edges are called **parallel edges.**

In Fig. 2.3(a) there are two parallel edges namely e_2 and e_3 between the vertices v_2 and v_3, and there are two parallel loops e_5 and e_6 at v_1. In Fig. 2.3(b) there is one loop at v_2 and three parallel edges between v_1 and v_2.

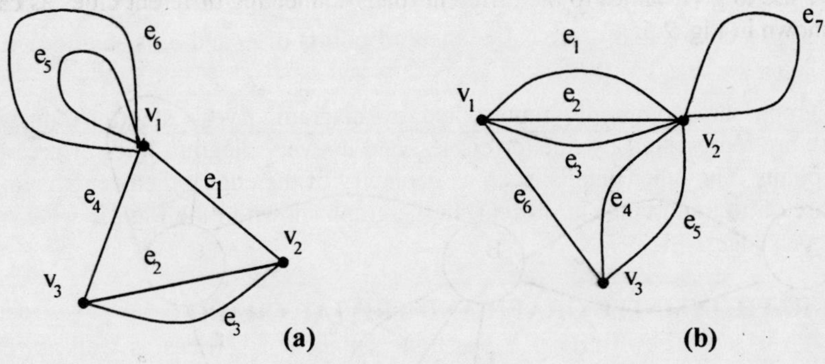

Fig. 2.3

2.4 SIMPLE GRAPH AND MULTIGRAPH

A **simple graph** is one for which there is no more than one edge between a pair of vertices.

Any graph, which contains some parallel edges, is called a **multigraph**. The graph shown in Fig. 2.3(a) and (b) are multigraph graphs while Fig. 2.4(b) is a simple graph.

A graph, in which loops and parallel edges are allowed, is called a **pseudo graph**. Graph shown in Fig.2.3 (a) is a Pseudo graph.

Example 2.1

An intercom in an institute can be easily described using a graph as shown in Fig.2.4 (a), which describes a network of six-telephone t_1, t_2, t_3, t_4, t_5 and t_6. An edge between two vertices (telephones) indicates that direct two-way communication is possible between the two persons. It is to be noted that edges of this graph is not directed. Similarly, we can take example of network of computers as shown in the Fig. 2.4(b). This describes a network of five-computer c_1, c_2, c_3, c_4, c_5 and c_6, which are directly or indirectly linked to each other. There are several situations for which this graph can serve as model.

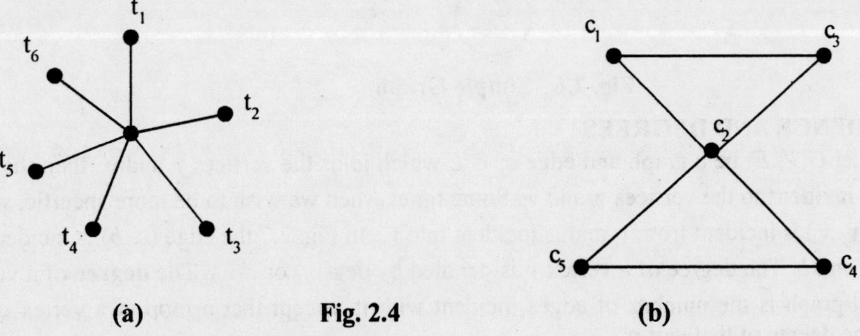

Fig. 2.4

Example 2.2

A road map is a common example of a multigraph. In a road map we can see there are many cities which are connected to other cities in more than one ways. In order to give clear travel directions, we use to give names to the different roads connecting different cities as can be seen in the multigraph shown in Fig. 2.5.

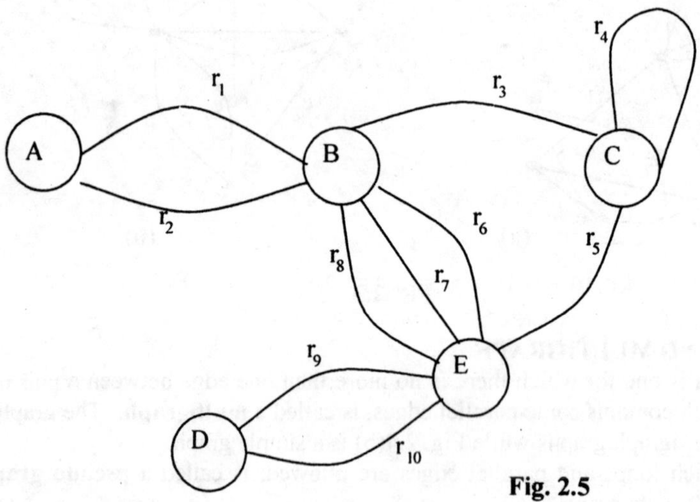

Fig. 2.5

Example 2.3

A single railway track connecting different cities is a very common example of a simple graph.

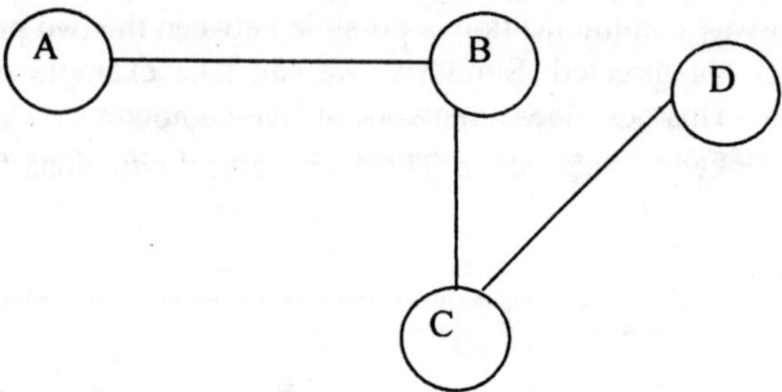

Fig. 2.6 Simple Graph

2.5 INCIDENCE AND DEGREES

Let $G(V, E)$ be a graph and edge $e_k \in E$ which joins the vertices v_i and v_j, then the edge e_k is said to be **incident** to the vertices v_i and v_j. Some times when we wish to be more specific, we say that the edge (v_i, v_j) is incident from v_i and is incident into v_j. In Fig. 2.7 the edge (a, b) is incident with the vertices a and b. The degree of a vertex v_i is denoted by $\deg(v_i)$ or $d(v_i)$. The **degree** of a vertex in an undirected graph is the number of edges incident with it, except that a loop at a vertex contributes twice to the degree of that vertex.

A vertex is said to be **even** or **odd** according as its degree is an even or odd number. Thus vertex v_1 and v_7 are even whereas v_2, v_3, v_4 and v_5 are odd in Fig. 2.8.

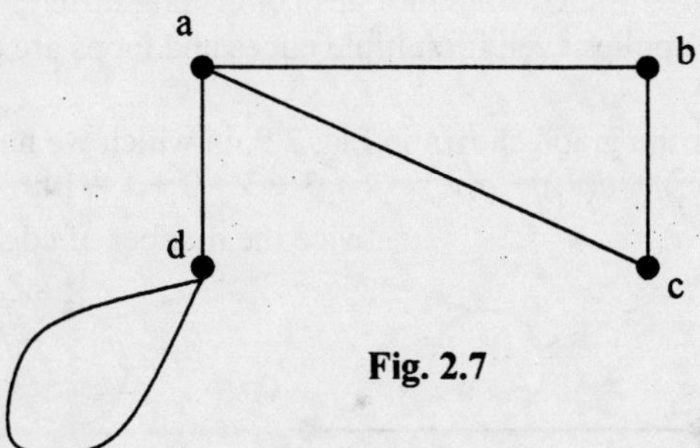

Fig. 2.7

2.6 ISOLATED VERTEX AND PENDANT VERTEX

A vertex of degree zero is called an **isolated vertex**. In other words, vertex having no incident edge is called an isolated vertex. In Fig. 2.8 vertices v_5 and v_6 are isolated vertices.

A vertex of degree one is called a **pendant vertex** or an **end vertex**. Vertex v_3 in Fig.2.8 is a pendant vertex.

Two adjacent edges are said to be in **series** if their common vertex is of degree two. In Fig.2.8 the two edges incident on v_7 are in series.

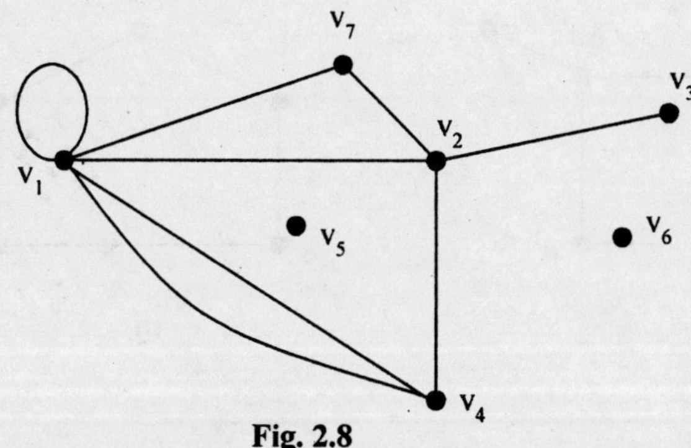

Fig. 2.8

THEOREM 2.1 (HANDSHAKING LEMMA)

The sum of the degrees of all vertices of a graph G is twice the number of edges in G, i.e.

$$\sum_{i=1}^{n} d(v_i) = 2e \text{, where } e \text{ is the number of edges in } G.$$

This result is due to Euler in 1736. This is called handshaking lemma because if several people shake hands, then the total number of hands shaking must be even, because just two hands are involved in each handshake.

PROOF

Consider a graph with m edges and n vertices $v_1, v_2, v_3, \ldots, v_n$. Since each edge contributes two degree, the sum of the degrees of all vertices in G is twice the number of edges in G. That is

$\sum_{i=1}^{n} d(v_i) = 2e$. Note that this applies even if multiple edges and loops are present.

For example, consider the graph shown in Fig. 2.9, in which we have

$d(v_1) + d(v_2) + d(v_3) + d(v_4) + d(v_5) = 4 + 3 + 3 + 3 + 1 = 14$

= twice the number of edges which is 7.

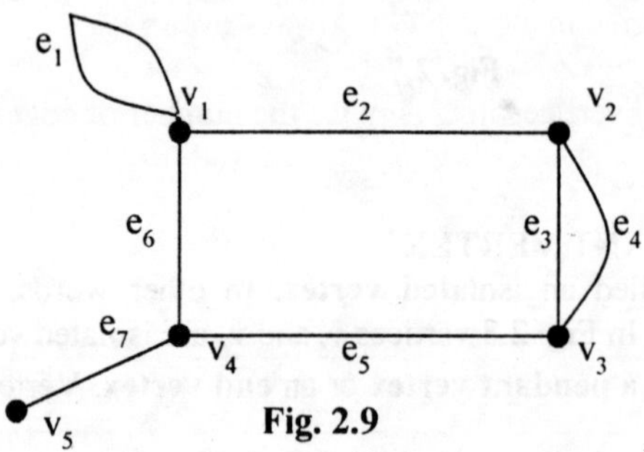

Fig. 2.9

Example 2.4

Verify handshaking lemma for the graphs shown in Fig. 2.10.

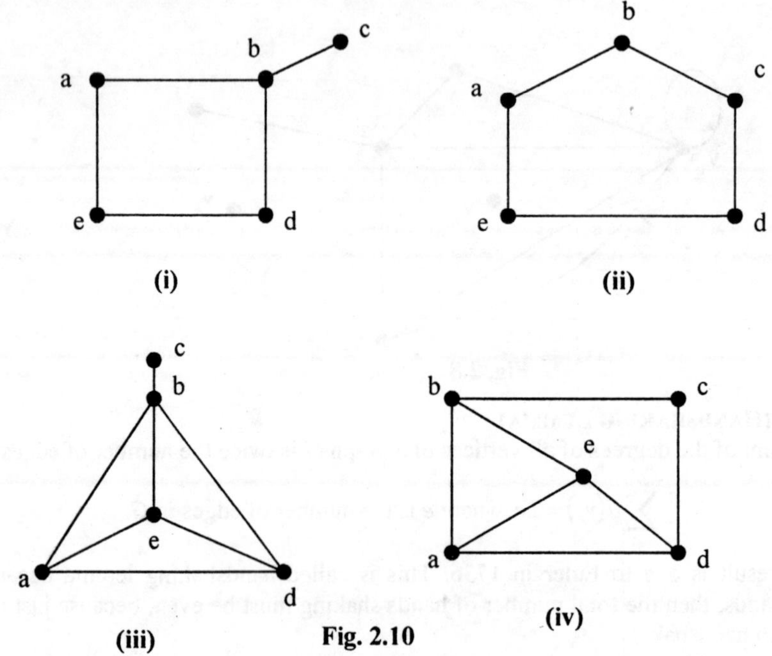

Fig. 2.10

Solution

(i) In Fig 2.10 (i)

$d(a) + d(b) + d(c) + d(d) + d(e) = 2 + 3 + 1 + 2 + 2 = 10 = 2 \times 5 = 2 \times$ no. of edges.

(ii) In Fig 2.10 (ii)

$d(a) + d(b) + d(c) + d(d) + d(e) = 2 + 2 + 2 + 2 + 2 = 10 = 2 \times 5 = 2 \times$ no. of edges.

(iii) In Fig 2.10 (iii)

$d(a) + d(b) + d(c) + d(d) + d(e) = 3 + 4 + 1 + 3 + 3 = 14 = 2 \times 7 = 2 \times$ no. of edges

(iv) In Fig 2.10 (iv)

$d(a) + d(b) + d(c) + d(d) + d(e) = 3 + 3 + 2 + 3 + 3 = 14 = 2 \times 7 = 2 \times$ no. of edges.

Hence, in all the graphs handshaking lemma is verified.

THEOREM 2.2

The number of vertices of odd degree in a graph G is always even.

PROOF

Since the sum of the degrees of all vertices in G is twice the number of edges in G, i.e.

$$\sum_{i=1}^{n} d(v_i) = 2e \qquad \ldots(1)$$

where e is the number of edges in G.

Let us consider the vertices with odd and even degrees separately. In equation (1) L.H.S. can be expressed as the sum of two sums that of vertices of even and odd degrees, i.e.

$$\sum_{i=1}^{n} d(v_i) = \sum_{\text{even}} d(v_j) + \sum_{\text{odd}} d(v_k)$$

$$\text{or} \quad \sum_{\text{even}} d(v_j) + \sum_{\text{odd}} d(v_k) = 2e$$

$$\text{or} \quad \sum_{\text{odd}} d(v_k) = 2e - \sum_{\text{even}} d(v_j) \qquad \ldots(2)$$

Here, we can observe that the expression in the R.H.S. is even. As in equation (2) each $d(v_k)$ is odd, the total number of terms in the sum must be even (Since the right hand side is even)to make the sum an even number. Hence the theorem.

2.7 DEGREE SEQUENCE

If $v_1, v_2, v_3, \ldots v_n$ are the vertices of graph G, then the sequence $(d_1, d_2, d_3, \ldots. d_n)$, where $d_i =$ degree(v_i), is the degree sequence of G. Normally, we order the vertices so that the degree sequence is monotone increasing, that is

$$\delta(G) = d_1 \leq d_2 \leq d_3 \leq \ldots. \leq d_n = \Delta(G).$$

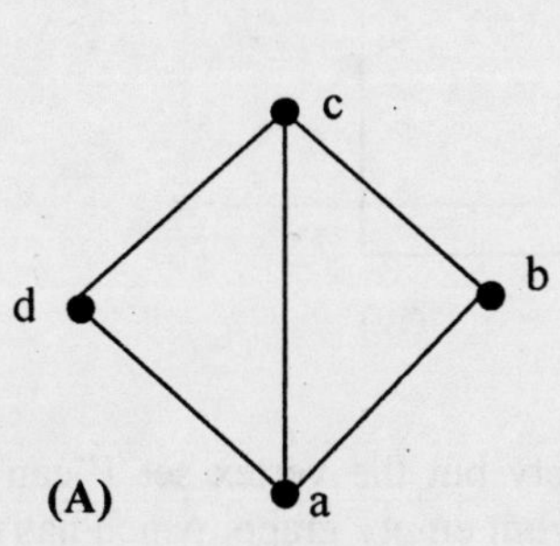

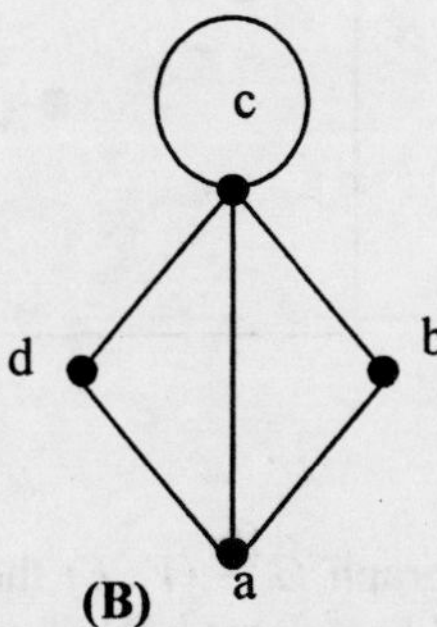

Fig. 2.11

where $\delta(G)$ = minimum of all degrees
and $\nabla(G)$ = maximum of all degrees.

Degree sequence of graph shown in the Fig.2.11 (A) and (B) is (2, 2, 3, 3) and (2, 2, 3, 5) respectively.

Example 2.5

Is there exist a non simple graph G with degree sequence (1, 1, 3, 3, 3, 4, 6, 7)?

Solution

Yes, a non-simple graph of degree sequence (1, 1, 3, 3, 3, 4, 6, 7) is possible, which is exhibited in the following graph.

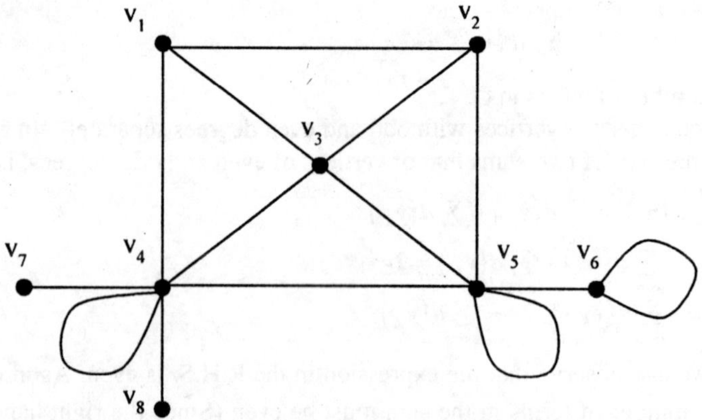

Fig. 2.12

2.8 NULL GRAPH

A graph containing only isolated vertices is called a **null graph**. In other words, the set of edges in a null graph is empty. We denote the null graph on n vertices by N_n. Graph shown in Fig. 2.13 is N_3.

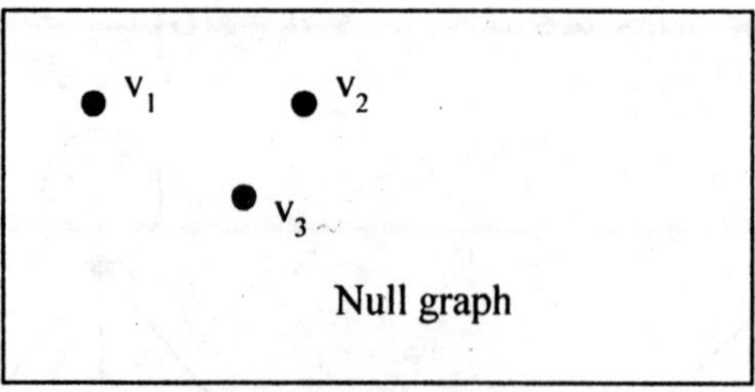

Fig. 2.13

NOTE

In a null graph $G = (V, E)$ the edge set is empty but the vertex set V can not be empty, otherwise there will be no graph. Null graph is different from empty graph, which has no vertices and no edges. A graph of N_6 class is shown in the Fig.2.14. In practice, an isolated vertex in a graph has very little importance.

Fig. 2.14

2.9 CONNECTED GRAPH

If there exist at least one path between any two vertices of a graph G, then it is called a **connected graph**. If graph is not connected then it is called a **disconnected graph**. Each of the connected part of the disconnected graph is called **components**. Graphs shown in the Fig.2.15 (a) is connected while that of (b) and (c) are disconnected and disconnected null graphs respectively.

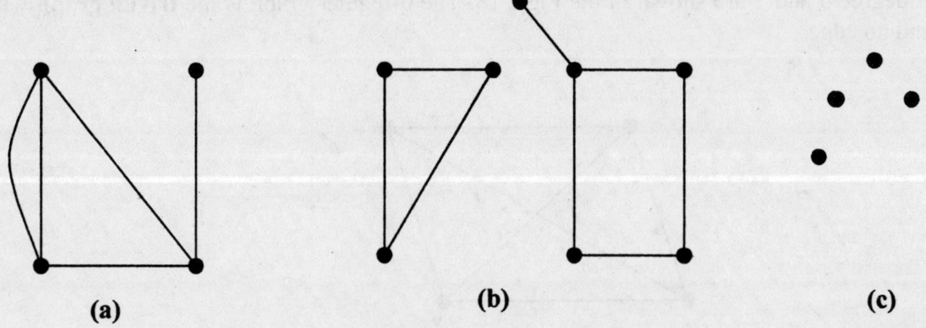

Fig. 2.15

The graph 2.15 (b) has two components since it has two connected parts.

2.10 COMPLETE GRAPH

A graph $G(V, E)$ is said to be **complete** if every vertex in G is connected to all other verities of G. We denote the complete graph on n vertices by K_n. Fig. 2.16 shows K_1, K_2, K_3, K_4 and K_5. One can note that in a complete graph of n vertices, degree of every vertex is n-1 as every vertex is joined with every other vertex through one edge and the total number of edges in G is $\frac{n(n-1)}{2}$.

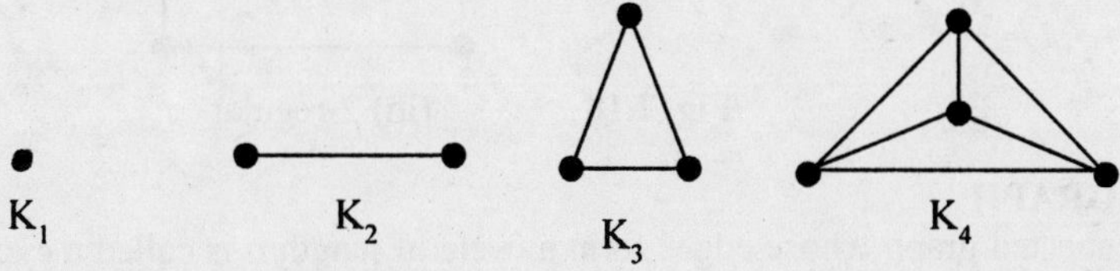

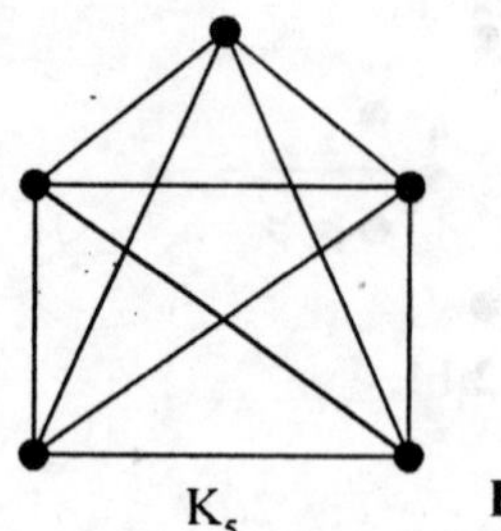

Fig. 2.16

2.11 REGULAR GRAPH

A graph in which all vertices are of equal degree is called a **regular graph** or simply a **regular**. A graph *G* is regular of degree *k* or ***k*-regular** if every vertex has degree *k*. The graph shown in Fig.2.17 is a regular of degree three. Usually a 3-regular graph is called a **cubic graph**. The regular graph of degree 0 and 1 are shown in the Fig.2.18. The 0-regular graph is the **trivial graph** with one vertex and no edge.

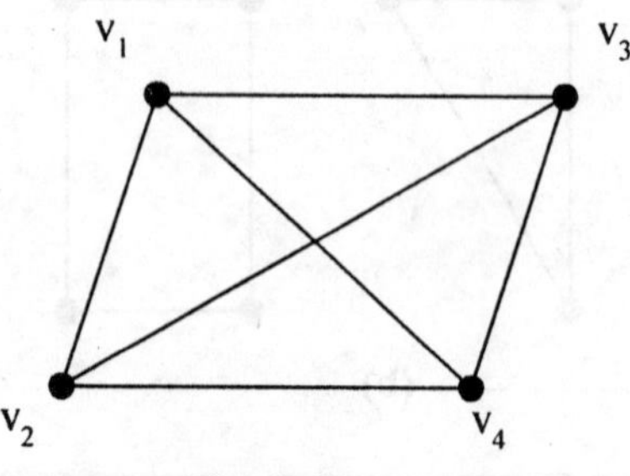

Fig. 2.17

REMARK

The *3*-regular graphs must have an even number of vertices, since the sum of degrees of the vertices is an even number. The same can be seen in Fig.2.17 for a 3-regular graph with four vertices. Note that the complete graph with *n* vertices i.e., K_n is regular of degree $n - 1$, null graph N_n is regular of degree 0.

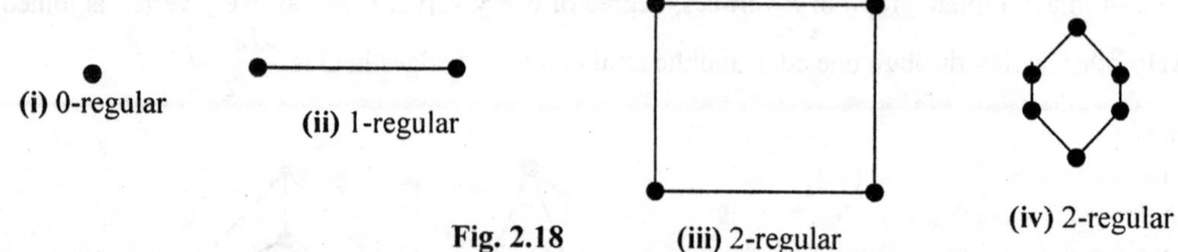

Fig. 2.18

2.12 CYCLE GRAPH

A connected graph whose edges form a cycle of length *n* is called a **cycle graph** of order *n*. Here length *n* means graph consists of *n* edges. Cycle graphs of length *n* is denoted by C_n. For a cycle graph $n \geq 3$.

Graphs of class C_6 is shown in the Fig.2.19. Note that the cycle graph C_6 is regular graph of degree 2.

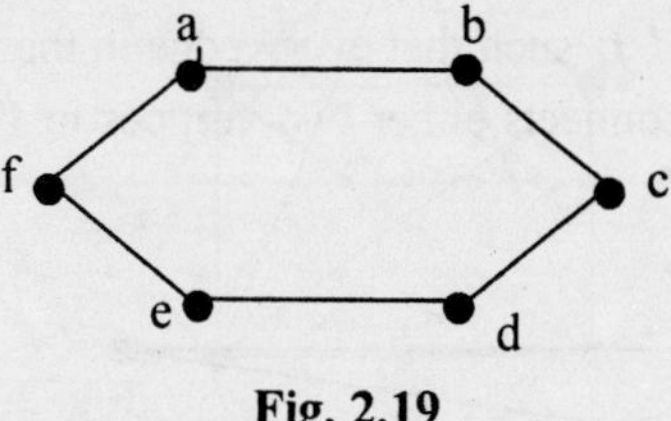

Fig. 2.19

2.13 WHEEL GRAPH

A **wheel** of order n is a graph obtained by joining a single new vertex to each vertex of cycle graph of order (n-1). Wheels of order n are denoted by W_n. A graph of class W_5 is shown in the Fig.2.20. Here W_5 is obtained by adding vertex e to cycle graph C_4. Cycle graph C_4 consists of vertices a, b, c, d.

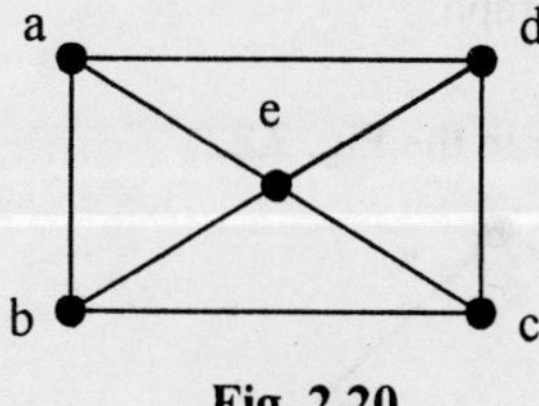

Fig. 2.20

2.14 PATH GRAPH

Removing an edge from a C_n graph, a **path graph** of order n is obtained. A path graph of order n is denoted by P_n. A graph of class P_5 is shown in the following figure.

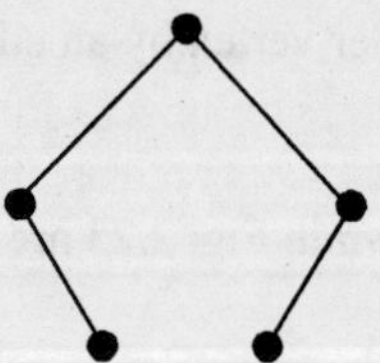

2.15 BIPARTITE GRAPH

A **bipartite graph** is a graph $G = (V, E)$ whose set of vertices can be partitioned into two nonempty subsets G_1 and G_2 such that there is no edge in E joining two vertices in G_1 or two vertices in G_2, i.e. its vertex set can be partitioned into two disjoint non-empty sets V_1 and V_2 such that every edge in the graph connects a vertex in V_1 and a vertex in V_2. Here (V_1, V_2) is called **partition**. In other words we can say that in a bipartite graph vertices can be coloured black and white in such a way that each way edge joins a black vertex and a white vertex.

Example 2.6

Show that C_6 is a bipartite graph.

Solution

C_6 is a bipartite graph can be shown by the fact that it can be partitioned into the two sets of vertices $V_1 = \{a, c, e\}$ and $V_2 = \{b, d, f\}$ such that every edge in the graph connects a vertex in V_1 and a vertex in V_2 so that no edge in G connects either two vertices in V_1 or two vertices in V_2 as shown in the Fig. 2.21.

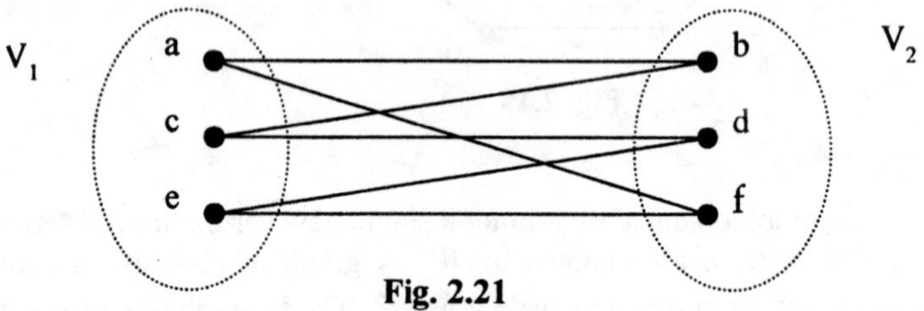

Fig. 2.21

Example 2.7

Show that K_3 is not a bipartite graph.

Solution

Let us take a K_3 graph as shown in the Fig. 2.22.

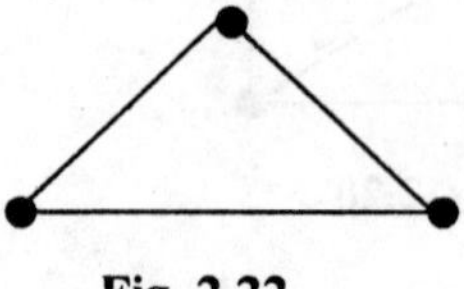

Fig. 2.22

Here, if we divide K_3 into two disjoint sets, one of the set will be containing two vertices. For this graph to be bipartite, these two vertices must not be connected by an edge. But this is not the case with K_3 as each vertex is connected to every other vertex by an edge. Hence K_3 is not bipartite.

Example 2.8

Are the graphs G_1, G_2, G_3 and G_4 shown in Fig. 2.23 are bipartite?

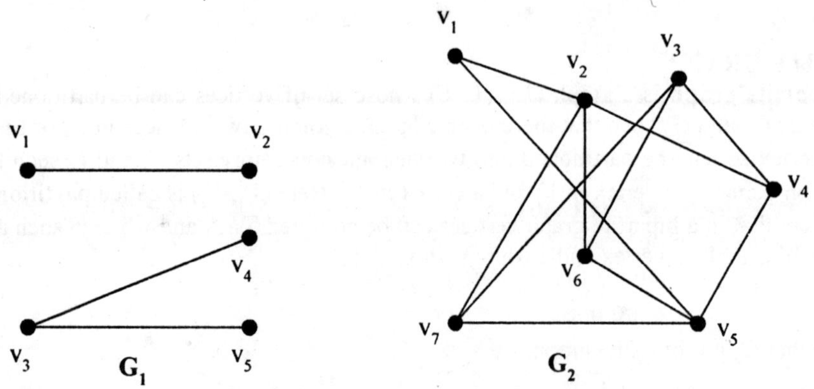

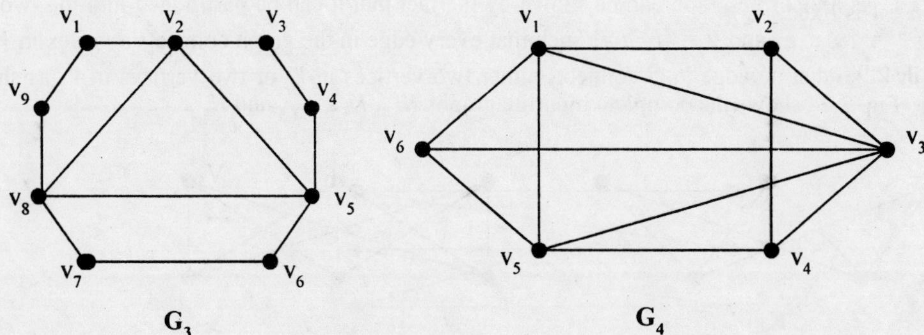

Fig. 2.23

Solution

(i) If we refer to our definition, we find that when we partition the graph into two portions say vertex sets V_1 and V_2, then we find one edge is incident on one vertex in V_1 and one vertex in V_2. It is to be noted here that if v_1 is a vertex in V_1 and v_2 is a vertex in V_2, it is not necessary that there is an edge between V_1 and V_2. If we see the graph G_1 of Fig.10.23, we find that if $V_1 = \{v_1, v_3\}$ and $V_2 = \{v_2, v_4, v_5\}$, then there is an edge between V_1 and V_2. However, not all edges between vertices in V_1 and V_2 are in graph. For instance $(v_2, v_3\}$ is not there. Hence G_1 is bipartite.

(ii) Graph G_2 is bipartite as it can be partitioned into two disjoint sets of vertices $V_1 = \{ v_2, v_3, v_5 \}$ and $V_2 = \{v_1, v_4, v_6, v_7\}$ such that each edge connects a vertex of V_1 to vertex of V_2. Here again we can note that every vertex of V_1 is not adjacent to every vertex of V_2. For instance v_1 and v_3 are not adjacent.

(iii) The Graph G_3 shown in Fig. 2.23 is not bipartite. Let us prove this arguing by contradiction. If graph G_3 had been bipartite it could be partitioned into two disjoint sets of vertices V_1 and V_2 such that $V_1 \cup V_2 = V$ (Refer 2.26-1) and each edge is incident on one vertex in V_1 and one vertex in V_2. Now consider the vertices v_2, v_5 and v_8. Since v_2 and v_8 are adjacent; one should be in V_1 and other should be in V_2. Again, v_2 and v_5 are adjacent and v_8 is in V_2, v_5 is in V_1. Further, v_2 and v_5 are adjacent and v_2 is in V_1, hence v_5 should be in V_2. But now we see that v_5 is both in V_1 and V_2, which is a contradiction since V_1 and V_2 are disjoint. Therefore, G_3 is not bipartite.

(iv) Graph G_4 is not bipartite because it can not be partitioned into two disjoint sets V_1 and V_2 such that every edge in the graph connects a vertex in V_1 and a vertex in V_2. It can be verified by taking the vertices v_1 , v_2 and v_3.

2.16 COMPLETE BIPARTITE GRAPH

A **complete bipartite graph** is a bipartite graph in which every vertex of G_1 is adjacent to every vertex of G_2, i.e. a complete bipartite graph is a bipartite graph in which each vertex in G_1 is joined to each vertex in G_2 by just one edge. A complete bipartite graph is denoted by $K_{m,n}$ **if it can be** partitioned into sets G_1 and G_2 in such a way that the number of vertices of G_1 and G_2 are ***m*** **and** ***n*** respectively. The normal order of m and n is such that $m \le n$. Clearly the graph $K_{m,n}$ **has** $m \cdot n$ **edges and**

$m + n$ vertices. Any graph that is $K_{1,n}$ is called a **star graph**. In Fig. 2.24 $K_{1,5}$ is a star graph. In a star graph $m = 1$ is fixed.

Example 2.9

Fig. 2.24 shows the complete bipartite graphs $K_{2,3}$, $K_{3,3}$, $K_{1,5}$ and $K_{2,4}$.

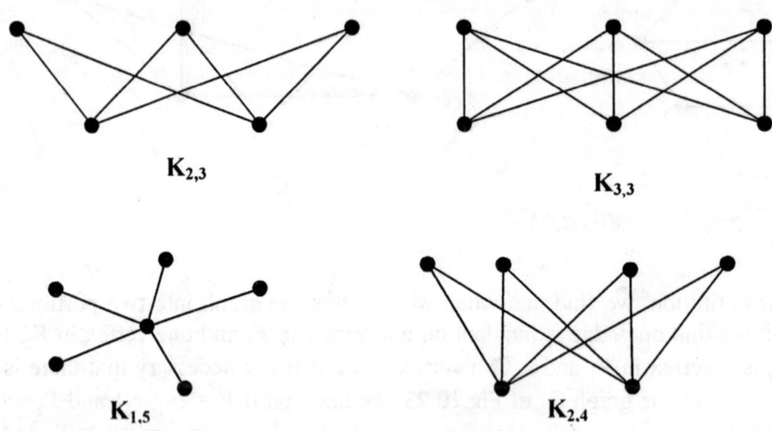

Fig. 2.24

Example 2.10

Draw the graph $K_{2,5}$.

Solution

Bipartite graph $K_{2,5}$ consists of seven vertices. These vertices are partitioned into a set G_1 of 2 vertices say a and b and a set G_2 of 5 vertices say p, q, r, s, and t, and all possible edges from vertices p, q, r, s, t. Thus there are 5×2=10 edges.

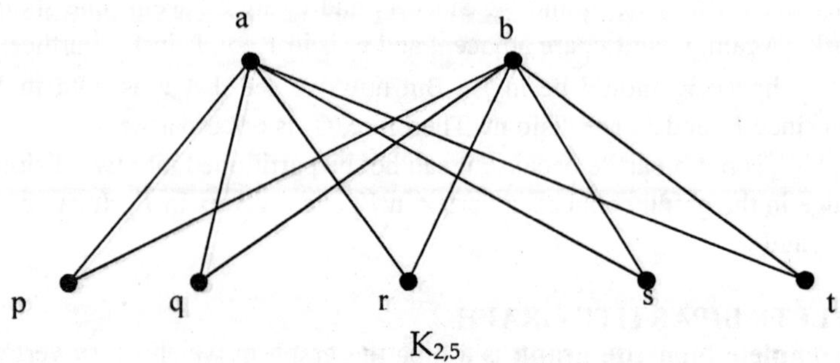

Fig. 2.25

Example 2.11

Which connected graphs can be both regular and bipartite?

Solution

Let us take a bipartite graph $K_{m,m}$. Since every complete bipartite graph is bipartite hence this graph is regular of degree *m* as each vertex is connected to *m* other vertices and thus its degree is *m*. If *m* disjoint edges (edges connected to distinct pair of vertices) are deleted from $K_{m,m}$, then the subgraph (Refer 2.23) obtained in this way will also be regular. For example, the subgraph of $K_{4,4}$ shown in Fig.2.26 (a) is 3-regular. If we continue to delete *m* (1,2,3,…,*m*) disjoint edges and each time we obtain a regular graph of one less degree. It is to be observed here that continuously doing so at last we get a disconnected graph, but in any case their connected components have the desired properties. Graph in Fig. 2.26 (b) is a graph obtained after deleting 1- disjoint edge from $K_{4,4}$.

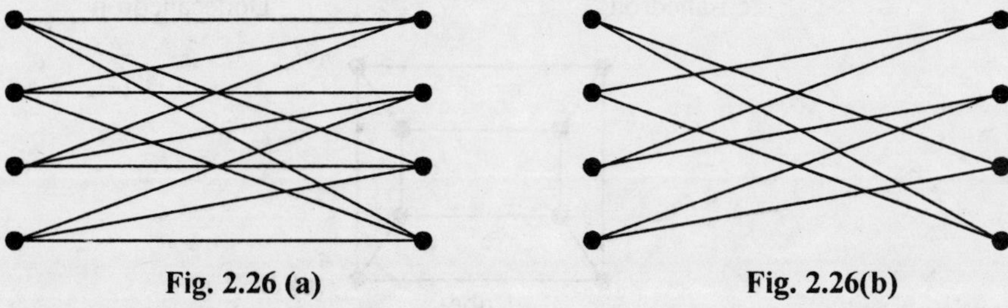

Fig. 2.26 (a) **Fig. 2.26(b)**

2.17 PETERSEN GRAPHS

The Petersen graph is drawn in the following Fig. (a). The graphs shown in Fig. (b) and (c) are the drawings of the Petersen graphs. The Petersen graph has a simple description using the set S of two element subsets of five-element set. Consider a graph G with vertex set S in which two pairs form an edge iff they are disjoint as sets. All the graphs shown in below are isomorphic(see article 2.26). This follows from the labelling the vertices of each graph with the members of 5 such that the adjacency relation is disjointness.

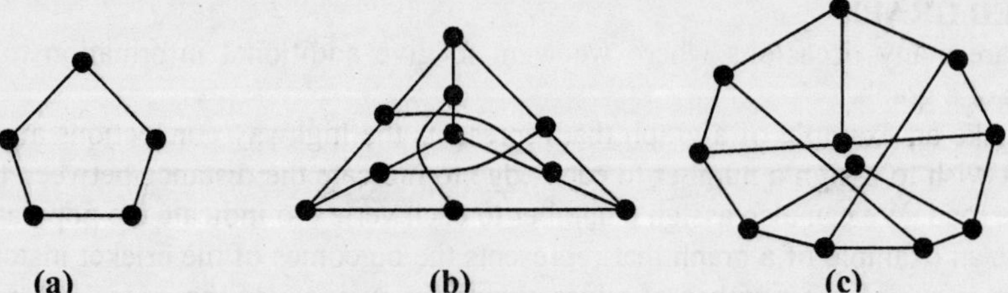

(a) **(b)** **(c)**

2.18 PLATONIC GRAPH

Other interesting graphs are shown in Fig. 2.27. The graphs formed by the vertices and edges of the five regular platonic solids are shown below.

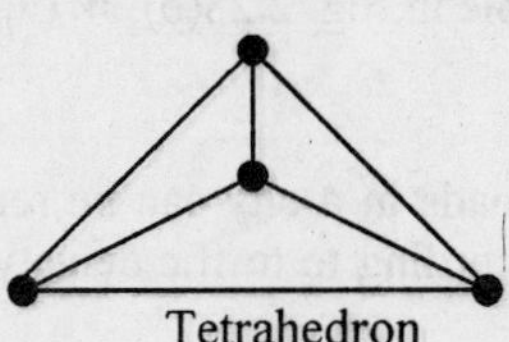

Tetrahedron

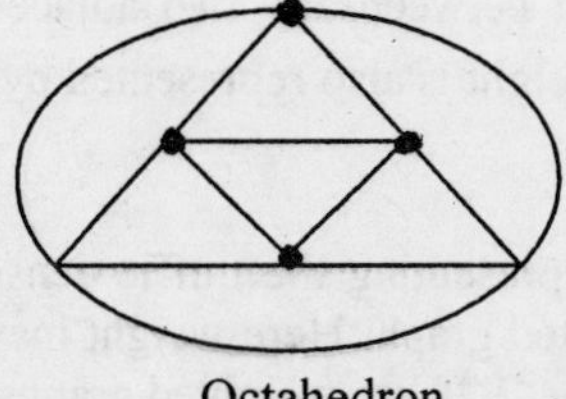

Octahedron

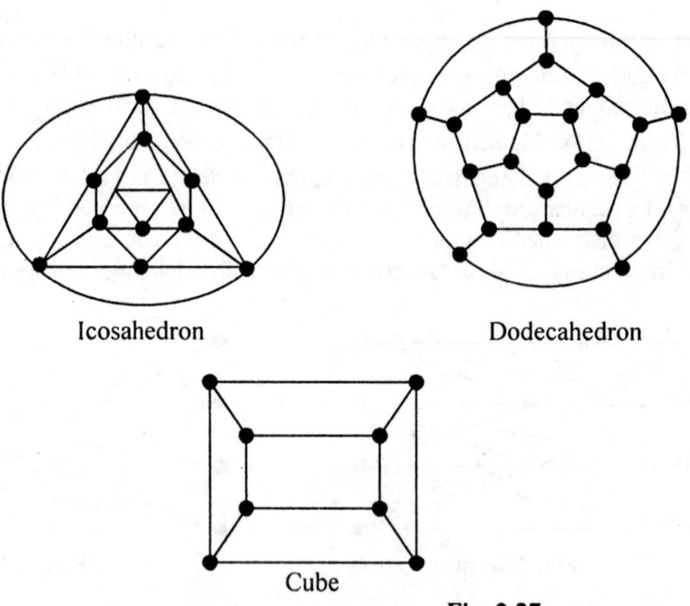

Fig. 2.27

2.19 LABELLED GRAPH

A graph G is called a labelled graph if its edges or vertices are assigned of one kind or another. The difference between labelled and unlabelled graphs is more apparent when we try to count them.

2.20 WEIGHTED GRAPH

There are many occasions where we want to give additional information to the vertices and/or the edges of a graph.

Let us take an example of a graph that represents the highway connections among different cities. We might wish to assign a number to each edge to indicate the distance between the two cities connected by the road. We can also assign a number to each vertex to indicate the population of a city. We can also take an example of a graph that represents the outcomes of the cricket matches. We may thus have graphs in which the number of edges shows the weights of the edges. A graph in which weights like capacities, distance, probabilities, etc. are assigned to every edge is called a **weighted graph**. The graphs in Fig.2.28 (a), (b) and (c) in which number on edges shows the weights of the edges. Ofcourse all weighted graphs are labelled graphs.

The weight between any two adjacent vertices v_i and v_j is written as $W(v_i, v_j)$ or if edge $e = v_i v_j$, then the weight is also represented by $W(e)$. For example in Fig. 2.28(b), $W(v_1, v_2) = 1.2$.

Example 2.12

A graph representing the traffic density of different roads in a city can be represented with the help of a weighted graph. Here weight may be assigned according to traffic density of each road. Graphs shown in Fig. 2.28 are weighted graphs.

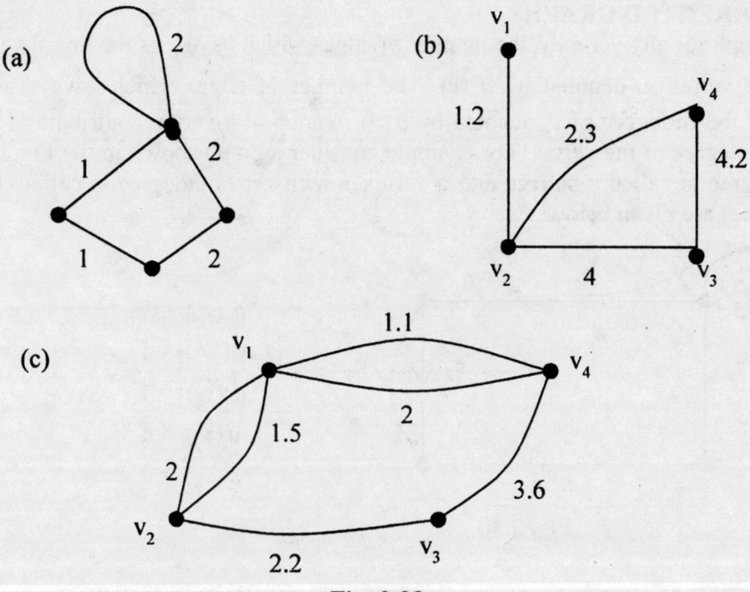

Fig. 2.28

2.21 DIRECTED GRAPH

In a graph $G = (V, E)$, an edge which is associated with an ordered pair of $V \times V$ is called a **directed edge** of G, while an edge which is associated with an unordered pair of vertices is called an **undirected edge**. A graph in which every edge is directed is called a **directed graph** or **digraph**. A graph in which every edge is undirected is called an **undirected graph**. A graph in which some edges are directed and some are undirected, is called a **mixed graph**.

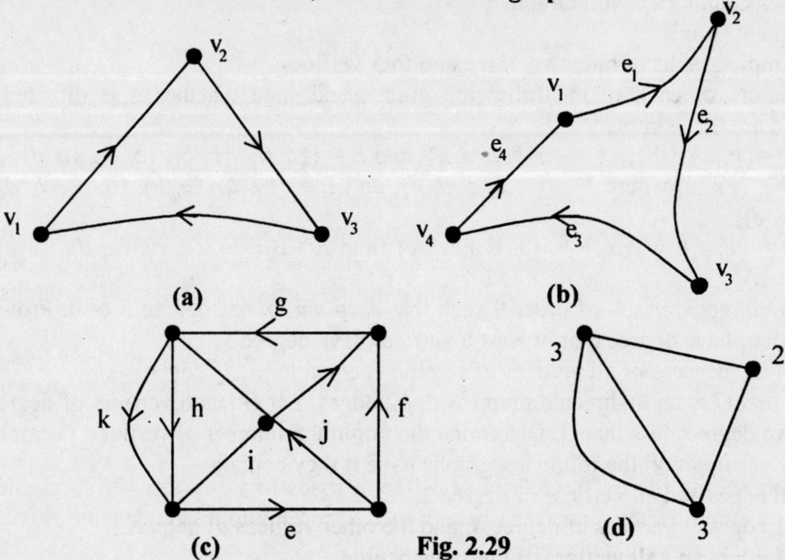

Fig. 2.29

2.22 DEGREE OF A DIRECTED GRAPH

In a directed graph for any vertex v_i the number of edges which have v as the initial vertex is called **outdegree** of the vertex v, denoted by $d^+(v)$. The number of edges which have v as their terminal vertex is called the **indegree** of v, denoted by $d^-(v)$. A loop at a vertex contributes 1 to the indegree and 1 to the outdegree of the vertex. For example consider a graph shown in the Fig.2.30. A vertex v with zero indegree is called a **source** and a vertex v with zero outdegree is called a **sink**. Degrees of various vertices are given below.

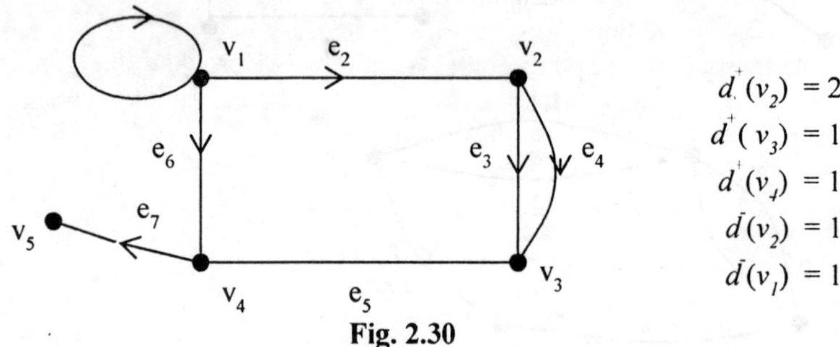

$d^+(v_2) = 2$

$d^+(v_3) = 1$

$d^+(v_4) = 1$

$d^-(v_2) = 1$

$d^-(v_1) = 1$

Fig. 2.30

EXERCISE 2.1

1. Define
   (i) Simple graph (ii) Multiple graph and give an example for each case.
2. Define
   (i) Directed graph (ii) Undirected graph.
3. Define
   (i) Regular Graph (ii) Null Graph
   (iii) Complete Graph
4. Draw all simple graphs of one, two, three and four vertices.
5. Draw a picture of each of the following graphs and state whether it is directed or non directed and are they simple
   (i) $G = (V, E)$, where $V = \{a, b, c, d, e\}$, and $E = \{\{a, b\}, \{b, c\}, \{a, c\}, \{a, d\}, \{d, e\}\}$.
   (ii) $G = (V, E)$, where $V = \{a, b, c, d, e\}$, and $E = \{(a, a), (a, b), (b, c), (c, d), (e, d), (d, e)\}$.
   (iii) $G = (V, E)$, where $V = \{a, b, c, d, e\}$, and $E = \{(a, b), (a, c), (a, d), (a, e), (e, c), (c, a)\}$.
6. Let G is nondirected graph of order 9 such that each vertex has degree 5 or 6. Prove that at least 5 vertices have degree 6 or at least 6 vertices have degree 5.
7. The sum of the degrees of all vertices of a graph is equal to ...... edges.
8. Suppose that G is an undirected graph with 12 edges. Let G has 6 vertices of degree 3 and the rest have degrees less than 3. Determine the minimum number of vertices G can have.
9. How many vertices will the following graphs have if they contain:
   (i) 16 edges and all vertices of degree 2,
   (ii) 21 edges, 3 vertices of degree 4, and the other vertices of degree 3,
   (iii) 24 edges and all vertices of the same degree.

10. For any simple graph of n vertices, prove that the number of edges of G is less than or equal to $\frac{1}{2}n(n-1)$.

11. A sequence $d = (d_1, d_2, d_3, \ldots d_n)$ of nonnegative integers is said to be **graphical** if there exists a simple non directed graph with no self-loop with degree sequence d. Show that
   (a) (2, 3, 3, 4, 4, 5) is not graphical
   (b) (4, 3, 2, 2, 1) is graphical
   (c) (2, 3, 3, 4, 5, 6, 7) is not graphical
   (d) (2, 2, 4) is not graphical

12. (i) Three married couples on a journey come to a river where they find a boat which can not carry more than two persons at a time. The crossing of the river is complicated by the fact that husbands are all very jealous and will not permit their wives to be left without them in a company where there are other men present. Construct a graph to show how the transfer can be made.
   (ii) Prove that the puzzle in part (i) cannot be solved if there are four couples.
   (iii) Prove that the puzzle in part (i) can be solved if there are four couples and the boat holds three persons.

13. Name five situations (games, activities, real life problems, etc.) that can be represented by means of graphs. Explain what the vertices and the edges denote.

14. Draw the graphs of the following chemical compounds
   (a) Methane CH_4
   (b) Ethane C_2H_6
   (c) C_6H_6
   (d) N_2O_3.

   (**Hint** Represent atoms by vertices and chemical bonds between them by edges.)

15. Define cycle graph, bipartite graph, path graph and star graph. Give one example of each of the graph.

16. Consider the following graph.
(i) Describe formally the graph G in the diagram, that is, find the set $V(G)$ of vertices of G and the set $E(G)$ of edges of G
(ii) Find the degree of each vertex.

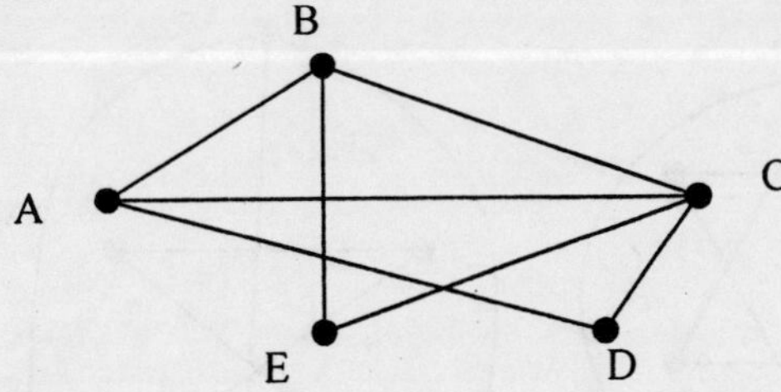

17. Draw the following graphs:
   (i) the complete graph K_6
   (ii) the complete bipartite graph $K_{5,7}$
   (iii) the complement of the cycle graph c_4.

18. Determine whether the following graphs are bipartite?

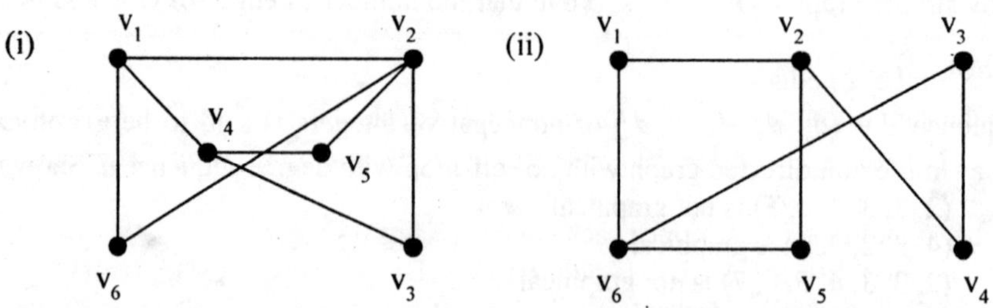

19. Determine the value of m for which the following graphs are regular.
    (i) C_m (ii) W_m (iii) K_m (iv) $K_{m,n}$
20. How many vertices does a regular graph of degree 4 with 10 edges have?
21. Consider the complete graph K_n
    (a) Find the number m of edges in K_n.
    (b) Find the degree of each vertex in K_n.
    (c) Find those values of n for which K_n is regular.
22. Show that a simple graph with $|V| = n$ is not bipartite if $|E| = \lfloor n^2/4 \rfloor$.
23. A **tripartite graph** is a graph where the set of vertices may be partitioned into three subsets so that no edge has incident vertices in the same subset.
    (i) Draw examples of $K_{2,3,3}$.
    (ii) Give an example of a tripartite graph.
24. Determine the number of vertices and edges in (i) K_n (ii) C_n (iii) P_n.
25. Draw a connected graph that becomes disconnected when any edge is removed from it.
26. Consider the following multigraph G.
    (i) Which of them are not connected?
    (ii) If they are not connected, find the number of connected components.
    (iii) Which are cycles free (without cycle)? If there are any, find the number of cycles.
    (iii) Which are loops free (without loops)?
    (iv) Which are (simple) graphs?

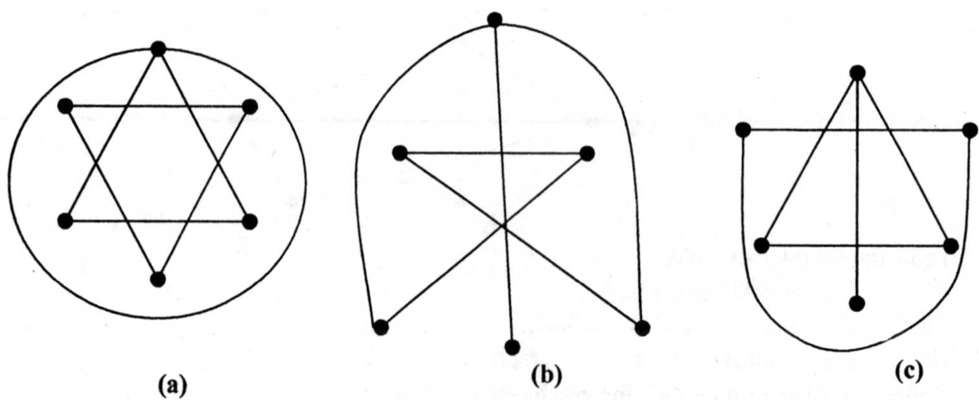

ANSWERS

7. Twice the number of edges
8. Equal to the sum of outdegrees of ½ the number of edges.
9. (i) 16 (ii) 13 (iii) solve $n|v|=48$ for all divisors n of 48, where v denotes the number of vertices.
10. The number of edges ≤ $C(n, 2)$ since each edge connects 2 vertices.
16. (i) Seven pairs; E(G) = [{A,B}, {A,C}, {A,D}, {B, C}, {B,E}, {C,D}, {C,E}
(ii) deg (A) = 3, deg (B) = 3, deg (C) = 4, deg (D) = 2, deg (E) = 2, The sum of the degrees of the vertices is 3 + 3 + 4 + 2 + 2 = 14. Which does equal twice the number of edges.
18. (i) Not bipartite
(ii) Not bipartite
19. (i) for all $m \geq 3$ (ii) for $m = 3$ (iii) for all $m \geq 1$
20. 5
21. (a) $m = c(n,2) = \dfrac{n(n-1)}{2}$
(b) $n - 1$
(c) all n
24. (i) n vertices and $\dfrac{n(n-1)}{2}$ edges.
(ii) n vertices and n edges.

2.23 COMPLEMENT OF A GRAPH

The complement $\overline{G}$ of a simple graph $G = (V, E)$ is the simple graph with vertex set $V(G)$ defined by the edge $v_i v_j \in E(G)$ iff $v_i v_j \notin E(\overline{G})$. In other words we can say $\overline{G}$ is a simple graph with vertex set $V(G)$ in which two vertices are adjacent iff they are not adjacent in G. One can observe here that complement of a complete graph is null graph. Graphs shown in the Fig.2.31 are complement to each other.

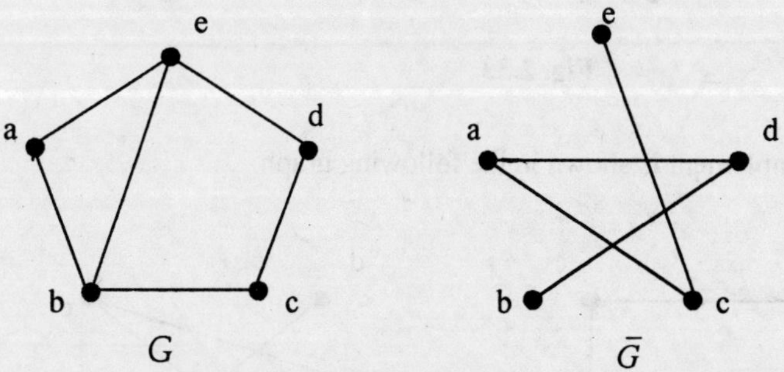

Fig. 2.31

Example 2.13

Graph $\overline{G}$ shown in Fig. 2.32 is a complement of a graph G.

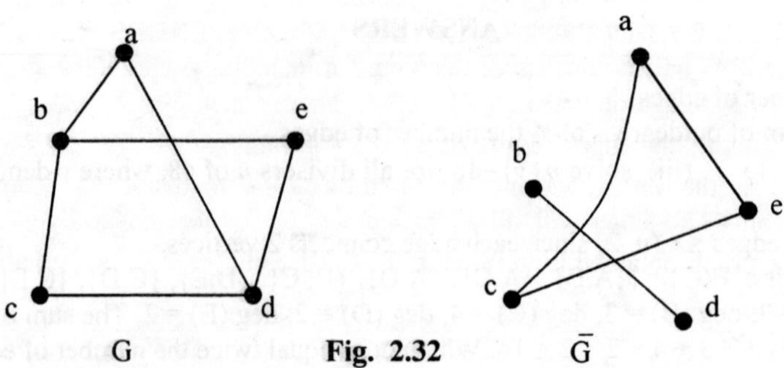

Fig. 2.32

Example 2.14

Graphs and their compliments are shown in Fig. 2.33.

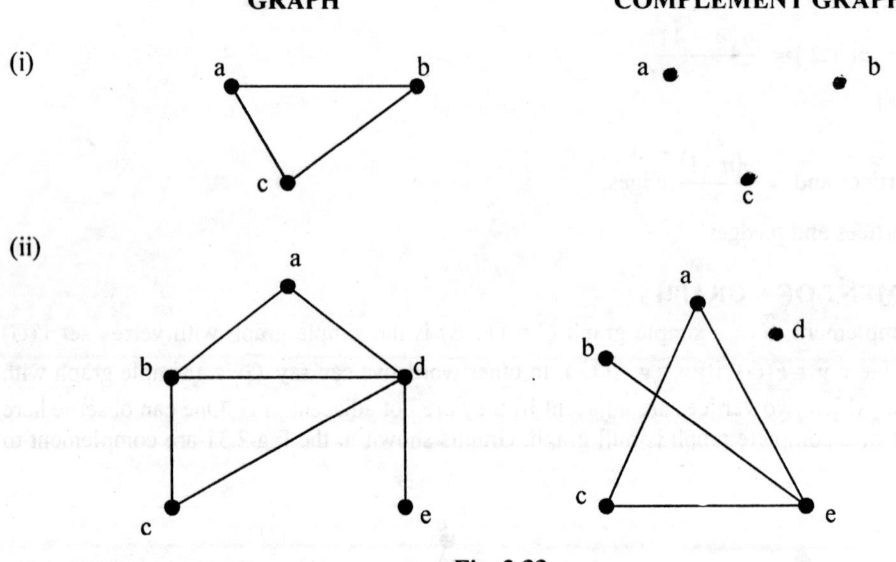

Fig. 2.33

Example 2.15

A graph and its complement is shown in the following graph.

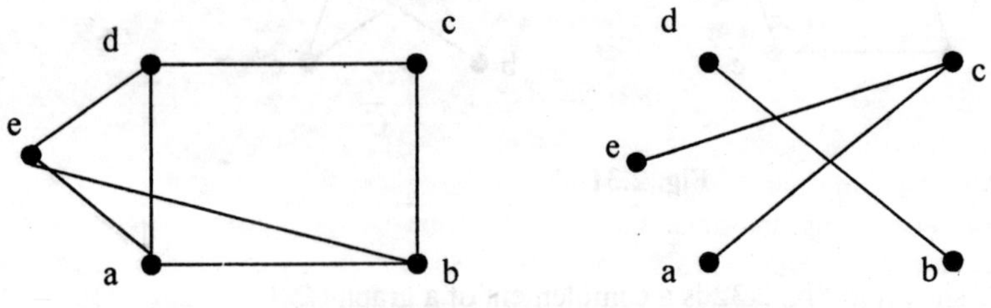

2.24 GRAPH OBTAINED BY DELETION OF VERTICES AND EDGES

Consider a graph G which consists of many edges including edge e. The graph obtained by deletion of the edge e from the graph is denoted by $G - e$. More generally, $G - \{e_1, e_2, e_3, \ldots, e_n\}$ is the graph obtained by deleting the edges $\{e_1, e_2, e_3, \ldots, e_n\}$ from G.

Similarly, if v is the vertex, we use the notation $G - v$ to denote the graph obtained by deleting the vertex v from G, together with all edges incident on v. More generally, we write $G - \{v_1, v_2, v_3, \ldots, v_n\}$ for the graph obtained by removing the vertices $v_1, v_2, v_3, \ldots, v_n$ and all the edges incident on any one of them.

2.25 SUBGRAPHS

Consider a graph $G = (V, E)$. A graph $g = (v, e)$ is said to be a subgraph of G if e is a subset of E and v is a subset of V such that edges in e are incident only with the vertices in v. For instance, the graph in Fig. 2.34(b) and (c) are a subgraph of Fig. 2.34(a).

The concept of subgraph is similar to the concept of subset in set theory. As a subset is contained in a set similarly a subgraph is thought of as being contained in (or a part of) another graph.

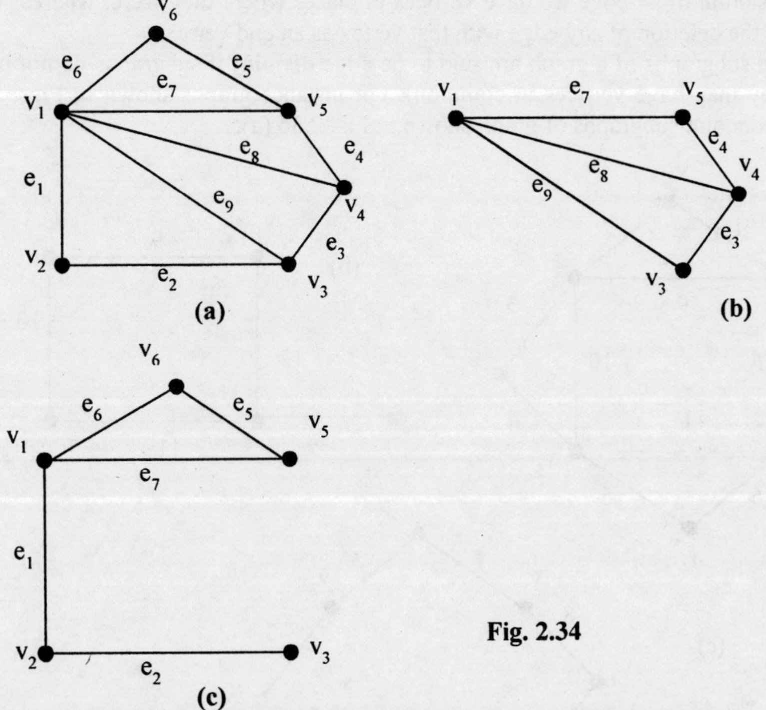

Fig. 2.34

Now one can draw following conclusion from the observations of subgraphs:

1. A graph G is a subgraph of itself. A null graph is also a subgraph of G.
2. A subgraph of a subgraph of G is a subgraph of G.
3. A single vertex in a graph G is a subgraph of G.
4. A single edge in G, together with its end vertices, is also a subgraph of G.

2.25 -1 Spanning Subgraph

A subgraph of G is said to be a **spanning subgraph** if it contains all the vertices of G. Graph shown in Fig. 2.35(b) is a spanning subgraph of Fig.2.35 (a).

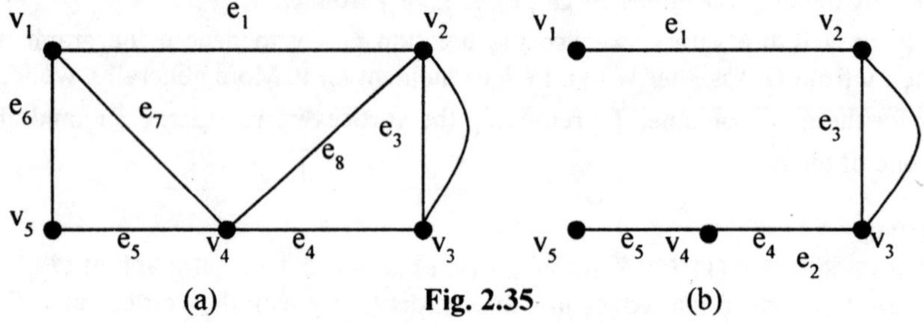

(a) **Fig. 2.35** (b)

2.25-2 Edge-Disjoint and Vertex Disjoint Subgraphs

Any subgraph of a graph can be obtained by deleting certain edges and certain vertices. It is to be noted that by deletion of an edge we have vertices in places where they were, whereas deletion of vertex necessitates the deletion of any edge with that vertex as an end vertex.

Two or more subgraphs of a graph are said to be **edge disjoint** if subgraphs do not have any edge in common (they may have vertices in common). For instance graphs shown in Figs. 2.36 (b) and 2.36 (c) are edge disjoint subgraphs of graph shown in Fig. 2.36 (a).

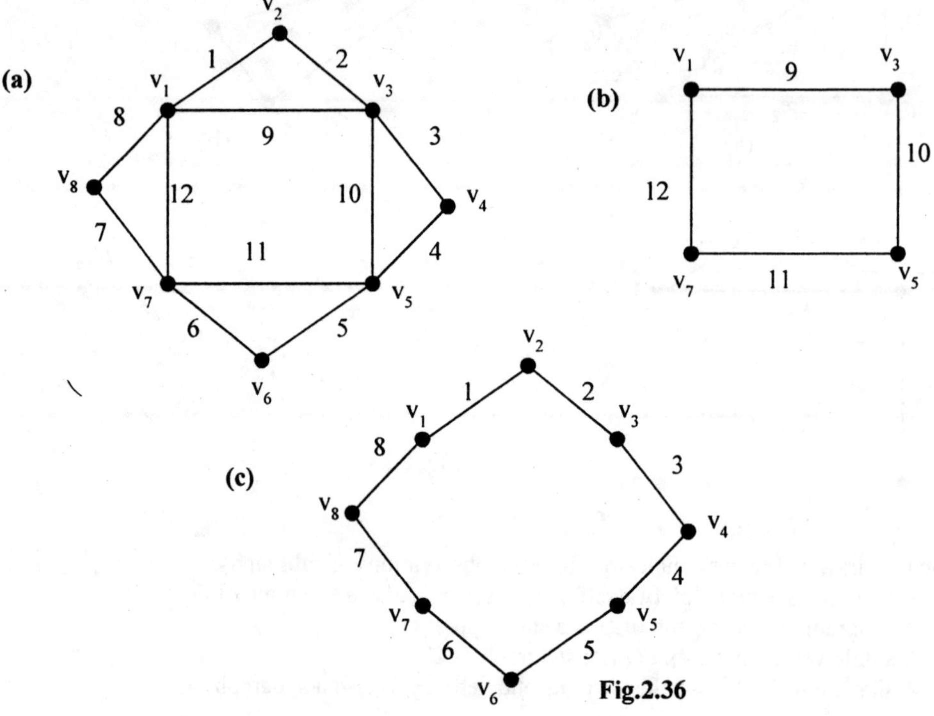

Fig.2.36

Subgraphs that do not have vertices in common are said to be **vertex disjoint**. One can observe that the graphs that have no vertices in common cannot possibly have edges in common.

For instance graphs show in Fig.2.37(b) and (c) are vertex disjoint graphs of graph Fig. 2.37(a).

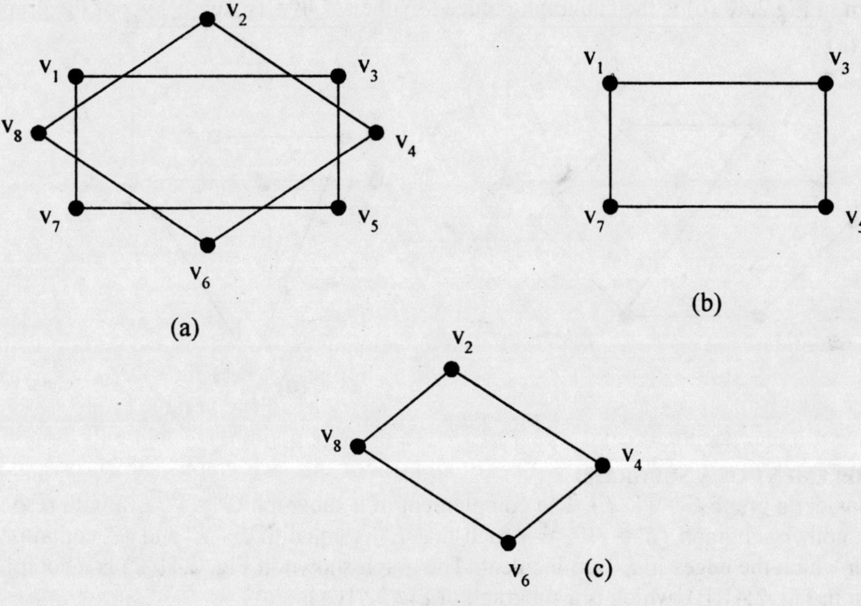

Fig. 2.37

Example 2.16

Find the vertex disjoint subgraph of graph shown in Fig.2.38.

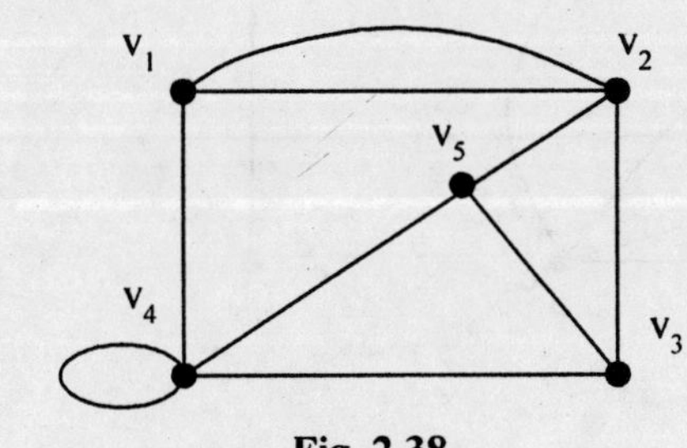

Fig. 2.38

Solution

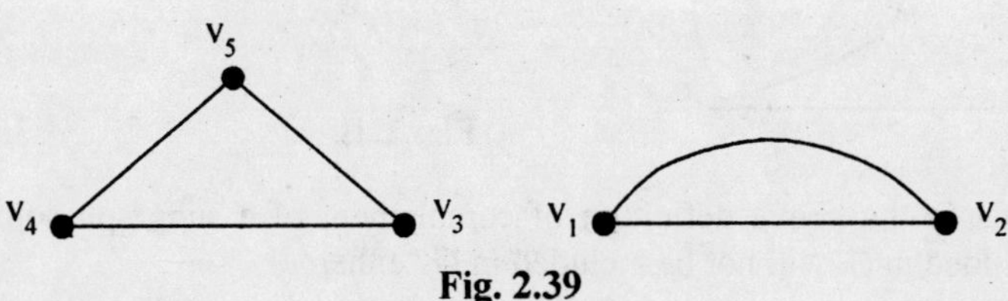

Fig. 2.39

2.25-3 Sub Graph Induced By A Set

Consider a graph $G = (V, E)$ and let $g = (v, e)$ be a subgraph of G. If $g = (v, e)$ be a subgraph of graph $G = (V, E)$ and W is any subset of vertices of G, then the subgraph induced by W is the subgraph g of G obtained by taking $v = W$ and e be those edges of G that join pairs of vertices in W. Graph shown in Fig.2.40 (b) is the subgraph induced by the set $W = \{v_1, v_2, v_3, v_4\}$ of the graph shown in Fig. 2.40 (a).

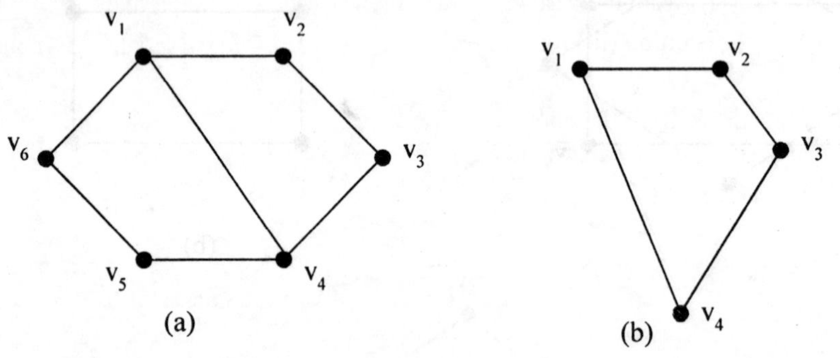

Fig. 2.40

2.25-4 Complement of a Subgraph

Consider a graph $G = (V, E)$. The complement of a subgraph $G' = (V', E')$ with respect to the graph G is another subgraph $G'' = (V'', E'')$ such that E'' is equal to $E - E'$ and V'' contains only the vertices with which the edges in E'' are incident. The graph shown in Fig. 2.41(C) is a complement of graph shown in Fig. 2.41(B), which is a subgraph of Fig.2.41(A).

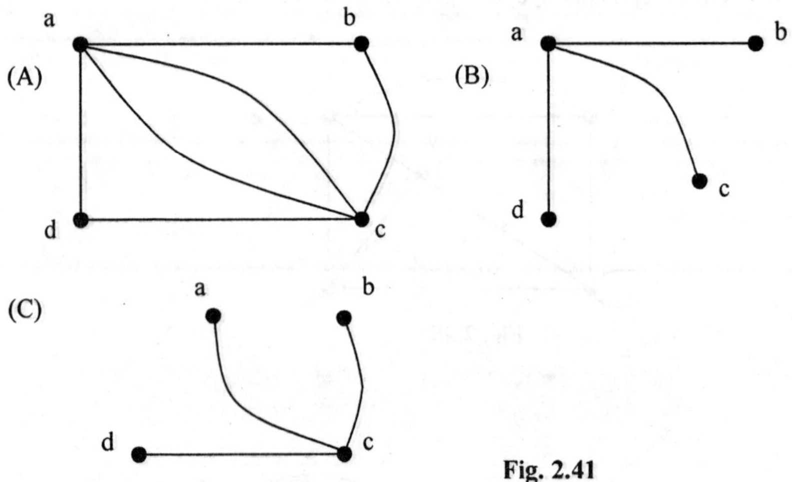

Fig. 2.41

It is to be noted that in the above definition of complement of a subgraph some isolated vertices in G that are not included in G' will not be included in G'' either.

2.26 ISOMORPHISM

In geometry if two figures resemble in terms of geometric properties and have identical behaviour then the figures are called congruent. Likewise in graph theory two graphs are thought of as equivalent and called **isomorphic** if they have identical behaviour in terms of graph-theoretic properties. More precisely it can be defined as follows:

Two graphs $G = (V, E)$ and $G_1 = (V_1, E_1)$ are **isomorphic** if there exist a function

$f: V(G) \to V_1(G_1)$ from the vertices of G to the vertices of G_1 such that

(i) f is one to one , (ii) f is onto, (iii) for each of vertices v_i and v_j of G; $\{v_i, v_j\} \in E$ if and only if $\{f(v_i), f(v_j)\} \in E_1$.

Any function f satisfying all the above mentioned properties is called an **isomorphism** from graph G to G_1. The condition (iii) says that function f preserves adjacency, i.e. v_i and v_j are adjacent in G iff $f(v_i)$ and $f(v_j)$ are adjacent in G_1.

In otherwords two graphs are said to be isomorphic if there is a one-to-one correspondence between their vertices and between their edges such that incidences are preserved. The graphs given in the Fig.2.42(a) and (b); (c) and (d); (e) and (f); (g) and (h) are isomorphic to each other.

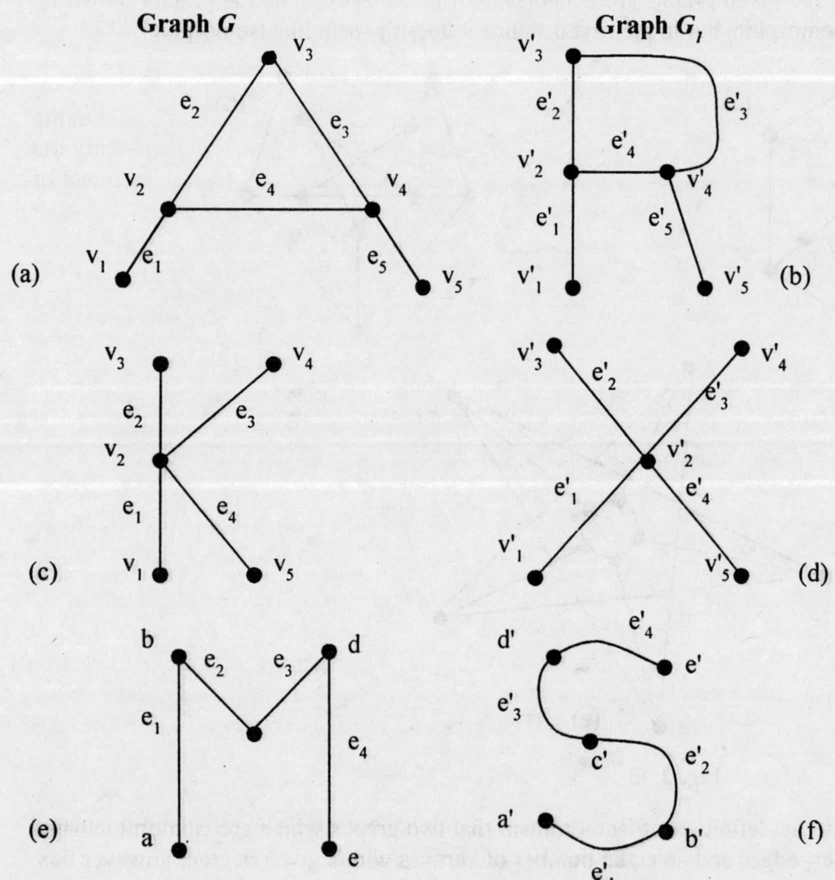

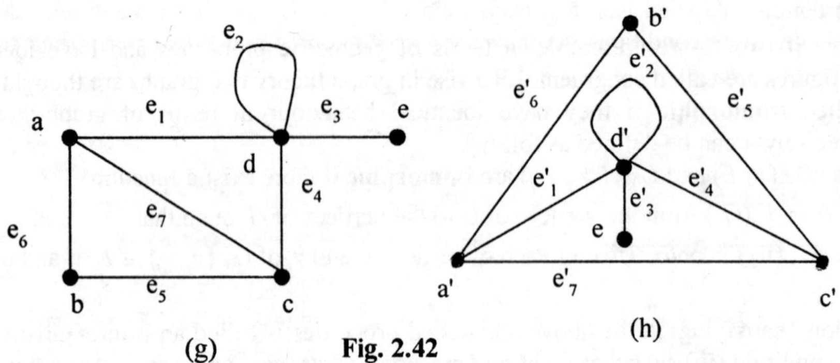

Fig. 2.42

From the Fig.2.42 one can see the correspondence between the graphs as: vertices v_is correspond to vertices v_i's and edges e_is correspond to edges e_i's of graphs G and G_1 respectively.

One can conclude from the above discussion that except for the labels (i.e. names) of their vertices and edges, isomorphic graphs are the same graph, may be drawn differently. It is not always easy to determine whether the given graphs are isomorphic or not. For example the graphs shown in the Fig. 2.43(*a*, *b*, *c*) are isomorphic, but in the first instance it doesn't seem like isomorphic.

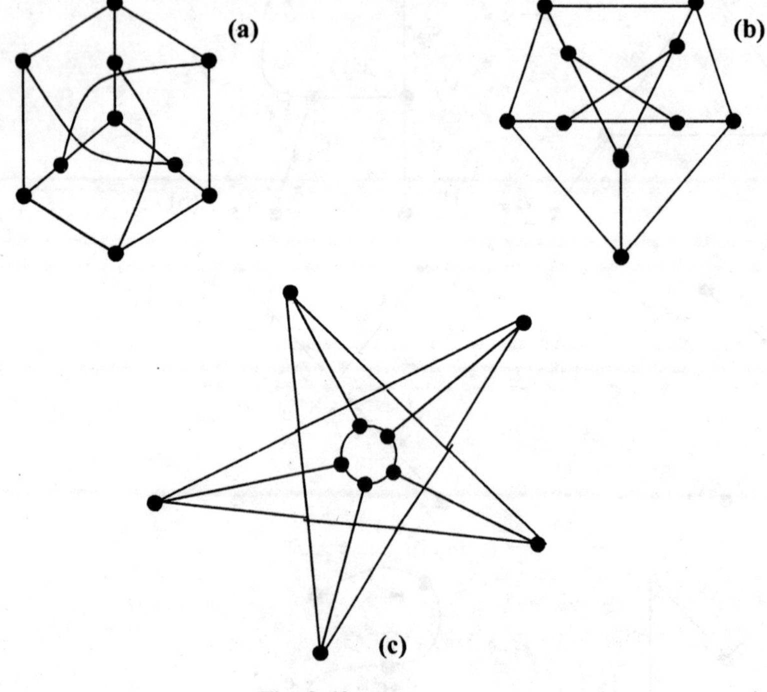

Fig. 2.43

It is apparent from the definition of isomorphism that two graphs which are isomorphic have the same number of vertices, edges and an equal number of vertices with a given degree; however this is not a sufficient condition for an isomorphism to exist, as can be seen from the graphs shown in Fig.2.44 satisfy all the three conditions yet they are not isomorphic and it can be verified from the following facts.

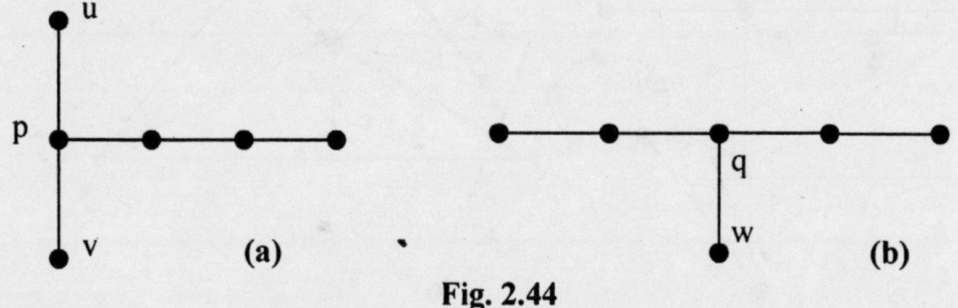

Fig. 2.44

(i) If graph in Fig. 2.44 (a) is to be isomorphic to Fig. 2.44 (b) then vertex p must correspond to vertex q as there is no other vertices of degree three.

(ii) In Fig. 2.44 (a) there are two pendant vertices u and v adjacent to p while in Fig. 2.44 (b) there is only one pendant vertex w adjacent to q.

2.26-1 DETERMINING WHEN GRAPHS ARE NOT ISOMORPHIC

Here are some tips, which may prove to be helpful in determining when two graphs are not isomorphic.

(i) To find some property that isomorphic graphs must share but the two graphs in question are not sharing that. For example they have different number of vertices, edges, degrees (degree sequences).

(ii) Some times two graphs may have the same degree (hence the same number of vertices and edges) and still not be isomorphic. For this, attempt to classify the vertices into classes according to some property preserved by isomorphism. For example, we classify into a single class the vertices of degree two in each graph and check these classes form a graph in their own right and we could ask if these graphs are isomorphic.

Example 2.17

Show that the following three graphs are isomorphic to each other.

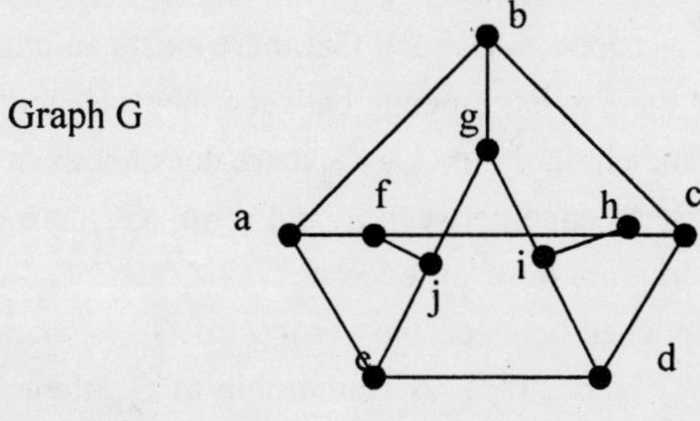

Fig. 2.45

Graph G_1

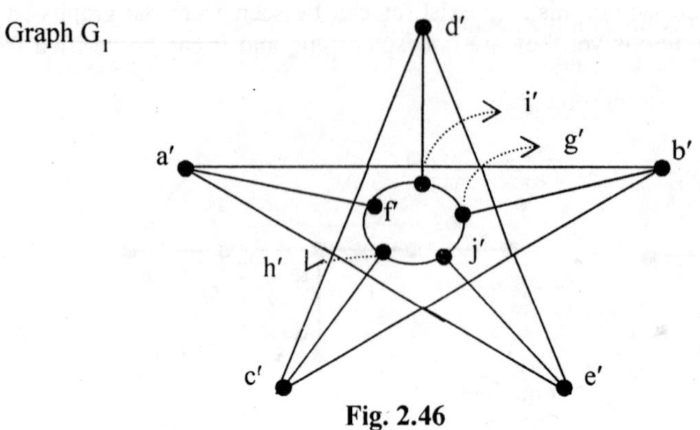

Fig. 2.46

Graph G_2

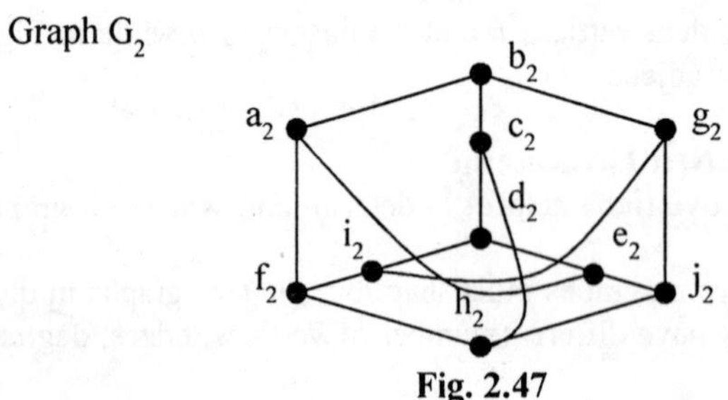

Fig. 2.47

Solution

All the three graphs have 10 vertices and 14 edges. The degree of each vertex is 3. Further, the correspondence between the vertices is $a - a' - a_2$, $b - b' - b_2$ $\ldots, j - j' - j_2$.

Example 2.18

Prove that two graphs are isomorphic if and only if their complements are isomorphic.

Solution

Let $\bar{G}_1$ and $\bar{G}_2$ be the complements of graphs of G_1 and G_2 respectively. Suppose G_1 is isomorphic to G_2. Then if u_1 and v_1 are 2 vertices in $\bar{G}_1$ such that there exists an edge between them, then there exist u_2 and v_2 in $\bar{G}_2$ such that there exists an edge between them. Hence, in $G_1 - \bar{G}_1$ there does not exist an edge between u_1 and v_1 and similarly in $G_2 - \bar{G}_2$ there does not exist an edge between u_2 and v_2. Similarly, if there does not exist an edge between u_1 and v_1 in $\bar{G}_1$ there does not exist an edge between u_2 and v_2 in $\bar{G}_2$ while these pairs have an edge in $G_1 - \bar{G}_2$ and $G_2 - \bar{G}_2$ respectively. Thus, there exists a one to one correspondence between the vertices of $G_1 - \bar{G}_2$ and $G_1 - \bar{G}_2$, and hence in $(G_1 - \bar{G}_2) \cup \bar{G}_1$ and in $(G_2 - \bar{G}_2) \cup \bar{G}_2$. If G_1 is isomorphic to G_2, their complements $\bar{G}_1$ and $\bar{G}_2$ have one to one correspondence as indicated and are hence isomorphic. The converse is true similarly from the fact that if G_1 and $\bar{G}_2$ are isomorphic, so are G_1 and G_2. Fig. 2.48 illustrates this.

Let us check the isomorphism by taking complements of the given graphs. Let $\bar{G}_1$ and $\bar{G}_2$ of Fig.2.48 (b) are complement of G_1 and G_2 (shown in Fig. 2.48 (a)) respectively. Now readers are advised to verify that G_1 and G_2 are not isomorphic.

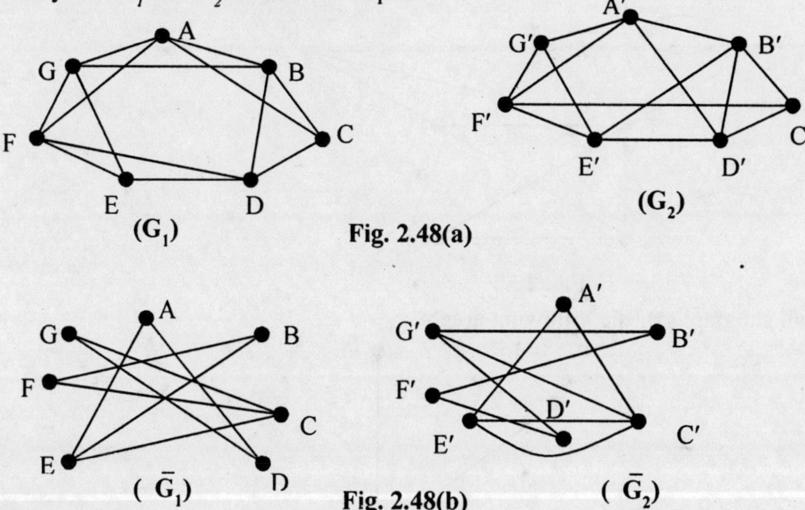

Fig. 2.48(a)

Fig. 2.48(b)

Example 2.19

If a graph of n vertices is isomorphic to its complement how many vertices it must have?

Solution

The total number of edges in a complete graph is $n(n - 1)/2$. Therefore, G and its complement must each have $n(n - 1)/4$ edges. Hence n or $(n-1)$ be a multiple of 4.

Example 2.20

Can a graph with seven vertices be isomorphic to its complement?

Solution

Since 7 is not a multiple of 4, a graph of 7 vertices cannot be isomorphic to its complements.

Example 2.21

Show that $K_{3,3}$ and G in Fig. 2.49 are not isomorphic.

Solution

In graph $K_{3,3}$ and G cycles of length 4 have different vertices, e.g. $(a\ e\ b\ f\ a)$, $(a'\ c'\ e'\ f'\ a')$. Further vertices in $K_{3,3}$ can be partitioned into 2 blocks (abc) (fed) such that there exists an edge from each vertex of one block to any vertex of other block and none within the block. G does not have such blocks.

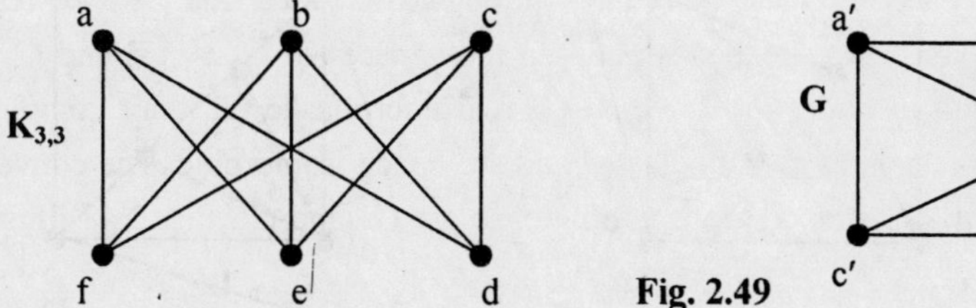

Fig. 2.49

EXERCISE 2.2

1. Find all connected subgraphs of the following graph.

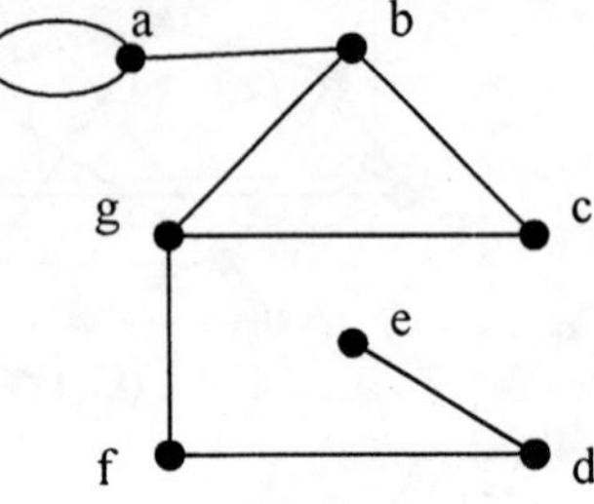

Fig. 2.50

2. Draw all subgraphs of the following graph.

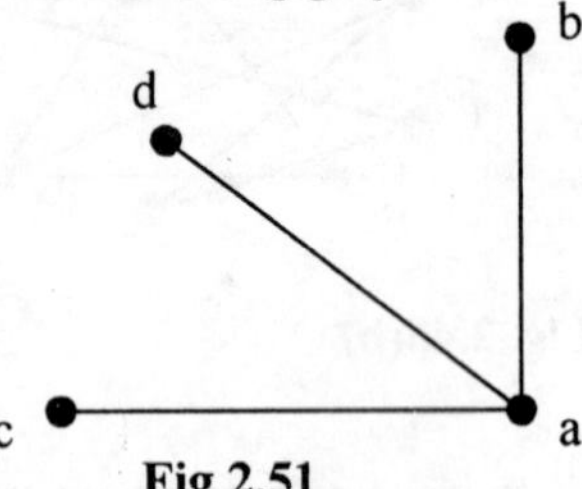

Fig 2.51

3. How many subgraphs does K_3 have?
4. Which of the following pairs of non-directed graphs in following figures are isomorphic?

(i)

(ii)

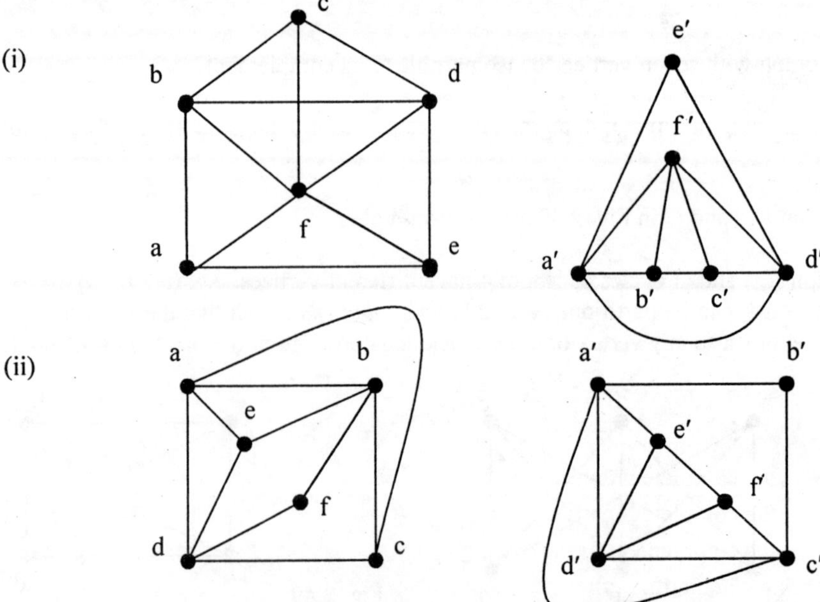

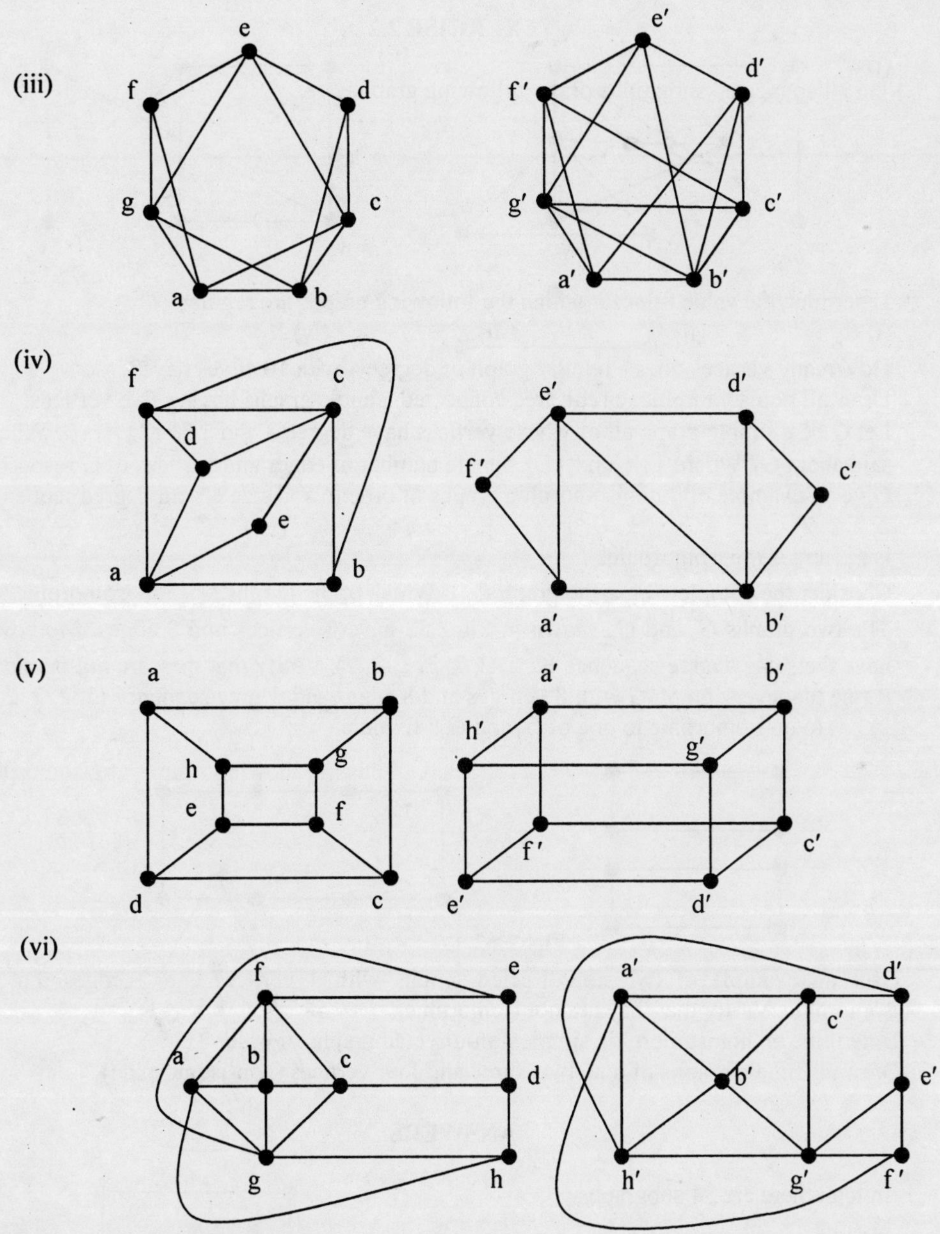

5. Draw the following graphs:
   (iv) the complete graph K_6
   (v) the complete bipartite graph $K_{5,7}$
   (vi) the complement of the cycle graph c_4.
6. Determine whether the following graphs are bipartite?

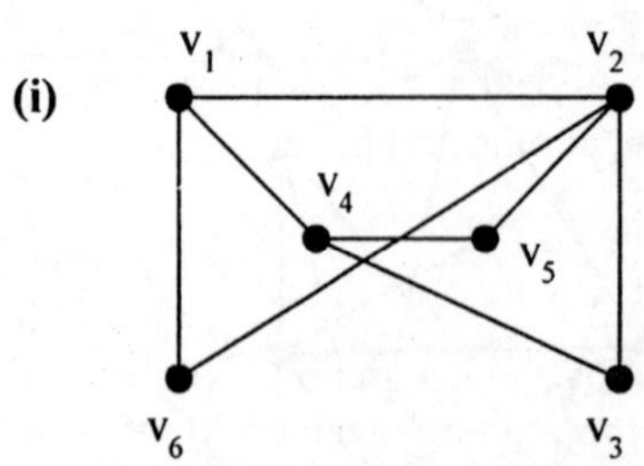

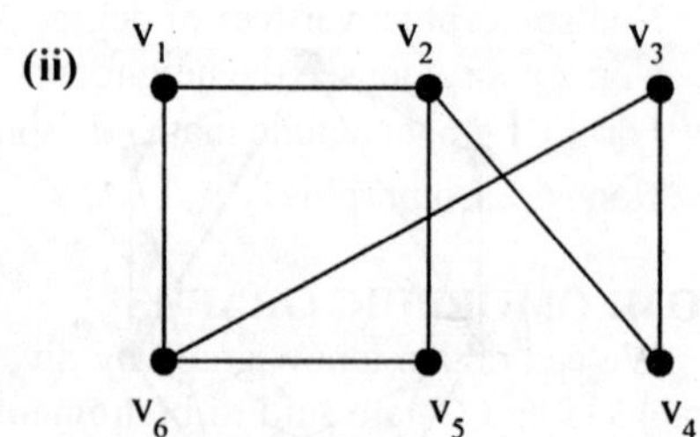

7. Determine the value of *m* for which the following graphs are regular.
   (i) C_m (ii) W_m (iii) K_m (iv) $K_{m,n}$
8. How many vertices does a regular graph of degree 4 with 10 edges have?
9. Draw all nonisomorphic, circuit free connected simple graphs having five vertices.
10. Let G be a simple graph all of whose vertices have degree 3 and $|E| = 2|V| - 3$. What can be said about G? Where $|V|$ and $|E|$ denote number of edges and vertices of G respectively.
11. Give an example of 2 nonisomorphic graphs of order 6 and size 6 with degree sequence (2, 2, 2, 2, 2, 2).
12. Is $k_{3,3}$ and k_6 are isomorphic.
13. Consider the complete bipartite graph $K_{m,n}$. Which of the graphs $K_{m,n}$ are isomorphic?
14. The two graphs G_1 and G_2 shown in Fig.2.52 have 8 vertices and 8 edges. Moreover, they have the same degree sequence (2, 2, 2, 2, 2, 2, 2, 2). Verify that they are not isomorphic. Is it true that every graph G with 8 vertices and 8 edges and degree sequence (2, 2, 2, 2, 2, 2, 2, 2) has to be isomorphic to one of G_1 and G_2? Explain

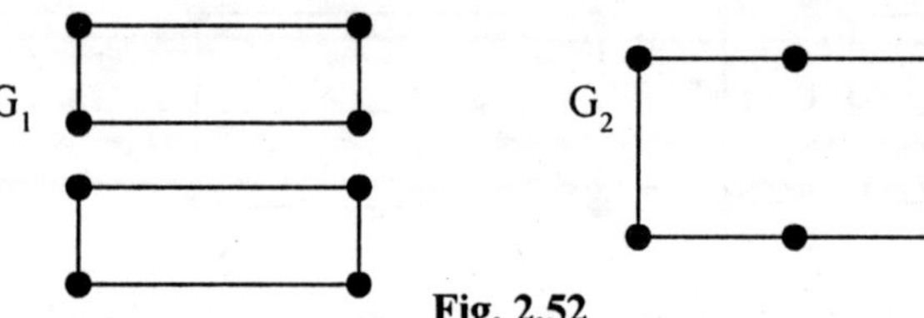

Fig. 2.52

15. Give an example of two non-directed graphs with 4 vertices and 2 edges that are not isomorphic. Verify that they are not isomorphic.
16. Determine all nonisomorphic simple non-directed graphs of order 3.
17. Draw all simple graphs of one, two, three and four vertices (non-isomorphic).

ANSWERS

2. In total there are 34 subgraphs.
3. 17.
4. (i) Isomorphic; by removing the vertices of degree 5 and their incident edges from each graph, observe the isomorphism for remaining graph and extend.

   (iii) Nonisomorphic, since vertices of degree 3 are adjacent in one graph and non-adjacent in the other.

   (iv) Nonisomorphic, consider their complements.

(v) Nonisomorphic, vertices of degree 3 are adjacent in one graph, non-adjacent in the other.
(vi) Nonisomorphic; see the neighbours of the vertex of degree 5 in each graph.

10. For $|V| = 6$, conclude that G is isomorphic to $K_{3,3}$ or the graph.

13. None are isomorphic.

2.27 HOMEOMORPHIC GRAPHS

We can obtain a new graph by dividing an edge of a given graph G with additional vertices. Two graphs G and G_l are said to be **homeomorphic** if they can be obtained from the same graph or isomorphic graphs by this method. For instance, graphs shown in Fig. 2.53(b) and (c) are not isomorphic but they are homeomorphic since they can be obtained from the graph shown in Fig. 2.53(a) by adding appropriate vertices.

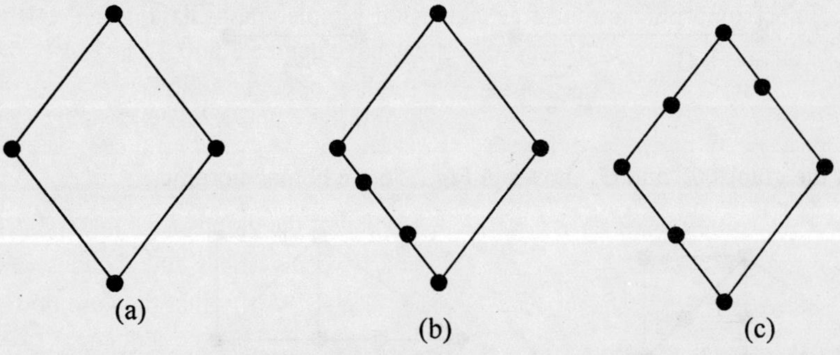

Fig. 2.53

Example 2.22

Show that the following figures are distinct, i.e. no two of them are isomorphic. Also show that G_2 and G_3 are homeomorphic.

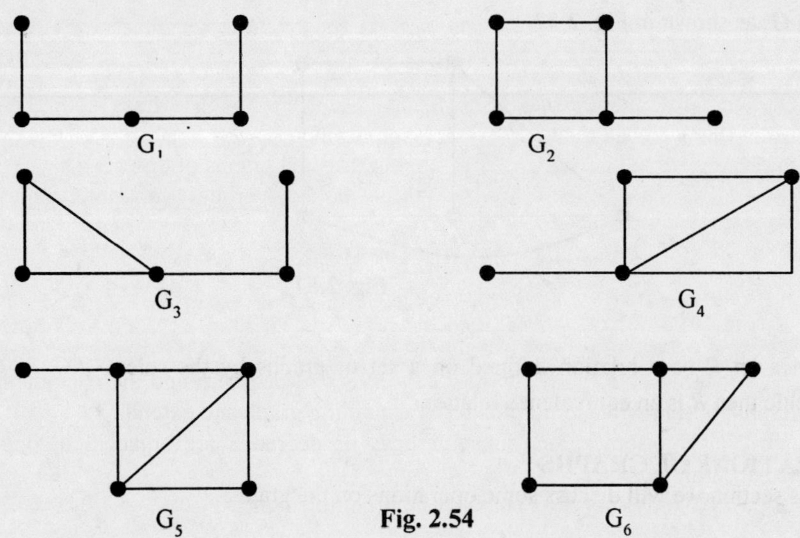

Fig. 2.54

Solution

In the given graphs the degrees of vertices of only G_2 and G_3 are same and the other graphs are having different degrees of vertices. Hence none of the graphs is isomorphic except possibly G_2 and G_3. But in G_2 and G_3 if we delete the vertices of degree 3, we obtain distinct subgraphs. Therefore G_2 and G_3 are also nonisomorphic. Hence all the graphs are distinct.

But graphs G_2 and G_3 are homeomorphic as it can be obtained from the following isomorphic graphs by adding appropriate vertices.

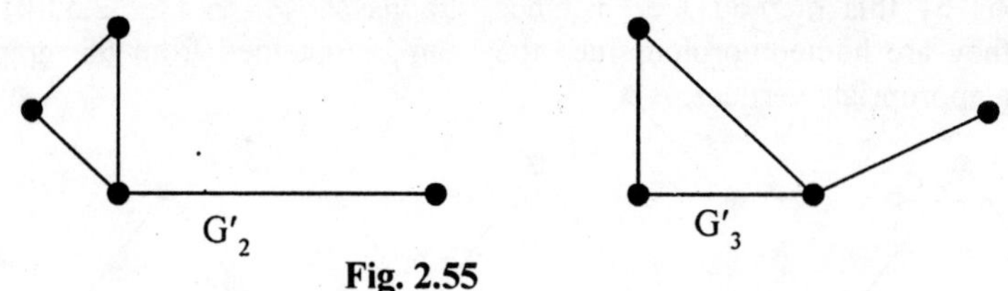

Fig. 2.55

Example 2.23

Show that the graphs G_1 and G_2 shown in Fig.2.56 are homeomorphic.

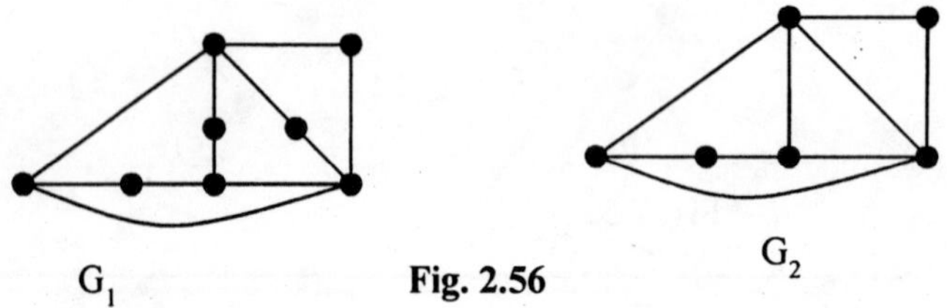

Fig. 2.56

Solution

The graphs G_1 and G_2 shown in Fig. 2.56 are homeomorphic as both of them can be reduced to the graph G_3 as shown in Fig. 2.57.

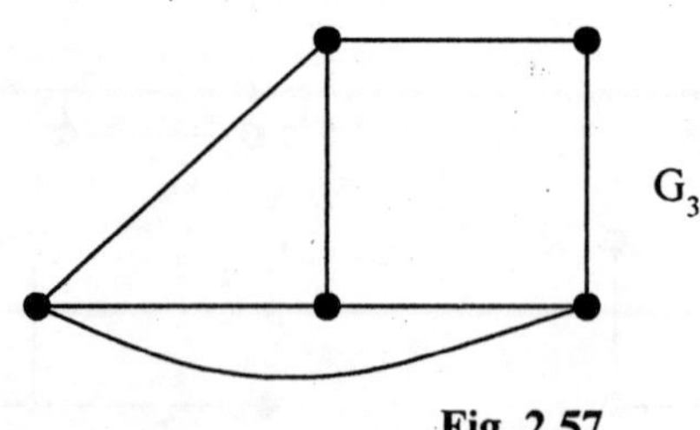

Fig. 2.57

NOTE

If we let R be a relation defined on a set of graphs by the rule G_1RG_2 if G_1 and G_2 are homeomorphic then R is an equivalence relation.

2.28 OPERATIONS OF GRAPHS

In this section we will discuss some operations on the graphs.

2.28-1 Union of Graphs

Consider two graphs G_1 and G_2. Let $V(G_1)$ and $V(G_2)$ denote the vertices of graph G_1 and G_2 and $E(G_1)$ and $E(G_2)$ denote the edges of graph G_1 and G_2 respectively. The union of graph G_1 and G_2 be a graph such that

$$V(G_1 \cup G_2) = V(G_1) \cup V(G_2)$$

and $$E(G_1 \cup G_2) = E(G_1) \cup E(G_2)$$

where $V(G_1 \cup G_2)$ and $E(G_1 \cup G_2)$ denote the vertices and edges of union of graphs G_1 and G_2. For example graph shown in Fig.2.58, $G_1 \cup G_2$ is union of G_1 and G_2.

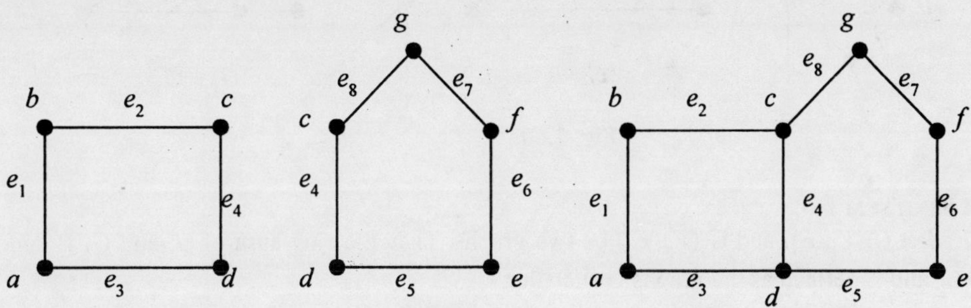

Fig 2.58

2.29-2 Intersection of Graphs

Let G_1 and G_2 be two graphs with at least one vertex in common then the intersection of these two graphs will be a graph such that

$$V(G_1 \cap G_2) = V(G_1) \cap V(G_2)$$

and $$E(G_1 \cap G_2) = E(G_1) \cap E(G_2)$$

For example graph shown in Fig.2.59, $G_1 \cap G_2$ represents the intersection of G_1 and G_2.

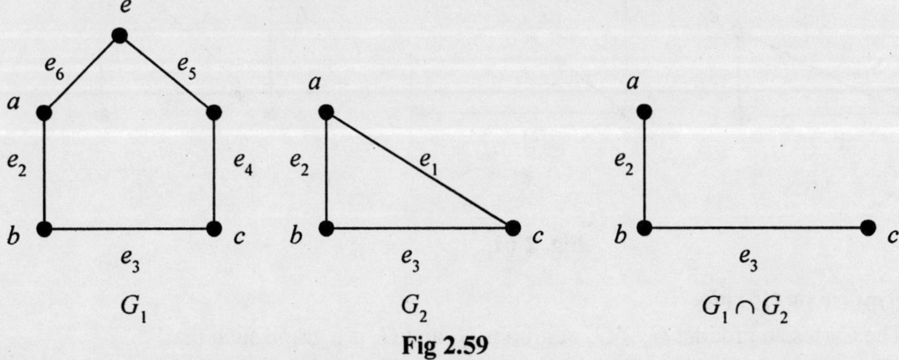

Fig 2.59

2.28-3 Sum of Two Graphs

Consider two graphs G_1 and G_2. Let $V(G_1)$ and $V(G_2)$ denote vertices of graph G_1 and G_2 respectively. $V(G_1 \cap G_2) = \phi$, then the sum of graphs G_1 and G_2: $G_1 + G_2$ is defined as the graph whose vertex set is $V(G_1) + V(G_2)$ and the edge set of $G_1 + G_2$ consist of those edges which are in G_1 and G_2 and the edges obtained by joining each vertex G_1 to each vertex of G_2. For example see the graphs shown in Fig. 2.60.

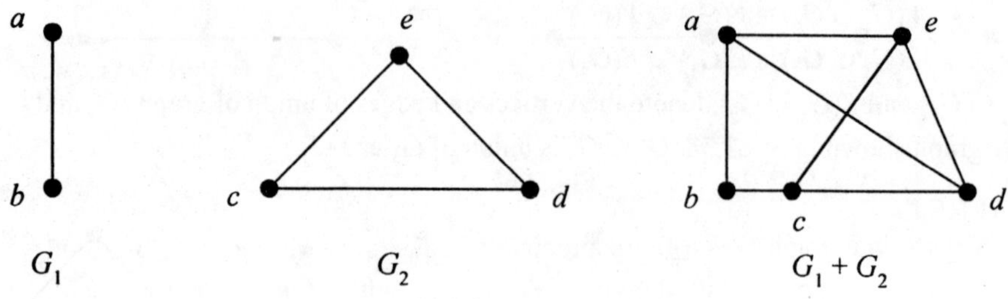

Fig. 2.60

2.28-4 RING SUM

Let $G_1(V_1, E_1)$ and $G_2(V_2, E_2)$ be two graphs. Then the **ring sum** of G_1 and G_2 is denoted by $G_1 \oplus G_2$ and is defined as the graphs G such that

(i) $V(G) = V(G_1) \cup V(G_2)$

(ii) $E(G) = E(G_1) \cup E(G_2) - E(G_1) \cap E(G_2)$.

The point (ii) says the graph G consists of those edges, which are either in G_1 or G_2 but not in both. The graphs shown in Fig.2.61 represent ring sum.

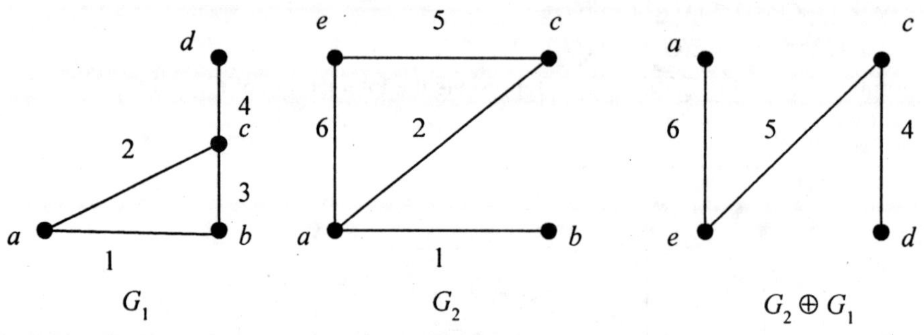

Fig. 2.61

2.28-5 PRODUCT OF GRAPHS

The Cartesian product $G_1 \times G_2$ of graphs G_1 and G_2 is a graph such that

(i) the vertex set of $G_1 \times G_2$ is the Cartesian product $V(G_1) \times V(G_2)$ and

(ii) any two vertices (u_1, u_2) and (v_1, v_2) are adjacent in $G_1 \times G_2$ if and only if either $u_1 = v_1$ and u_2 is adjacent with v_2 or u_1 is adjacent with v_1 and $u_2 = v_2$.

For example:

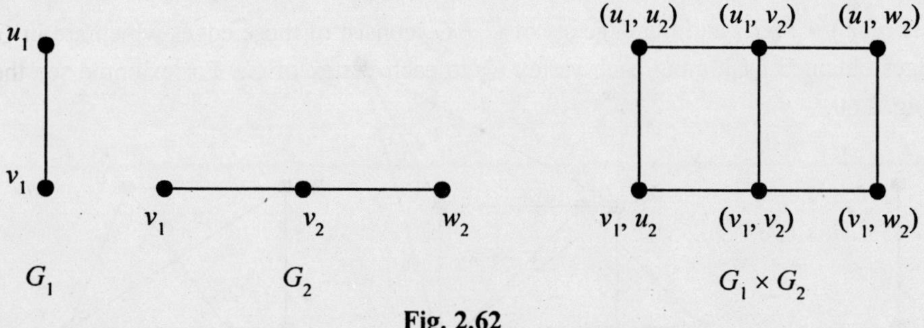

Fig. 2.62

2.29 WALKS

A **walk** in a graph G is a finite alternating sequence of vertices and edges of the form v_0, e_1, v_1, e_2, ..., e_{n-1}, v_{n-1}, e_n, v_n, (starting and terminating with vertices) such that each edge is incident with the vertices preceding and following it.

For instance $v_1 e_1\, v_2\, e_2\, v_3\, e_3 v_3\, e_4\, v_4\, e_5\, v_2\, e_6\, v_5$ is a walk in the Fig.2.63.

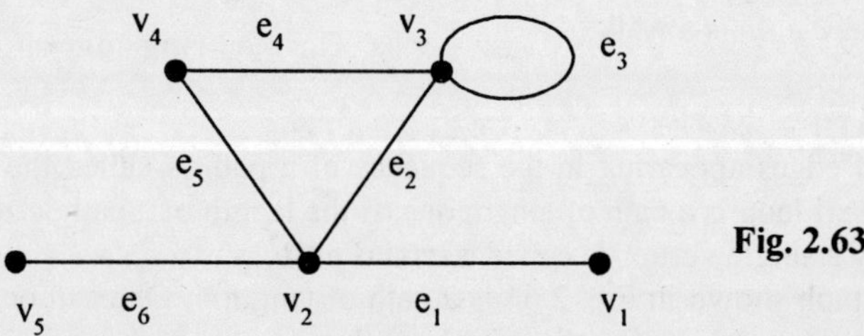

Fig. 2.63

A walk is also called **chain** or an **edge train**. It is to be noted that set of vertices and edges of a given walk in a graph is a subgraph of G. The vertices and edges in a walk need not be distinct.

A walk is said to traverse through the vertices in the sequence originating in the initial vertex of the first edge and ending at the terminal vertex of the last edge in the sequence. Vertices with which a walk begins and ends are called **initial** and **final vertex** respectively or simply **terminal vertices**. In the graph shown in Fig. 2.63, vertices v_1 and v_5 are the terminal vertices and the walk is $v_1\, e_1\, v_2\, e_2\, v_3\, e_3\, v_3\, e_4\, v_4\, e_5\, v_2\, e_6\, v_5$.

A **close walk** is a walk, which begins and ends at the same vertex. A walk, which is not close, is called an **open walk**. For example graph shown in the Fig. 2.63 is an open walk.

2.29 -1 TRAIL

If the edges of a walk are distinct then we call it a **trail** or we can say a trail is a walk with no repeated edge. A trail may visit same vertex more than once, but only if it comes and goes from a different edge each time.

2.29 -2 PATH

A path is a walk that does not include any vertex twice except that its initial vertex may be same as final vertex. In other words if in a trail all the vertices are distinct except possibly terminal vertices, then it is called a **path**. Let us consider the graph shown in Fig. 2.64.

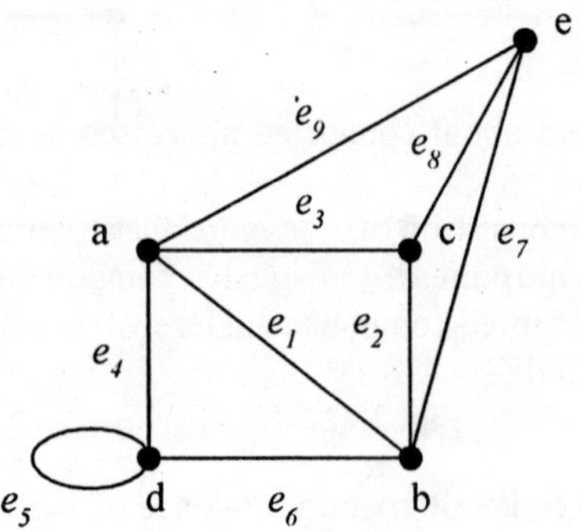

Fig. 2.64

Here, we see that $e\, e_8\, c\, e_2\, b\, e_1\, a\, e_4\, d\, e_5\, d\, e_6\, b$ is a trail, whereas $e\, e_8\, c\, e_2\, b\, e_1\, a\, e_4\, d$ is a path and $e\, e_8\, c\, e_3\, a\, e_9\, e\, e_7\, b\, e_2\, c\, e_3\, a$ is a walk.

2.29-3 Length of a Path

The number of edges appearing in the sequence of a path is called the **length of the path,** viz. an edge if its not a self loop is a path of length one. If the length of a path is zero then the path has no edge, it contains only a single vertex, is called a **trivial path**.

For example graph shown in Fig. 2.64 is a path of length 4. One can observe that the length of a path is one less than the number of vertices in the path.

2.29-4 Cycle and Circuit

A **cycle** is a closed path in which all the vertices are distinct except for the initial and final vertex. A **circuit** is a trail, which begins and ends on the same vertex. It means a circuit may have internal repeated vertices other than the end vertices. A cycle of length k is called a k-cycle. In a graph any cycle must have length two or more. Two different cycles are shown in the Fig.2.65.

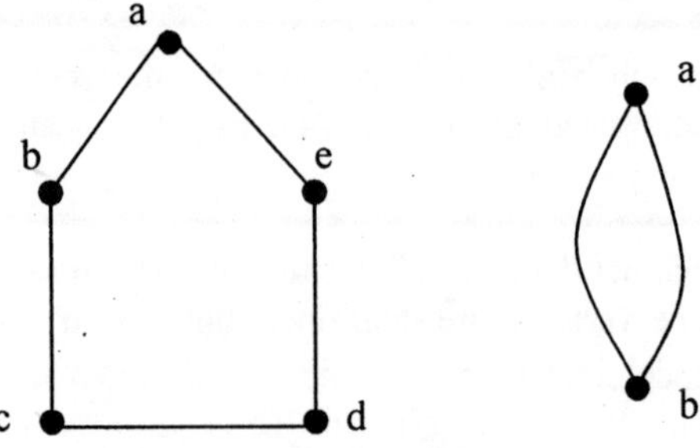

Fig. 2.65

For example, consider the graph shown in Fig. 2.64. Here

Cycle: $e\, e_7\, b\, e_2\, c\, e_3\, a\, e_9\, e.$

Circuit: $e\, e_9\, a\, e_3\, c\, e_2\, b\, e_1\, a\, e_4\, d\, e_6\, b\, e_7\, e.$

Caution! Many authors call walks as paths or simple paths and some of them call path chains.

2.30 CONNECTIVITY

We have already discussed about connected and disconnected graphs. Here we take some theorems related to these graphs.

As a disconnected graph consist of two or more than two connected graphs. Each of these connected subgraphs is called a **component.** Obviously, component itself is a graph. For example graph shown in Fig. 2.66 consist of three components. Here we can say that the graph G is the union of three disjoint components A, B and C.

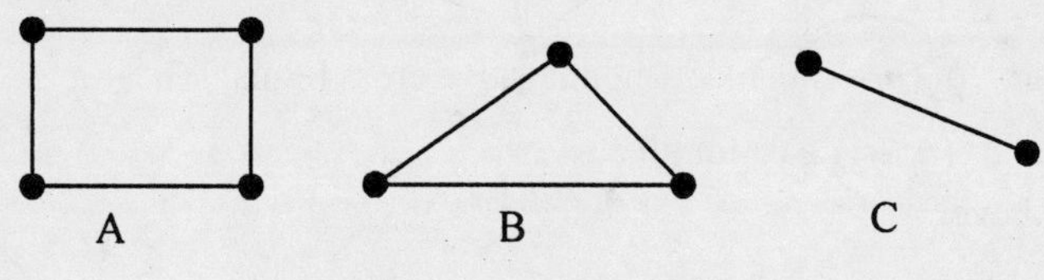

Fig. 2.66

THEOREM 10.6

A graph G is disconnected iff its vertex set V can be partitioned into two nonempty, disjoint subsets S_1 and S_2 such that there exists no edge in G whose one end vertex is in subset S_1 and the other is in subset S_2.

PROOF

To prove this theorem let us assume that such a partitioning exists. Consider two arbitrary vertices v_1 and v_2 of graph $G = (V, E)$ such that $v_1 \in S_1$ and $v_2 \in S_2$. As per our assumption no path can exist between vertices v_1 and v_2, otherwise there would be at least one edge whose one end vertex would be in S_1 and other in S_2. Hence, if a partition exists, graph G is not connected.

Conversely, let us assume G is a disconnected graph. Let v_1 be a vertex in G. Consider S_1 be the set of all vertices that are joined by paths to v_1. Since graph G is disconnected, S_1 does not include all vertices of G. And the remaining vertices will form a nonempty set S_2. Now it is clear that no vertex in S_1 is joined to any in S_2 by an edge. Hence the partition.

THEOREM 2.7

If a graph, which may be connected or disconnected, has exactly two vertices of odd degree, then there must be a path joining these two vertices.

PROOF

Consider a graph $G = (V, E)$ with all even vertices (i.e. vertices with even degrees) except vertices v_1 and v_2, which are odd. From above theorem which is applicable for every graph (directed and undirected) therefore for every component of a disconnected graph, no graph can have an odd number of odd vertices. This implies that vertices v_1 and v_2 in graph G must belong to the same component and hence must have a path between them.

THEOREM 2.8

A graph G without parallel edges or self loop (i.e. simple graph) with n nodes and k components can have at most $(n - k)(n - k + 1)/2$ edges.

PROOF

Let $n_1, n_2, ..., n_k$ be the number of vertices in each of the k components of a graph G. Then

$n_1 + n_2 + ..., n_k = n$, where $n_i \geq 1$

Before proving the theorem let us take the inequality

$$\left(\sum_{i=1}^{n}(n_i - 1)\right) = n - k$$

$$\left(\sum_{i=1}^{n}(n_i - 1)\right)^2 = n^2 + k^2 - 2nk$$

$$\text{or} \quad \sum_{i=1}^{n}(n_i^2 - 2n_i) + k + \text{nonnegative cross terms} = n^2 + k^2 - 2nk$$

$$\therefore \quad (n_i - 1) \geq 0, \text{ for all } i$$

Squaring both sides, we have

$$\therefore \quad \sum_{i=1}^{n} n_i^2 \leq n^2 + k^2 - 2nk - k + 2n = n^2 - (k-1)(2n-k)$$

$$\Rightarrow \quad \sum_{i=1}^{n} n_i^2 \leq n^2 - (k-1)(2n-k) \quad ...(1)$$

As we know the maximum number of edges in the ith component of graph G is $(1/2)n_i(n_i-1)$. Therefore, the maximum number of edges in G is

$$\frac{1}{2}\sum_{i=1}^{k}(n_i - 1)n_i = \frac{1}{2}\left(\sum_{i=1}^{k} n_i^2\right) - \frac{n}{2}$$

$$\leq \frac{1}{2}\left[n^2 - (k-1)(2n-k)\right] - \frac{n}{2} \quad \text{(from 1)}$$

$$= \frac{1}{2}(n-k)(n-k+1)$$

2.31 DISTANCE AND DIAMETER

Consider a connected graph $G = (V, E)$. The distance between vertices v_i and v_j denoted by $d(v_i, v_j)$ or $d_G(v_i, v_j)$ is the minimum number of edges between any two points in G. For example in Fig.2.67(a), $d(v_8, v_3) = 2$. The **diameter** of the graph $G = (V, E)$ written as diam(G) or Δ, is the maximum distance between any two vertices, i.e.

$$\Delta = \max_{u,v \in V} d(u,v)$$

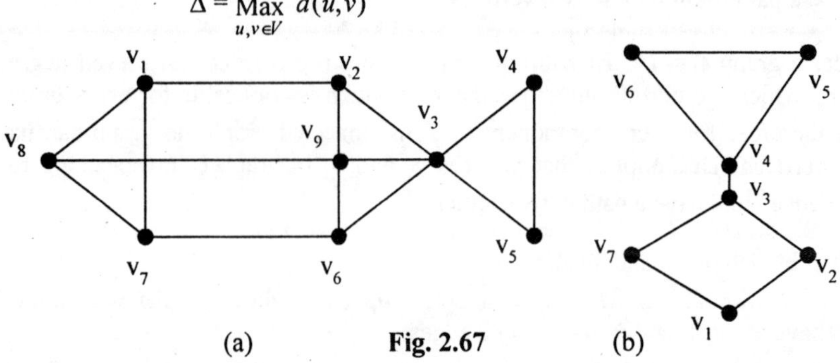

(a) **Fig. 2.67** (b)

Diameter of the graph shown in Fig. 2.67(b) is 4, i.e. $\Delta = 4$.

Example 2.24

The paths between the vertices u and v in the graph shown in the Fig. 2.68 are (a, e), (a, c, f), (b, c, e), (b, f), (b, g, h) and (b, g, i, k). Out of these (a, e) and (b, f) are two shortest paths, each of length 2, i.e. $d(u, v) = 2$.

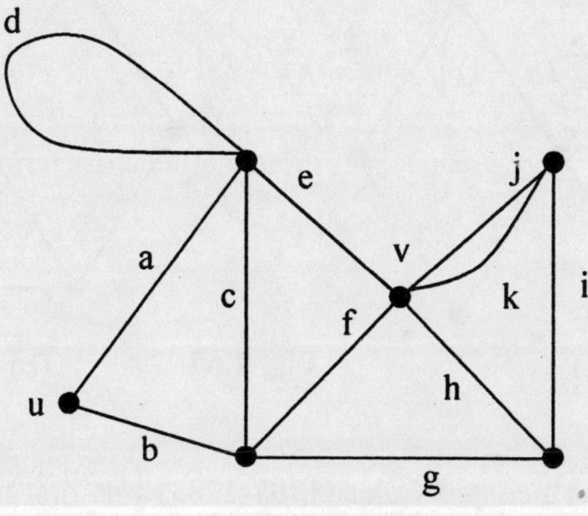

Fig. 2.68

Example 2.25

Under what condition is the equation $d(v_1, v_2) + d(v_2, v_3) = d(v_1, v_3)$ satisfied for $v_1, v_2, v_3 \in V$ in a simple graph $G = (V, E)$.

Solution

The given equation holds if there is a unique path of minimum length between the vertices v_1, v_2 and v_3.

2.32 REACHABILITY

A vertex v_i of a simple graph is said to be **reachable** or **accessible** from v_j if there is a path from v_i to v_j. If a vertex v_i is reachable from a vertex v_j then the path of minimum length from v_i to v_j is called **geodesic**.

The length of geodesic from v_i to v_j is called the **distance** from v_i to v_j denoted by $d(v_i, v_j)$. It is assumed that $d(v_i, v_i) = 0$. We assume that every vertex is reachable from itself. Here, it is to be noted that reachability is a relation, which is reflexive, and transitive. Reachability need not be symmetric or anti-symmetric.

2.33 CUTPOINTS AND BRIDGES

Let $G = (V, E)$ be a connected graph. A vertex v in a graph G is called a **cutpoint** (or **cut vertex** or **articulation point**) if $G - v$ is disconnected. It is to be recalled here that the $G - v$ is the graph obtained from G by deleting v and all edges incident on v.

An edge e of graph G is called a **bridge** (or **cut edge**) if $G - e$ is disconnected. Again it is to be noted here that $G - e$ is the graph obtained from G by simply deleting the edge e. In the Fig.2.69(a) vertex v_3 is a cutpoint and there are no bridge.

In Fig.2.69 (b), the edge (v_4, v_5) is a bridge and its end points v_4 and v_5 are cutpoints.

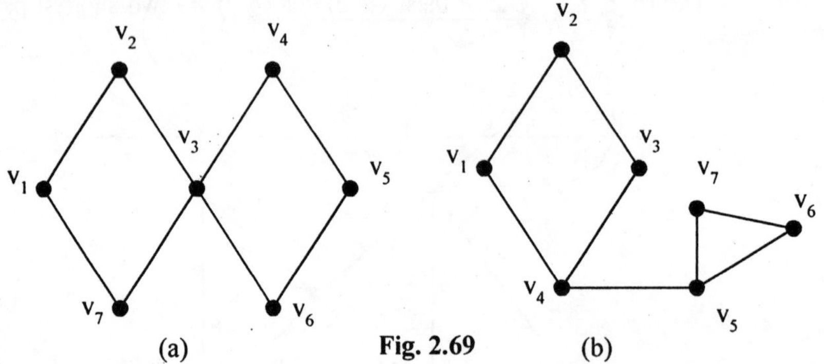

Fig. 2.69

2.34 EULER# GRAPHS

Swiss mathematician Leonard Euler (1707-1783) was first mathematician whose concept of using graph had been recorded. He in 1736 settled a famous unsolved problem of eighteenth century, which is known as Konigsberg Bridges problem. Konigsberg (now called Kaliningrad and part of the Russia) was located on the banks of Pregil River included two islands and seven bridges as shown in the Fig.2.70(a). The problem was beginning at the starting point and ending at the starting point can a person walk through town crossing all seven bridges but not crossing any bridge twice?

The problem of Konigsberg referred to the celebrated Swiss Mathematician L. Euler. To prove that such a walk is impossible, Euler replaced each of the areas A, B, C and D by vertices and

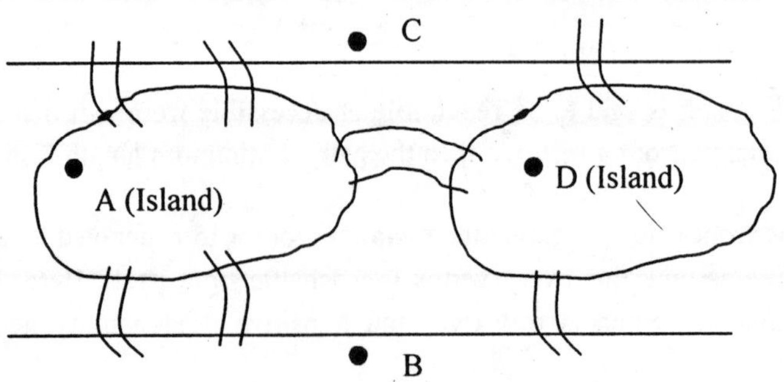

(a) The Konigsberg Bridges

Leonhard Euler (1707-1783) was an enormously creative Swiss mathematician. He studied in Basel under Johann Bernoulli and in 1730 became a professor of physics and later of mathematics in St. Petersburg, Russia. He contributed to almost all branches of mathematics and its application to physical problems even after he became totally blind in 1766.

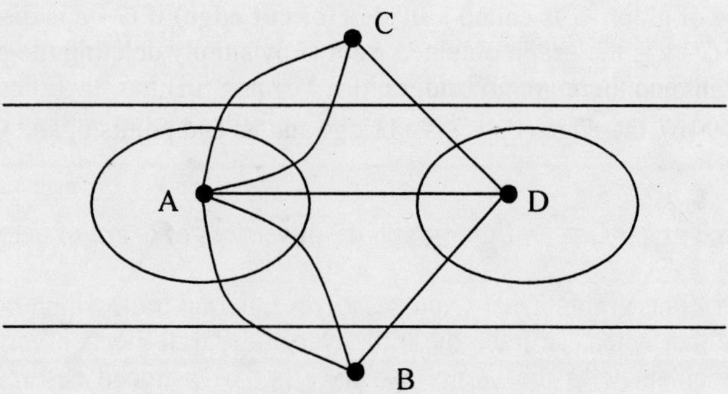

(b) Euler's graphical representation

Fig. 2.70

each bridge by an edge joining the corresponding vertices, thereby producing the graph. On the pattern of this problem Euler discussed more general problems such as in what type of graph G is it possible to find a closed walk running through every edge of G exactly once? Later on such a walk is called **Euler line** and a graph that consists of an Euler line is called an **Euler graph**. More specifically a connected graph G is **Eulerian** if there exist a closed trail containing every edge of G. Such a trail is called **Eulerian trail** or **Eulerian line**. Some authors also refer this as Euler path. Note that this definition requires each edge to be traversed only once. Similarly, we define an **Eulerian circuit** in a graph to be a circuit that traverses each edge in a graph once.

Since walk is always connected and Euler line contains all the edges of the graph. Hence, here after it is assumed that Euler graphs do not have any isolated vertices hence are always connected.

Example 2.26

Which of the graphs shown in Fig.2.71 have an Euler circuit. If it does not contain Euler circuit, which have an Euler trail?

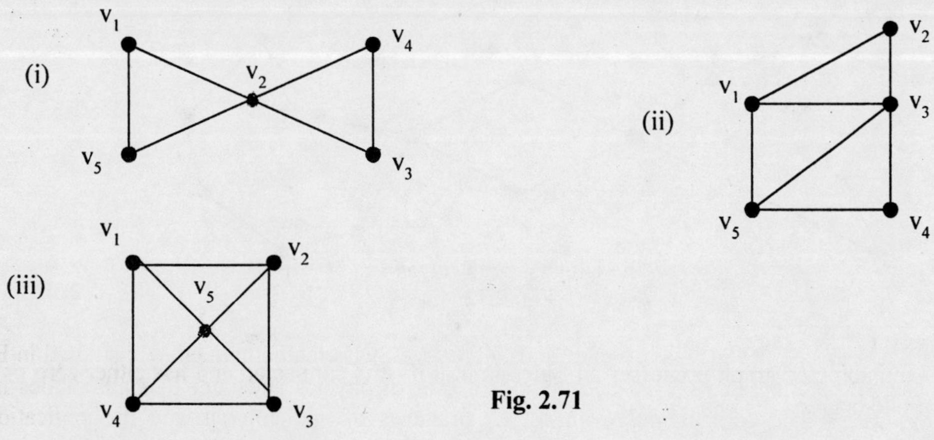

Fig. 2.71

Solution

The graph (i) has an Euler circuit $v_1, v_2, v_3, v_4, v_2, v_5, v_1$. We can observe here that all edges are traversed once. Readers can verify that neither (ii) nor (iii) has an Euler circuit. But (ii) has an Euler trail $v_5, v_4, v_3, v_2, v_1, v_3, v_5 v_1$. (iii) Doesn't have an Euler trail as well as Euler path.

THEOREM 2.10

A connected graph G is an Euler graph iff all vertices of G are of even degree.

PROOF

Let G be a Euler graph. Then G possesses an Eulerian trail, which begins and ends at vertex a, say. We observe that when we trace the trail, every time trail meets a vertex, it goes through two edges, which are incident with the vertex and have not been traced before. Since any trail (hence Eulerian trail) uses any edge once, each traversal of any vertex, say b represents a contribution of two to its degree. Hence deg (v) is even.

Conversely, suppose that graph G is connected and every vertex of G is even. Let us try to construct an Eulerian trail T_1 by beginning at an arbitrary edge e, say. We go on extending the trail T_1 by adding an edge after the other. During the tracing if the trail is not closed at any step (means if tracing started at vertex a and ended at b; where $a \neq b$) then last vertex b encountered must be of odd degree (i.e. odd number of edges incident b). Hence, T_1 can be incident on b. Continuing extensions of T_1 in this way we must have reached the initial vertex a, i.e. trail T_1 closed. If in this process T_1 includes all the edges of G then T_1 is Eulerian.

Let us assume that trail T_1 does not include all the edges of graph G. Then we remove those edges of G that have been traced by T_1 and for the remaining edges obtain a graph H (H is a subgraph of G). The degree of all vertices of H is even since T_1 contains vertices of even degree. Since G is connected let us assume that there exist an edge e' of H that has an end point a in T_1. Here, again we construct a trail T_2 in H starting with the vertex a' and using e'. As all the vertices of H are of even degree, this trail ultimately returns to the vertex where it started i.e. a' (see Fig. 2.72). We can join T_1 and T_2 to form a larger closed trail in G. We continue this process until all the edges of G are covered. In this way we obtain a Eulerian trail and hence G is Eulerian.

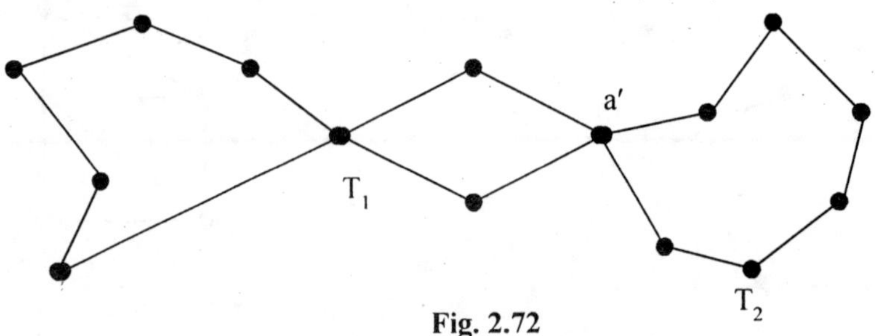

Fig. 2.72

COROLLARY 1

An undirected graph possesses an Eulerian trail iff it is connected and has either zero or two vertices of odd degree.

PROOF

Let us assume that the graph possesses an Eulerian path. We know that this graph is connected. We can observe that when we trace the Eulerian trail, every time trail meets a vertex, it goes through two edges, which are incident with the vertex and have not been traced before. Now, one can conclude very easily that except for the two vertices at the two ends of the path, the degree of any vertex in the graph must be even. If there exist two distinct vertices at two ends of the Eulerian trail, then they are the only two vertices with odd degree. If they coincide, all vertices have even degree, and the Eulerian trail become an Eulerian circuit. This proves the necessary condition.

To prove the sufficient condition let us construct an Eulerian trail by starting at one of the two vertices that are of odd degree (if there is no vertex of odd degree we start at an arbitrary vertex) and going through the edges in such a way that no edge is traced more that once. And for a vertex of even degree whenever the trail enters through an edge, it will always have the vertex through another edge that is never been covered during tracing. Hence, when the tracing is finally ended we must have reached the other vertex of odd degree. Tracing all the edges of the graph in this way, we will have an Eulerian trail. If doing so some of the edges of the graph are left, we shall remove those edges that have been traced and for the remaining edges obtain a subgraph. The degrees of the vertices of this subgraph are all even. As the trail path is connected, this subgraph must touch the path that we have traced at least at one vertex. Here, again we can construct a path that passes through the edges beginning from one of these vertices. As vertices are of even degree, this trail ultimately returns to the vertex at which it started. We can join this trail with the trail we have already constructed to obtain one which begins and ends at the two vertices of odd degree.

COROLLARY 2

A connected multigraph has an Euler path but not an Euler circuit iff it has exactly two vertices of odd degree.

COROLLARY 3

If G is a connected graph and every vertex has even degree, then G has an Euler circuit.

Example 2.27

Show that the graph shown in Fig. 2.73 is not Eulerian.

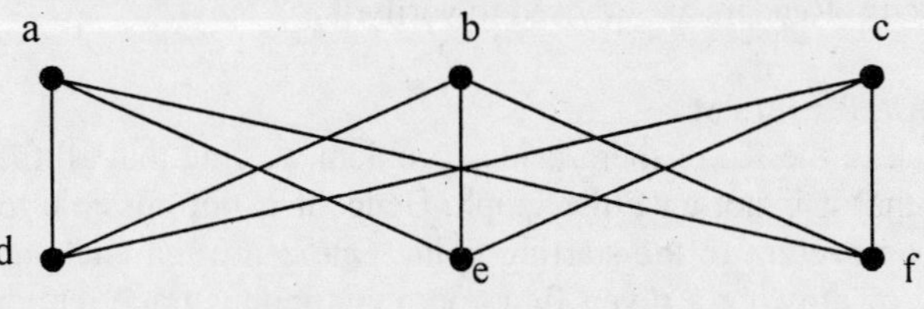

Fig. 2.73

Solution

Since all the vertices of graph shown in Fig. 2.73 are of degree 3 (i.e. odd). Therefore from corollary 2, the graph is not Eulerian.

Example 2.28

Consider a graph G shown in Fig. 2.74. Show that G has an Euler cycle. Find also an Euler circuit.

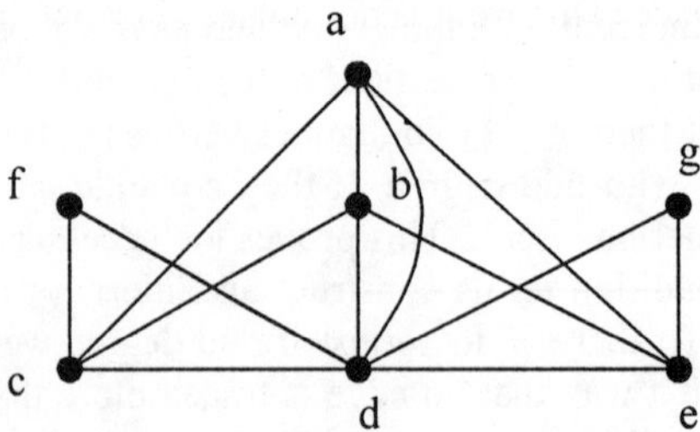

Fig. 2.74

Solution

Let us verify that the graph shown in Fig.2.74 has an Euler cycle with the help of corollary 3. We can note that the graph is connected and all the vertices are of even degree. Therefore G has an Euler cycle. By inspection, we find one Euler circuit as $(f, d, g, e, a, c, d, a, b, e, d, b, c, f)$.

Example 2.29

Whether graph shown is Fig. 2.75 contains Euler circuit?

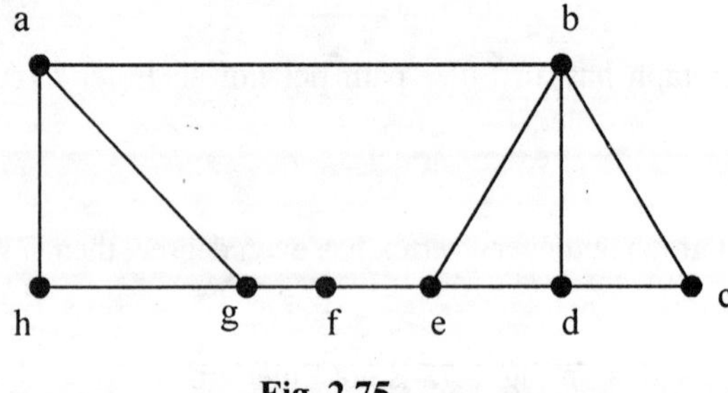

Fig. 2.75

It doesn't contain Euler circuit. Readers are advised to verify it.

2.34-1 Konigsberg Bridge Problem

Again looking back to the Konigsberg bridge problem we find that all the vertices are not of even degree. This implies that it is not an Euler graph. Hence it is not possible to walk over each of such bridges for one time and return to the starting point. One can often encounter this in puzzle of determining the possibility of drawing a given figure in a continuous trace without repetition of any part.

Example 2.30

The problem of drawing a multigraph on a paper with a pen without lifting the pencil or repeating any lines is clearly a matching with a problem of finding an Euler path in the multigraph. Here it is to be noted that multigraph can be drawn in this way if and only if it has an Euler trail.

2.34-2 UNICURSAL LINE (OPEN EULER LINE), UNICURSAL GRAPH

In Fig.2.76 the walk (v_1 v_3 v_4 v_1 v_2 v_4 v_5 v_2) includes all the edges of the graph and doesn't passes through any edge more than once. This walk is not closed since the initial vertex is v_1 and final

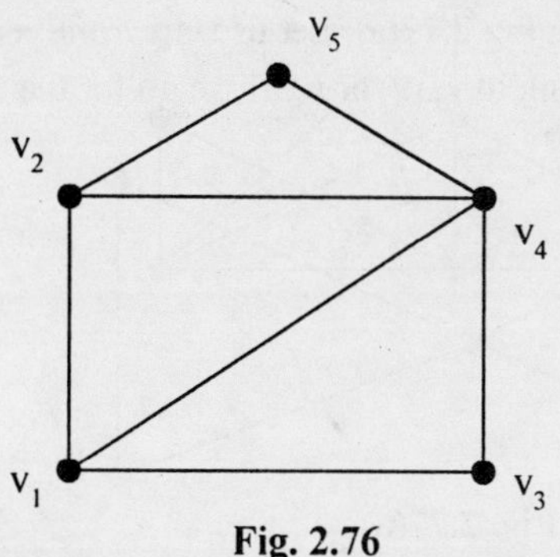

Fig. 2.76

vertex is v_2. The open walk that covers all the edges of a graph without retracing any edge is called an **open Euler line** or **unicursal line**. A connected graph that has a unicursal line is called a **unicursal graph**. One can see here that by adding an edge between initial and final vertices of an open Euler line we shall get an Euler line. We conclude that a connected graph is unicursal iff it has exactly two vertices of odd degree.

THEOREM 2.11

In a connected graph G with exactly $2n$ odd vertices, there exist n edge disjoint subgraphs such that they together contain all edges of G and that each is a unicursal graph.

PROOF

Let the odd vertices of the given graph G be named arbitrarily v_1, v_2, v_3 ... v_n, v_1', v_2', v_3', v_4', ... v_n'. Let us form a new graph G' by adding n edges to G between the vertex pairs (v_1, v_1') (v_2, v_2'), ..., (v_n, v_n'). As every vertex of G' is of even degree, G' consists of an Euler line λ. Removing the n edges just added from λ we find that λ will split into n walks, each of which is a unicursal line. It is to be noted here that first removal will leave a singular unicursal line, the second removal will split that into two unicursal lines and each successive removal will split a unicursal line into two unicursal lines until there are n of them. Hence the theorem.

THEOREM 2.12

A connected graph G is an Euler graph if and only if it can be decomposed into circuits.

PROOF

Let us decompose graph G into circuits i.e. G is a union of edge disjoint circuits. As we know degree of every vertex in a circuit is two, hence every vertex of G is of even degree. Thus G is an Euler graph. Conversely, Suppose G is an Euler graph. Consider a vertex v_1. There are at least two edges incident at v_1. Out of these two edges let us consider that one of the edges be between v_1 and v_2. As vertex v_2 is also of even degree, it must have at least another edge, say between v_2 and v_3. Continuing doing so we arrive at a vertex from where we have started, thus forming a circuit C. Now, if we remove C from graph G. Graph of remaining vertices must also be of even degree. But this graph is not necessarily connected. Proceed in this fashion until no edge is left. Hence the theorem.

2.35 ARBITRARILY TRACEABLE GRAPHS

Consider a Euler graph shown in Fig.2.77(a). Let us start from vertex v_1 and trace the path v_1 v_2 v_3. Here, at v_3 we have an option of going to v_1, v_4 or v_5. If we go for the first option then we would

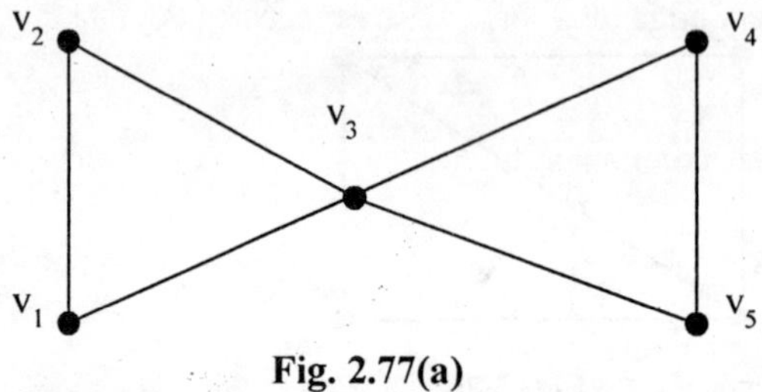

Fig. 2.77(a)

trace the circuit v_1 v_2 v_3 v_1, which is not a Euler line. From here we conclude that starting from vertex v_1 it is not always that the entire Euler line can be traced by moving along any edge that has not traced earlier. Now following question arises:

Is there any property that a vertex has in an Euler graph such that an Euler line is always obtained when one follows any walk according to a particular rule that whenever one reaches at a vertex one should select any edge, which is not previously traversed?

Graph of this type is called **arbitrarily traceable graph** from vertex v. The graph shown in Fig. 2.77(a) is an arbitrarily traceable graph from vertex v_3 but not from any other vertex. The graph shown in the Fig.2.77(b) is not arbitrarily traceable from any vertex.

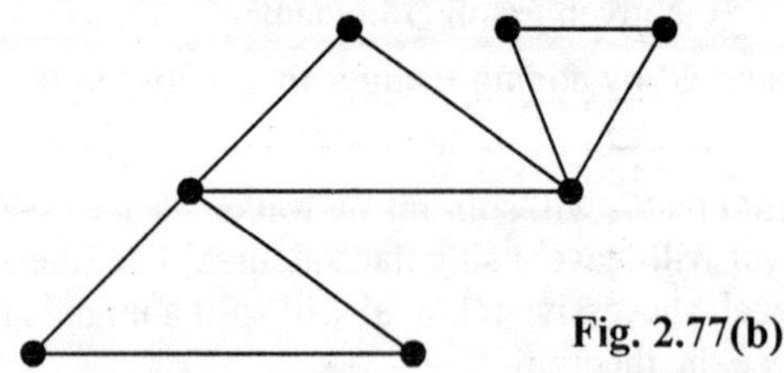

Fig. 2.77(b)

THEOREM 2.13

An Euler graph G is arbitrarily traceable from vertex v in G iff every circuit in G contains v.

PROOF

Proof of the theorem is beyond the scope of this book however readers interested in the proof are advised to refer Ore, O. "A problem regarding the tracing of graphs," Rev. Elementary Math. Vol.6, 1961, 49-53.

2.36 FLEURY'S ALGORITHM

The algorithm constructs Euler tour (moving on a graph by covering all the edges of the graph and returning to the original vertex) in an Euler graph. Its approach is to construct a trail that grows to the desired Eulerian tour. Following steps construct Fleury's algorithm.

Step 1: Start at a vertex v_0 that has odd degree unless graph G is even in which case start at any vertex and set $W_0 = v_0$

Step 2: If the trail $W_i = v_0, e_1, v_1, \ldots, e_i, v_i$ has been chosen , choose an edge e_{i+1} different from e_1, $e_2, \ldots, e_i$ such that:

(i) e_{i+1} is incident with v_i and

(ii) If there is more than one such edge, select one that is not a bridge.

Step 3: Stop, if we contain every edge of G, otherwise repeat step 2.

THEOREM 2.14

If G has one nontrivial component and at most two odd vertices, then Fleury's algorithm constructs an Eulerian trail.

Example 2.31

Use Fleury's algorithm to construct an Euler circuit for the graph in Fig. 2.78.

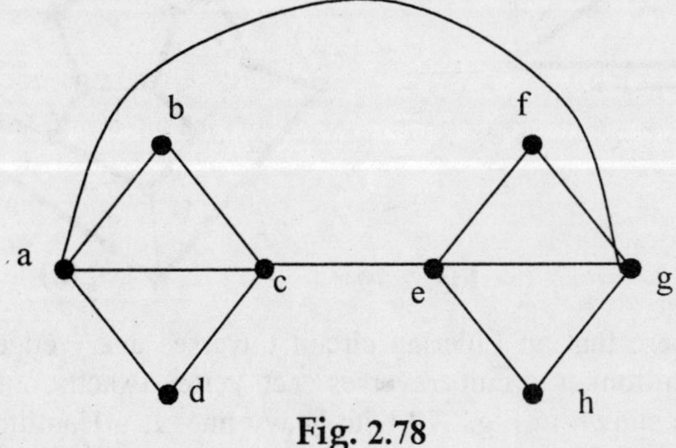

Fig. 2.78

Solution

Let us start with vertex a. Apply step 2 repeatedly. The results obtained in step 2 are generalized in the following table:

Current Path	Next edge	Reasoning
$p : a$	$\{a, b\}$	No edge from a is a bridge. Choose any one.
$p : a, b$	$\{b, c\}$	Only one edge from b remains.
$p : a, b, c$	$\{c, a\}$	No edge from c is a bridge. Choose any one.
$p : a, b, c, a,$	$\{a, d\}$	No edge from a is a bridge. Choose any one.
$p : a, b, c, a, d$	$\{d, c\}$	Only one edge from vertex d remains.
$p : a, b, c, a, d, c$	$\{c, e\}$	Only one edge from vertex c remains.
$p : a, b, c, a, d, c, e$	$\{e, g\}$	No edge from e is a bridge. Choose any one.
$p : a, b, c, a, d, c, e, g$	$\{g, f\}$	$\{a, g\}$ is a bridge. Choose $\{g, f\}$ or $\{g, h\}$.
$p : a, b, c, a, d, c, e, g, f$	$\{f, e\}$	Only one edge from f remains.
$p : a, b, c, a, d, c, e, g, f, e$	$\{e, h\}$	Only one edge from e remains.
$p : a, b, c, a, d, c, e, g, f, e, h$	$\{h, g\}$	Only one edge from h remains.
$p : a, b, c, a, d, c, e, g, f, e, h, g$	$\{g, a\}$	Only one edge from g remains.
$p : a, b, c, a, d, c, e, g, f, e, h, g, a$		

Here readers may note that at several vertices we could have other choices then what we have done here. In other words we can say that, if a graph has an Euler circuit, it is quite possible to have several different Euler circuits.

2.37 HAMILTONIAN CIRCUIT AND PATH

A Hamiltonian* circuit is defined as a closed walk that traverses every vertex of a connected graph G exactly once except for the end vertices, which is also the terminal vertex. A circuit in a connected graph G is said to be **Hamiltonian** if it contains Hamiltonian circuit. It means a Hamiltonian circuit of a graph having n vertices consists of exactly n edges.

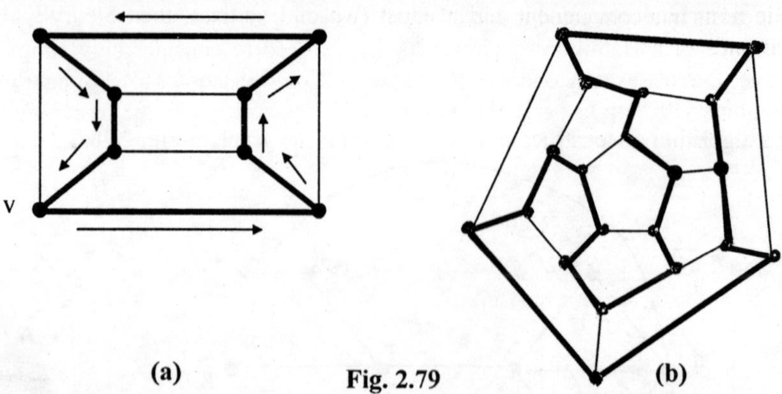

(a) Fig. 2.79 (b)

It is to be noted here that an Eulerian circuit traverses every edge exactly once, but may repeat vertices while a Hamiltonian circuit traverses each vertex exactly once and hence miss some edges altogether. The graph shown in Fig.2.79 with heavy lines is a Hamiltonian circuit. Here when one traverses along the heavy lines he passes through each vertex exactly once.

Hamiltonian circuit is always a connected graph, but every connected graph does not have a Hamiltonian circuit. The graph shown in the Fig.2.77(a) and (b) have a Hamiltonian circuit whereas the graph shown in Fig. 2.80 is non-Hamiltonian.

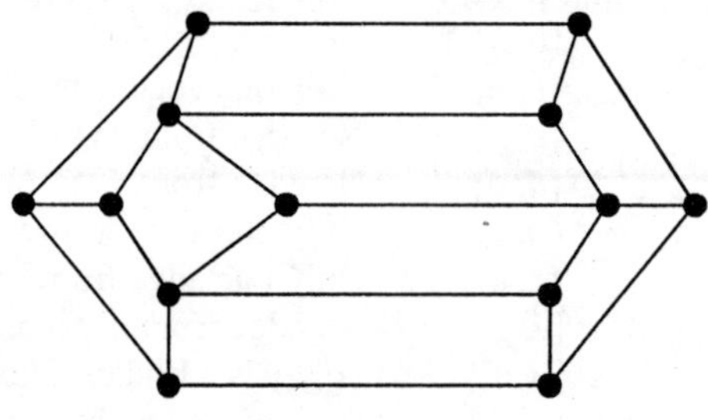

Fig. 2.80

* The name "Hamiltonian" is derived from the Irish mathematician Sir William Rowan Hamilton, who invented a game in 1857 consisting of a solid regular dodecahedron.

Now from the definition of Hamiltonian circuit we can have following observations:

1. A complete graph with more than two vertices is Hamiltonian.
2. Every cycle graph is Hamiltonian.
3. Every platonic graph is Hamiltonian.

2.37-1 HAMILTONIAN PATH

The path obtained by removing any one edge from a Hamiltonian circuit is called a **Hamiltonian path**. Hamiltonian path in a graph G visits every vertex of graph G. It is to be noted here that a Hamiltonian path is a subgraph of a Hamiltonian circuit, it implies that graph having a Hamiltonian circuit also has a Hamiltonian path. But the converse is not always true. The length of a Hamiltonian path if it exists in a connected graph of n vertices is $n - 1$.

For the existence of a Hamiltonian circuit (path) we need to consider only simple graphs because it traverses every vertex exactly once, it should not include self-loop or a set of parallel edges. If a general graph contains self-loop or parallel edges it can be made simple graph by removing cycle or parallel edges.

THEOREM 2.15

A graph G with n vertices always has a Hamiltonian path if the sum of the degrees of every pair of vertex (v_i, v_j) in G satisfies the condition

$$d(v_i) + d(v_j) \geq n - 1.$$

PROOF

To prove this theorem first of all we show that G is a connected graph. Let us assume that G has two or more than two disconnected components. Let the graph G consists of two components, one having n_1 vertices while other n_2. Suppose v_1 is a vertex in one component and v_2 is vertex in another component. Since degree of v_1 and v_2 be at most $n_1 - 1$ and $n_2 - 1$ respectively, hence the sum of their degree is at most $n_1 + n_2 - 2$ which is less than $n - 1$. But this is a contradiction. Hence the graph is disconnected.

Let us show how a Hamiltonian path can be constructed starting with a path containing single edge. Consider a path of $m - 1$ edges such that $m < n$ in graph G, which meets the sequence of vertices say $v_1, v_2, v_3, \ldots, v_m$. Obtain a path of m edges if either v_1 or v_m is adjacent to a vertex that is not in the path by extending the path to include this vertex. If this is not the case than both v_1 and v_m are adjacent only to the vertices that are in the path. Now we will show that in that case there is a circuit containing exactly the vertices $v_1, v_2, \ldots, v_m$. In case v_1 is adjacent to v_m then the circuit $(v_1, v_2, \ldots, v_m, v_1)$ will suffice, so let us assume here that v_1 is adjacent only to $v_{i_1}, v_{i_2}, \ldots v_{i_m}$ where $2 \leq i_j \leq m - 1$. And if v_m is adjacent to one of $v_{i_1-1}, v_{i_2-1}, \ldots, v_{i_m-1}$, say v_{j-1}, then from Fig.2.81(a), one can see that the circuit $(v_1, v_2, v_3, \ldots, v_{j-1}\, v_m, v_{p-1}, \ldots v_j, v_1)$ contains exactly the vertices $v_1, v_2, \ldots, v_m$. Suppose v_m is not adjacent to any one of $v_{i_2-1}, v_{i_2-1}, \ldots, v_{i_k-1}$ then v_m is adjacent to at most $m - k - 1$ vertices. Hence, the sum of the degree of v_1 and v_m is at most $n - 2$ which is a contradiction.

Now we have got a circuit containing all the vertices $v_1, v_2, \ldots v_m$. Let us take one vertex, say v_2 which is not in the circuit. Since G is connected there is a vertex v that is not in the circuit with an edge between v_j and v_k for some v_k in $\{v_1, v_2, \ldots v_m\}$ as shown in Fig. 2.81(b). Now we have got the path $(v_z, v_k, v_{k-1}, \ldots, v_m, v_{m-1}, \ldots, v_j, v_1, v_2, v_{k-1})$ which contains m edges as shown in Fig. 2.81(c), This construction is repeated until we get a path of n -1 edges.

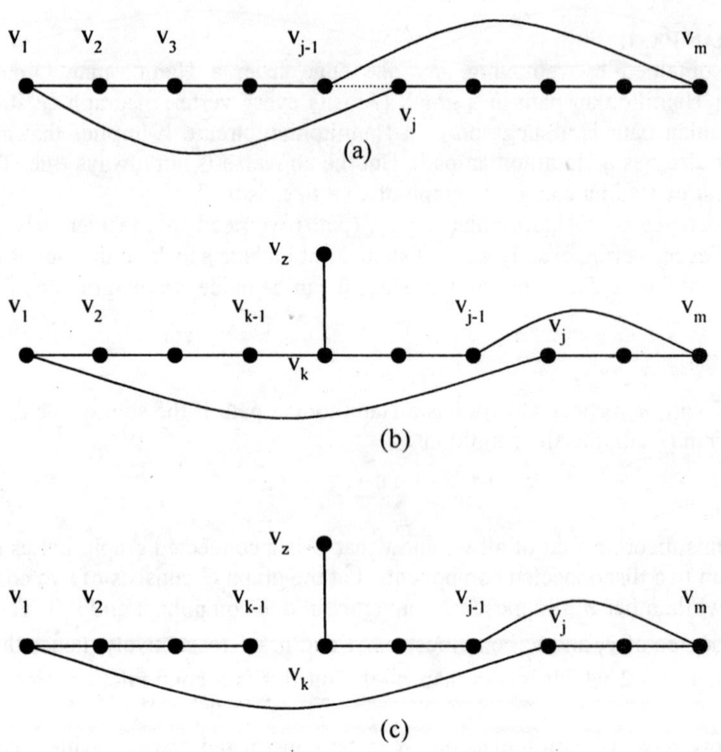

Fig. 2.81

THEOREM 2.16 (Ore, 1960).

If G is a simple graph with n (≥ 3) vertices, and if

$$\deg(v) + \deg(w) \geq n$$

for each pair of non-adjacent vertices v and w, then G is Hamiltonian.

PROOF

Let us assume that the theorem is not true and derive a contradiction. Let G be a non-Hamiltonian graph with n vertices, satisfying the given condition on the vertex degrees. Let us suppose that graph G is 'only just' non-Hamiltonian in the sense that addition of any further edge gives a Hamiltonian graph. It is to be noted here that addition of an extra edge does not violate the condition on the vertex degrees. From this we conclude that graph G contains a path $v_1, v_2, \ldots, v_n$ passing through every vertex. But we have assumed that graph G is non-Hamiltonian, hence vertices v_1 and v_n are not adjacent, and so $\deg(v_1) + \deg(v_n) \geq n$. This implies that there exist some vertex v_i adjacent to

v_1 with the property that v_{i-1} is adjacent to v_n (See Fig. 2.82). But this gives us the required contradiction since $v_1, v_2, \ldots, v_{i-1}, v_n, v_{n-1}, v_{i+1}, v_i, v_1$ is then a Hamiltonian cycle.

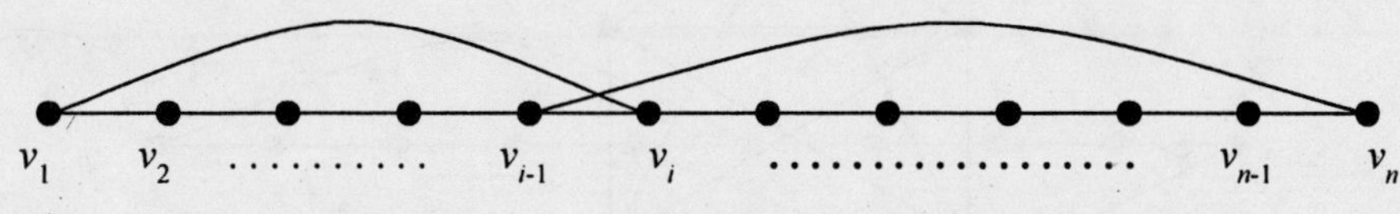

Fig. 2.82

Corollary (Dirac, 1962)

Let G be a simple graph with n (≥ 3) vertices then graph G is Hamiltonian if deg (v) > $n/2$ for each vertex v.

PROOF

This result follows immediately from theorem 2.16. Since deg (v) + deg (w) $\geq n$ for each pair of vertices v and w irrespective of adjacency.

Example 2.32

Verify hypothesis of theorem 2.16 for the following graph.

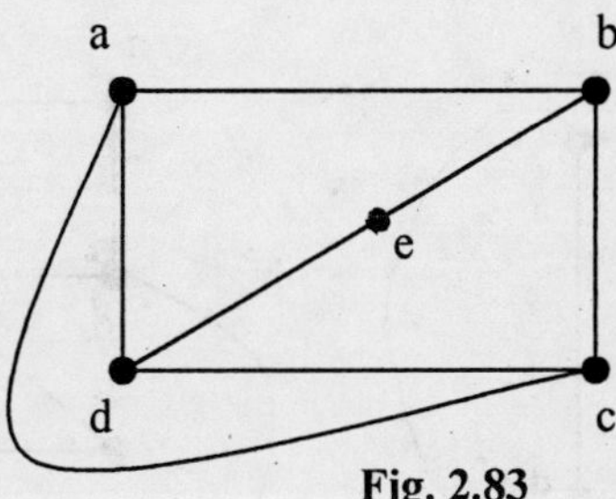

Fig. 2.83

Solution

Here number of vertices is 5, i.e. $n = 5$. We can observe form the figure there exist three pair of non-adjacent vertices. Now for pair of vertices (b, d)

deg (b) + deg (d) = 3 + 3 = 6 > 5

For pair (a, e), deg (a) + deg (e) = 3 + 2 = 5 = 5.

For pair (e, c), deg (e) + deg (c) = 2 + 3 = 5 = 5.

In all the three cases sum of the degrees of nonadjacent vertices is either equal to or greater than number of vertices (n = 5).

Example 2.33

Show that K_n has a Hamiltonian circuit whenever $n \geq 3$.

Solution

In graph K_n, we can form a Hamiltonian circuit beginning at any vertex. Such a circuit can be formed by visiting vertices in any order, as long as initial and final vertices are same and visits each vertex exactly once. All this is possible because there exist edges in K_n between any two vertices.

Particular Case k_5:

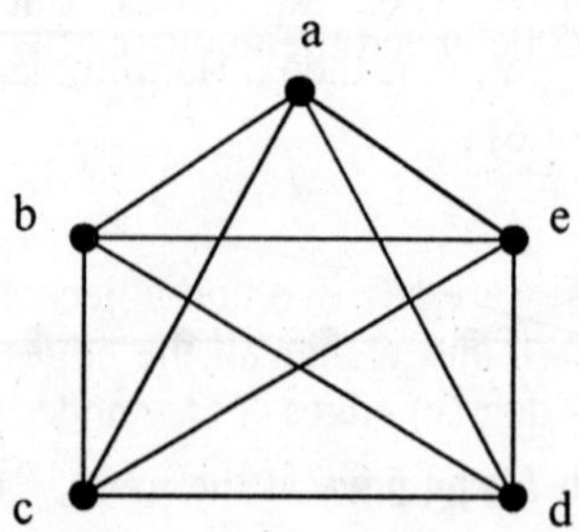

Solution

There exist a Hamiltonian circuit (a, b, c, d, e) is k_5 and k_5 is Hamiltonian. We can observe that each vertex of k_5 is of degree 4. And number of vertices of k_5 is 5, i.e. $n = 5$. Since $n = 5 \geq 3$ and $\deg(v) = 4 > \frac{n}{2} = \frac{5}{2}$ for each vertex. Hence according to hypothesis of corollary of theorem 2.16 it should be Hamiltonian.

Example 2.34

Which of the following graphs shown in Fig.2.84 have a Hamiltonian circuit? If not, why?

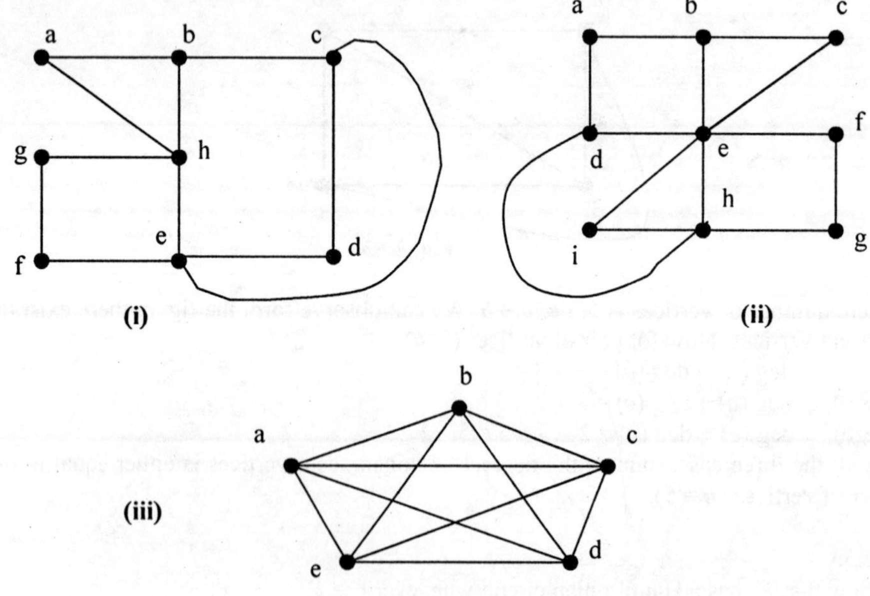

Fig. 2.84

Solutions

Graph (i) and (iii) have Hamiltonian circuits but (ii) has no Hamiltonian circuit (Readers are advised to verify it).

THEOREM 2.17

Let the number of edges of graph G be m and the number of vertices be n respectively. Then G has Hamiltonian circuit if $m \geq \frac{1}{2}(n^2 - 3n + 6)$.

PROOF

Let G be a graph and suppose u and v are any two non-adjacent vertices of G. Let H be a sub graph of G obtained by eliminating vertices u and v, and all the edges incident on u and v from G. Then H has n - 2 vertices and m - deg (u) - deg (v) edges (If u and v were adjacent then one fewer edge would have been removed). The graph H can have at the most ${}^{n-2}C_2$ number of edges. This is possible when there is an edge connecting every pair of vertices. Thus the maximum possible number of edges of H is

$$n-2_{C_2} = \frac{(n-2)(n-3)}{2} = \frac{1}{2}(n^2 - 5n + 6)$$

$$\Rightarrow \quad m - \deg(u) - \deg(v) \leq \frac{1}{2}(n^2 - 5n + 6)$$

$$\Rightarrow \quad \deg(u) + \deg(v) \geq m - \frac{1}{2}(n^2 - 5n + 6)$$

By the hypotheses of the theorem

$$\deg(u) + \deg(v) \geq \frac{1}{2}(n^2 - 3n + 6) - \frac{1}{2}(n^2 - 5n + 6)$$

$$\deg(u) + \deg(v) \geq n$$

Example 2.35

Give an example of the graph, which is Hamiltonian but not Eulerian, and vice-versa.

Solution

Graph shown in the Fig.2.85 is Hamiltonian but not Eulerian. The Hamiltonian circuit is

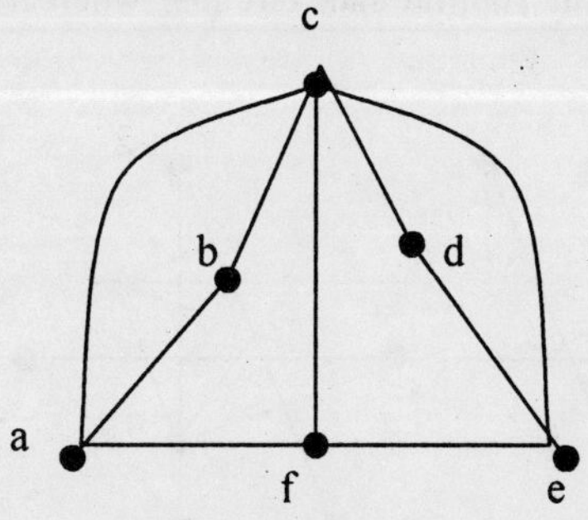

Fig. 2.85

a, b, c, d, e, f, a. But in this graph degree of each vertex is not even hence the graph is non-Eulerian.

The graph shown in Fig. 2.86 is Eulerian but not - Hamiltonian.

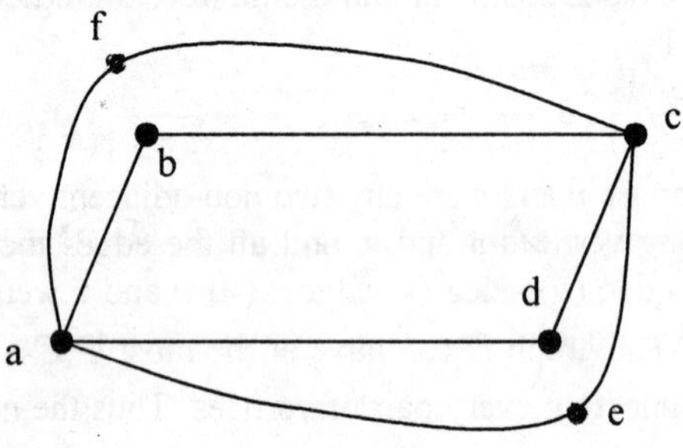

Fig. 2.86

In this graph we can observe that the degree of each vertex is even hence it is Eulerian. The graph is not - Hamiltonian because we can't find any circuit in the graph, which visits all the vertices once except initial and final vertex.

2.37-2 Number of Hamiltonian Circuits in a Graph

A given graph may contain more than one Hamiltonian circuit. As far as all the edge disjoint Hamiltonian circuits in a graph is concerned, the exact number (of edge disjoint-Hamiltonian) is still an unsolved problem but the following theorem gives the number of edge-disjoint Hamiltonian circuit in a complete graph with odd number of vertices.

Theorem 2.18

In a complete graph with n vertices there are $(n-1)/2$ edge disjoint Hamiltonian circuits, if n is an odd number ≥ 3.

Proof

Consider a complete graph G with n vertices. It has $n(n-1)/2$ edges and a Hamiltonian circuit in G consists of n edges.

$\therefore$ the number of edge-disjoint Hamiltonian circuit in G will not be greater than $(n-1)/2$.

There are $(n-1)/2$ edge disjoint Hamiltonian circuits, when n is odd. It can be shown as follows:

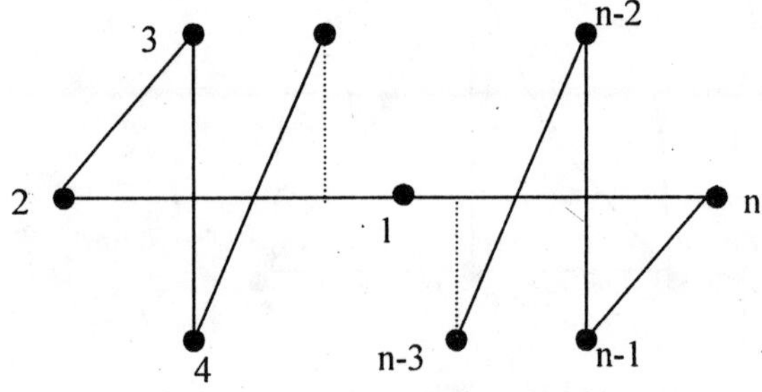

Fig. 2.87

The subgraph of a complete graph of n vertices as shown in the Fig.2.87 is a Hamiltonian circuit. If we keep the vertices fixed on a circle and rotate the polygonal pattern clockwise by

$$\frac{360}{(n-1)},\ 2.\frac{360}{(n-1)},\ 3.\frac{360}{(n-1)},\ \ldots,\ \frac{(n-3)}{2}.\frac{360}{(n-1)} \text{ degrees.}$$

One can see here that in every rotation we are getting a new Hamiltonian circuit without any edge common to previous ones.

⇒ We have $(n - 3)/2$ all edge disjoint new Hamiltonian circuits and also edge disjoint among themselves. Hence the theorem.

2.37-3 Some Basic Rules For Constructing Hamiltonian Paths and Cycles

1. If given graph G has n vertices, then Hamiltonian path and Hamiltonian circuit must contain $n - 1$ and n edges respectively.
2. If a vertex v in graph G has degree m, then a Hamiltonian path must contain at least one edge incident on v and at most two edges incident on v. But Hamiltonian cycle contains exactly two edges incident on v.

NOTE

In particular, both edges incident on a vertex of degree two will be contained in every Hamiltonian cycle. We can say that there can not be three or more edges incident with one vertex in a Hamiltonian cycle.

3. While building a Hamiltonian path or cycle no cycle that does not contain all the vertices of G can be formed.
4. During the construction of a Hamiltonian circuit if we have passed through a vertex v, then all other unused edges incident on v can be deleted because only two edges incident on v can be included in a Hamiltonian cycle.

2.37-4 Nonexistence of a Hamiltonian Path

In order to prove the nonexistence of a Hamiltonian path or cycle first of all we try to form parts of a Hamiltonian path and show that construction must always fail. Showing that we cannot visit all the vertices without visiting some vertices at least twice can show failure.

Example 2.36

Consider the problem of assigning duties to seven drivers in seven days so that two duties given by the same driver are not scheduled on consecutive days. If no driver serves more than four days show that it is always possible to schedule the duties.

Solution

Let G be a graph with seven vertices corresponding to the seven duties. The edge between any two vertices corresponds to two duties performed by different drivers. One can see here that degree of each vertex is at least 3, the sum of the degrees of any two vertices is at least 6. Hence, G always contains a Hamiltonian path, which corresponds to a suitable schedule for the seven duties.

Example 2.37

In the Fig.2.88 which is formed in the order of appearance of English alphabet. Does it form Hamiltonian path and Hamiltonian cycle.

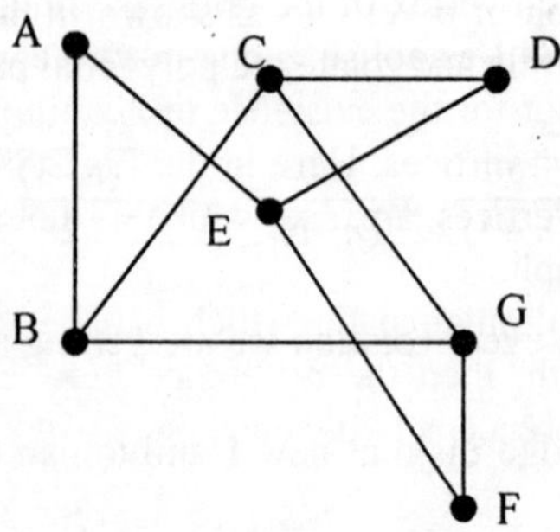

Fig. 2.88

Solution

Going through the graph one can conclude that the path through the vertices of graph shown in the Fig.2.88 in the order of appearance in the English alphabet forms a Hamiltonian path. But graph has no Hamiltonian cycle.

Example 2.38

Show that there is no Hamiltonian cycle and Hamiltonian path for the graph shown in Fig.2.89.

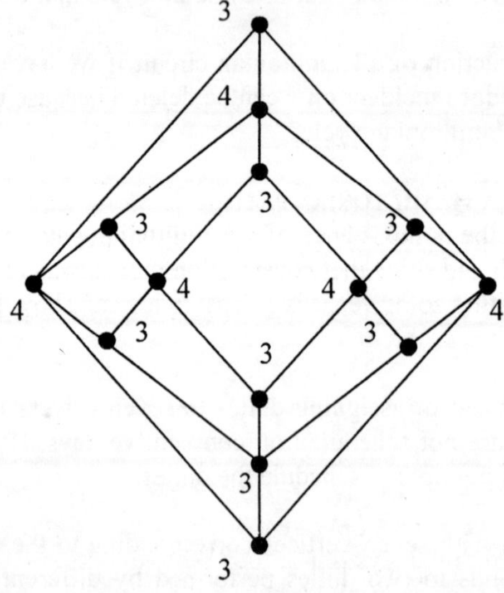

Fig. 2.89

Solution

From the Fig.2.89 it is clear that each vertex of the graph G has degree three or four. We can also observe that every edge connects a vertex of degree three with one of degree four. If we had a

Hamiltonian cycle it would visit each vertex of the graph passing alternately through a 3-vertex and a 4-vertex. Hence in this cycle we will establish a one-to-one correspondence between the set of 3-vertices and the set of 4-vertices. But for the existence of Hamiltonian cycle in this graph there would be same number of 3-vertices and 4-vertices. Here in the Fig.2.89 we can see that total number of 3-vertices are eight and that of 4-vertices are six, which is not equal. Therefore, there can be no Hamiltonian cycle for the given graph.

Further, there is no Hamiltonian path, which can be verified with the help of the above reasoning. If there exit such a path, then the difference between the number of 3-vertices and 4-vertices must be one. But here difference is two, hence there is no Hamiltonian path.

Example 2.39

Is it always possible to rank players in a round robin badminton tournament?

Solution

Let us arrange the rank of the players in round-robin badminton tournament in following way such that player *A* will be ranked higher than player *B* if

1. *A* beats *B* or
2. *A* beats a player who beats *B*
3. *A* beats a player who beats another player who beats *B* and so on

Now as the outcome of the matches can be represented with the help of a directed graph which implies that existence of Hamiltonian path. Hence, it is always possible to rank the players.

Example 2.40

Show that the graph shown in Fig. 2.90 does not contain a Hamiltonian cycle.

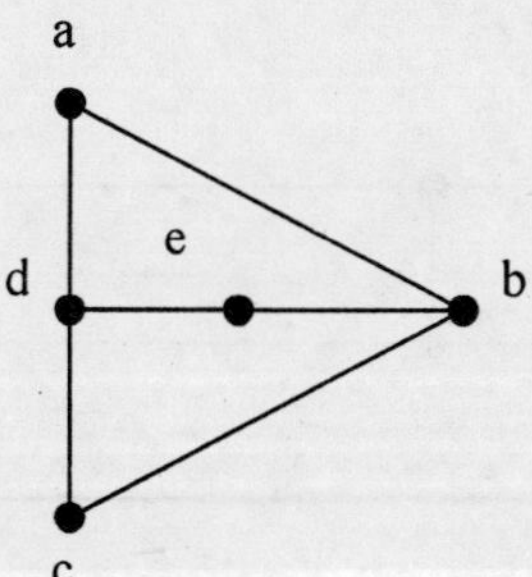

Fig. 2.90

Solution

We can observe that there are five vertices in the graph shown in Fig. 2.90. As it contains five vertices, a Hamiltonian cycle must have five edges. Let us try to eliminate some of the edges of the graph, leaving just a Hamiltonian cycle. For this, we would have to eliminate one edge incident at *b* and one edge incident at *d* because each vertex in a Hamiltonian cycle must have a degree 2. But this removal leaves only four edges, which is not sufficient for a Hamiltonian cycle of length 5. Therefore, this graph does not have a Hamiltonian cycle.

Example 2.41

Prove that there is no Hamiltonian cycle in each of the following graphs:

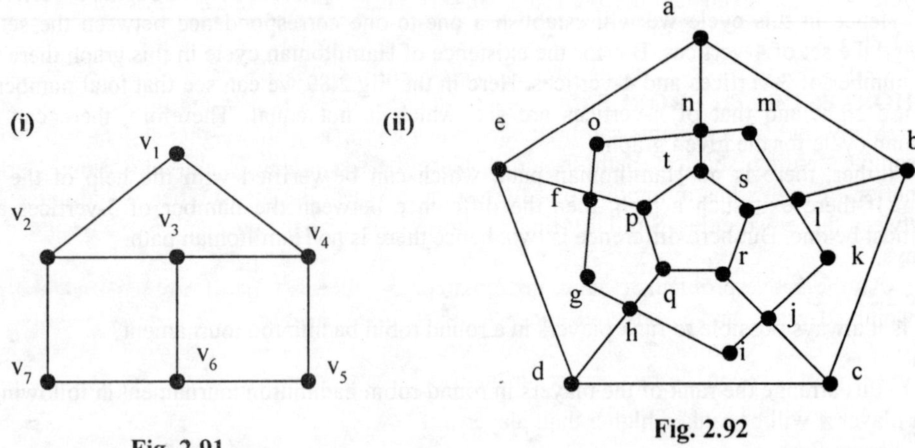

Fig. 2.91

Fig. 2.92

Solution

(i) If we consider the vertices of degree 2, we see that the edges (v_1, v_2), (v_1, v_4), (v_2, v_7), (v_7, v_6), (v_5, v_6) and (v_4, v_5) must be included in a Hamiltonian cycle. But this forms a proper subcycle of G.

(ii) Similar to the above proof (i), consider the vertices o, g, i, k and m.

Example 2.42

Show that any Hamiltonian cycle which contains $(v_3\ v_4)$ in the graph in Fig. 2.93 must also contain $(v_8\ v_3)$.

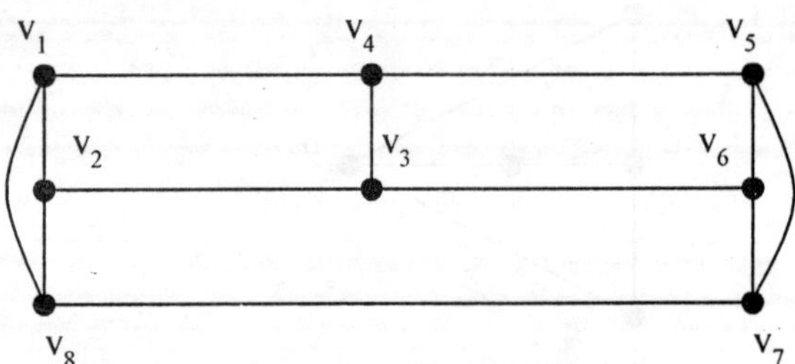

Fig. 2.93

Solution

To verify this let us first include $(v_3\ v_4)$ in the Hamiltonian cycle (H-cycle). Then, at v_4 we have to remove $(v_4\ v_1)$ or $(v_4\ v_5)$; at v_3 remove $(v_3\ v_2)$ or $(v_3\ v_6)$. Now the following cases are possible:

(i) If we remove $(v_4\ v_1)$ and $(v_3\ v_2)$ then, the H- cycle must include $(v_1\ v_2)$ and $(v_2\ v_8)$. But here observe that unless $(v_8\ v_7)$ is included, the path $(v_1\ v_2\ v_8)$ gets disconnected with the rest of the graph. But if we include $(v_8\ v_7)$, we get H-path $(v_1\ v_2\ v_8\ v_7\ v_6\ v_3\ v_4\ v_5)$. But this is not a Hamiltonian cycle.

(ii) If we remove $(v_4\ v_1)$ and $(v_3\ v_6)$ and include cycles $(v_4\ v_3)$ and $(v_2\ v_3)$, then, at v_2 we can remove $(v_2\ v_8)$, which will give the path $(v_4\ v_3\ v_2\ v_1\ v_8)$ and $(v_4\ v_5\ v_6\ v_7)$. But, unless $(v_7\ v_8)$ is included in H-cycle, these two paths get disconnected. But including $(v_7\ v_8)$, we get the H-cycle $(v_3\ v_4\ v_5\ v_6\ v_7\ v_8\ v_1\ v_2\ v_3)$.

2.38 SHORTEST PATH PROBLEM

Many real life problems can be modeled using graphs with weights assigned to their edges. For example finding shortest path in air distance between New Delhi and New York. What combination of flights has minimum total flight time between these two cities. What is the cheapest fare between these two cities. Similarly finding shortest distance between any two cities connected by highways.

All the above problems (Many move matching with this type of problems) can be modeled using weighted graph and is covered under **'shortest path problem'**. It is to be recalled that the length of a path in a weighted graph is the sum of weights of the edges of this path and the minimum length of the path between the two vertices is the shortest path between them.

There are many algorithms to find the shortest distance between any two vertices in a weighted graph. Dijkstra's algorithm is one of them named after E.W. Dijkstra born in 1930 in the Netherlands who has given this algorithm.

Let us first have an example, which will prove to be helpful in understanding Dijkstra's algorithm.

Find the shortest distance of the graph shown in Fig.2.94 from the vertex v_1 to successive vertices until v_6 is reached.

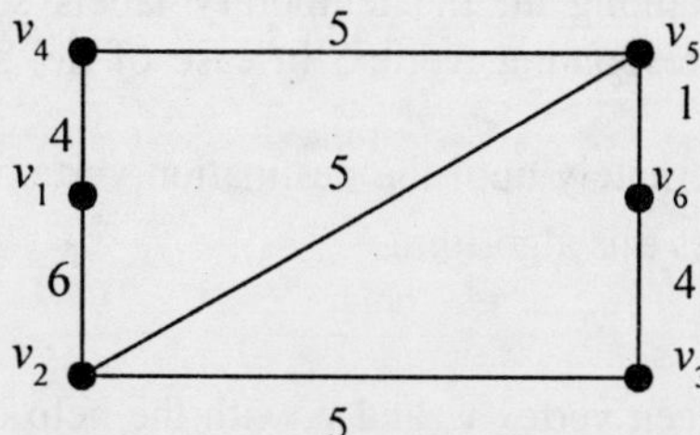

Fig 2.94

From vertex v_1 we have two paths namely (v_1, v_2) and (v_1, v_4) until final vertex v_6 is reached. Here the length of edge (v_1, v_2) and (v_1, v_4) is 6 and 4 respectively. Hence vertex v_4 is closest. Again we search for the paths that go through only v_1 and v_4 until the final vertex v_6 is reached. The shortest path to v_2 is still (v_1, v_2) having length 6 whereas the shortest path to v_4 is (v_1, v_4, v_5) having length 9. Therefore next closest vertex to v_1 is v_2.

Now to find the third closet vertex to v_1 we will examine only paths that go through v_1, v_2 and v_4 until the final vertex v_6 is reached. Here we can observe two paths (v_1, v_4, v_5, v_6) of path length 10 to v_6 and path (v_1, v_2, v_3) to v_3 of path length 11. Hence v_6 is the next closet vertex to v_1. Consequently (v_1, v_4, v_5, v_6) is the shortest path form v_1 to v_6 with length 11.

Now we enquire how to obtain a shortest path between v_1 to v_6 of an undirected connected simple weighted graph. Dijkstra's algorithm is based on finding the shortest length from vertex v_1 to first vertex and then length of the shortest path from v_1 to a second vertex and so on. This process in continued till we reach the final vertex v_6 with shortest path from v_1.

Dijkstra's algorithm is based on iterations. Adding one-vertex in each iteration forms distinguished set of vertices. Simultaneously, a labeling is done in each iteration. In the labeling procedure, assign a label to vertex v_i, with the length of the shortest path form v_1 to v_i that consist of only vertices already in the distinguished set. The new vertex added to the distinguish set should be that one which has minimal label among those vertices not already in the set.

Let v_1 and v_n be initial and final vertices between which the shortest path is to be determined.

2.38-1 DIJKSTRA'S ALGORITHM

Following is the **Dijkstra's algorithm** for finding shortest path from one specified vertex v_l to another specified vertex v_n.

Step 1 : Assign a permanent label zero to the starting vertex v_l and temporary labels ∞ to the remaining vertices.

Step 2 : Each vertex v_j that is not got permanently labeled gets a new temporary label whose value is given by

$$\text{Min}\left\{\text{old label of } v_j,\ \text{old labelof } v_i + d(v_j, v_j)\right\}$$

where v_i is the latest vertex permanently labeled in the previous iteration and $d(v_i, v_j)$ is the direct distance between vertex v_i and vertex v_j. If direct distance is not given between vertex v_i and v_j then assign $d(v_i, v_j) = \infty$.

Step 3 : Select the smallest value among all the temporary labels so found and this becomes a permanent label of the corresponding vertex. In case of tie, select any one for permanent label.

Step 4 : Repeat step (2) and (3) alternately until the destination vertex v_n gets permanent label. The following example illustrates this algorithm.

Example 2.43

Find the shortest path between vertex v_l and v_7 with the help of Dijkstra's algorithm for the Fig.2.95.

Solution

By applying Step 1 of Dijkstra's algorithm assign permanent label zero to the starting vertex v_l and ∞ to remaining all the vertices.

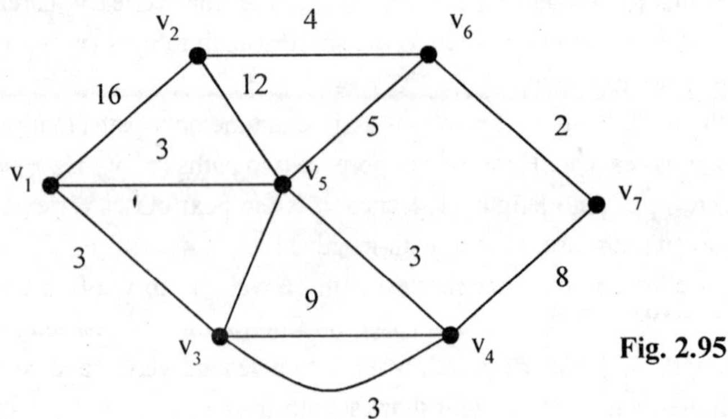

Fig. 2.95

Vertex	v_1	v_2	v_3	v_4	v_5	v_6
Distance	0	∞	∞	∞	∞	∞

For vertex v_2

$$\text{Min}[\text{old } of\ v_2,\ \text{ old of } v_1 + d(v_1, v_2)]$$
$$\text{Min}[\infty,\ 0+16] = \text{Min}[\infty, 16] = 16$$

Applying Dijkstra's algorithm finally we have following table

Vertex	v_1	v_2	v_3	v_4	v_5	v_6	v_7
Distance	0	∞	∞	∞	∞	∞	∞
	0	16	∞	∞	3	∞	∞
	0	15	5	6	3	8	∞
	0	15	5	6	3	8	∞
	0	15	5	6	3	8	14
	0	15	5	6	3	8	10

Thus shortest path between vertex v_1 and v_7 is 10.

2.38-2 Travelling Salesperson Problem

Travelling salesman problem is one of the problems considered as puzzles by the mathematicians. Suppose a salesman wants to visit a certain number of cities allotted to him. He knows the distance (or cost or time) of journey between every pair of cities. His problem is to select such a route that starts from his home city, passes through each city once and only once and return to his home city in the shortest possible distance (or at the least cost or in least time).

This problem can closely be related to Hamiltonian circuits. If we represent the cities by vertices and road connecting two cities by edges we get a graph. Here, we can see with every edge e_i a number (distance or cost or time) is associated say w_i. This type of graph is called weighted graph with $w(e_i)$ being the weight of edge e_i.

Given a weighted graph, the problem of finding a circuit (e_1, e_2,.... e_n) that visits every vertex exactly once and minimizes the sum of weights is called **travelling salesperson problem**, which is equivalent to determining Hamiltonian circuit for the graph. If the students are looking for a circuit which must visit each vertex at least once and if the task asked them to visit each vertex once and only once, then the problem would require them to find a Hamiltonian circuit.

In the graph if we have n vertices (cities) then there are (n-1)! possible edges (routes). The total number of Hamiltonian circuits in a complete graph of n-vertices is (n - 1)!/2 (As out of total (n-1)! number of choices, each Hamiltonian circuit has been counted twice, hence the number is divided by 2).

2.38-3 Nearest Neighbour Method

Here we present a procedure, which is called the **nearest neighbour Method**, that gives results close to the shortest routes.

Step 1. Choose any arbitrary vertex v and find the vertex nearest to this starting vertex to form an initial path of one edge.

Step 2. Add new edges to the path already obtained by selecting new vertices in the following manner:

Let p denote the latest vertex that is most recently added to the path, then eliminate from consideration all edges incident with p that lead to the vertices that have already been chosen (including v). Now among the rest of the vertices, which have not been selected so far, pick the closest one to p and add to the path, the edge-connecting p and the new vertex. Go on doing so until all the vertices of graph G is included in the path.

Step 3. Join initial and the final vertex by an edge and form a circuit.

NOTE

This algorithm produces a directed cycle in the complete graph but not necessarily the shortest one and different solution may be obtained with respect to different initial vertex.

Example 2.44

In the Fig.2.96 (a) starting with vertex v_1 construct a Hamiltonian circuit according to the nearest neighbour method.

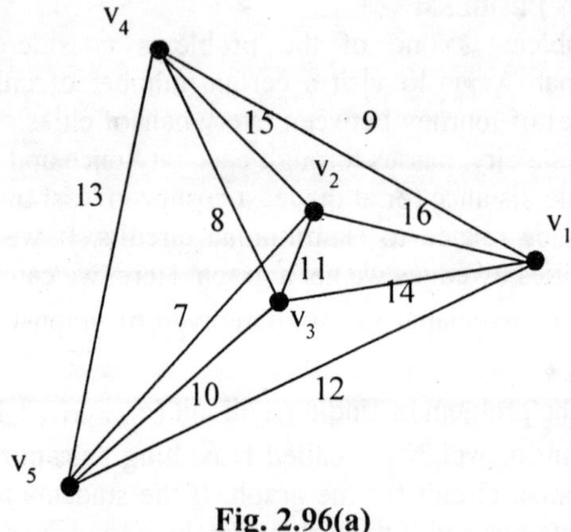

Fig. 2.96(a)

Solution

As it has been asked to start with vertex v_1. Vertex nearest to v_1 is v_4 (because v_4 has the least weight of the four edges incident with v_1), hence select the edge $(v_1\, v_4)$. The next edge must have v_4 as an end vertex, and $(v_4\, v_1)$ is not allowed (one cannot return to v_1, it has already been used), so smaller weight of the remaining edges is chosen namely $(v_4\, v_3)$. The weight with smallest edge originating at v_3 is $(v_3\, v_5)$. The next edge with smallest weight originating at v_5 is $(v_5\, v_2)$. The next edge with smallest weight originating at v_2 is $(v_2\, v_3)$. So algorithm finds route $(v_1\, v_4\, v_3\, v_5\, v_2\, v_1)$ as shown in Fig.2.96 (b). It is to be noted here that a different route may be obtained if one starts with different vertex.

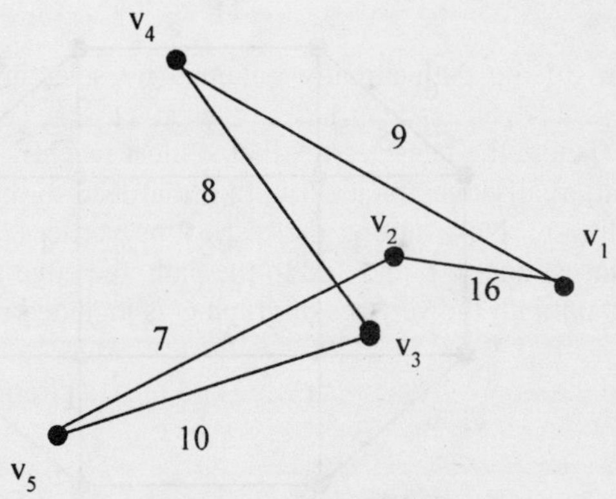

Fig. 2.96(b)

2.38-4 THE CHINESE POSTMAN PROBLEM

The Chinese mathematician Guan Meigee [1962] proposed this problem. Suppose a mail carrier traverse all edges in a road network starting and ending at the same vertex. If the edges have been assigned weights representing distance (or time). Our aim is to find a closed walk of minimum total length that uses all the edges. This is called the **chinese postman problem** in honour of **Guan Meigee.**

PROOF

To prove this theorem let us use the principle of mathematical induction on size (number of edges) $e(G)$. The result is true for $e(G) = 1$. Consider the case when $e(G)>1$ and suppose that the claim holds for graphs with $e(G)$ - 1 edges.

If graph G is even, then G has no cut-edge, since deleting it would leave subgraphs with one odd vertex*. Hence starting from some vertex v_i along an edge v_j leaves a graph with two odd vertices. By the induction principle the algorithm completes the circuit of G by completing a (v_i, v_j) trail in $G - v_i v_j$.

Suppose G has two odd vertices v_i and v_j. If every vertex is even, the graph G will be Eulerian and then sum the weights gives us the answer. Otherwise, we must repeat edges. Every traversal is an Eulerian circuit of a graph obtained by duplicating edges of graph G. Here searching the shortest traversal is same to finding the minimum total weight of edges whose duplication** will make all vertices even. If we use an edge three or more times in making all vertices even, then deleting two of those copies will have all vertices even. There may be more than one ways to choose the duplicated edges.

Example 2.45

In the graph given in the Fig.2.97 the eight outer vertices are odd. If we match them around the outside to make the degrees even, the extra cost is 6 + 6 + 6 + 6 = 24 or 1 + 11 + 11 + 1 = 24.

---

* A graph with every vertex degree even has no cut edge.

** We say duplication because we need not use an edge more than twice.

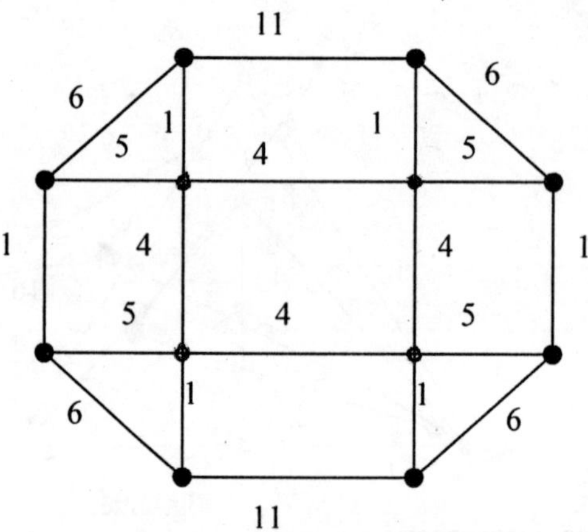

Fig. 2.97

We can do better by using all the vertical edges, which total only 18. Here it is to be observed that adding an edge from an odd vertex to an even makes the even vertex odd. We must continue adding edges until we complete a trail to an odd vertex. The duplicate edges must consist of collection of trails that pair up the odd vertices. We may restrict our attention to paths pairing up the odd vertices, though the path may need to intersect.

EXERCISE 2.3

1. Consider the graph shown in Fig.2.102. Find
   (i) all cycles, if any
   (ii) all cut points if any
   (iii) all bridges, if any
   (iv) the degree of each vertex
   (v) all paths from *a* to *g*
   (vi) all trails (distinct edges) from *b* to *c*
   (vii) the distance from *a* to *c*, *d*(*a*, *c*)
   (viii) diam (*G*), the diameter of *G*.
2 (a) Consider a connected graph *G*. Prove that if *G* contains a cycle *C*, which contains an edge *e*; then *G-e* is still connected.
   (b) Let *G* be a connected graph. Prove that if $e = \{u, v\}$ is an edge such that *G* - *e* is disconnected, than *u* and *v* belong to different components of *G* -*e*.
3. List all the paths between vertices v_5 and v_6 in Fig.2.98. Also give the length of each of these paths.

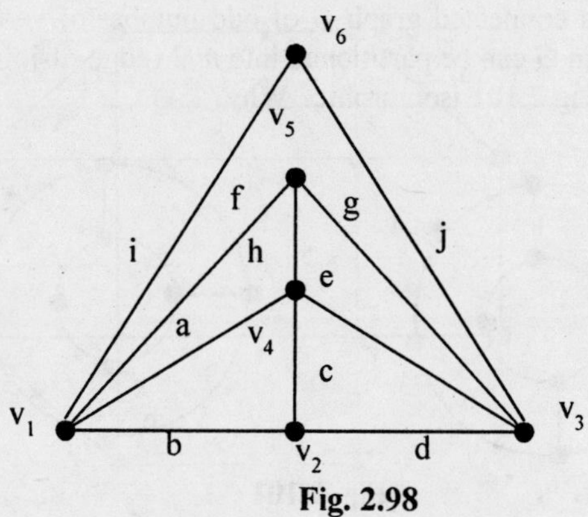

Fig. 2.98

4. (i) What is the diameter of the graph shown in Fig.2.99.

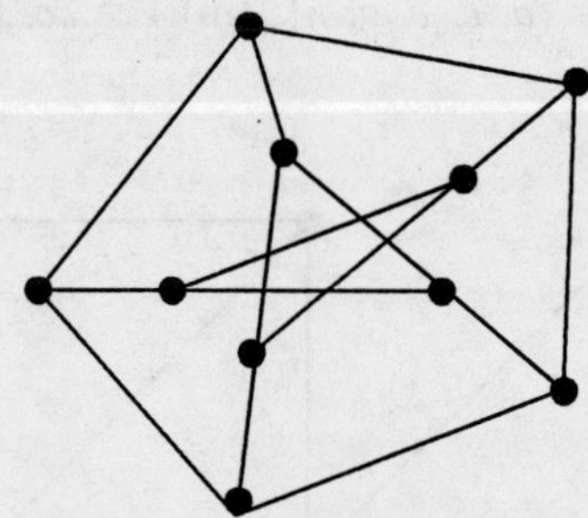

Fig. 2.99

(ii) If the vertices of the graph represent computers and edges represent data communication links between computers. What is the physical significance of the graph in this case?

5. Prove that a simple graph with n vertices must be connected if it has more than $[(n-1)(n-2)]/2$ edges.

6. Prove that any two simple connected graphs with n vertices, all of degree two are isomorphic.

7. If the intersection of two paths is a disconnected graph. Show that the union of the two paths has at least one circuit.

8. In the graph shown in the Fig. 2.100 edges can be partitioned into two edge-disjoint paths.

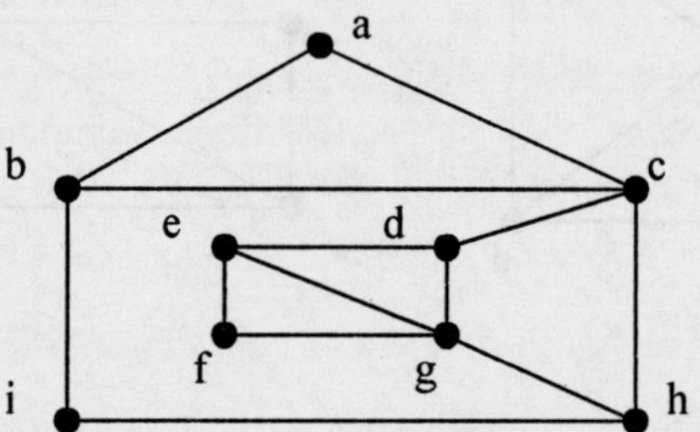

Fig. 2.100

(a) Show one such partition.

(b) Consider a connected graph G of odd number of vertices say m ($m>0$). Show that the edges in G can be partitioned into $m/2$ (edge-disjoint) paths.

9. Are two graphs in Fig.2.101 isomorphic? Why?

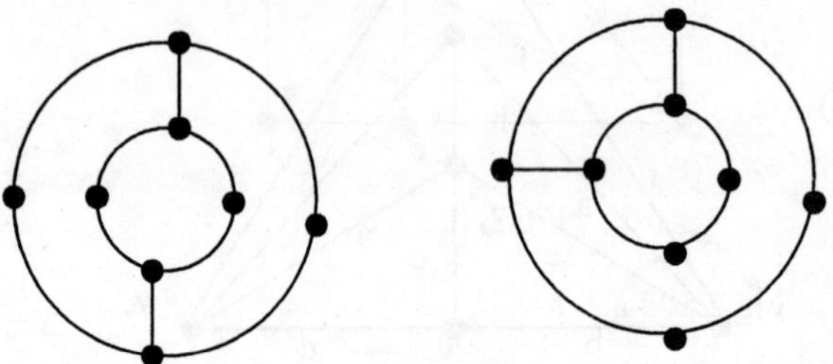

Fig. 2.101

10. Consider the graph given in Fig.2.102 and find the subgraph $M(V', E')$ generated by: (i) $V' = \{b, c, d, e, f\}$; (ii) $V' = \{a, c, e, g, h\}$; (iii) $V' = \{c, f, g, h\}$. Which of them are homeomorphic or Isomorphic?

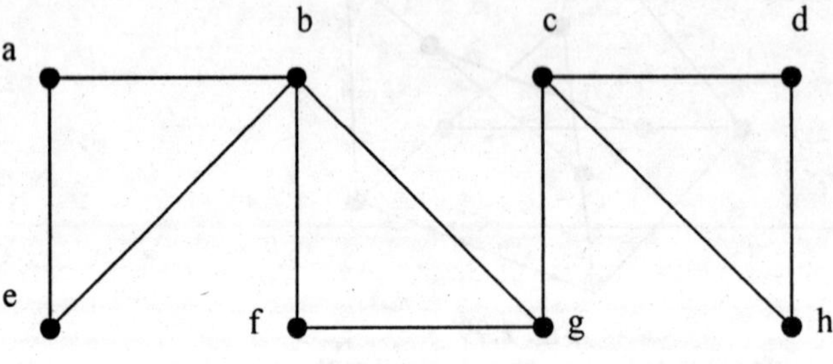

Fig. 2.102

11. Which of the graphs in the Fig.2.103 have an Euler circuit, an Euler path but not an Euler circuit or neither.

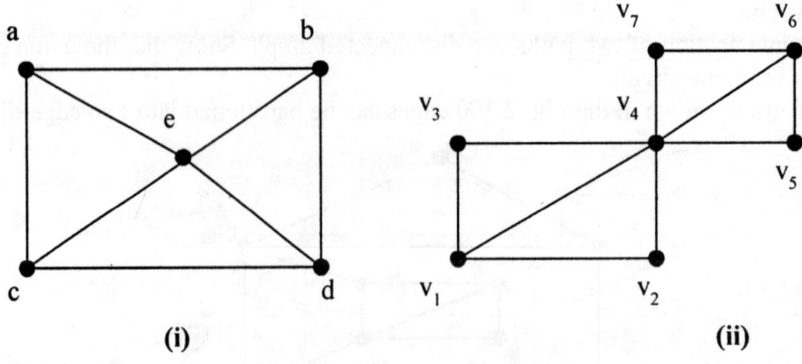

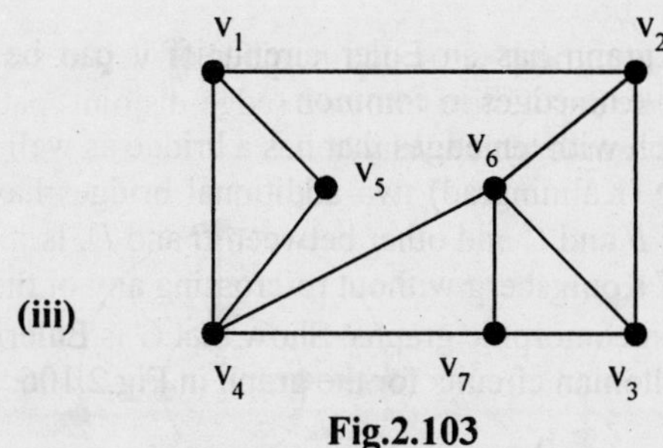

Fig.2.103

12. Find an Euler circuit in Fig.2.104.

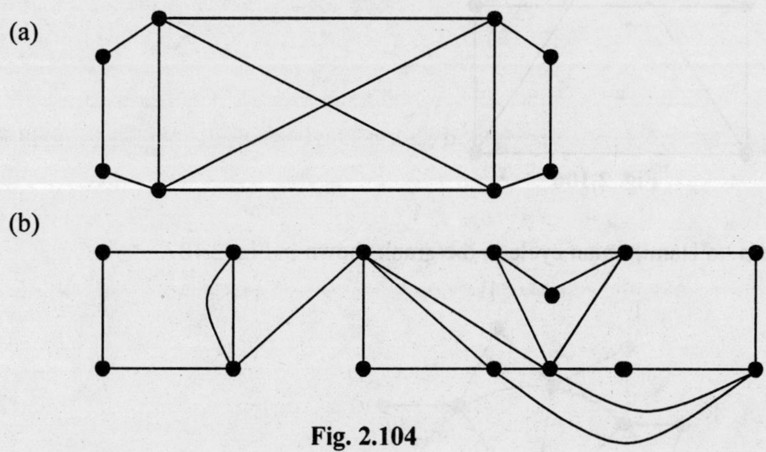

Fig. 2.104

13. Which of the multigraph in Fig. 2.105 have Euler paths, circuits or neither.

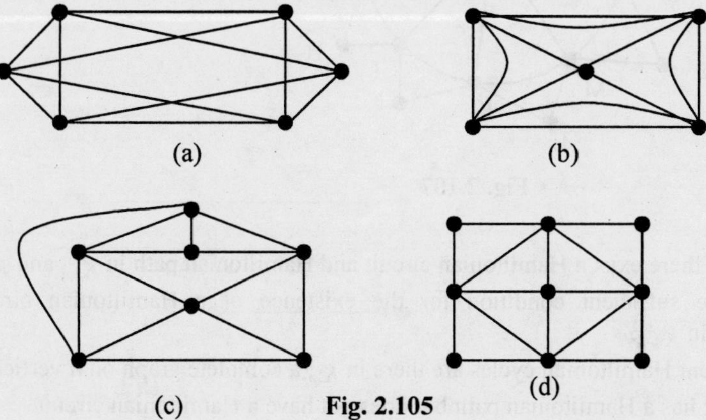

Fig. 2.105

14. Prove that a connected graph has an Euler circuit iff it can be decomposed to a set of elementary cycles that have no edges in common.
15. Give an example of a graph with ten edges that has a bridge as well as an Euler path.
16. At present in Konigsberg (Kaliningrad) two additional bridges have been constructed. One bridge is between regions B and C and other between B and D. Is it now possible to construct a route over all bridges of Konigsberg without re-crossing any of them.
17. Suppose G and G' are Homeomorphic graphs. Show that G is Eulerian iff G' is Eulerian.
18. Find the number of Hamiltonian circuits for the graph in Fig.2.106.

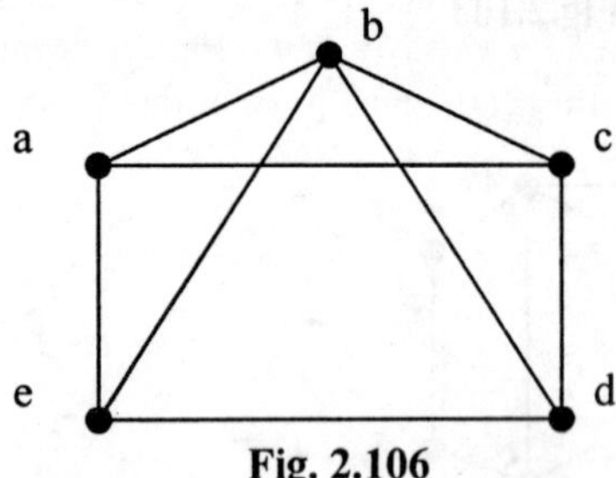

Fig. 2.106

19. Prove that there is no Hamiltonian cycle in the graph shown in Fig. 2.107.

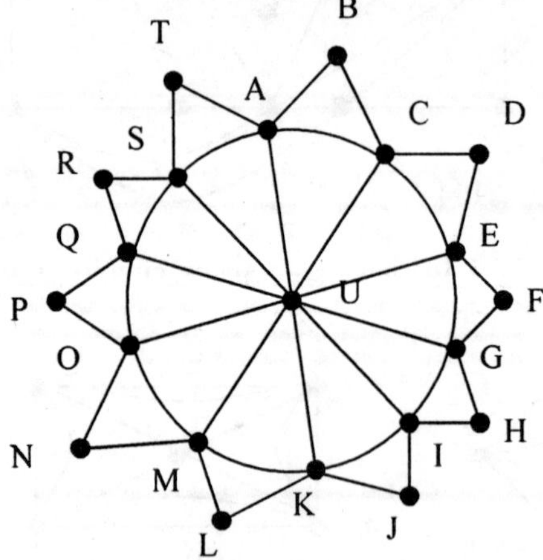

Fig. 2.107

20. (a) Test whether there exist a Hamiltonian circuit and Hamiltonian path in $K_{4,4}$ and $K_{4,6}$?

    (b) What is the sufficient condition for the existence of a Hamiltonian circuit and Hamiltonian path in $K_{m,n}$.
21. How many different Hamiltonian cycles are there in K_n, a complete graph on n vertices?
22. Draw a graph that has a Hamiltonian path but does not have a Hamiltonian circuit.

23. Show that the any Hamiltonian cycle in the graph shown in Fig.2.108 which contains the edge (c, d) must also contain the edge (g, h).

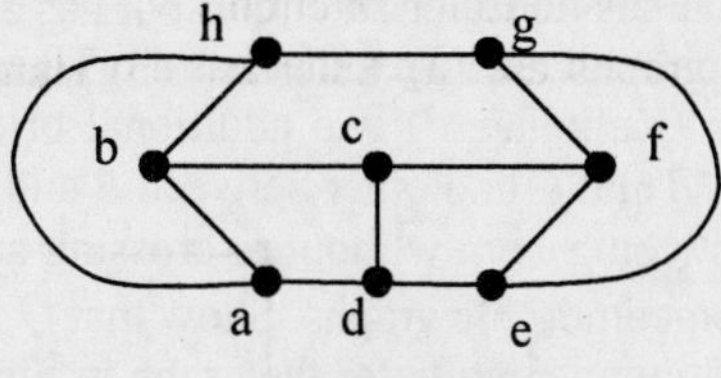

Fig. 2.108

24. Show that any Hamiltonian cycle in the graph shown in the Fig.2.109 must contain exactly two edges (v_1, v_8), (v_3, v_4), (v_9, v_{10}). Further show that any Hamiltonian cycle that contains both edges (v_4, v_5), (v_5, v_{10}) cannot also contain (v_1, v_8).

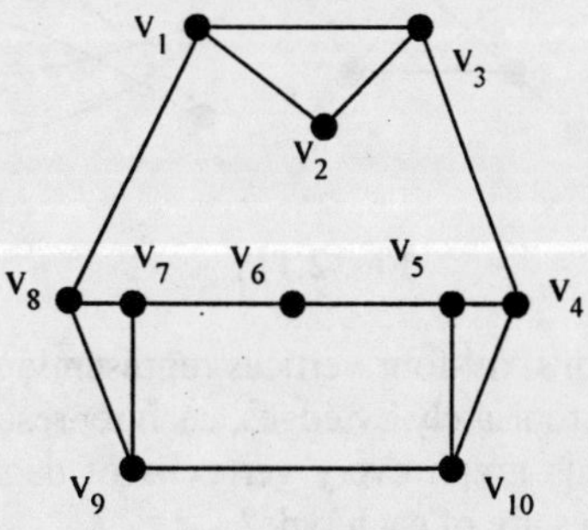

Fig. 2.109

25. Find a Hamiltonian cycle in each of the following graph.
(i) The dodecahedron
(ii) The octahedron
(iii)
(iv)

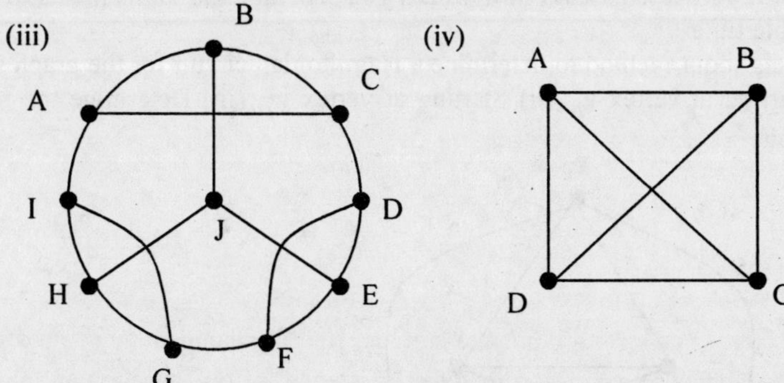

Fig. 2.110

26. Show that there can be no path longer than Hamiltonian path if it exist in a graph.

27. (a) Draw a graph that has both an Eulerian circuit and a Hamiltonian circuit.
    (a) Draw a graph that has neither an Eulerian circuit nor a Hamiltonian circuit.
    (b) Draw a graph that has an Eulerian circuit but has no Hamiltonian circuit.
    (c) Draw a graph that has no Eulerian circuit but has a Hamiltonian circuit.
28. Which of the following graphs are (a) Eulerian (b) Hamiltonian. Determine the Eulerian path (or circuit).

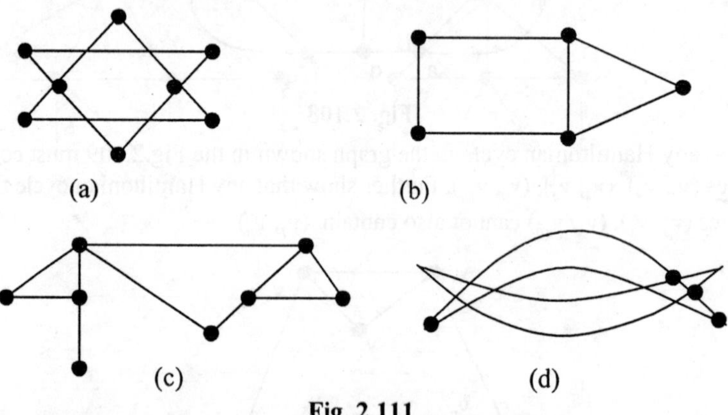

Fig. 2.111

29. (a) Draw a graph with sixty-four vertices representing the squares of a chessboard. Join these vertices appropriately by edges, each representing the move of the knight. One can see that in this graph every vertex is of degree two, three, four, six or eight. How many vertices are of each type?
    (b) Is it possible, starting from any sixty-four squares of the chessboard to move a knight such that it occupies every square exactly once and return to the original position? If so give one such tour?
    [Hint Look for a Hamiltonian circuit]
30. A round-robin tournament among n players (n being an odd number) can be represented by a complete graph of n vertices. Discuss how would you schedule the tournaments to finish in the shortest possible time.
31. Use the nearest-neighbour method to determine a Hamiltonian circuit for the graph shown in Fig.2.112 (i) Starting at vertex v_1 (ii) Starting at vertex v_4 (iii) Determine the minimum Hamiltonian circuit.

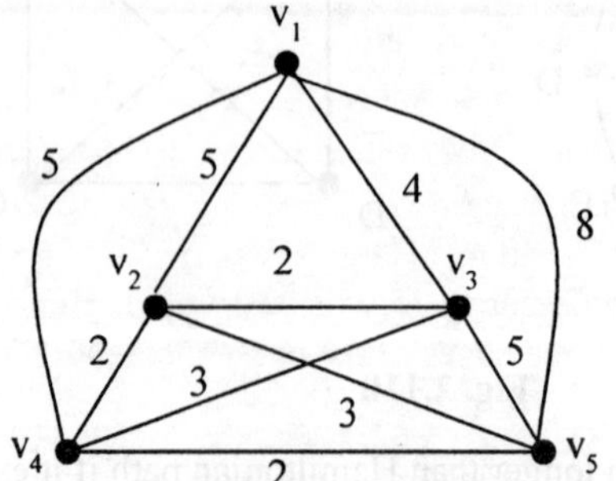

Fig. 2.112

32. Determine the shortest path between the vertex A and P of the graph shown in Fig.2.113(a) and (b) where the numbers associated with the edges are the distance between the vertices.

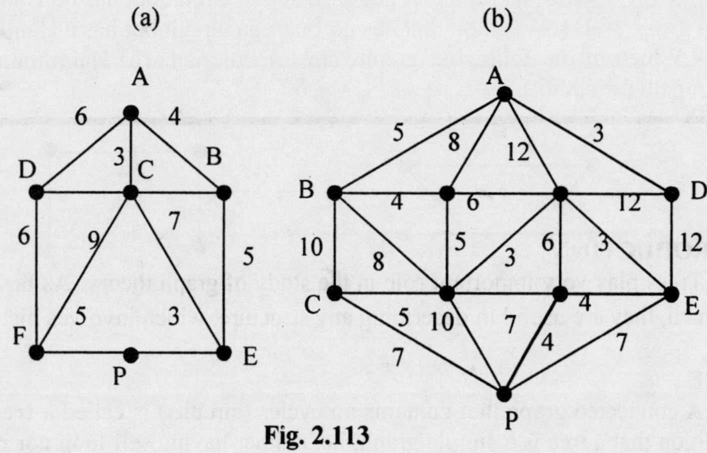

Fig. 2.113

ANSWERS

1. (i) *abea, bfgb, cdhc,*
   (ii) *b,c,g,*
   (iii) only {*c, g*},
   (iv) 2, 4, 3, 2, 2, 2, 3, 2,
   (v) *abg, abfg, aebg, aebfg,*
   (vi) *bge, bfgc, baebgc, baebfgc,*
   (vii) 3,
   (viii) 4.

10. (i) E' = {*be, bf, cd*}, (ii) E' ={*ae,fg,gc*}, (iii) E' = {*fg, gc, ch*}. Further (i) and (ii) are isomorphic, and (i) and (ii) are homeomorphic.

20. Hint: Adding a vertex by dividing an edge doesn't change the degree of the original vertices and simply adds a vertex of even degree.

24. In case two of $\{v_1, v_8\}$, $\{v_3, v_4\}$, and $\{v_9, v_{10}\}$, are not used (let us assume $\{v_3, v_4\}$, and $\{i, j\}$ are not used), then $\{v_4, v_5\}$, $\{v_5, v_{10}\}$ and $\{v_4, v_{10}\}$ must be used, giving a triangle subcycle, impossible. This implies that at least two of $\{v_1, v_8\}$, $\{v_3, v_4\}$ and $\{v_9, v_{10}\}$ must be used. But all three should not be used because in that case (i) Suppose $\{v_4, v_{10}\}$ is used, then $\{v_4, v_5\}$ and are $\{v_5, v_{10}\}$ not used. It means two of the three edges at *e* are not used; impossible.

28. Let us assume that {*g, h*} is not used but {*e, d*}, is used. In this case edges {*a, h*} {*b, h*}, {*g, f*} and {*e, g*} must be included in a Hamiltonian cycle. But then {*a, b*} and {*e, b*} have to be deleted otherwise a triangle subcycle will be formed. Hence, {*b, c*} and {*e, f*} must be included. But then there are three edges of the cycle incident on *C*. contradict.

3 TREES

3.0 INTRODUCTION

Trees play very important role in the study of graph theory. As far as application of the trees is concerned, they are useful in describing any structure, which involves hierarchy.

3.1 TREE

A connected graph that contains no cycles (circuits) is called a **tree**. One can conclude from the definition that a tree is a simple graph, i.e. neither having self loop nor parallel edges as they both form cycles (circuits). Example of a tree is shown in the Fig.3.1. A tree T with only one vertex is called a **trivial** or **degenerate** or **empty** tree, otherwise T is a **non-trivial tree**. The number of vertices in a tree is called **order** of the tree.

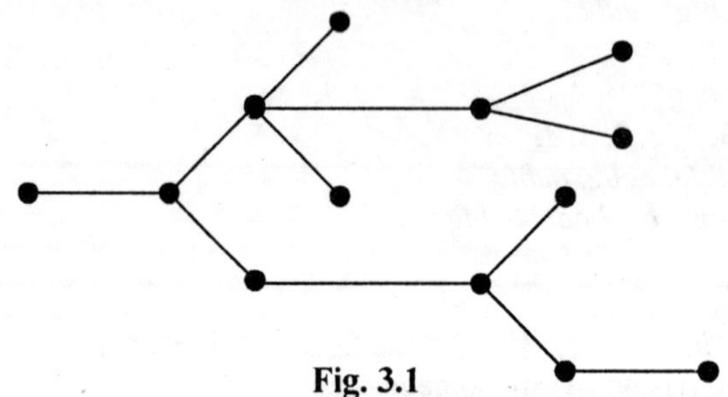

Fig. 3.1

Examples of tree

(i) Chain reaction
(ii) Administrative set up in an organization
(iii) Flow of electricity
(iv) Water/ L.P.G./Petroleum supply pipeline
(v) Postal Flow
(vi) Decision Tree.

3.2 FOREST

A **forest** G is a graph with no cycles, hence the connected components of a forest G are trees. A vertex of degree 1 in a tree is called a **leaf** or a **terminal node** or a **pendant vertex** and a vertex of degree larger than 1 is called a **branch node** or an **internal node**. In the graph shown in the Fig. 3.2. a, c, d, f and i are leaves and b, e, h and g are branch nodes.

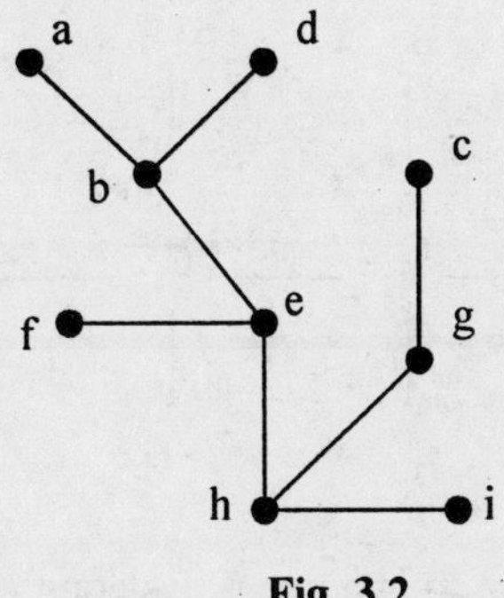

Fig. 3.2

Example 3.1

(a) Graphs (i), (ii) and (iii) in Fig. 3.3 are trees while graphs (iv) and (v) are not tree.

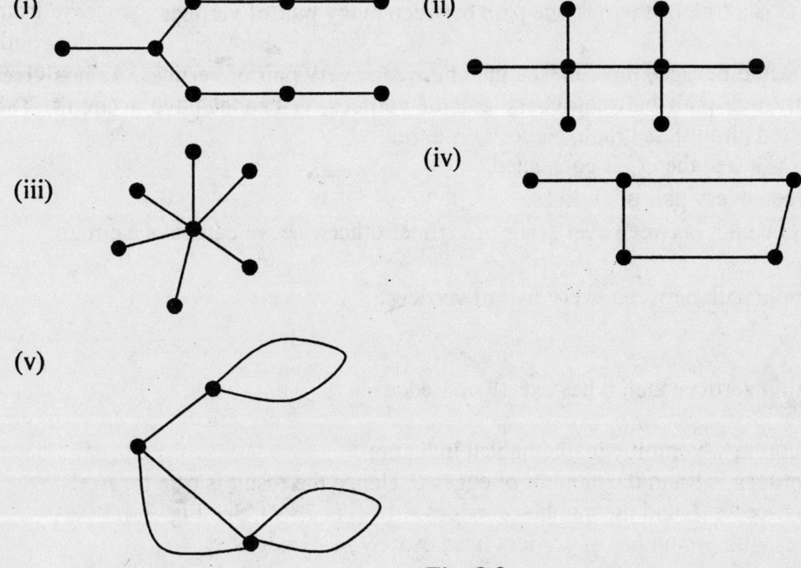

Fig. 3.3

(b) In a tree T there exist one simple path between two vertices of T; otherwise the two paths would form a cycle.

(c) A K_2 is a tree, but for $n \geq 3$, K_n is not a tree.

(d) If a tree T doesn't contain an edge $e = (u, v)$ and we add the edge e to T, then the simple path from u to v in T and e will form a cycle, thus T will no longer be a tree.

(e) Some chemical compound structures are modeled by a tree, e.g. Butane consists of four carbon molecules and ten hydrogen molecules where an edge between two molecules represents a bond between them and vertices represent molecules.

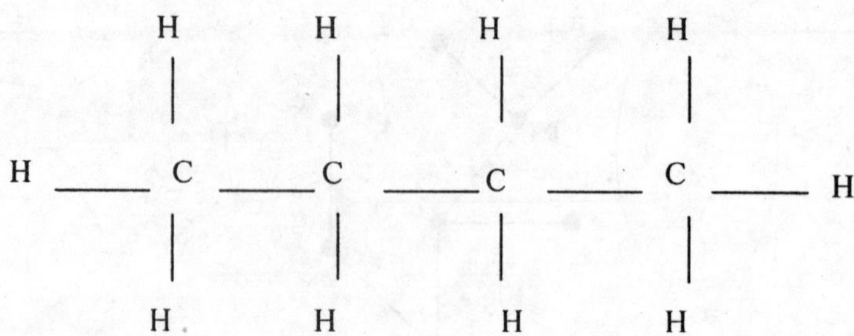

(f) Suppose there is an edge $e = (u, v)$ in a tree T. Deleting e from T, the tree will no longer be connected as there cannot be a path from u to v and hence T is no longer a tree.

In the following articles we discuss some important properties of trees.

THEOREM 3.1

A simple graph G is a tree iff there is one path between every pair of vertices.

PROOF

Graph G is connected because there exist a path between every pair of vertices. As it is given that there is one and only one path between every pair of vertices, G cannot have a circuit. This implies that G is a connected circuitless graph. Hence G is a tree.

Conversely, if G is a tree then G is connected.

$\Rightarrow$ There is a path between every pair of vertices.

$\Rightarrow$ There must be only one path between every pair of vertices otherwise we can have a circuit

$\therefore$ G is a tree.

$\Rightarrow$ There is a unique simple path between every pair of vertices.

THEOREM 3.2

If G is a tree with n vertices then it has exactly n-1 edges.

PROOF

To prove this theorem we employ mathematical induction.

If $n = 1$, the number of vertices = 1 and the number of edges 0. Hence the result is true for $n = 1$.
If $n = 2$, the number of vertices = 2, and the number of edges = 1 (i.e. $2 - 1$). See Fig. 3.4.
Assume $n = m$, i.e. all trees with m number of vertices have exactly $m - 1$ edges.

Fig. 3.4 Trees with one and two vertices

Let us assume that e_m be an edge connecting vertices v_i and v_j. As G is a tree, there exist only one path i.e. e_m between the vertices v_i and v_j. Hence, if we delete e_m, it will disconnect tree G into two components, say G_1 and G_2 as one can see in Fig. 3.5. It is easy to see that each such components has less than $m+1$ vertices and since there are no circuits in G, each of these components is a tree.

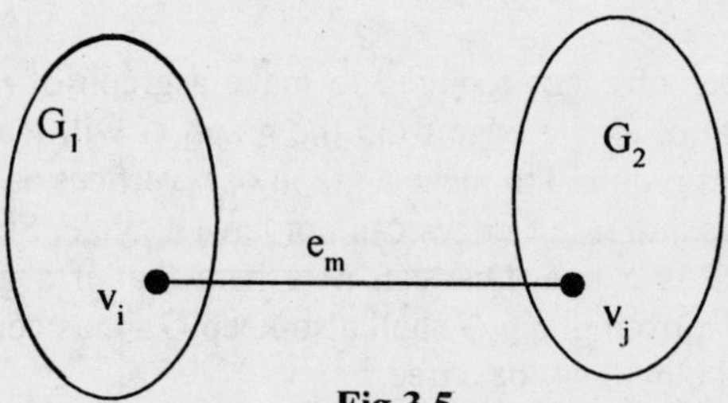

Fig.3.5

Let the two components G_l and G_2 have m_l and m_2 vertices respectively. Now, the total number of edges will be

$$m_1\text{-}1 + m_2\text{-}1 + 1 = m_1 + m_2\text{-}1$$
$$= m + 1 - 1$$
$$= m.$$

Thus the theorem is true for $n = m + 1$ if it is true for $n = m$. Hence by mathematical induction tree G has exactly $n - 1$ edges.

COROLLARY

If G is a nontrivial graph then G contains at least two vertices of degree 1.

PROOF

Let the number of vertices of graph G be n. Then from the sum of degrees formula,

$$\sum_{i=1}^{n} \deg(v_i) = 2e = 2(n-1) = 2n-2.$$

where e is the number of edge.

If we have only one vertex of degree 1, then

$$\deg(v_i) \geq 2 \text{ for } i = 2,3, \ldots, n.$$

and

$$\sum_{i=1}^{n} \deg(v_i) = 1 + \sum_{i=2}^{n} \deg(v_i)$$
$$\geq 1 + 2n - 2$$
$$= 2n - 1$$

But then $2n - 2 \geq 2n - 1$ or $-2 \geq -1$

which is a contradiction. Hence, if G is a nontrivial tree then G contains at least 2 vertices of degree one.

THEOREM 3.3

Every tree with at least two vertices has a pendant vertex.

PROOF

Consider a tree T in such a way that degree of all vertices are greater than or equal to 2. Now define a walk from an arbitrary vertex say u of T. Since T has finite vertices and every vertex has more than one degree, this implies the walk is a cycle. But this is a contradiction. Hence every tree with at least two vertices has a pendant vertex.

THEOREM 3.4

Any connected graph G with n vertices and $(n - 1)$ edges is a tree.

PROOF

We know minimum number of edges required to make a graph of n vertices connected is $n - 1$. We can observe that removal of single edge from the graph G will make it disconnected. So $n-1$ is the minimum number of edges required to make a graph of n vertices as connected and acyclic. Thus a connected graph of n vertices and $n - 1$ edges can not have a cycle. Since connected graph G is not having cycle, hence graph G is a tree. One can note here that if a graph is not minimally connected, then deletion of an edge e from graph G shall also keep $G - e$ as connected. Therefore this e shall be an edge in some cycle and G will not be a tree.

3.3 MINIMALLY CONNECTED GRAPH

A connected graph is said to be **minimally connected** if removal of any one edge from it disconnects the graph. It is obvious that a minimally connected graph cannot have a circuit. Because, if it contains a cycle then even after removal of an edge from the graph, it will remain connected. Hence a minimally connected graph is a tree.

Conversely, let G is a tree and G is not minimally connected. It means removal of an edge (say e) from G leaves the graph connected. It means edge e is in some circuit. But this contradicts the assumption that G is a tree, hence the converse. From here, we can conclude that a graph is a tree iff it is minimally connected.

THEOREM 3.5

A graph G with n vertices, $n-1$ edges and no circuits is connected.

PROOF

Let G be a disconnected circuitless graph with n vertices and $n-1$ edges. Then G consists of two or more components. Let us assume that G consists of two component T_1 and T_2 and let v_1 and v_2 are two vertices in T_1 and T_2 respectively. Add an edge e connecting v_1 and v_2. As there is no path between v_1 and v_2 in G, addition of e will not create a circuit. This implies that the graph $G \cup e$ will be a circuitless, connected graph with n vertices and n edges. But this contradicts the fact that a circuitless, connected graph with n vertices can have only $n-1$ edges (see theorem 3.2).

REMARK

A graph G is a tree iff G has no cycles and $|E| = |V| - 1$, where $|V|$ denotes number of vertices and $|E|$, number of edges.

Remember

A graph G with n vertices is called a tree if

(i) G is circuitless and is connected or
(ii) G is connected and has $n - 1$ edges or
(iii) G is circuitless and has $n - 1$ edges or
(iv) There is exactly one path between every pair of vertices in G or
(v) G is minimally connected graph.

Example 3.2

Draw six non-isomorphic trees with six vertices.

Solution

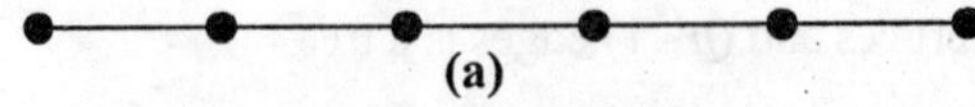

(a)

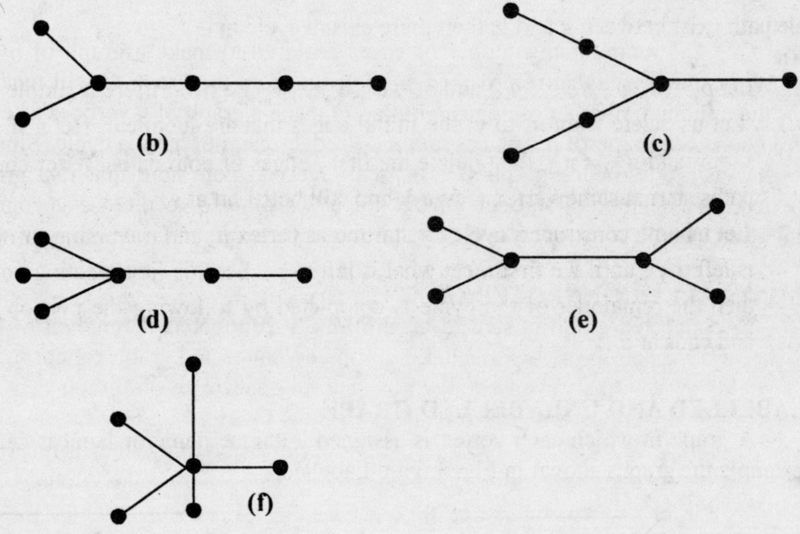

It is to be noted here that tree (*a*) has diameter 5, (*b*) and (*c*) have diameter 4, (*d*) and (*e*) have diameter 2 and (*f*) has diameter 2 (Refer 3.7-3).

Example 3.3

Let G be a graph with k-components, where each component is a tree. Obtain a formula for $|E(G_j)|$ in terms of $|V(G_j)|$ and k.

Solution

Let G_1 and $G_2, \ldots, G_k$ are components of G and $|E(G_j)|$ and $|V(G_j)|$ denote the number of edges and vertices of component G_j respectively.

$$|E(G_j)| = |V(G_j)| - 1 \qquad \forall j$$

$$\Rightarrow \quad |E(G)| = \sum_{j=1}^{k} |E(G_j)| = \sum_{j=1}^{k} \left(|V(G_j)| - 1\right)$$

$$\text{or} \quad |E(G)| = |V(G)| - k.$$

Example 3.4

Consider a tree with n_1 vertices of degree 1; 2 vertices of degree 2; 4 vertices of degree 3 and 3 vertices of degree 4. Find n_1.

Solution

Since $|E| = |V| - 1$. And from the sum of degree formula, we have

$$n_1 \cdot 1 + 2 \cdot 2 + 4 \cdot 3 + 3 \cdot 4 = 2\,|E|$$

$$n_1 + 4 + 12 + 12 = 2\,(|V| - 1)$$

$$\text{or} \quad n_1 + 28 = 2\,(n_1 + 2 + 4 + 3 - 1)$$

$$\text{or} \quad n_1 + 28 = 2\,n_1 + 16$$

$$\text{or} \quad n_1 = 12.$$

THEOREM 3.6

Consider an undirected graph $G(V, E)$ with no self-loops and let $u, v \in V$. If two different simple path exist between u and v, then there exist a cycle in G.

PROOF

Let $p_1 = (u_1, u_2, u_3, \ldots, u_m)$ and $p_2 = (v_1, v_2, \ldots, v_n)$ be two different simple paths from u to v.

Step 1 Let us delete from u_1 to v_1 the initial edges that are identical. Here, if $u_1 = v_1$, $u_2 = v_2$, . . . , $u_i = v_i$ and $u_{i+1} \neq v_{j-1}$, then delete the first i edges of both paths. After completion of this, both paths start at same vertex (say u_c), and still both end at v.

Step 2 Let us now construct a cycle by starting at vertex u_c and traversing through the vertices what is left to p_1 until we first meet what is left to p_2. Let this first meeting occurs at vertex v_f, and then the remainder of the cycle is completed by following the portion of p_2 that starts at v_f and ends at u_c.

3.4 LABELLED AND UNLABELLED GRAPH

A graph in which each vertex is assigned a unique name or label is called **Labelled graph**. For example the graphs shown in Fig. 3.6 are Labelled.

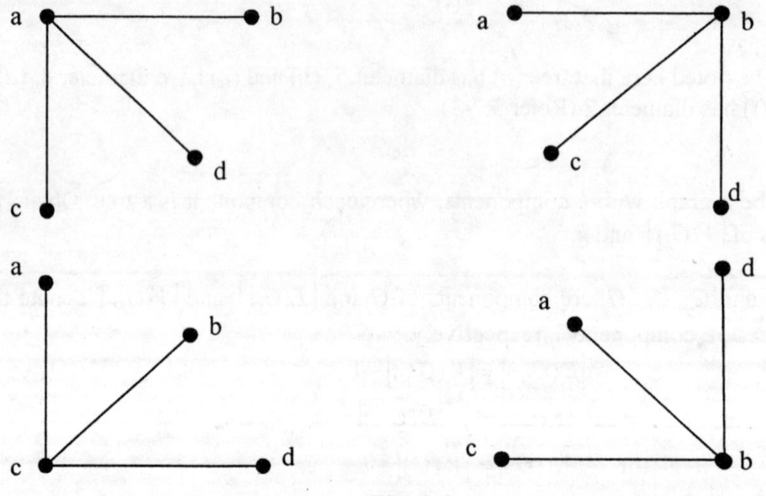

Fig. 3.6

It is to be noted here that the four graphs shown in Fig. 3.6 are treated as four different trees even though they are isomorphic because they are labelled. If it had not been labelled there were no distinction made between vertices a, b, c or d and these four trees would be treated as one.

THEOREM 3.7

Every tree T with at least two vertices has at least two leaves. Deleting a leaf from a tree with n vertices produces a tree with n - 1 vertices.

PROOF

We know that every connected graph with at least two vertices has an edge. Further, we know that the end points of a maximal nontrivial path in an acyclic graph have only one neighbour on the path, hence have degree 1. Therefore, the end points of maximal path provide the two desired leaves.

Let us consider a leaf l of a tree T, and let $T' = T - 1$. If $v_1, v_2 \in V(T')$, then there exist no v_1, v_2-path p in T which can pass through the vertex 1 of degree 1. This implies path p is also present in G'. Hence G' is connected. Since removal of a vertex never create a cycle, thus T' is also acyclic. Hence T' is a tree with n -1 vertices.

THEOREM 3.8

Consider a tree T with m edges and let G is a simple graph with minimum degree $(\delta(G)) \geq m$, then T is a subgraph of G.

PROOF

Here, we use mathematical induction on m to prove the theorem. When $m = 0$, note that every simple graph contains K_1. Let $m > 0$ and supposing theorem holds. Refer theorem 3.7, it allows us to choose a leaf l with neighbour v in T. Let us consider a smaller tree $T' = T - l$.

By mathematical induction T' is a subgraph of G, since $\delta(G) \geq m > m - 1$. Let y be the vertex in this copy of T that corresponds to v, y has a neighbour x, say in G that doesn't appear in this copy of T'. As T' has only m -1 vertices other than v, y has a neighbour x in G that doesn't appear in this copy of T'. Adding the edge (y, x) to correspond to (u, v) expands this copy of T' in G into a copy of T. See Fig. 3.7.

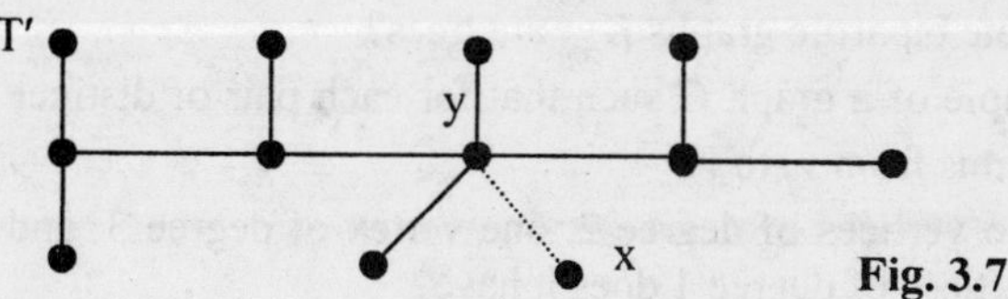

Fig. 3.7

EXERCISE 3.1

1. (a) What is the smallest complete undirected graph (with more than one vertex) which is a tree.
   (b) Characterise all trees that have exactly two leaves.
2. (a) Draw all trees of n labelled vertices for n =1, 2, 3, 4 and 5
   (b) Draw all trees of n unlabelled vertices for n =1, 2, 3, 4 and 5.
3. Write down the number of different (pair-wise non-isomorphic) trees there are of orders (number of vertices in a graph is called **order of a graph**).
   (i) 2 (ii) 3 (iii) 4 (iv) 5.
4. List all pair-wise non-isomorphic trees with degree sequences
   (i) (1, 1, 1, 1, 2, 2, 2, 3, 3)
   (ii) (1, 1, 1, 1, 2, 2, 3, 3)
   (iii) (1, 1, 1, 1, 1, 1, 2, 3, 5).
5. Prove that in a tree G, the sum of degrees equals to $2|v| - 2$.
6. Mention any three applications of trees with specific examples.
7. Show that there are only three different trees of five vertices.
8. Consider a tree T with n_1 vertices of degree 1, n_2 vertices of degree 2, n_3 vertices of degree 3,...., n_k vertices of degree k. Find n_1 in terms of n_2, n_3,.... and n_k.

9. Let $G = (V, E)$ be an undirected graph with no self-loops and $|V| = n$. Consider the following propositions:
   P_1: G is a tree
   P_2: For each pair of distinct vertices in v_1 there exist a unique simple path between them.
   P_3: G is connected and if $e \in E$, then $(V, E - \{e\})$ is disconnected.
   P_4: G contains no cycles, but by adding one edge, exactly one circuit is created.
   Show that $P_1 \Rightarrow P_2 \Rightarrow P_3 \Rightarrow P_4 \Rightarrow P_1$
10. Find the number of trees with seven vertices.
11. Prove that if $G = (V, E)$ is a tree and $e \in E$, then $(V, E-\{e\})$ is a forest of two trees.
12. Draw all pair-wise non-isomorphic trees with $|V|=7$ and at least one vertex of degree ≥ 4. **[Hint** Determine the degree sequence first]
13. Show that a graph G is a tree iff G is connected and contains no circuits.
14. Show that there is no tree with degree sequence
(i) (1,1,1,1,1,1,2,3,7)
(ii) (1,1,2,2,3,3).
15. Show that any tree is a bipartite graph.
16. Which complete bipartite graphs $K_{m,n}$ are trees?
17. Give an example of a graph G such that for each pair of distinct vertices v_1 and v_2 there are exactly two paths from v_1 to v_2.
18. A tree has two vertices of degree 2, one vertex of degree 3, and three vertices of degree 4. How many vertices of degree 1 does it have?

ANSWERS

3. (i) 1 (ii) 1 (iii) 2 (iv) 3
8. $n_1 - 2 = n_3 + 2\,n_4 + 3\,n_5 + \ldots + (k - 2)\,n_k$.
10. Ten
16. $m = 1$ 17. Any cycle graph c_n.

3.5 DISTANCE IN A TREE

The distance between two vertices in a tree is defined exactly as it was defined for a graph. The **distance** between two vertices u and v in a tree $G = (V,E)$, written $d_G(u, v)$ or simply $d(u, v)$, is defined to be the minimum number of edges between them (i.e. shortest one). There may be several shortest paths. If G has no such path, then $d(u, v) = \infty$.

Example 3.5

The paths between the vertices u and v in the graph shown in the Fig. 3.8 are (a, e), (a, c, f), (b, c, e), (b, f), (b, g, h) and (b, g, i, k). Out of these (a, e) and (b, f) are two shortest paths, each of length 2, i.e. $d(u, v) = 2$.

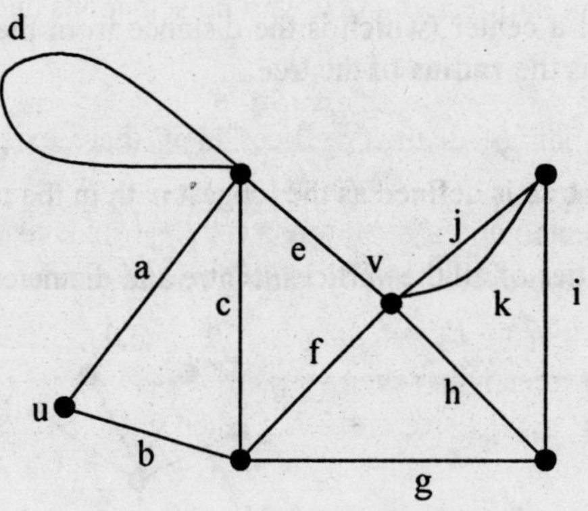

Fig. 3.8

Example 3.6

Determination of distance in a tree is much easier as there exist exactly one path between any two vertices. In the tree shown in the Fig. 3.9.

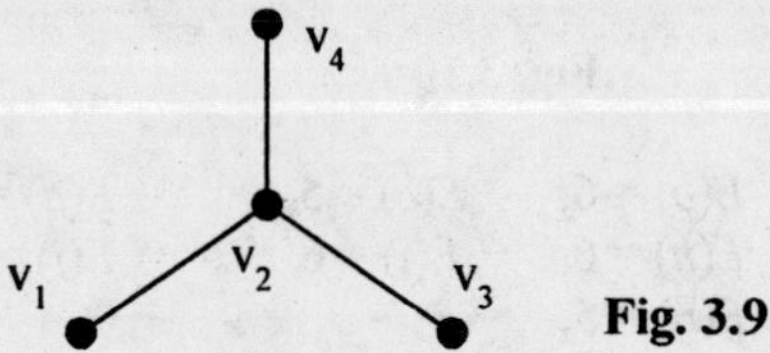

Fig. 3.9

$d(v_1, v_2) = 1$, $d(v_1, v_3) = 2$, $d(v_3, v_2) = 1$, and so on.

3.6 METRIC

A function $f(x, y)$ which satisfy following conditions:

(i) $f(x, y) \geq 0$, and $f(x, y) = 0$ iff $x = y$

(ii) $f(x, y) = f(y, x)$

(iii) $f(x, y) \leq f(x, z) + f(z, y)$ for any z

is said to be a **metric**. It is obvious that the distance $d(u, v)$ (distance function) satisfies (i) and (ii) conditions. And as $d(u, v)$ is the shortest path between u and v, this path cannot be longer than any other path which goes through a vertex w. Hence $d(u, v) \leq d(u, w) + d(w, v)$. Hence the distance between vertices of a connected graph is a metric.

3.7 ECCENTRICITY

Eccentricity of a vertex v of a graph G, denoted by $E(v)$, is the distance from v to the vertex farthest from v in G, i.e.

$$E(v) = \operatorname*{Max}_{v_i \in G} d(u, v_i)$$

Eccentricity is also referred to as **associated number** or **separation**.

3.7-1 CENTRE

A vertex of graph G with minimum eccentricity is called the **centre** of G.

3.7-2 RADIUS

The eccentricity of a center (which is the distance from the center of the tree to the farthest vertex) in a tree is defined as the **radius** of the tree.

3.7-3 DIAMETER

The **diameter** of a tree is defined as the longest path in the tree. See example 3.2.

Example 3.7

Find the eccentricities of all the vertices, centre and diameters of the following trees:

(i)

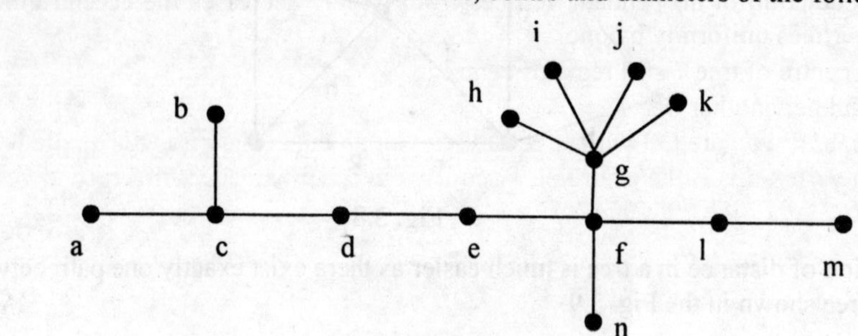

Fig. 3.10

Solution

$E(a) = 6$, $E(b) = 6$, $E(c) = 5$, $E(d) = 4$, $E(e) = 3$, $E(f) = 4$,
$E(g) = 5$, $E(h) = 6$, $E(i) = 6$, $E(j) = 6$, $E(k) = 6$, $E(l) = 5$,
$E(m) = 6$, $E(n) = 5$,

e is the centre, radius = 3, diameter = 6.

(ii)

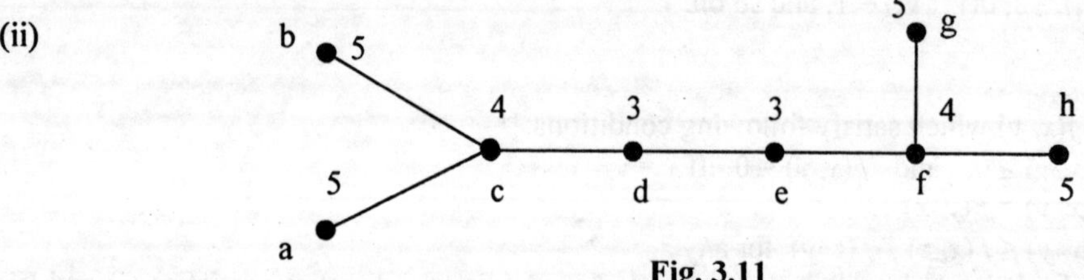

Fig. 3.11

Centres are *d* and *e*; radius = 3, diameter = 5. The eccentricities are written with vertices in the Fig. 3.11.

REMARK

A graph can have more than one centres. Every vertex of a graph that consists of a circuit (a polygon), is a centre. See Fig. 3.12

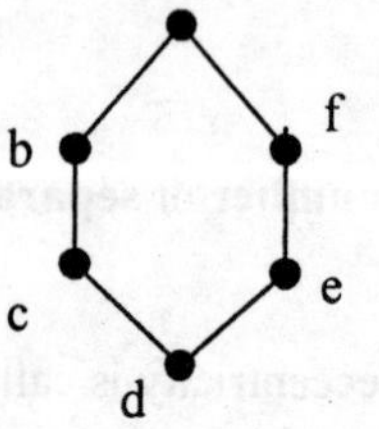

Fig. 3.12

THEOREM 3.9

Every tree has either one or two centres.

PROOF

We know that the maximum distance between two vertices v and v_i (i.e. max $d(v, v_i)$) is possible only when v_i is a pendant vertex. Let us consider a tree T having more than two vertices. We know that tree T must have two or more pendant vertices. Now, to prove the theorem :

(i) Remove all the pendant vertices from T, the resulting graph T' is still a tree.

(ii) We can see deletion of all pendant vertices from tree T, decreases the eccentricities of the remaining vertices uniformly by one.

(iii) We see that centre of tree T still remains centre in T'.

(iv) Repeat (i) and get another tree T'.

(v) Continue (i), until you are left with either a vertex or an edge. Here it is to be noted that if we are left with vertex this is the centre of T and if we left with an edge then its end vertices are the two centres of T. See the same in Fig.3.13.

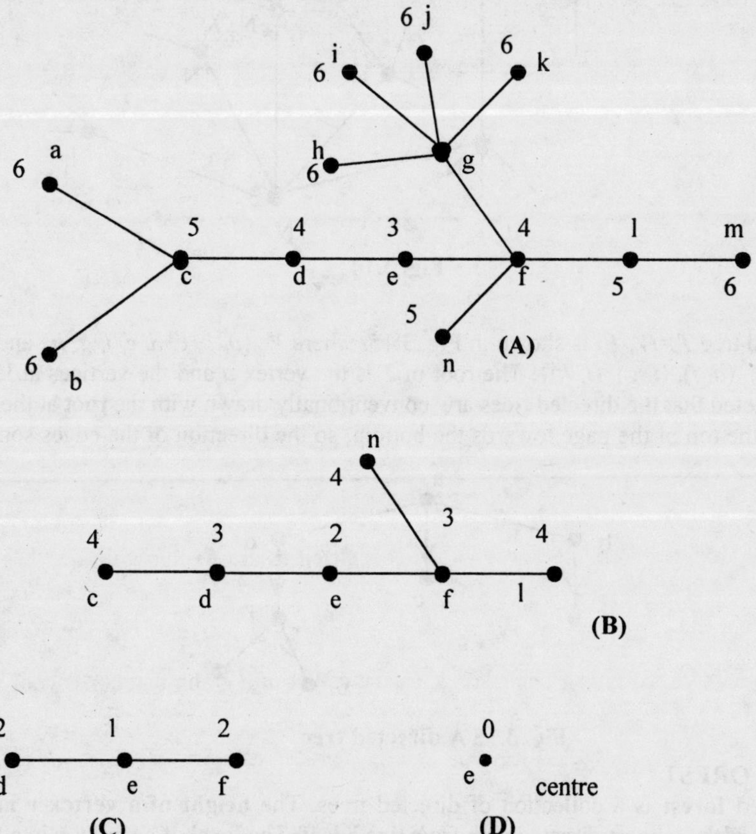

Fig. 3.13

COROLLARY

We can see that if a tree T has two centres, then they are adjacent.

3.8 DIRECTED TREE

A **directed tree** is an acyclic digraph which is a tree as an undirected graph and which has one vertex called its **root** with indegree 0, while all other vertices have indegree 1. An isolated vertex is also a directed tree. In a directed tree any vertex which has out degree zero is called a **terminal vertex** or a **leaf**. All other vertices are called **branch vertices** (or nodes). The **level** of any vertex is the length of its path from the root.

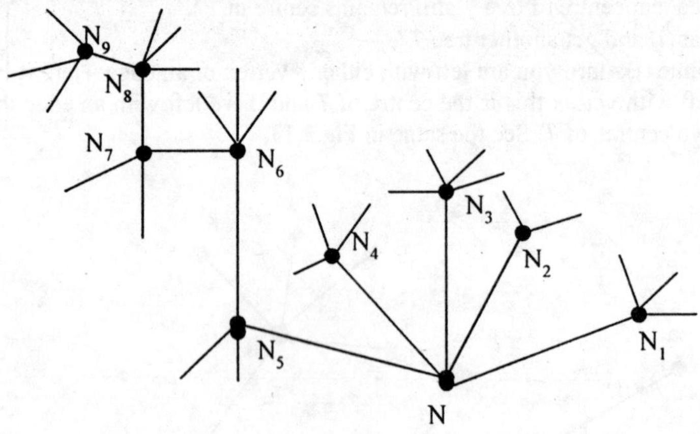

Fig. 3.14

Example 3.8

A directed tree $T=(V, E)$ is shown in Fig. 3.15, where $V=\{a, b, c, d, e, f, g, h\}$ and $E=\{(a, b), (a, c), (a, d), (b, e), (d, f), (f, g), (f, h)\}$. The root of T is the vertex a and the vertices at level 2 are e and f. It is to be noted that the directed trees are conventionally drawn with the root at the top and all edges going from the top of the page towards the bottom, so the direction of the edges sometimes not explicitly shown.

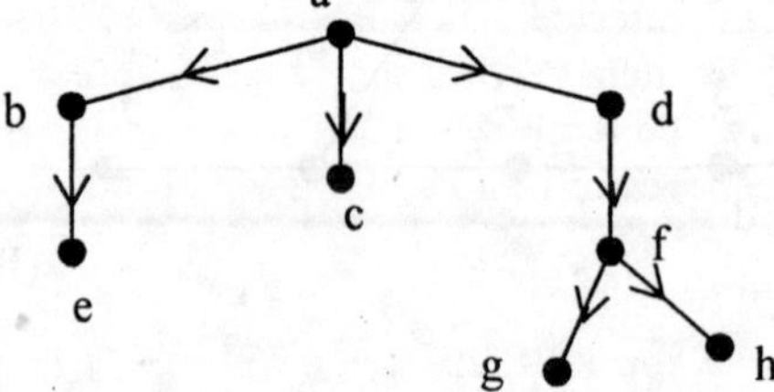

Fig. 3.15 A directed tree

3.9 DIRECTED FOREST

A **directed forest** is a collection of directed trees. The **height of a vertex** v in a directed forest is the length of the longest directed path from v to a leaf. The level of a vertex v in a forest is the length of the path v from the root of the tree to which it belongs. A directed tree T is said to have degree m if m is the maximum of the out degrees of all the vertices in T.

Example 3.9

The forest shown in Fig. 3.16 consists of three trees, T_1, T_2 and T_3.

$T_1 = (V_1, E_1)$, where $V_1=\{A, B, C\}$, $E_1= \{(A, B), (A, C)\}$

$T_2 = (V_2, E_2)$, where $V_2 =\{D, \phi\}$ and

$T_3 = (V_3, E_3)$, where $V_3 = \{E, F, G, H, I, J, K\}$,

$E_3= \{(E, F), (E, G), (G, H), (H, I), (H, J), (H, K)\}$.

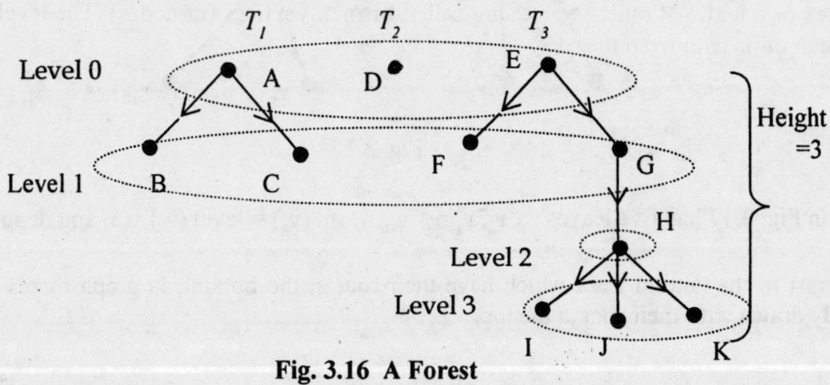

Fig. 3.16 A Forest

THEOREM 3.10

There are between 0 and m^n vertices at level n in any directed tree of degree m.

THEOREM 3.11

There are between $h+m$ and $(m^{h+1})/(m-1)$ vertices in a directed tree of height h and degree m.

3.10 ROOTED AND BINARY TREES

A **rooted tree** is a tree in which there is one designated vertex called the **root**. The **level** of a vertex v in a rooted tree is the length of the path to v from the root and the maximum vertex level is called the **depth** or **height** of the tree. Those vertices with degree 1, other than root r are called **leaves** or **terminal vertices** of T and the directed path from a vertex to a leaf is called a **branch.** The vertices other than leaves are called **branch vertices** or **nodes**. Note that in a tree any vertex may be designated as a root. Generally, the term tree means trees without any root. However, for emphasis they are sometimes called **free trees** or **non-rooted trees** to differentiate them from the rooted kind. Tree shown in Fig.3.17 is a rooted tree with root r and 10 other vertices.

In a rooted tree T with root $r = v_0$ if $r = v_0, v_1, v_2, \ldots, v_{i-1}, v_i, v_{i+1}, \ldots, v_{n-1}, v_n$, then

(i) v_{i-1} is the **parent** of v_i for $i = 1, 2, 3, \ldots, n-1$.

(ii) $v_1, v_2, \ldots, v_{i-1}, v_i, v_{i+1}, \ldots, v_{n-1}$ are **ancestors** of v_n.

(iii) v_i is a **child** of v_{i-1} for $i = 1, 2, 3, \ldots, n-1$. For example v_2 is a child of v_3.

(iv) v_i is a **descendant** of v_{i-1}. For example v_3 is a descendant of v_2.

(v) If v_i has no child then v_i is a **terminal vertex.**

(vi) If a vertex v_i has one or more children then it is called an **internal vertex.**

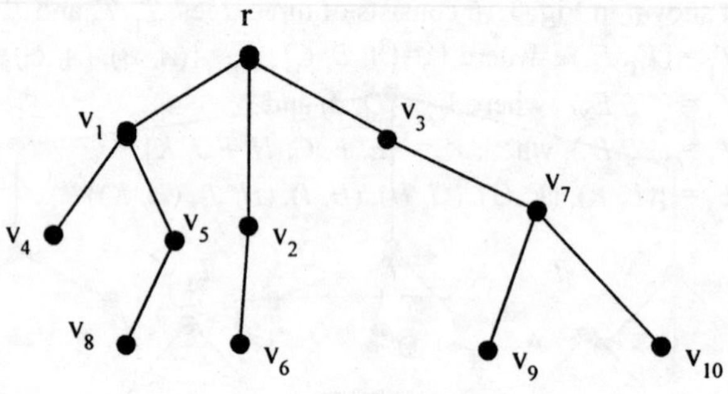

Fig. 3.17

The tree shown in Fig. 3.17 has five leaves v_4, v_6, v_8, v_9, v_{10}, level (v_8) = level $(v_9) = 3$, and depth of the tree is 3.

In contrast to the natural trees which have their root at the bottom, in graph theory rooted trees are typically drawn with their root at the top.

Example 3.10

Consider the tree in Fig. 3.18. Let vertex a has been designated as the root of the tree.

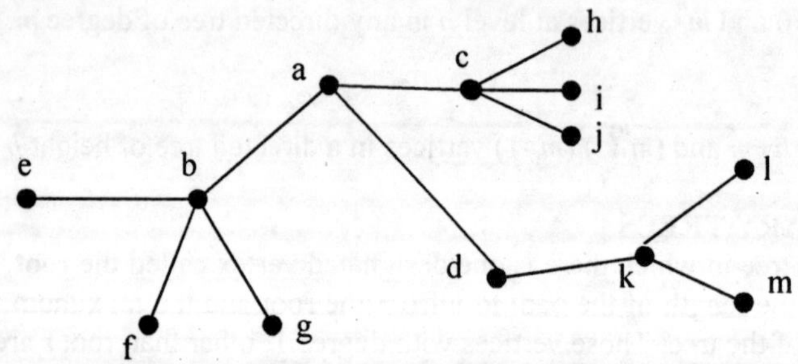

Fig. 3.18

It is to be noted here that if we choose any vertex other then root, say m in the tree, then the path from a to m will be unique. The vertices on this path (a, d, k, m) are described in genealogical terms as follows:

(i) m is a child (son or daughter) of k (so is l).

(ii) k is m's parent (father).

(iii) a, d and k are m's ancestors.

(iv) d, k and m are descendants of a.

As from the Fig.3.18 one can see that genealogical relationships are often easier to visualize if we rewrite the tree in a way so that children are positioned below their parents.

From the Fig.3.19 one can think each vertex as the root of a tree that contains in addition to itself, all of its descendants, e.g. c is the root of a tree that contains h, i and j. And finally h, i and j are roots of the tree that contain only themselves.

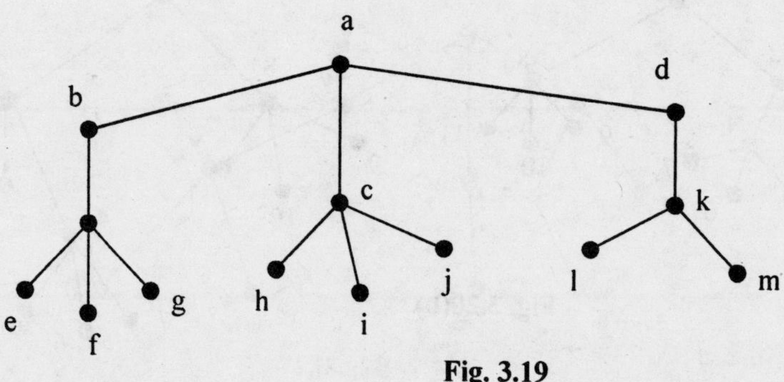

Fig. 3.19

Example 3.11

Graph shown in Fig. 3.14 is a rooted tree, with *N* as the root, which is an example of a **decision tree.**

3.11 SUBTREE

Let *a* be a branch node in a tree *T*. Then a **subtree** with *v* as the root is a subgraph say *T* '= (*V*', *E*') of *T* such that *V* ' contains *v* and all its descendants, and *E* ' contains the edges in all paths emanating from *v*. When we say a subtree of *v* we mean a subtree that has a son of *v* as root. For example graph shown in Fig. 3.20(a) is a subtree of graph shown in Fig. 3.19.

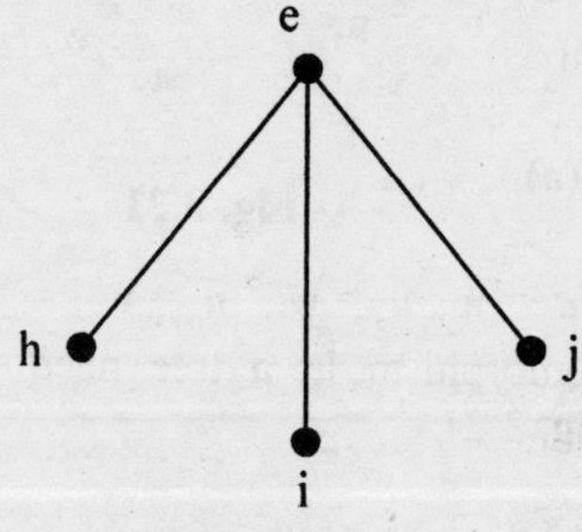

Fig. 3.20(a)

3.12 ORDERED ROOTED TREE AND ISOMORPHIC TREE

An **ordered rooted tree** is a rooted tree whose subtrees are put into a definite order and are themselves ordered trees. An empty tree (a tree with no vertices) and a single vertex are ordered rooted trees.

The trees shown in the Fig.3.20 (b) represents the same tree but they are of different order.

In an ordered tree the edges incident from each branch node labelled with integers 1, 2, 3, . . ., *n*. As a result of this the subtrees of a branch node can be referred to as the 1st, 2nd, 3rd,..... subtrees of the branch node corresponding to the labels of the edges incident from it. In an ordered tree it is not necessary that orders of tree are consecutive.

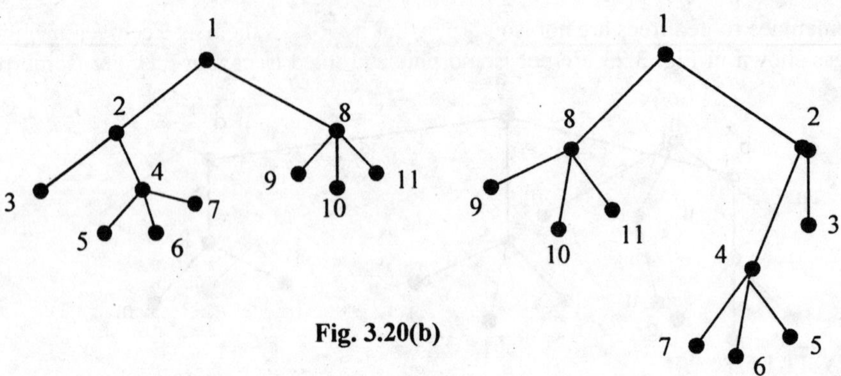

Fig. 3.20(b)

Two ordered trees are said to be **isomorphic** if there is one to one correspondence between their vertices and edges that preserves the incidence relation and the labels of the corresponding edges match.

Example 3.12

The rooted tree shown in Fig 3.21(a) and (b) are isomorphic.

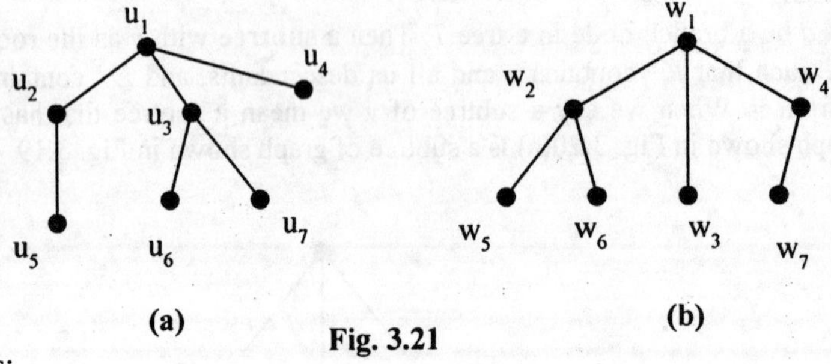

Fig. 3.21

Here the isomorphism is:

$u_1 = w_1,\ u_2 = w_4,\ u_4 = w_3,\ u_3 = w_2,\ u_5 = w_7,\ u_6 = w_5,\ u_7 = w_6$

One can observe that isomorphism in a tree need not unique. Can you find another isomorphism of the tree shown in Fig. 3.21.

Example 3.13

The rooted trees shown in Fig. 3.22 are not isomorphic since the root of the tree T_1 is 3 but that of T_2 is 2.

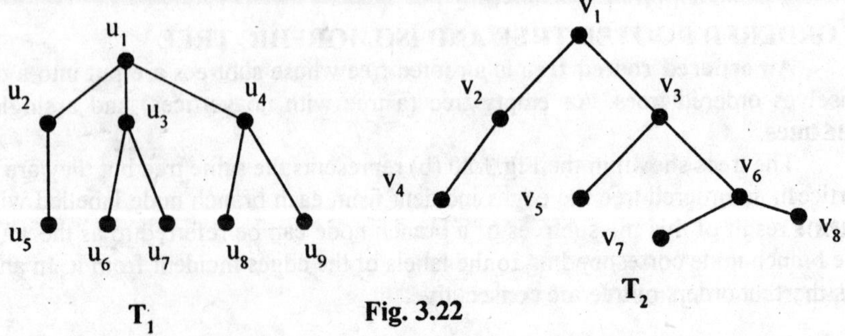

Fig. 3.22

Sometimes rooted trees are not isomorphic but if we take them as free tree, then they are. For example trees shown in Fig. 3.23 are not isomorphic as rooted trees, but they are isomorphic as free trees.

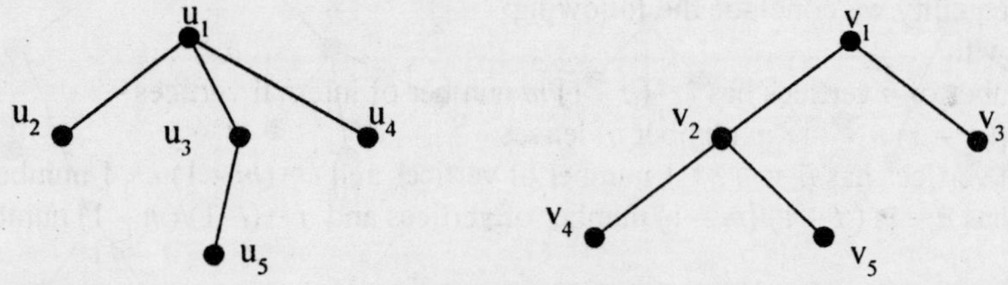

Fig. 3.23

3.13 *M*-ARY TREE

A tree is said to be an ***m*-ary** tree if every parent (internal vertex) has at most *m*-children. An *m*-ary is said to be **complete** or **full *m*-ary** tree if every parent has exactly m children.. If $m = 2$, the trees are called **binary**. An *m*-ary tree is said to be **regular** if each of its branch nodes has exactly m children.

THEOREM 3.11

Let $T = (V, E)$ be a rooted tree then:

(i) There is no cycle in T.

(ii) If r is the root of the tree then r is the only root of T.

(iii) Each vertex in T, other than r has indegree one.

PROOF

(i) Consider a cycle beginning and ending at a vertex r. We know from the definition of a tree that any vertex v cannot be equal to r and there must exist a path from r to v. But this implies that there is more than one path from r to v which contradicts the definition of a tree. Hence there are no cycles in tree T.

(ii) Let us assume that there exist a root r' of T other then root r, then there must exist a path p_l from r to r' and a path p_2 from r' to r since r' is a root.

$\Rightarrow$ There exist a cycle from r to r'. But this is not possible. Therefore the vertex r is unique.

(iii) Consider a vertex a in T other than r. Then there exist a unique path from r to w. We can conclude from here that a has indegree at least one. In case the indegree of a is more than one then there must be distinct vertices b and c such that the branch ba, ca are both in T. Let $a \neq r$, $c \neq r$, and there are paths p_2 from r to b and p_3 from r to c by definition of a tree.

$\Rightarrow$ There are two different paths from r to a. But this is a contradiction (by the definition of a tree with root r). Therefore, the indegree of a is one and r has indegree zero.

If possible, let the indegree of r is 1 and let branch $(a\ r)$ be in T. But there is a path from r to a by definition and a path from a to r. This implies that there exist a cycle in T which is a contradiction. Therefore, r has indegree zero.

THEOREM3.12

If T is a full *m*-ary tree with i internal vertices then there are $n = m.i + 1$ total number of vertices.

PROOF

Since a full *m*-ary tree contains exactly *m*-children and here given number of internal vertices is i, therefore total number of vertices excluding root is $m.i$. Hence the tree contains $m.i + 1$number of vertices including root.

REMARK

If number of leaves in a *m*-ary tree is l then number of vertices $n = i + l$. Using above theorem and this equality we conclude the following:

A full *m*-ary tree with

(a) total number of n vertices has $i = (n - 1)/m$ number of internal vertices and $l = [(m - 1)\, n + 1] / m$ number of leaves.

(b) i internal vertices has $n = m\, i + 1$ number of vertices and $l = (m - 1)\, i + 1$ number of leaves.

(c) l leaves has $n = m\, (l - 1)/(m - 1)$ number of vertices and $i = (l - 1)/(m - 1)$ number of internal vertices.

THEOREM 3.13

An *m*-ary tree of height h has at most m^h leaves.

PROOF

We use mathematical induction on height to prove the theorem.

For $h = 1$:

When height of *m*-ary tree is 1, then tree consists of a root with at most *m*- children, which are leaves. Therefore there are at most $m^1 = m$ leaves in an *m*-ary tree of height 1.

For height less than h:

Let us assume that the result is true or all *m*-ary tree of height less than h.

For height equal to h:

Now assume that T be an *m*-ary tree of height h. If we delete the edges from the root to each of the vertices of level 1 we obtain a subtree. The leaves of T are the leaves of subtree so obtained. Each of this subtree has height less than or equal to $h - 1$. Hence by induction hypothesis each of these has at most m^{h-1} leaves. Since at most m such subtrees are possible with a maximum of m^{h-1} leaves. Hence there are at most $m.m^{h-1} = m^h$ leaves.

3.14 BINARY TREE

A **binary tree** is a rooted tree in which every vertex has at most two children, and each child of a vertex is designated as its **left child** or **right child**. The subtrees rooted at the children of the root are the **left subtree** and the **right subtree** of the tree. In a more formal way, we can say a binary tree is a rooted tree in which each vertex has either one child or two children.

A binary tree T is frequently presented by a diagram in the plane called a **picture** of T.

The difference between binary trees and ordered trees is that every vertex of a binary tree has exactly two subtrees (one or both of which may be empty), while a vertex of an ordered tree may have any number of subtrees. The Fig.3.24 represents a binary tree.

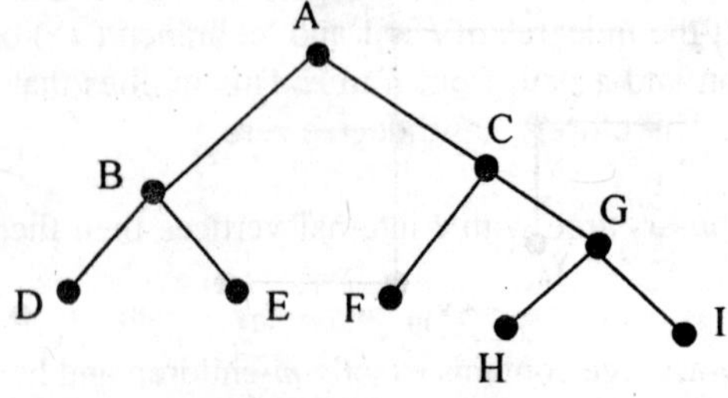

Fig. 3.24

(i) Tree T consists of 9 vertices.

(ii) The root of T is the vertex A at the top of the graph.

(iii) A left downward slanted line at a vertex, say v indicates a left successor of v and a right downward slanted line v indicates a right successor of v.

Further in Fig.3.24, we find that

(a) B is a left successor (child) and C in a right successor (right child) of the root A.

(b) the left subtree of the root A consists of the vertices B, D and E, and the right subtree of A consists of the nodes C, F, G, H, I.

(c) the vertices A, B, C and G have two successors and the vertices D, E, F, H, I are terminal vertices.

Example 3.14

Four distinct binary trees are shown in the Fig.3.25.

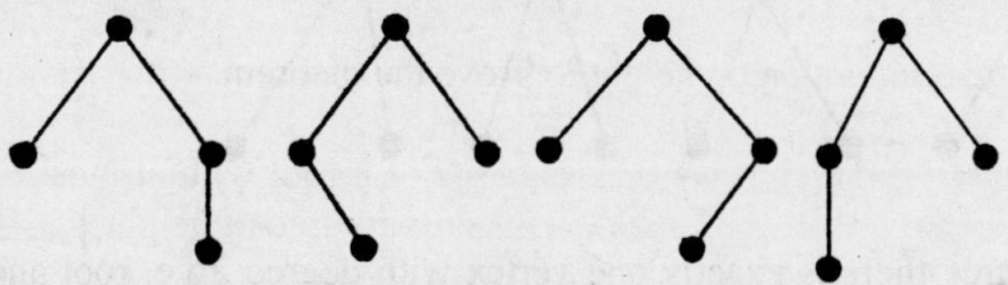

Fig. 3.25

Example 3.15

Obtain binary tree corresponding to the tree given below.

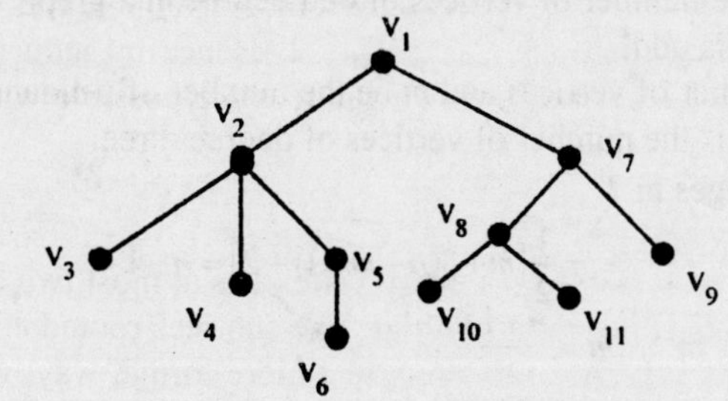

Fig. 3.26

Solution

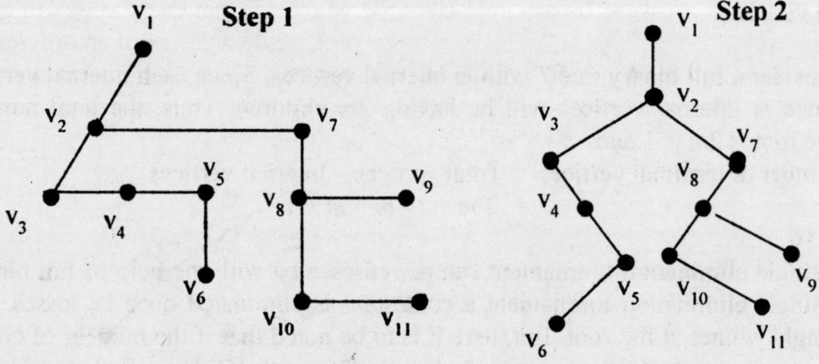

Fig. 3.27

3.14-1 FULL BINARY TREE

A tree in which every vertex has either zero or two subtrees is said to be a **full binary tree.** In other words, each vertex has either two or zero children. Tree shown in following figure is a full binary tree.

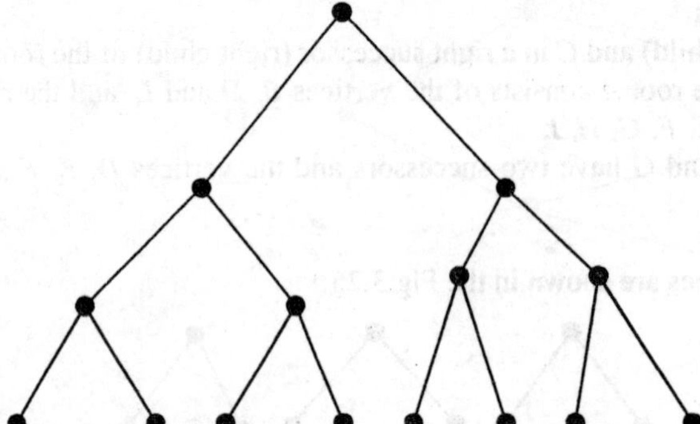

NOTE

(i) In a full binary tree there is exactly one vertex with degree 2 i.e. root and all other vertices have degree 1 or 3.

(ii) A full binary tree will always have odd number of vertices, say n.
Since there is exactly one vertex (root) of even degree, and the remaining $n - 1$ vertices are of odd degrees. But the number of vertices of odd degree in a graph is always even, therefore $n - 1$ is even. Hence n is odd.

(iii) Let n be the total number of vertices and m be the number of pendant vertices in a full binary tree T. Then $n - m - 1$ is the number of vertices of degree three.
$\therefore$ The number of edges in T

$$= \frac{1}{2}[m + 3(n - m - 1) + 2] = n - 1$$

$$\Rightarrow \quad m = \frac{n+1}{2}$$

THEOREM 3.14

If T is a full binary tree with m internal vertices, then there are $m + 1$ terminal vertices and $2m + 1$ total vertices.

PROOF

Consider a full binary tree T with m internal vertices. Since each internal vertex contains two children hence m internal vertices will be having $2m$ children. Thus, the total number of vertices including the root is $2m + 1$ and

Number of terminal vertices $=$ Total vertices – Internal vertices

$$= 2m + 1 - m = m + 1.$$

Example 3.16

A single elimination tournament can be represented with the help of full binary tree. As we know in a single elimination tournament a contestant is eliminated once he losses. And eventually there is a single winner at the root. But, here it is to be noted that if the number of contestant is not a power of 2, some contestants receive byes. In Fig. 3.28 consistent 7 has a first round bye.

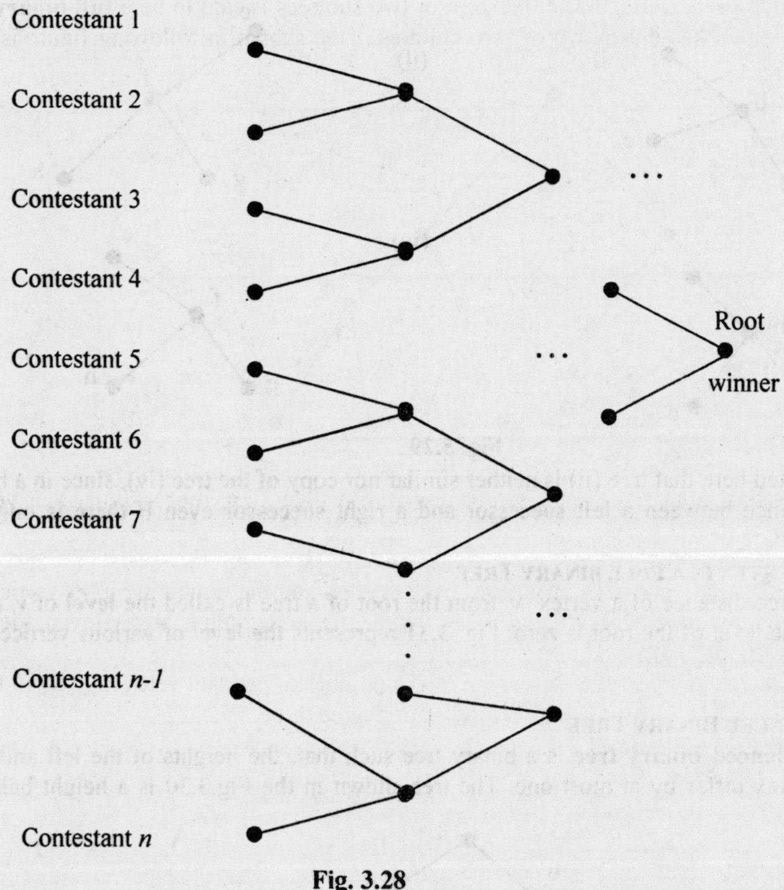

Fig. 3.28

We can show that if there are n contestant in single elimination tournament, then there are $(n-1)$ matches to be played.

Here the number of contestants is same as number of terminal vertices. And the number of matches m to be played is same as number of internal vertices. Then from the above theorem

$$n + m = 2m + 1$$

or

$$m = n - 1.$$

3.14-2 SIMILAR BINARY TREES

Two binary trees T and T' are said to be **similar** if they have the same structure or shape. The trees are said to be **copies** if they are similar and have the same contents as corresponding vertices.

Example 3.17

Four binary trees are shown in Fig.3.29. The three trees (i), (iii) and (iv) are similar and the trees (i) and (iii) are the copies as they also have same data at corresponding vertices.

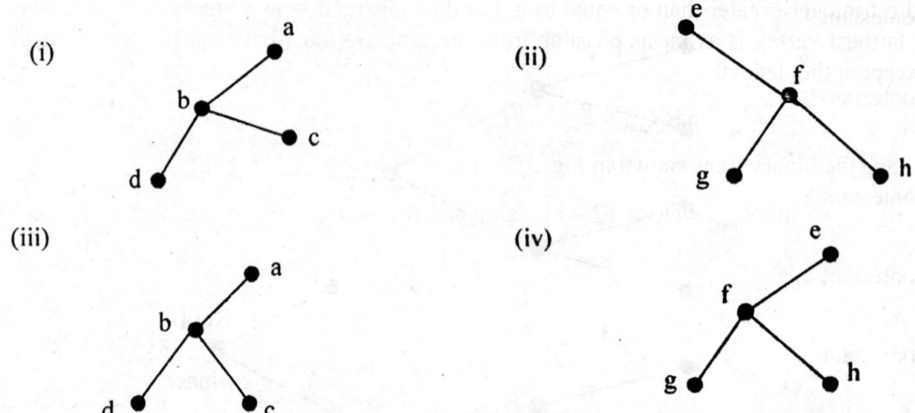

Fig. 3.29

It is to be noted here that tree (ii) is neither similar nor copy of the tree (iv), since in a binary tree there is a difference between a left successor and a right successor even if there is only one successor.

3.14-3 Level of a vertex in a full binary Tree

In a binary tree distance of a vertex v_i from the root of a tree is called the **level of** v_i and is denoted by l_i .Thus the level of the root is zero. Fig. 3.31 represents the level of various vertices of a rooted tree.

3.14-4 Height Balanced Binary Tree

A **height balanced binary tree** is a binary tree such that. the heights of the left and right subtrees of every vertex differ by at most one. The tree shown in the Fig.3.30 is a height balanced binary tree.

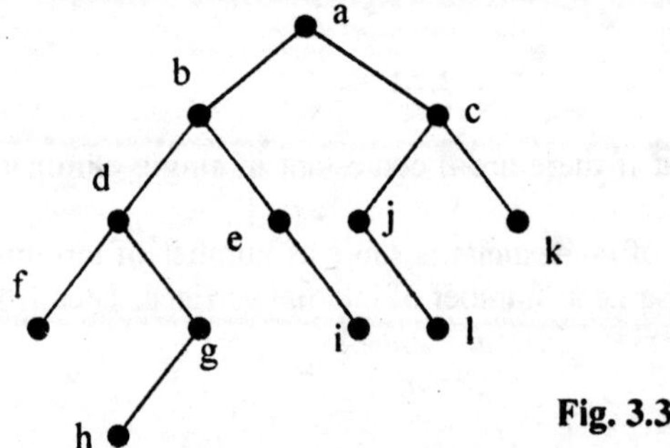

Fig. 3.30

Example 3.18

Consider a binary Tree $T = (V, E)$ with n number of vertices. As there will be one root at level 0, at most two vertices at level 1, at most four vertices at level 2, and so on.

$\therefore$ The maximum number of vertices possible in a m-level binary-tree is

$$2^0 + 2^1 + 2^2 + \ldots + 2^m \geq n.$$

The minimum possible height of an n-vertex binary tree is

$$\min l_{max} = \lceil \log_2 (n+1) - 1 \rceil.$$

where l_{max} is the maximum level of any vertex in a binary tree, i.e. height of the tree, and $\lceil n \rceil$ denotes the smallest integer greater than or equal to n. For the construction of a binary tree for a given n such that the farthest vertex is as far as possible from the root, we must have exactly two vertices at each level, except at the level 0.

$$\max l_{\max} = \frac{n-1}{2}$$

See the binary trees shown in Fig. 3.28. For $n = 11$.

$$\min l_{\max} = \lceil \log_2 12 - 1 \rceil \quad \text{and} \quad \max l_{\max} = \frac{11-1}{2} = 5.$$

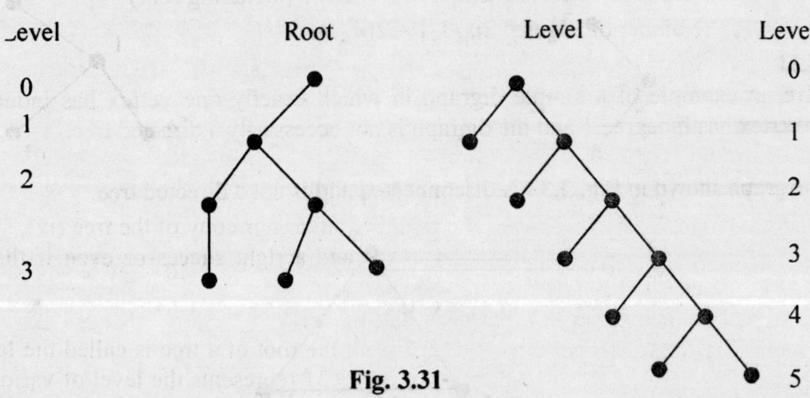

Fig. 3.31

Example 3.19

In the three trees T_1, T_2 and T_3 shown in Fig.3.32, identify those that represent the same:

(i) rooted tree
(ii) ordered rooted tree
(iii) binary tree.

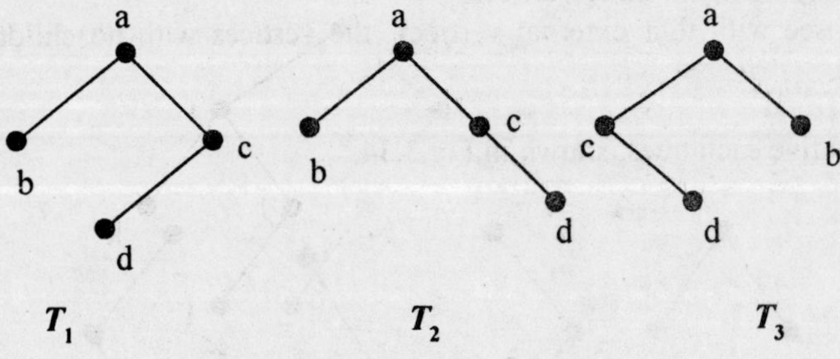

Fig. 3.32

Solution

(i) As all the three trees have same root a, they represent same rooted tree. Root a is the root with children b and c, and c has the single child d.

(ii) T_1 and T_2 are same ordered rooted tree but T_3 is different. As b is the first child of a in T_1 and T_2 but it is second child of a in T_3.

(iii) All three trees, T_1, T_2 and T_3 represent different binary trees. Specifically T_1 and T_2 are different since we distinguish between left and right successors even if there is only one successor (which are not true for ordered rooted trees). d is the left successor of c in T_1 but a right successor of c in T_2.

Example 3.20

Show that for n_l terminal vertices in a complete binary tree the total number of edges is given by $2(n_i - 1)$.

Solution

Given number of terminal vertices = n_i

$\therefore$ Number of internal vertices = n_i -1

Total number of vertices = $2(n_i-1) + 1 = 2n_i - 1$ (including root)

Number of edges = $2n_i - 1 - 1 = 2(n_i - 1)$.

Example 3.21

Give an example of a simple digraph in which exactly one vertex has indegree zero and every other vertex has indegree 1 and the digraph is not necessarily a directed tree.

Solution

The graph shown in Fig. 3.33 is disconnected and is not a directed tree.

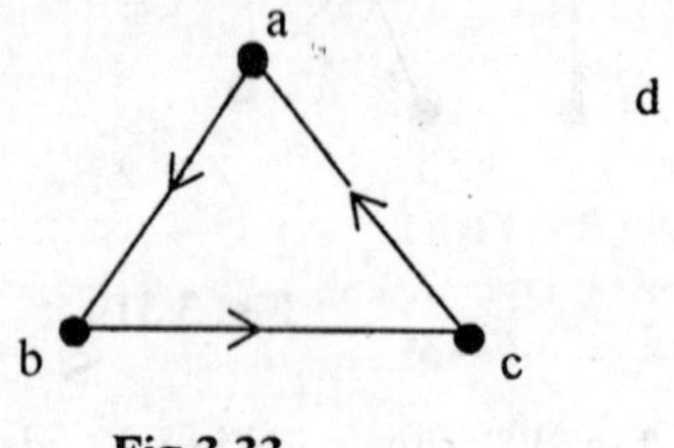

Fig.3.33

Example 3.22

Draw all possible non similar trees T where:

(i) T is a binary tree with three vertices.

(ii) T is a 2-tree with four external vertices (the vertices with no children are called **external vertices**).

Solution

There are five such trees, shown in Fig.3.34.

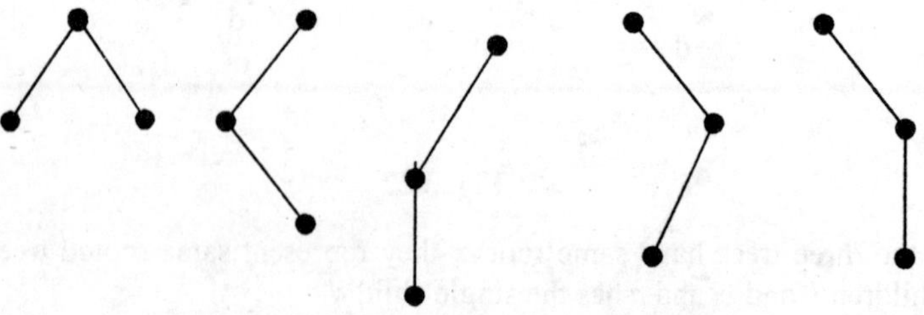

Fig. 3.34

(ii) Each 2-tree with four external vertices is determined by a binary tree with three vertices. Thus there are five such trees shown in Fig.3.35.

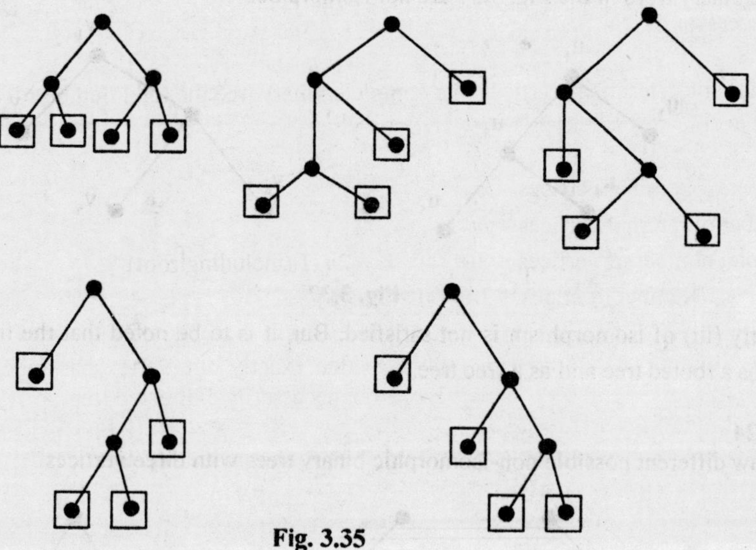

Fig. 3.35

3.15 ISOMORPHISM IN A BINARY TREE

Consider two binary trees T_1 and T_2 with their respective roots r_1 and r_2. The binary tree T_1 and T_2 are said to be **isomorphic** if there exists one-to-one, on to function f from the vertex set of T_1 to the vertex set of T_2 and satisfies the following:

(i) $f(r_1) = r_2$

(ii) If vertex v_i and v_j are adjacent in T_1 then $f(v_i)$ and $f(v_j)$ must be adjacent in T_2.

(iii) If v is a left child (right child) of u in T_1 then $f(v)$ must be left child (right child) of $f(u)$ in T_2.

The function f is called **isomorphism**.

Example 3.23

(i) The binary trees shown in the Fig.3.36 are isomorphic. The isomorphism is $f(u_i) = v_i$, where $i = 1, \ldots 5$.

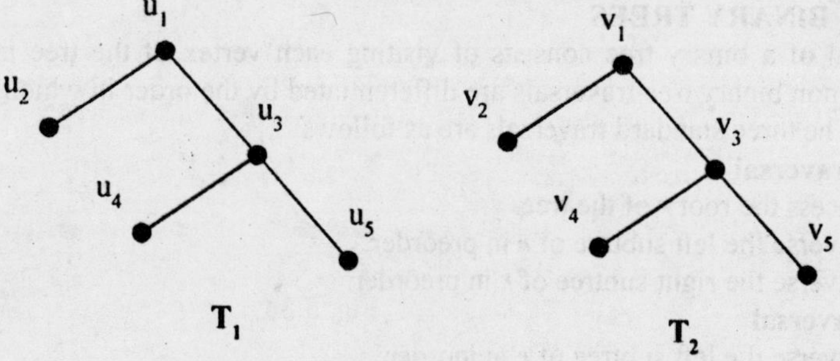

Fig. 3.36

(ii) The binary trees in the Fig. 3.37 are not isomorphic.

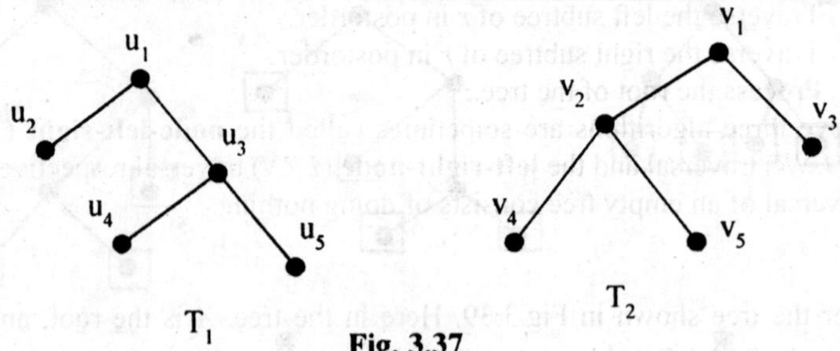

Fig. 3.37

Here, property (iii) of isomorphism is not satisfied. But, it is to be noted that the trees T_1 and T_2 are isomorphic as a rooted tree and as a free tree.

Example 3.24

Draw different possible non-isomorphic binary trees with three vertices.

Solution

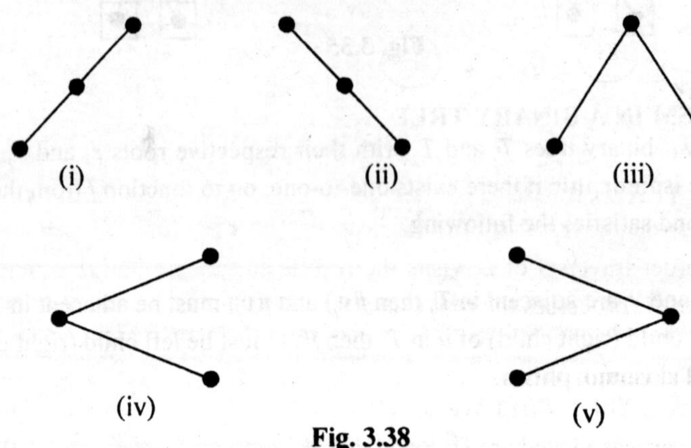

Fig. 3.38

There are five nonisomorphic binary trees with three vertices.

3.16 TRAVERSING BINARY TREES

The traversal of a binary tree consists of visiting each vertex of the tree in some defined order. The most common binary tree traversals are differentiated by the order in which the root and its subtrees are visited. The three standard traversals are as follows:

(i) **Preorder Traversal**

(a) Process the root r of the tree.

(b) Traverse the left subtree of r in preorder.

(c) Traverse the right subtree of r in preorder.

(ii) **Inorder Traversal**

(a) Traverse the left subtree of r in inorder.

(b) Visit the root r of the tree.

(c) Traverse the right subtree of r in order.

(iii) Postorder Traversal

(a) Traverse the left subtree of *r* in postorder.

(b) Traverse the right subtree of *r* in postorder.

(c) Process the root of the tree.

The above three algorithms are sometimes called the **node-left-right** (*NLR*) traversals, the **left-node-right** (*LNR*) traversal and the **left-right-node** (*LRN*) traversal respectively.

Any traversal of an empty tree consists of doing nothing.

Example 3.25

Consider the tree shown in Fig.3.39. Here in the tree, *A* is the root, and its left subtree L_T consists of vertices *B, D* and *E* and its right subtree R_T consists of vertices *C* and *F*.

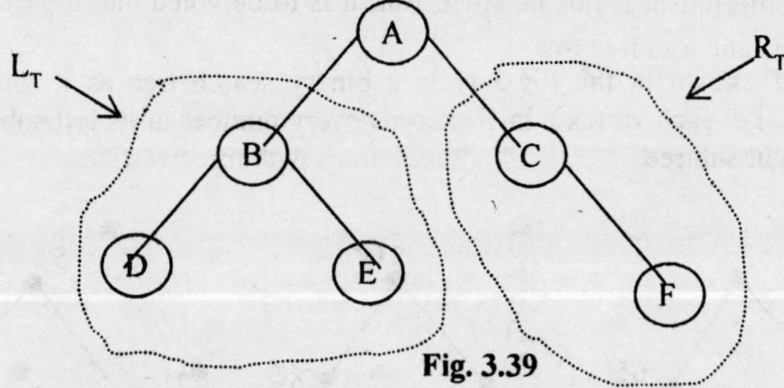

Fig. 3.39

(i) Preorder Traversal

In the preorder traversal of *T*, first process root *A*, then traverse L_T and then traverse R_T. However, the preorder traversal of L_T visits the root B and then *D*, and *E*, and in a similar way the preorder traversal of R_T processes the root *C* and then *F*. Hence *A-B-D-E-C-F* is the preorder traversal of *T*.

(ii) Inorder Traversal

In this traversal of *T*, first traverses L_T, visits *A* and then traverses R_T. Inorder traversal of L_T the order in which vertices visited are *D, B* and then *E*, and inorder traversal of R_T visits C and then *F*. Hence *D-B-E-A-C-F* is the inorder traversal of *T*.

(iii) Postorder Traversal

In this traversal of *T*, first traverses L_T then traverses R_T, and then visits *A*. *D-E-B-F-C-A* is the postorder traversal of *T*.

Example 3.26

For the tree in Fig.3.40, the different orders in which the nodes processed are

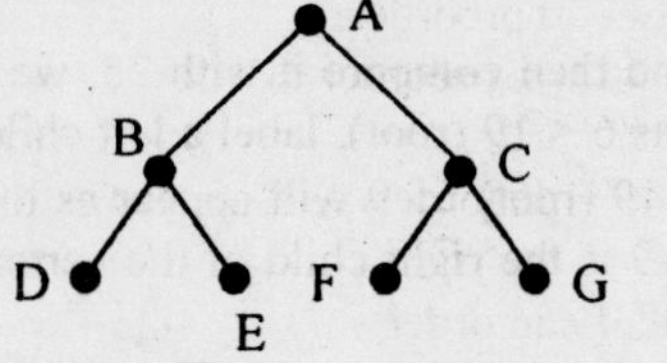

Fig. 3.40

(i) Preorder Traversal ABDECFG
(ii) Inorder Traversal DBEAFCG
(iii) Post order Traversal DEBFGCA

3.17 BINARY SEARCH TREES

A **binary search tree** is a binary tree T in which each vertex v of T satisfies the following properties:

(i) Each data item in the left subtree of v is less than the data item in v.

(ii) Each data item in he right subtree of v is greater than the data item in v.

One can observe that the above property guarantees that the inorder traversals of T will give a sorted listing of the elements of T. See Fig. 3.41.

Example 3.27

The binary tree T shown in the Fig.3.41 is a binary search tree as it satisfies both the conditions of above article, i.e. each vertex v in T exceeds every number in its left subtree and is less than every number in its right subtree.

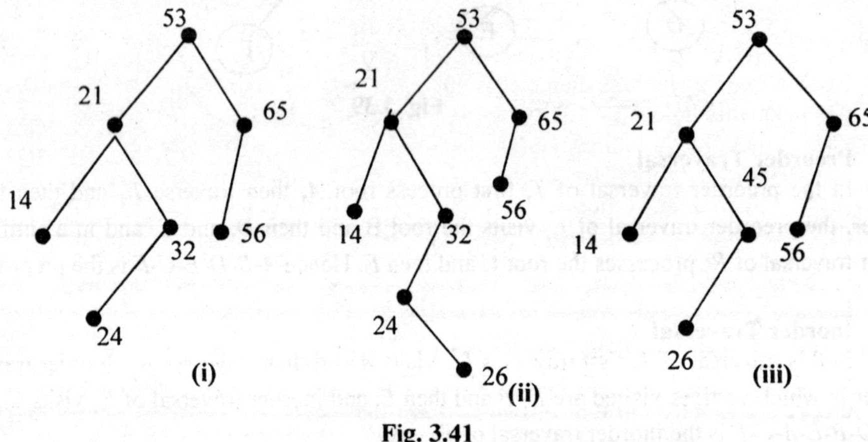

Fig. 3.41

Had we replaced 32 by 90 in (i) then T would not be binary search tree because 90 would be in the left subtree of 53 but 90 is greater than 53.

Example 3.28

Given a sequence of numbers 19, 25, 6, 9, 11, 21, 47, 8, 4, 39, 77. Build a binary search tree for the sequence.

Solution

First of all label the root with 19 and then compare it with 25, we see that 25 > 19, make a right child for the root and label it 25. Next as 6 < 19 (root), label a left child of the root with label 6. Continue this and go to next number 9. 9 < 19 (root) ⇒ 9 will appear as the label of some vertex in the left subtree of the root, but 9 > 6, hence 9 is the right child of the vertex. Continuing this we get the binary search tree as shown in the Fig.3.42.

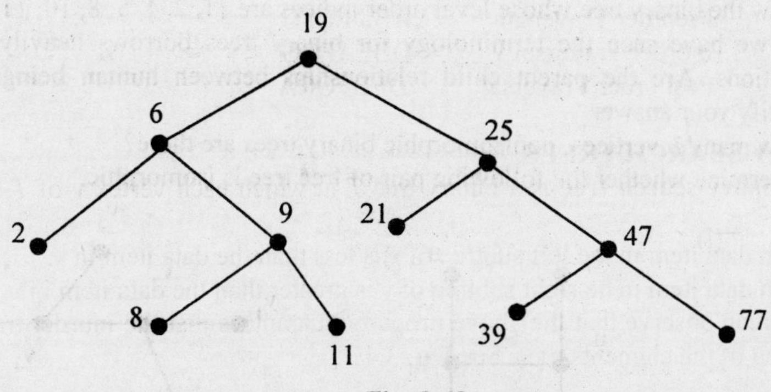

Fig. 3.42

NOTE

One of the good features of binary search tree is that additional numbers appended to the original sequence can be incorporated into the binary search tree just as if these had been in the sequence all along.

EXERCISE 3.2

1. Show a tree in which its diameter is not equal to twice the radius. Under what condition does this inequality hold?
2. Draw a binary tree with six vertices and only one leaf.
3. Draw all distinct binary trees
   (a) with three vertices
   (b) with four vertices.
4. Prove that the maximum number of vertices at level l of a binary tree is 2^l and that a tree with that many vertices at level l must have at least $2^{l+1}-1$ vertices.

   (Hint Prove it by induction on l, for $l = 0$, $n > 0$ $\exists$ one vertex with index $2^0 = 2^{0+1} - 1 = 1$. If $n < 2^l$, theorem holds vacuously. Otherwise invoke the definition of level. The vertices in level l are exactly those at distance l from the root and so in level l-1 are exactly those at distance l - 1 from the root. Proceed further to get the result.
5. How many different directed trees are there \ th three vertices? How many ordered trees are there with three vertices?
6. (i) Find the level of each vertex of the tree in Fig.3.43.
   (ii) What is the height, centre and diameter of this tree?

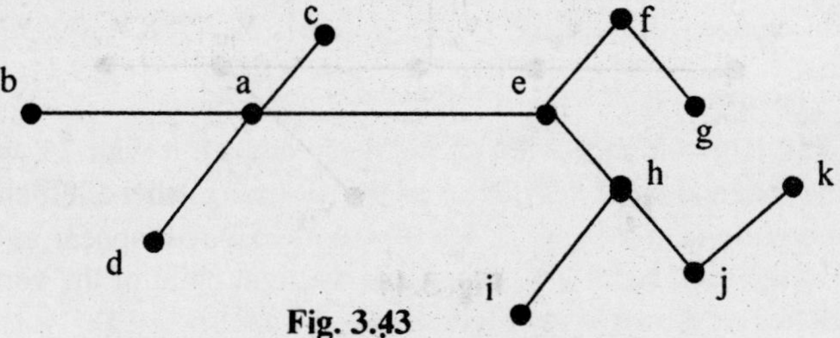

Fig. 3.43

7. Draw the binary tree whose level order indices are {1, 2, 4, 5, 8, 10, 11, 20}.
8. As we have seen the terminology for binary trees borrows heavily from human kinship relations. Are the parent child relationships between human beings really binary trees? Justify your answer.
9. How many *n* vertices, nonisomorphic binary trees are there?
10. Determine whether the following pair of free tree is isomorphic.

(i)

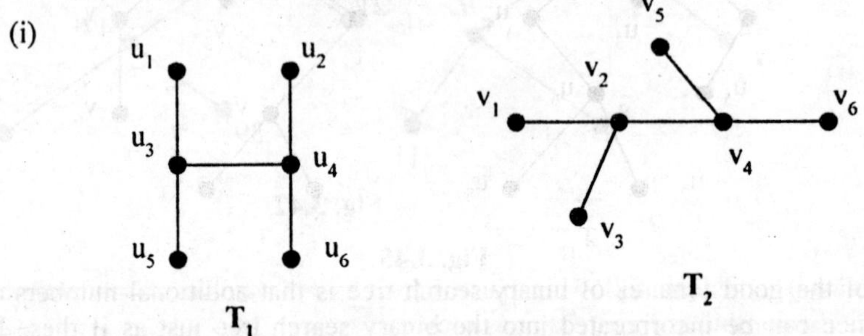

(ii)

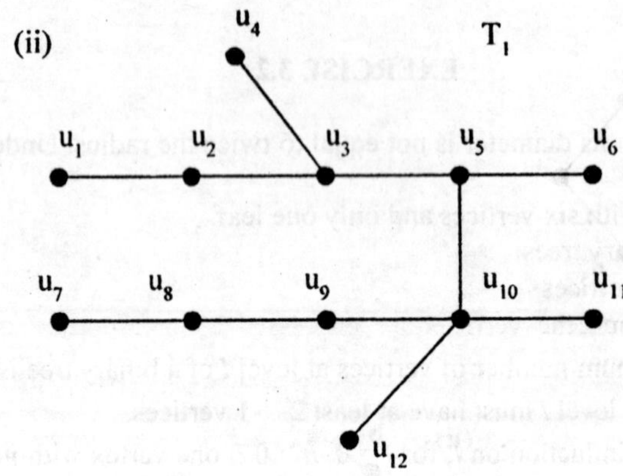

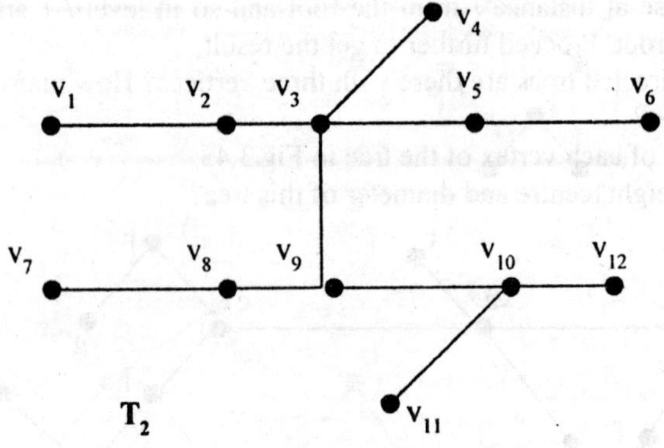

Fig. 3.44

11. Determine whether the following pair of rooted tree is isomorphic. If they are isomorphic, specify an isomorphism. If not give the invariant. Also, determine whether the trees are isomorphic as free trees.

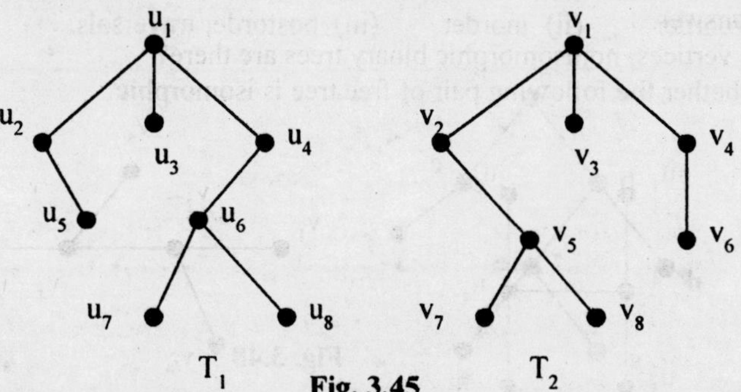

Fig. 3.45

12. Determine whether the following pair of trees is isomorphic. Also determine whether the trees are isomorphic as free trees or as rooted trees.

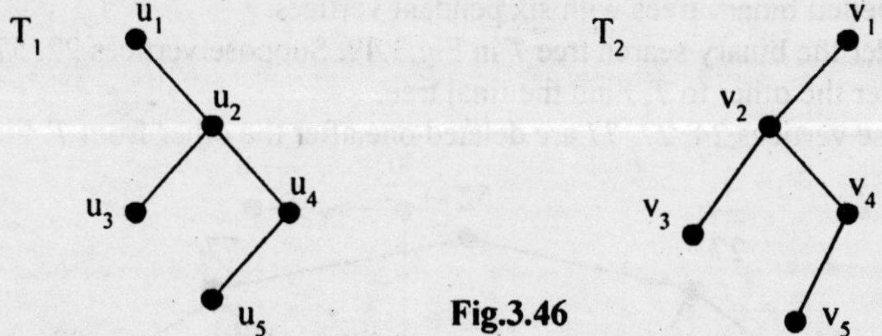

Fig.3.46

13. List the order in which the vertices are processed using preorder, inorder and postorder traversal for the following:

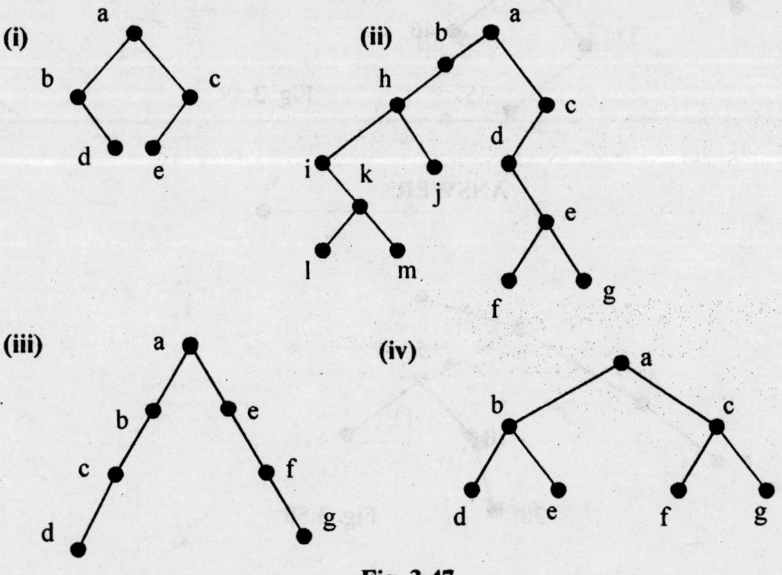

Fig. 3.47

14. Consider the binary tree T in the Fig.3.48.
 (a) Find (i) depth d (ii) the descendants of b
 (b) List the vertices of T in
 (i) preorder (ii) inorder (iii) postorder traversals.

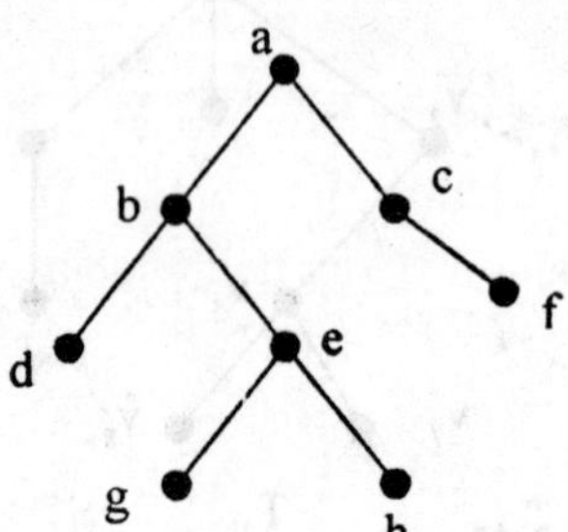

Fig. 3.48

15. Draw the final tree T if the following numbers are inserted into an empty binary search tree T: 52, 35, 46, 24, 79, 37, 62, 42.
16. Sketch all unlabelled binary trees with six pendant vertices.
17. (a) Consider the binary search tree T in Fig.3.49. Suppose vertices 22, 57, 90 are added one after the other to T. Find the final tree.
 (b) Suppose vertices 24, 27, 77 are deleted one after the other from T. Find the final tree T.

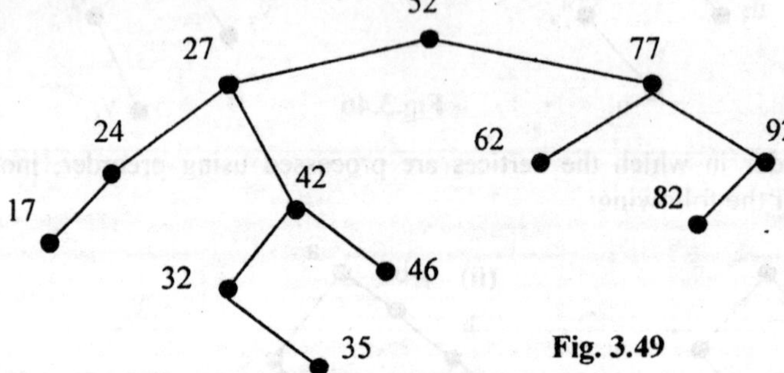

Fig. 3.49

ANSWERS

7.

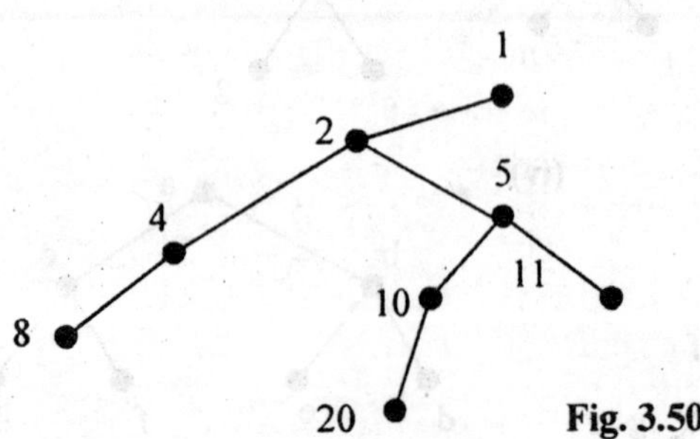

Fig. 3.50

8. First, they are not binary as a person can have more than two children. Second, they need not be trees, as a person may have a common ancestor on two sides of the family, i.e.

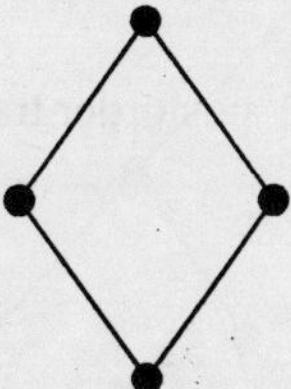

Fig. 3.51

But when viewed linguistically, human kinship terms do form binary trees, even in the above case, by ignoring cross breeding.

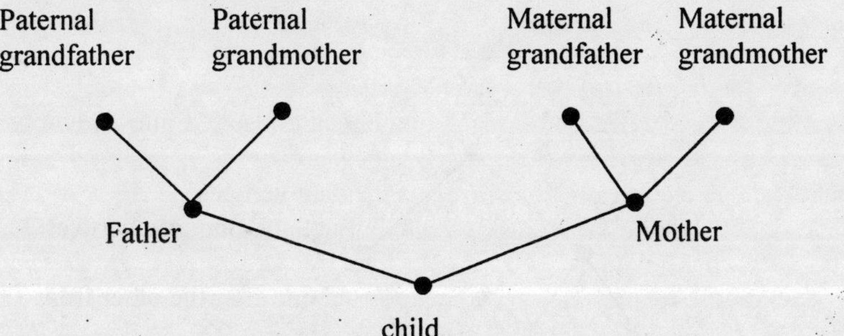

Fig. 3.52

9. $C(2n, n)/(n+1)$ 10(i). Isomorphic

11. Isomorphic as rooted tree. Isomorphism is $f(u_1) = v_1, f(u_2) = v_5, f(u_3) = v_3, f(u_4) = v_4, f(u_5) = v_2, f(u_6) = v_6$.

12. Not isomorphic as a binary tree. Because the root of T_2 has left child but root of T_1 does not. but they are isomorphic as rooted tree and free tree.

14. (a) (i) $d = 4$ (ii) d, e, g, h
(b) (i) $a\,b\,d\,e\,g\,h\,c\,f$ (ii) $d\,b\,g\,e\,h\,a\,c\,f$
(iii) $d\,g\,h\,e\,b\,f\,c\,a$.

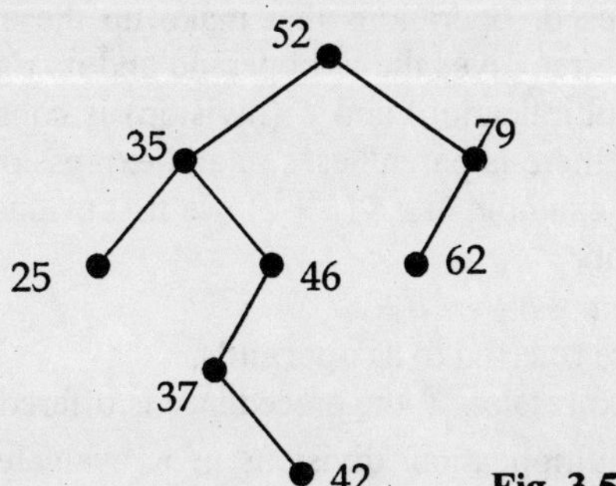

Fig. 3.53

16. **Hint** Distribute the 11 vertices (because $n = 6 + 5$) among different levels. Note that level 0 has exactly one vertex, level 1 has exactly two vertices, level 2 can have either two or four vertices; and so on.

17 (a)

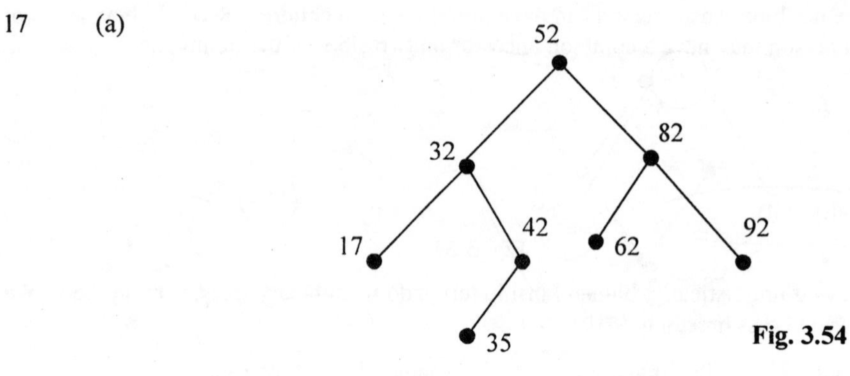

Fig. 3.54

(b)

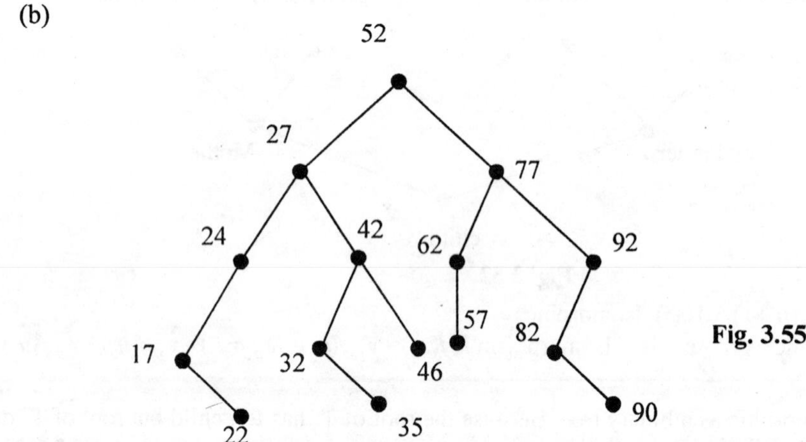

Fig. 3.55

3.18 EXPRESSION TREES

Algebraic expression can be represented by binary trees in which vertices of the trees are labeled with numbers, variables or operations that make up the expressions. For each vertex that contains an operator, its left subtree gives the left operand and its right subtree the right operand. The precedence of operators '*' (multiplication) and '/' (division) is considered to be of equal and higher value than that of '+' and '-'. If there is parentheses in and expression then the order of precedence is offered by the parentheses. For example in $a * (b + c)$, we first evaluate $(b + c)$ and then $a * (b + c)$.

Consider the expression

$$X = a * b - c / d + e \quad \text{... (1)}$$

The variables *a, b, c, d* and *e* are referred to as **operands**.

Now to evaluate the expression X, the precedence is offered to parentheses. According to our convention a precedence on multiplication/ divisions: X is evaluated as $((a * b) - (c/d)) + e$. Since multiplication/ divisions or addition / subtractions are evaluated from left to right, therefore, we can analyse by noting that it is sum of two simple expressions $(a * b) - (c/d)$ and e. The first of all these expressions can be decomposed down further into the difference of the expressions $a * b$ and c/d.

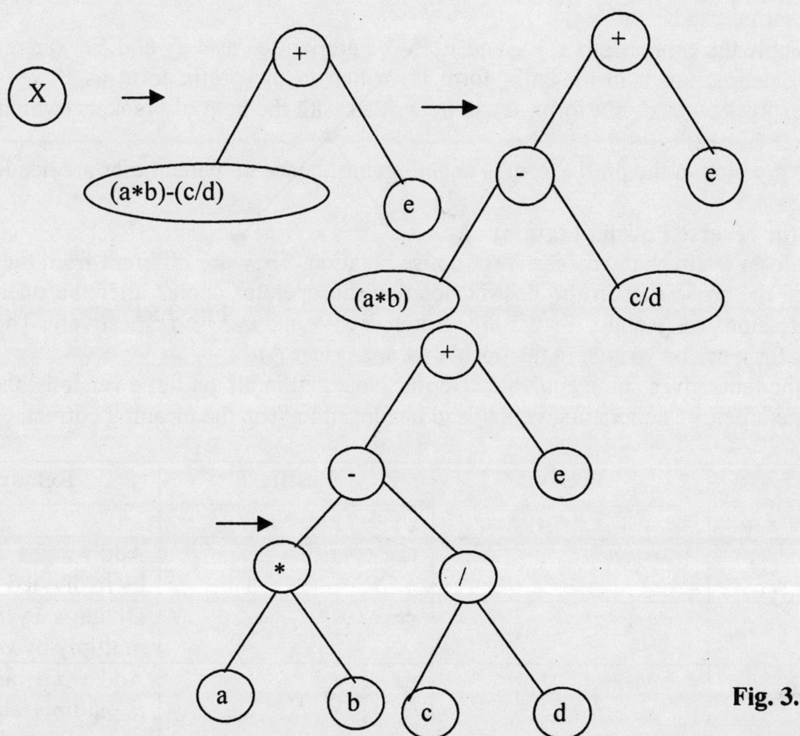

Fig. 3.56

When we decompose any expression into (left expression) (operation) (right expression), the expression tree of that expression is the binary tree whose root contains the operation and whose left and right subtrees are the trees of the left and right expressions, respectively. The evolution of the expression tree for the expression *X* is shown in the Fig.3.56.

A binary operation applied to a pair of numbers can be written in following three ways:
(i) Infix (ii) Prefix (iii) Postfix.

3.18-1 INFIX

In the **infix** form the operator is written in between operands. This is the usual way we write an expression. The in order traversals of an expression tree result in the infix form of an expression. Infix form gives the original expression with the element and operation in the same order as they originally occurred.

An expression of the form $x * (y + z) / u$ is usually taken to mean something like, first add y and z together then multiply the result by x then divide by u to get the final answer.

To make the infix forms of an expression unambiguous it is always advisable to include parenthesis in the in order traversal whenever we encounter an operation.

3.18-2 PREFIX (or Polish notation)

Prefix notation also known as polish notation, is a symbolic logic invented by polish mathematician Jan Lukasiwic in the 1920's. In the prefix notations as the name suggests, operator are written before their operands.

For example the expressions $x + y$ and $x + y * z$ are written as $+ xy$ and $* + xyz$ respectively in prefix form. Example taken in the infix form is written in the prefix form as $/* \ x + y \ zu$. To understand this expression in prefix form, it can be written with the help of brackets (which is totally unnecessary) as $(/* \ (x + yz) \ u)$.

Since expression in the prefix form is unambiguous, hence no parentheses are needed in such expressions.

3.18-3 POSTFIX (or reverse Polish notation)

Postfix form is also known as reverse polish notation. They are different from the infix and prefix notation in the sense that in the postfix notation the operator comes after the operands. For example the expressions $x + y$ and $x + y * z$ are written as $xy +$ and $xyz + *$ respectively. The example taken in the infix form can be written in the postfix form as $xyz + * \ u/$.

Following table gives some equivalent forms. Note that in all the three versions, the operand occurs in the same order, it is operators, which is to be moved to keep the meaning correct.

Infix	Prefix	Postfix	Remark
$x + y * z \ x * y + z / u$	$* + x \ y \ z + * xy / zu$	$x \ y + * z \ xy * zu / +$	
$x * (y + z) / u$	$/ * x \ yzu$	$yzu + * t /$	Add y and z, multiply by n, divvied by t
$x * (y + z / u)$	$* x + y / zt$	$xyz \ t / + *$	Divide z by t, add y, multiply by x
$(x + y) * (z + t)$	$* + xy + zt$	$xy + \text{zt} + x$	Add x *ady*, add z and t, multiply addition x and y by addition z and t.

The binary tree representation of the expression $((a + b) * c - (d/e)$ is shown in Fig.3.57.
The connection between traversals of an expression tree and these forms is simple:

The preorder traversal and postorder traversals of an expression tree will result in the prefix and postfix form of the expression respectively, however in order traversal of an operation tree will not in general yield the proper infix form of the expression.

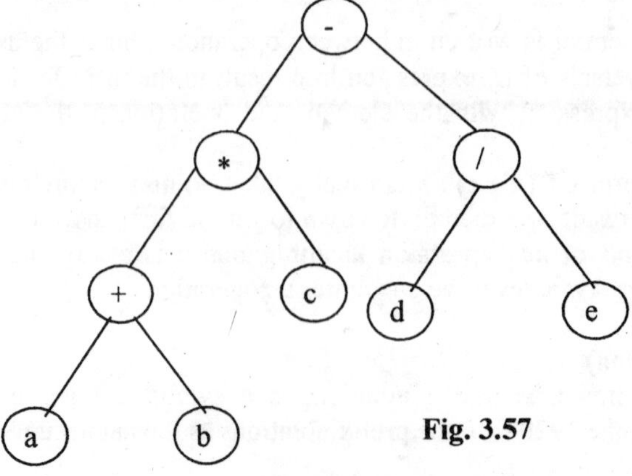

Fig. 3.57

Example 3.29

Draw expression tree for

(a) $a*(b-c)/(d+e)$ (b) $a^2 - b^2$

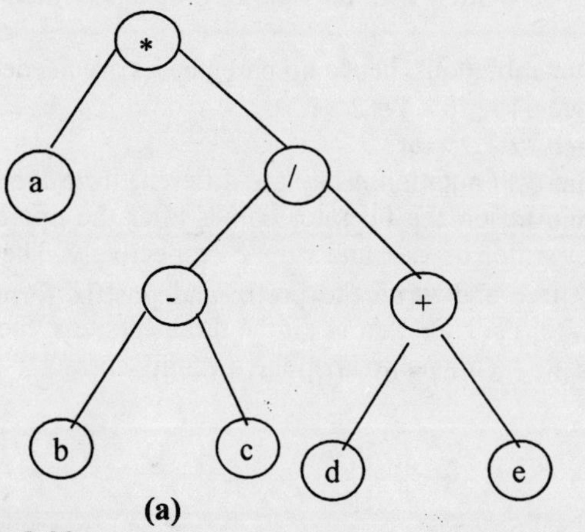

(a) Fig. 3.58 (b)

Example 3.30

List the order in which the vertices are processed using preorder, inorder, and postorder interval.

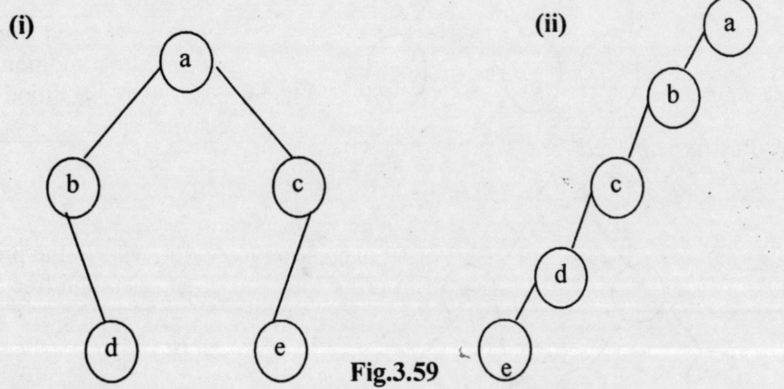

Fig.3.59

Solution

(i) Preorder *a b d c e*, Inorder *b d a e c,* Postorder *d b e c a*

(ii) Preorder *a b c d e,* Inorder e *d c b a,* Postorder *e d c b a*

Example 3.31

Find the value of

(a) Prefix expression * - 95 + 2 / 4.

(b) Postfix expression 22 * - 2182 / +.

Solution

(a) Let us discuss the evaluation step by step.

(i) * - 95 + 2 / 44

(ii) * 4 + 2 / 44 , since the first string is - 95 which is 9 - 5 = 4

(iii) * 4 + 21, since / 44 = 4 / 4 = 1
(iv) * 43, since + 21 = 2 + 1 = 3
(v) 12, since * 43 = 4 × 3 = 12.

(b) Following is the step by step evaluation:

(i) 622 * - 2 ↑ 82 / +
(ii) 64 - 2 ↑ 82 / + , since 22 * = 2 * 2 = 4
(iii) 22 ↑ 82 / + , since 64 - = 6 - 4 = 2
(iv) 482 / +, since 22 ↑ 2^2 = 4
(v) 44 + , since 82 / = 8 / 2 = 4
(vi) 8, since 44 + = 4 + 4 = 8.

Example 3.32

Represent the expression as a binary tree and write the prefix and postfix forms of the expression

(i) (a + b) ∗ (c-d) (ii) (((a + b) ∗ c + d) ∗ e) – ((a + b) ∗ c - d)

Solution

(i)

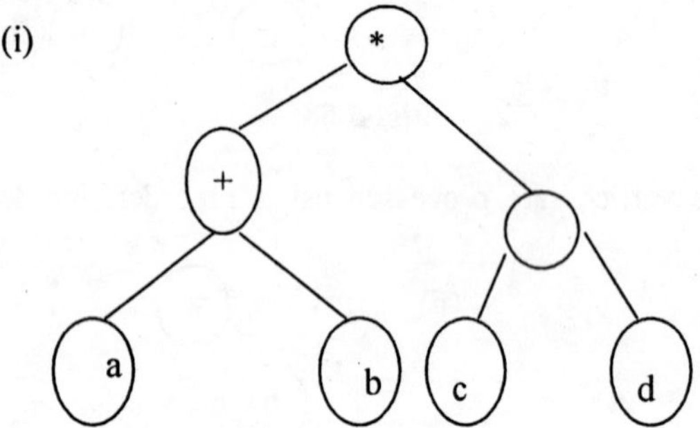

Fig.3.60

Prefix : ∗ + *ab* - *cd*, Postfix : *ab* + *cd* - ∗

(ii)

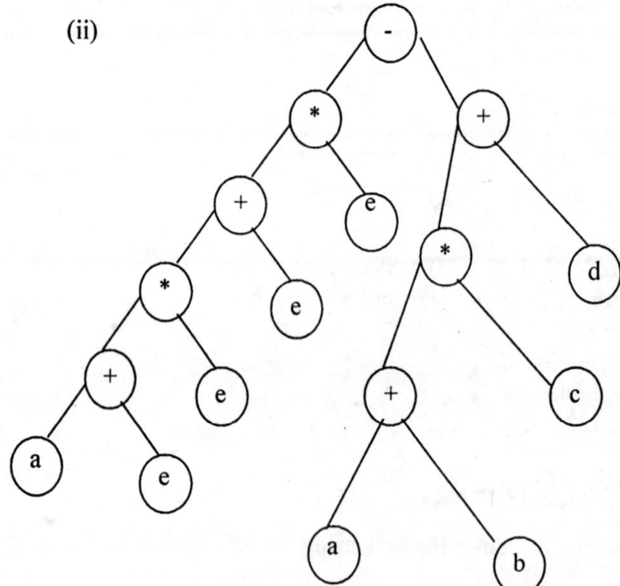

Fig. 3.61

Prefix : $- * + * + a\, b\, c\, d\, e + * + a\, b\, c\, d$

Postfix : $ab + c \;\; * d + e * ab + c \; *d + -$.

Example 3.33

Represent the postfix expression as a binary tree and write the prefix form, the infix form and the fully parenthesised infix form of the expression:

(i) $ab + c -$.

Solution

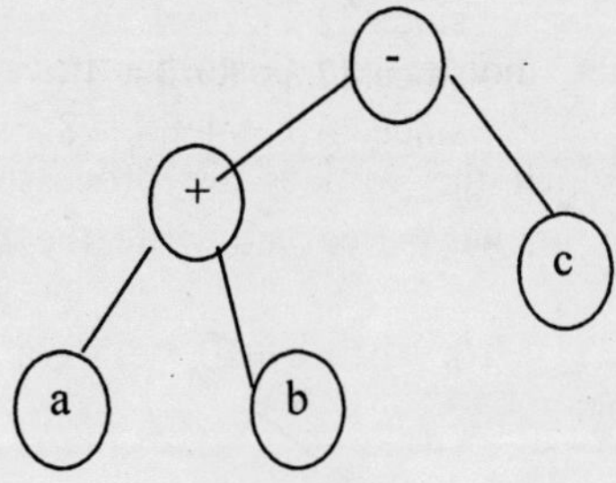

Fig. 3.62

Prefix : $- + abc$

Infix : $a + b - c$

Fully parenthesized infix form : $((a + b) - c)$.

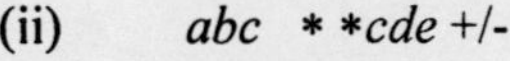

(ii) $abc \;\; * *cde +/-$

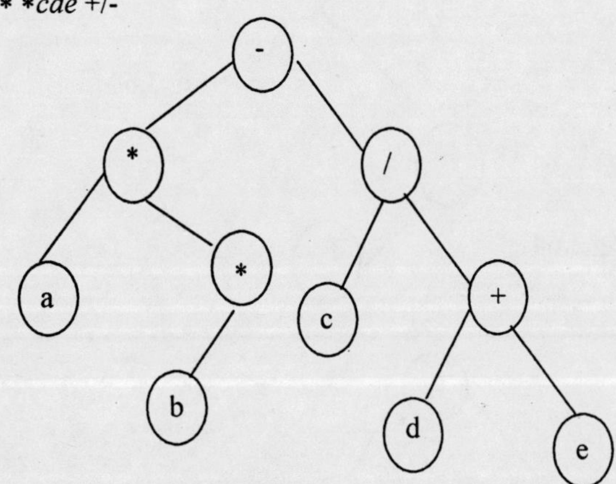

Fig.3.63

Prefix : $- * a *c/c + de$

Infix: $a * b * c - c/(d + e)$

Fully parenthesised infix form : $((a * (b *c)) - \; (c/(d + e)))$.

Example 3.34

Find the value of the expression $abc + -$ if $a = 1$, $b = c$, $c = 3$.

Solution

-4, try yourself.

EXERCISE 3.3

1. Draw the expression tree for the following expressions:

   (i) $x(y + z)$ (ii) $x^2 - 4yz$

   (iii) $\dfrac{a^3 + 1}{a^2 - 1}$ (iv) $((dz + c)z + b)z + a$

2 (a). List out the preorder, inorder and postorder traversal of the trees in question 1 above.

(b). List the order in which the vertices are processed using preorder, inorder and postorder traversal.

(i)

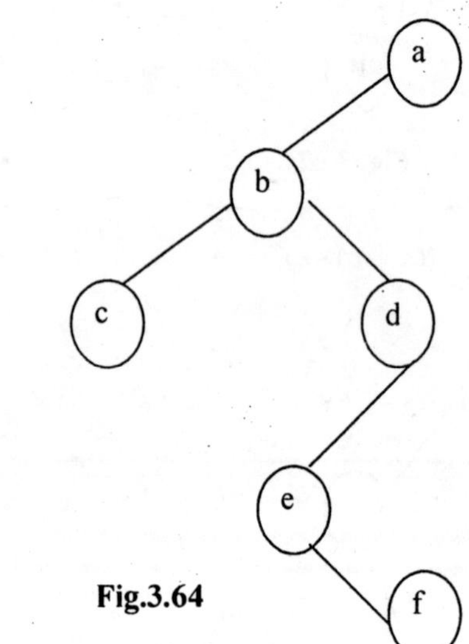

Fig.3.64

(ii)

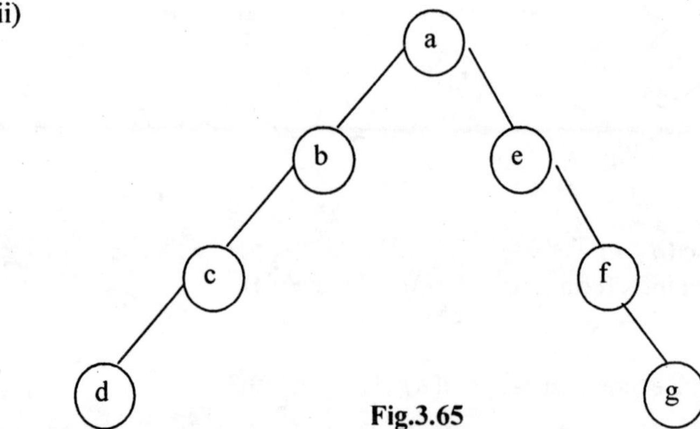

Fig.3.65

(iii)

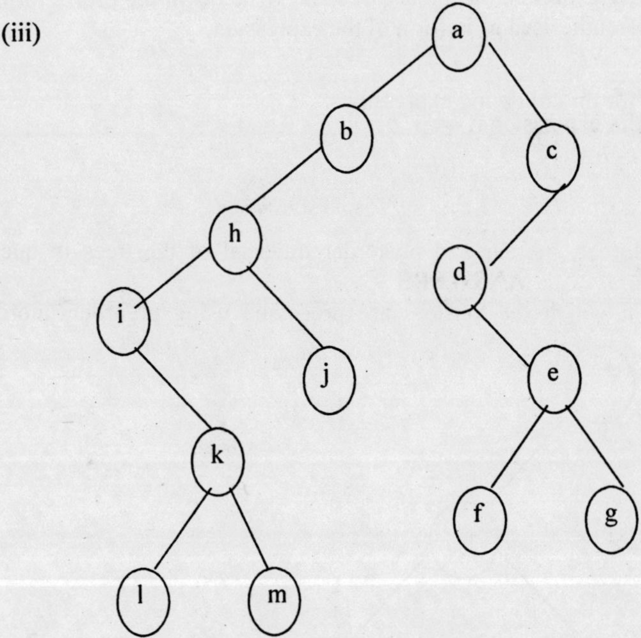

Fig.3.66

3 (i) Draw a binary tree with seven vertices and only one leaf.

(ii) Draw a binary tree with seven vertices and as many leaves as possible.

4. Draw a binary tree corresponding to the following expressions and write the prefix and postfix form of the expression.

(i) $(a * b + c * d) - (a/d\text{-}(d + e))$

(ii) $((a - c) *d)/ (a + (b + d))$.

5. In what order the vertices of the binary tree shown in the Fig. 3.67 is processed if postorder traversal is used.

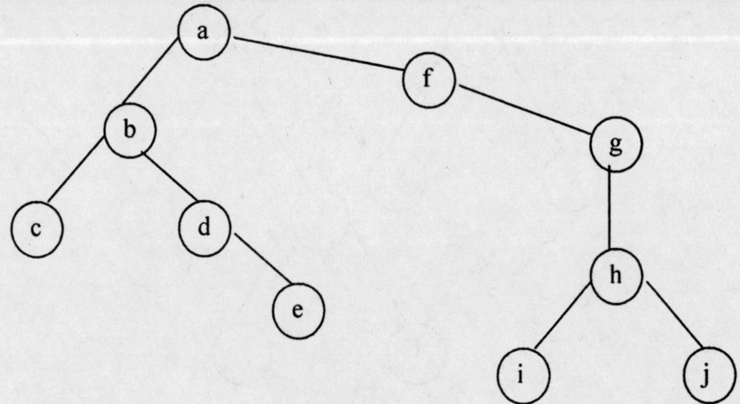

Fig. 3.67

6. Represent following postfix expression as binary tree and write down the prefix form, usual infix form and the fully parenthesised infix form of the expression.
   (i) *abcd* + ∗/*e*-
   (ii) *ab* + *cd* ∗ *ef*/- - *a* ∗.
7. Find the value of the postfix expression if $a = 3$, $b = 4$, $c = 9$ and $d = 3$
   (i) *ab* + *cd* ∗ + *aa*/ - - *b* ∗
   (ii) *adbcd* ∗ - + ∗.

ANSWERS

1. (i)

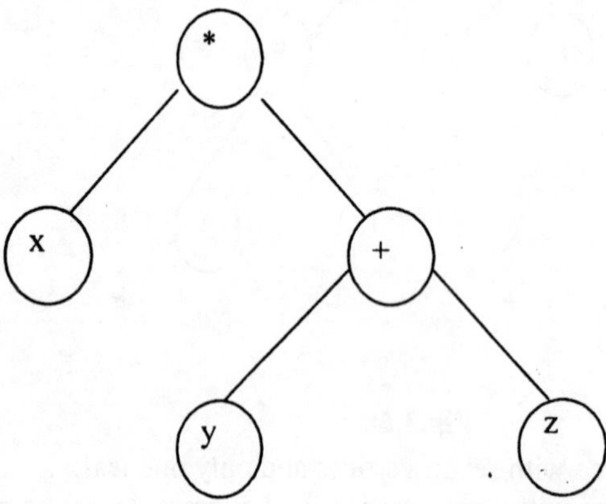

(ii)

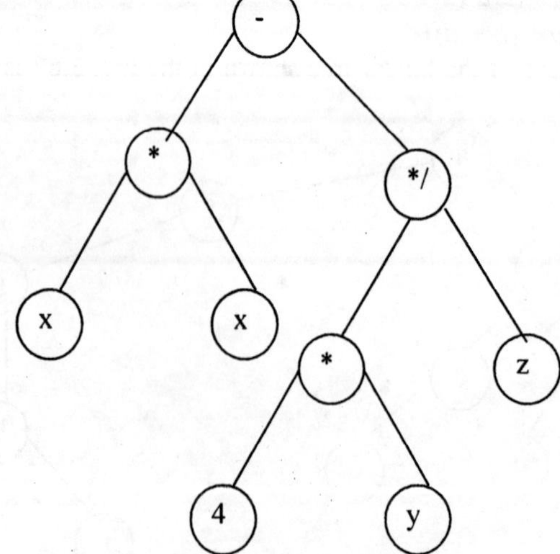

(iv)

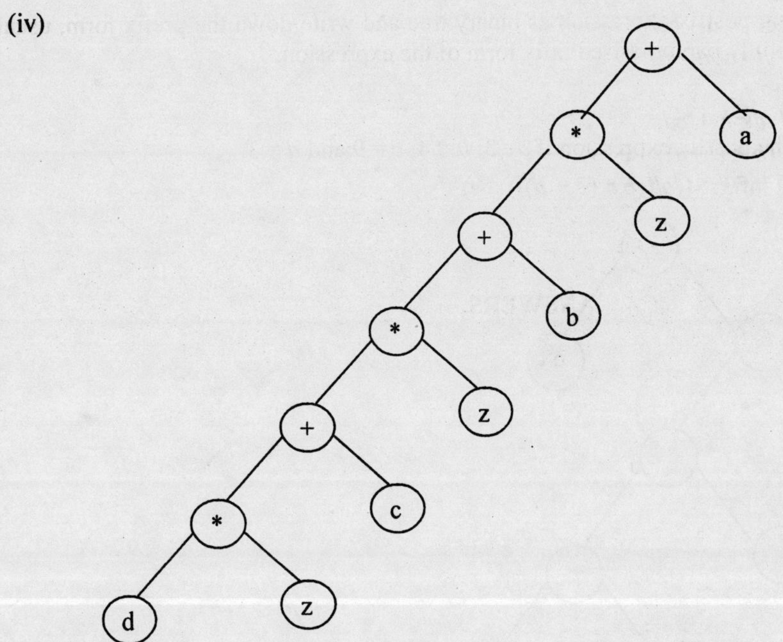

2(a)

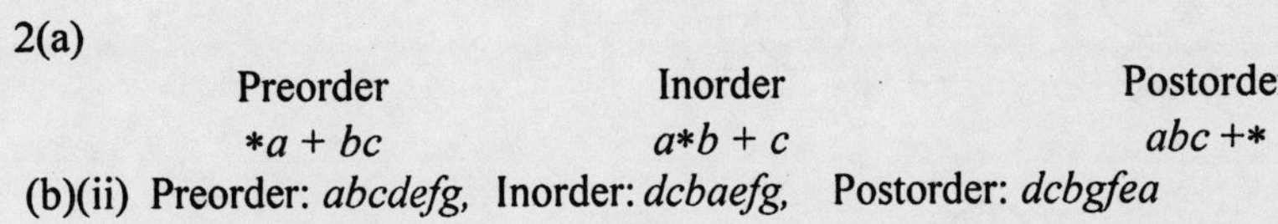

(b)(ii) Preorder: *abcdefg,* Inorder: *dcbaefg,* Postorder: *dcbgfea*

3 (i) (ii)

4 (ii)

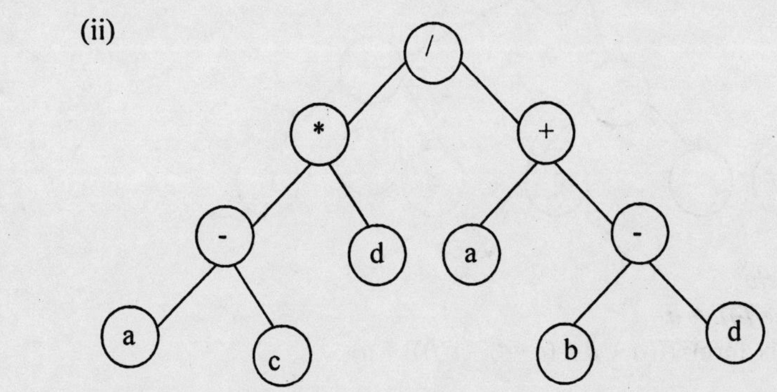

Prefix form: $/ * - acd + a - bd$

Postfix form: $ac - d * abd - + /$

5. $c\ e\ d\ b\ i\ j\ h\ g\ f\ a$

6 (i) Prefix: $- / a * b + cde$

Infix: $a / (b * (c+d)) - e$

Fully parenthesised infix: $((a/((b * (c + d))) - e)$

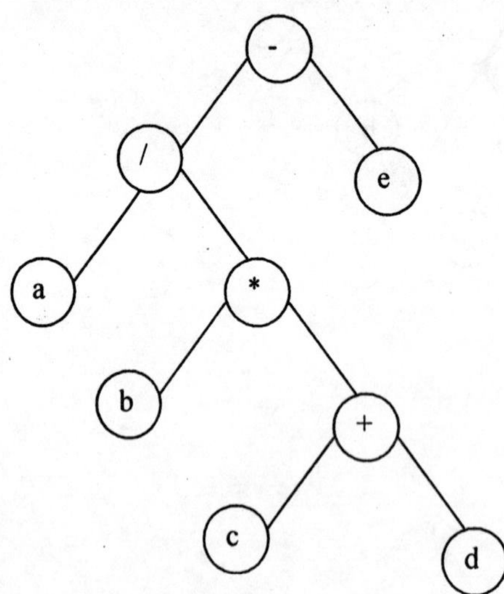

(ii)

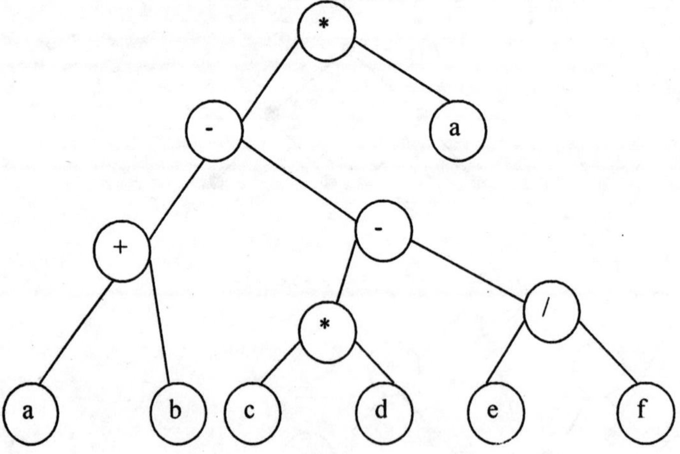

Prefix: $* - + ab - * cd / efa$

Infix: $(a + b - (c * d - e/f)) * a$

Fully parenthesized infix form: $(((a + b - (c * d) - e/f))) * a)$

7. (ii). - 6.

3.19 PATH LENGTH

The **path length** of a tree is defined as the sum of the path lengths from the root to all pendant vertices. The path length of the tree shown in the Fig.3.68 is 24 i.e. 1 + 3 + 4 + 4 + 4 + 4 + 4 = 24.

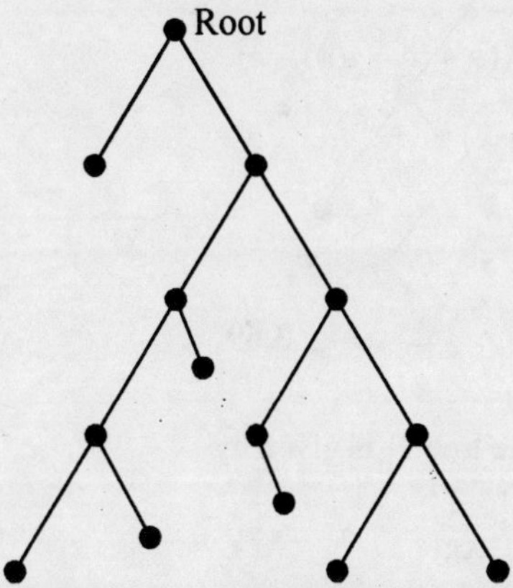

Fig. 3.68

3.19-1 WEIGHTED PATH LENGTH

Consider a binary tree T with n pendant vertices and suppose each pendant vertices n_i is associated with positive real number w_i (weight). The **weighted path length** P of tree T is defined to be

$$P = \Sigma W_i l_i = W_1 l_1 + W_2 l_2 + \ldots + W_n l_n$$

where l_i is the length of the path from the root (level) to an external vertex N_i.

Example 3.35

Three trees T_1, T_2 and T_3 are shown in Fig.3.69 each are having pendant vertices with same weight 4, 5, 7 and 13. Find weighted path length.

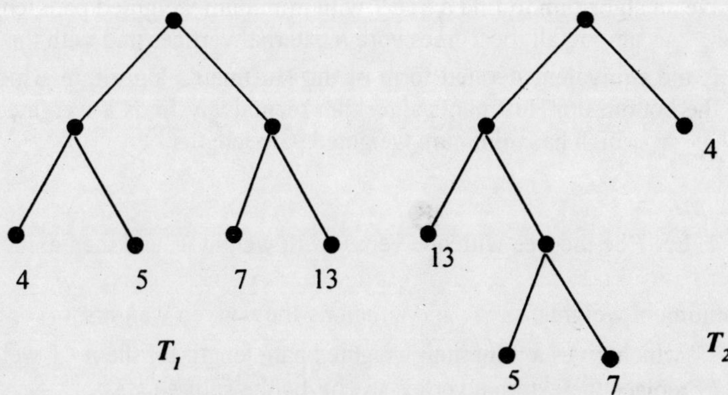

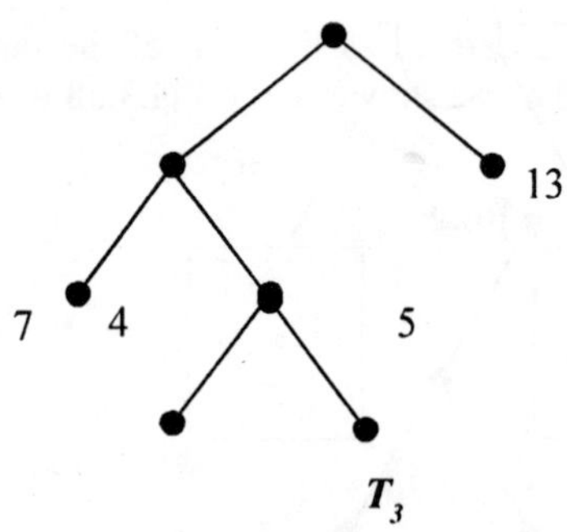

T_3

Fig. 3.69

Solution

The weighted path lengths of the tree T_1 is given by

$$P_1 = 4{\cdot}2 + 5{\cdot}2 + 7{\cdot}2 + 13{\cdot}2$$
$$= 8 + 10 + 14 + 26 = 58.$$

Here all the pendant vertices are at level 2.

Similarly, weighted path length of T_2 is given by

$$P_2 = 4{\cdot}1 + 5{\cdot}3 + 7{\cdot}3 + 13{\cdot}2 = 66$$

In tree T_3, there are one pendant vertex at level 1, one pendant vertex at level 2 and two pendant vertices at level 3. Hence weighted path length of tree T_3 is given by

$$P_3 = 4{\cdot}3 + 5{\cdot}3 + 7{\cdot}2 + 13{\cdot}1 = 54.$$

OBSERVATION

From the quantities of P_1, P_2 and P_3 one can observe that P_1 and P_3 indicate that the complete tree need not give a minimum path length and quantities P_2 and P_3 indicate that similar trees need not give the same path lengths.

3.20 HUFFMAN'S ALGORITHM

Huffman gave an algorithm to find a tree T with minimum weighted path length for a list of n weights say $w_1, w_2\, w_3..., w_n$ among all the 2-trees with n external vertices and with the given n weights.

Following is the equivalent iterated form of the Huffman's algorithm which constructs the desired tree T from the bottom up. Huffman's algorithm recursively finds a weighted 2-tree T with n given weights $w_1, w_2 ..., w_n$ which has minimum weighted path lengths.

ALGORITHM

Step 1 When $n = 1$. Let T be the tree with one vertex with weight w_1 and then exit.

Step 2 When $n > 1$

(i) Find two minimum weights, say w_i and w_j among the n given weights.

(ii) Find a tree T' which gives a minimum weighted path length for the n - 1 weights.

(iii) In the tree T' replace the external vertex $w_i + w_j$ by the subtree.

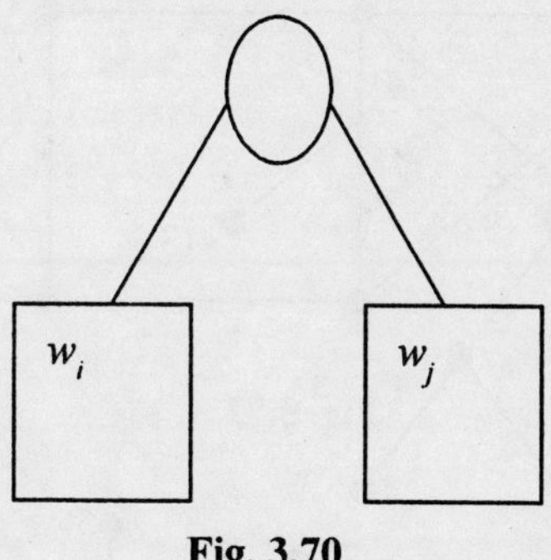

Fig. 3.70

(iv) Exit.

3.21 SPANNING TREES

3.21 -1 SPANNING SET

Let $G = (V, E)$ is a connected undirected graph. A **Spanning set** for G is defined as a subset F of E such that (V, F) is connected.

3.21-2 SPANNING TREE

A subgraph S of a connected graph $G = (V, E)$ is called a **spanning tree** of G if

(i) S is a tree.

(ii) S contains all the vertices of G.

Spanning trees some time also referred to as a **skeleton** or **scaffolding** of G. A spanning tree that is directed is called directed spanning tree of G. The subgraph in heavy lines in Fig.3.71 is the spanning tree of the graph G.

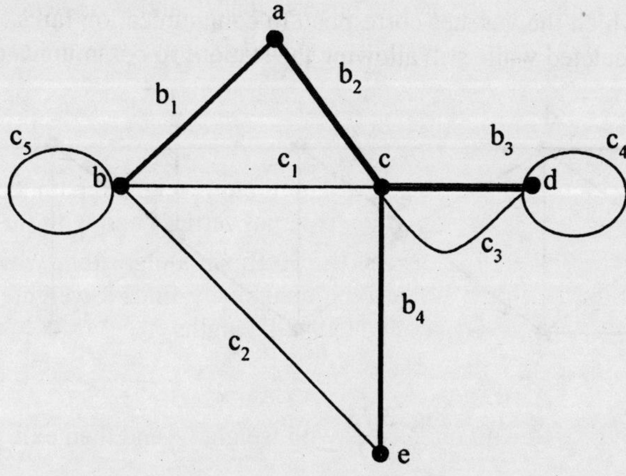

Fig. 3.71

Example 3.36

Draw three spanning trees of graph shown in Fig. 3.72.

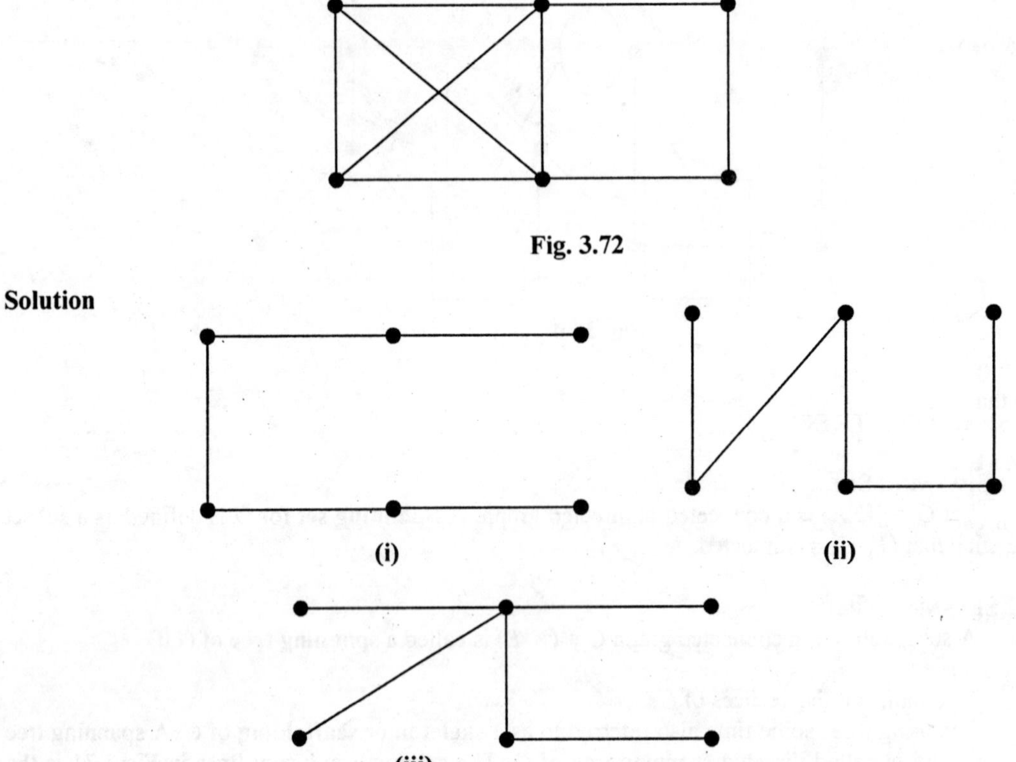

Fig. 3.72

Solution

Example 3.37

Consider the graph shown in the Fig.3.73. Suppose that this graph represents a communication network in which the vertices correspond to communication links. What is the largest number of edges that can be deleted while still allowing the stations to communicate with each other?

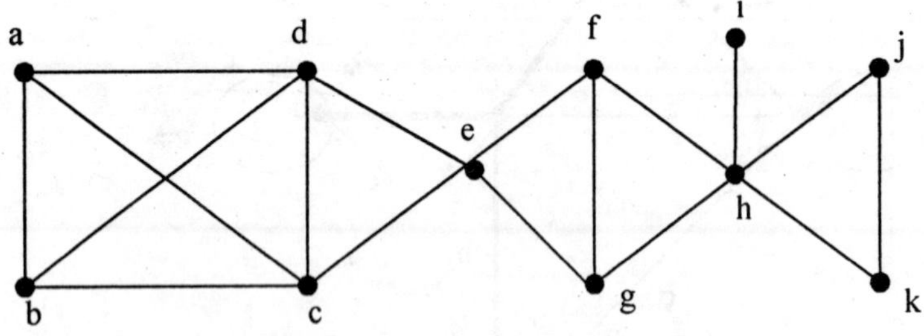

Fig. 3.73

Solution

As we know for communication, the two stations are linked in one way. Here the cycle *dced* gives two ways for *d* and *e* to communicate, namely, along the path *dce* or directly from *d* to *e*. Even if the edge (*d, e*) is deleted, *d* and *e* can still communicate via *c*. Likewise one edge of each cycle in *G* can be deleted. Continuing this we will get the spanning tree as shown in Fig.3.74.

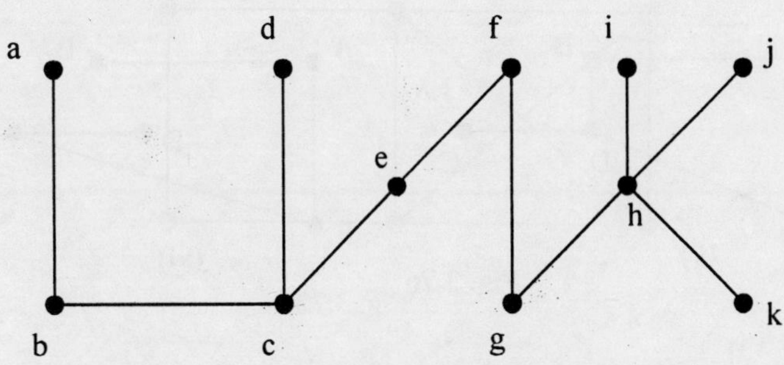

Fig. 3.74

Note that G has 15 edges and the spanning tree for G has 10 edges so 5 are to be deleted.

REMARK

If G is a connected graph with m edges, a spanning tree of G must have n - 1 edges (A tree with n vertices has exactly n - 1 edges). So, the number of edges that must be removed to obtain a spanning tree is $m - (n - 1) = m - n + 1$. This number is frequently called the **circuit rank** of G.

Example 3.38

Which of the following graph in Fig. 3.76 (i) to (vi) are spanning tree of the graph G shown in Fig.3.75.

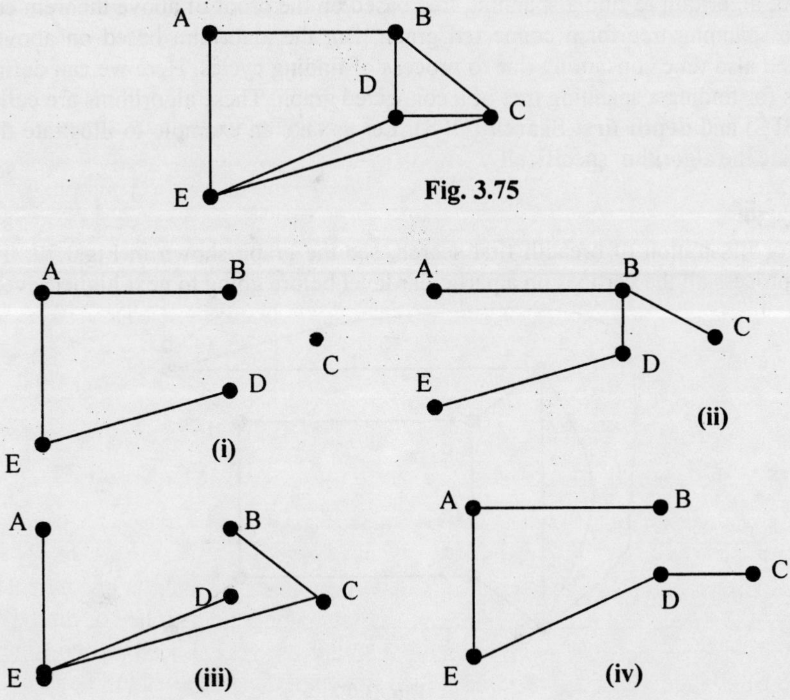

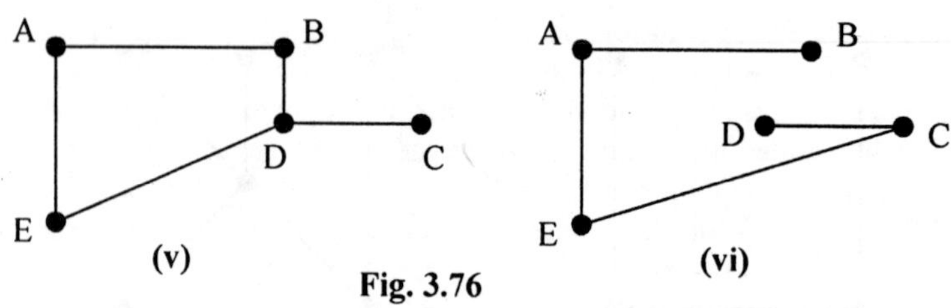

Fig. 3.76

Solution

(ii), (iii), (iv) and (vi). Readers are advised to enquire the reasons.

THEOREM 3.15

A graph *G* has a spanning tree if and only if *G* is connected.

PROOF

If *G* has a spanning tree *T*, there is a path between any pair of vertices in *G* along the tree *T*. Thus *G* is connected. If the graph *G* is connected and acyclic, then *G* is a tree.

Suppose that *G* is having a cycle. If we delete an edge (but no vertex), the graph will remain connected. If after removal it becomes acyclic, we stop here. If it contains a cycle, we delete an edge from this cycle. Continuing this way, we finally produce an acyclic, connected graph *T* which is a tree. Since *T* contains all the vertices of *G*, *T* is a spanning tree of *G*.

3.22 BREADTH-FIRST SEARCH AND DEPTH-FIRST SEARCH

An algorithm to find a spanning tree based on the proof of above theorem could be designed to find the spanning tree for a connected graph. But the algorithm based on above theorem is not efficient and also time consuming due to process of finding cycles. Here we can define other efficient algorithms for finding a spanning tree of a connected graph. These algorithms are called **breadth-first Search** (BFS) and **depth-first Search** (DFS). Let us take an example to illustrate the algorithm and then we state the algorithm specifically.

Example 3.39

For illustration of breadth first search, see the graph shown in Fig.3.77. The basic idea of BFS is to process all the vertices on a particular level before going to next higher level.

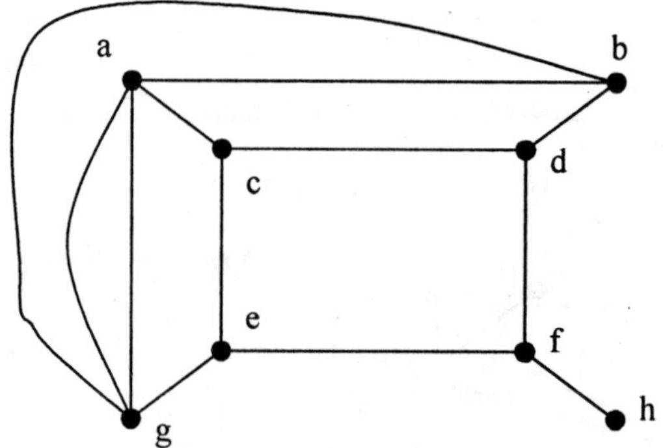

Fig. 3.77

Let us first select an ordering of the vertices say *abcdefgh* of graph *G* shown in Fig 3.77. Then we select *a* as first vertex and label it as the root. At this state *T* consists of the single vertex *a* and no edge. Add to *T* all edges (*a, x*) and vertices on which they are incident, for *x* = *b* to *h*, that do not produce a cycle when added to *T*. Thus, we add edges (*a, b*), (*a, c*) and (*a, g*) to *T* (We could use either of the parallel edges incident on *a* and *g*).

Now repeat the process for all vertices on level one from the root by examining each vertex in the designated order.

b: Include (*b, d*). (We reject (*b, g*) since its inclusion would produced a cycle in *T*).

c: Include (*c, e*). (We reject (*c, d*) since its inclusion would produce a cycle in *G*).

g: None (Since inclusion of any of the edge at this stage would produce a cycle).

Repeat this procedure with the vertices on level 2:

d: Include (*d, f*)

e: none

Repeat this procedure with the vertices on level 3:

f: Include (*f, h*).

As no edges can be added to the single vertex *h* on level 4, the procedure ends. The spanning tree obtained is shown in Fig. 3.78.

The edges which are included in BFS are called **tree edges** and the edges that are rejected in the BFS are called **cross edges.** Hence, BFS partitions the edges of the graph *G* into two sets of edges namely tree edges and cross edges. BFS can be used to test whether an arbitrary graph *G* with *n* vertices is connected or not. It can also be used to find minimum path lengths in an unweighted graph from a fixed vertex *v* to all other vertices. We note that the length of a shortest-path from *v* to a vertex on a level *i* of the spanning tree is *i*. Dijkstras shortest path algorithm for weighted graph (discussed later) can be considered as a generalization of BFS. Let us formalize the description of BFS.

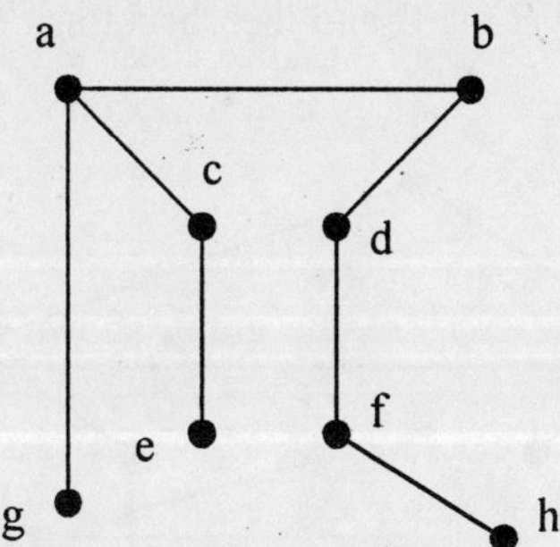

Fig. 3.78

3.22-1 Breadth-First Search For a Spanning Tree

Input: A connected graph *G* with vertices labelled $v_1, v_2, ..., v_n$.

Output: A spanning tree *T* for *G*.

Procedure

1. Start: Let v_1 be the root of *T*. Form the set $V = \{v_1\}$.

2. Add new edges: Consider the vertices of *V* in order consistent with original labeling. Then for each vertex $x \in V$, add the edge (x, v_i) to *T* where *i* is the minimum index such that addition of the edge (x, v_i) to *T* does not produce a cycle. If no edge can be added, then stop; *T* is a spanning tree for *G*. When all the vertices of *V* have been considered in order, go to step 3.

3. Update V: Replace V by all the children v in T of the vertices x of V where the edges (x, v) were added in step 2. Go back and repeat step 2 for the new set V.

There is an alternative to BFS called **depth-first search**, which proceeds to successive levels in a tree at the earliest possible opportunity.

Example 3.40

Illustration of Depth-First search on the graph of Fig. 3.79.

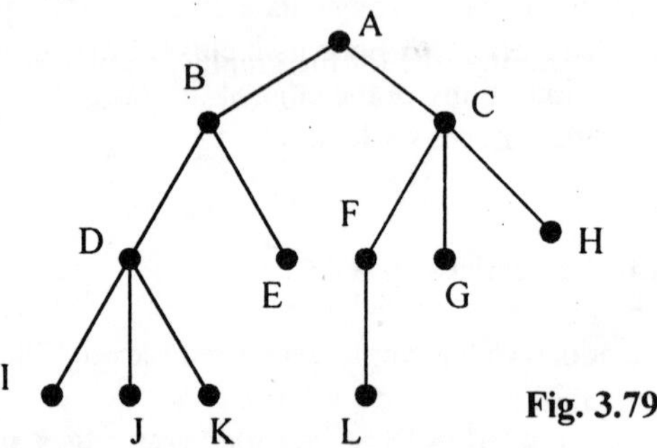

Fig. 3.79

Solution

In depth first Search basic concept is to penetrate as deeply as possible into a tree before fanning out to other vertices. For the illustration of this consider the tree of Fig. 3.79. In order to perform a depth first search, let us start at vertex A and move down to B, D and I. There is no scope for further penetration, we backtrack to D and then go down to J. Again we backtrack to D and go down to K. We now backtrack via D to B, from which we can go down to E. Again, backtrack to A then move to C, F and L. Then to G and H via C and then backtrack to A. This gives us the labelled tree as shown in Fig.3.80.

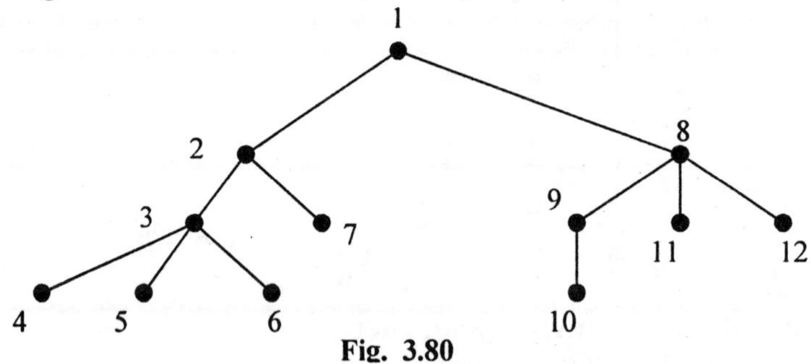

Fig. 3.80

DFS is also called **backtracking** since the algorithm returns to vertices previously visited to add paths.

3.22-2 Algorithm for Depth First Search for a Spanning Tree

Input: A connected graph G with vertices labeled $v_1, v_2, \ldots v_n$.

Output : A spanning tree T of G.

Procedure

(i) **Visit a vertex:** Let v_1 be the root of T, and set $L = v_1$. (The name L stands for the vertex last visited)

(ii) **Find an unexamined edge and an unvisited vertex adjacent to *L*:** For all vertices adjacent to L, choose the edge (L, v_i) where i is the minimum index such that adding (L, v_i) to T doesn't create a cycle. If no such edge exists, go to step (iii); otherwise, add edge (L, v_i) to T and set $L = v_i$; repeat step (ii) at the new value for L.

(iii) **Backtrack or terminate:** If x is the parent of L in T, set $L = x$ and apply step (ii) at the new value of L. If L has no parent in T (so that $L = v_i$) then DFS terminates and T is a spanning tree for G.

Example 3.41

Use a DFS to find a spanning tree for the graph G shown in Fig. 3.81.

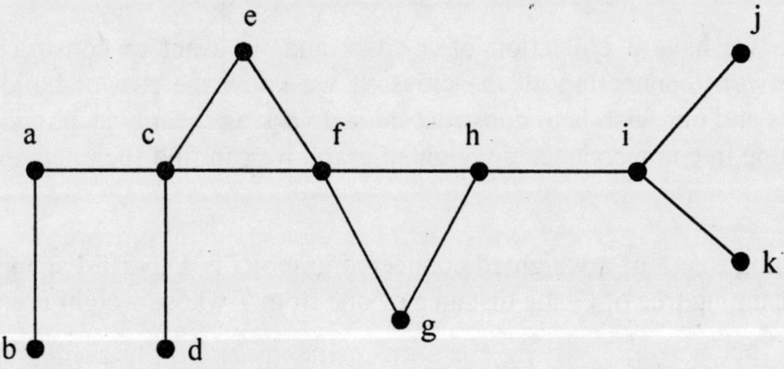

Fig. 3.81

Solution

Let us arbitrarily start with the vertex f (one can start with some other vertex also). Make a path by successively adding edges with vertices not already in the path, as long as possible. In this way, we are getting a path f, e, c, a, b (note that there is a possibility of building other paths). There is no further scope for penetration, we backtrack to a. But, there is no beginning at a containing vertices not in the path already visited. So backtrack to c. Now, form the path c, d. Again backtrack to e, and then to f. From f we can build a path f, g, h, i, j. Then backtrack to i and form the path i, k. This produces the spanning tree as shown in Fig. 3.82.

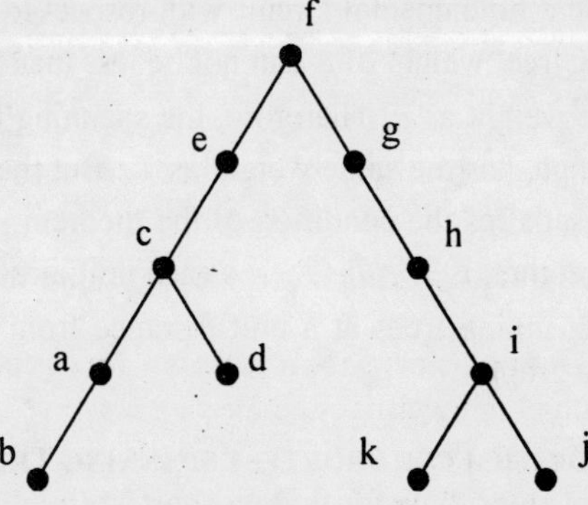

Fig. 3.82

3.23 MINIMAL SPANNING TREES

3.23-1 WEIGHT OF A SPANNING TREE

Consider a connected weighted graph G. The **weight of a spanning tree** T of G is defined as the sum of weights of all the branches in T. Here, one can observe that the different spanning trees will have different weights.

3.23-2 SHORTEST SPANNING TREE

A spanning tree in a weighted connected graph G with smallest weight is called a **shortest spanning tree** or **minimal spanning tree** or **shortest distance spanning tree.**

APPLICATIONS

Suppose we have a collection of n cities and we want to construct a communication or transportation network connecting all the cities. If we know the cost of building the links between each pair of cities and our wish is to construct the network as cheaply as possible. Then with the help of shortest spanning tree in a connected weighted graph we can find such network.

THEOREM 3.16

A spanning tree T of a weighted connected graph G is a shortest spanning tree of G iff there exists no other spanning tree of G at a distance of one from T whose weight is smaller than that of T.

PROOF

There is no need to prove the necessary condition (only if condition) as it is clear that otherwise we shall get another tree shorter than T by a cyclic interchange.

To prove the sufficient condition, consider a spanning tree T_1 in G and suppose there is no spanning tree at a distance of one from T_1, which is shorter than T_1 (hypothesis of the theorem). The proof will be completed if we show that there exist a shortest tree T_2 different from T_1 in G such that the weight of T_1 is also equal to that of T_2. If T_2 is a shortest spanning tree it will satisfy the hypothesis of the theorem. Let us consider an edge a in T_2 which is not in T_1. If we add a to T_1, it will form a fundamental circuit with branches in T_1. But it is to be noted here that not all of the branches in T_1 that form the fundamental circuit with a may also be in T_2. But because of the assumption on T_1, each of these branches in T_1 has a weight $\leq e$ (say). Out of the edges in this circuit which are not in T_2 at least one say e_i, must form some fundamental circuit with respect to T_2 containing a. As we have assumed T_2 to be shortest spanning tree, weight of e_i can not be less than that of a.

$\Rightarrow$ e_i must have the same weight as a. Therefore, the spanning tree $T_1' = (T_1 \cup e - e_i)$ obtained from T_1 through one cycle exchange, has the same weight as T_1. But the number of edges of T_1 is one more in common with T_2, and it satisfies the condition of the theorem. Repeating this argument and getting a series of trees of equal weights, $T_1, T_1, T_1, T_1, \ldots$ each unit closer to T_2, until we get T_2 itself.

$\Rightarrow$ if none of the spanning trees at a unit distance from T is shorter than T, no spanning tree shorter than T exist in the graph.

3.23-3 KRUSKAL'S ALGORITHM FOR SHORTEST SPANNING TREE

There are several algorithms for finding shortest (minimal) spanning tree. One algorithm due to Kruskal is as follows:

Step 1: List all the edges of the graph G in the nondecreasing order of weight.

Step 2: Select any edge of smallest weight that is not in a loop. This is the first edge of spanning tree *T*. In case there exist more than one edge of smallest weight, arbitrarily choose one of them.

Step 3: Select from all the remaining edges of *G* having smallest value that does not form a circuit with the edges already included in *T*.

Step 4: Continue step 3 until n -1 edges have been selected (where n is the number of vertices of G). These edges, will constitute the required shortest spanning tree.

NOTE

The weight of a minimal spanning tree is unique but the minimal spanning tree itself is not. It is quite possible that the different minimal spanning trees may result when two or more edges have the same weight. In such a case, the arrangement of the edges in step 1 is not unique and hence may result in different spanning trees.

Example 3.42

Determine the road network of minimal cost of the towns in Fig.3.83.

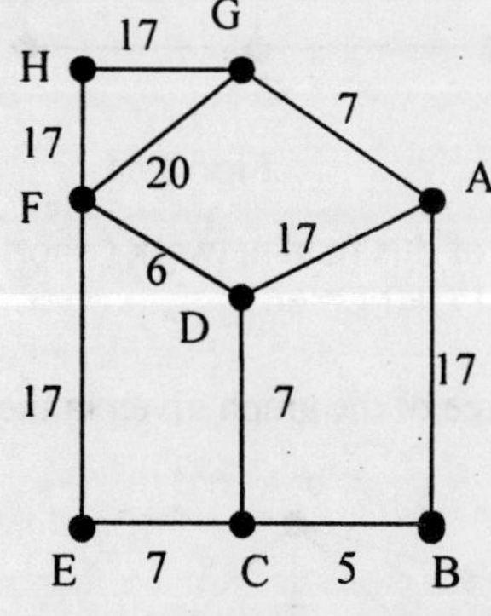

Fig. 3.83

Solution

Applying Kruskal's algorithm to find the shortest spanning tree.

(i) Let us collect the weight of edges in increasing order in the following table :

(ii) Choose the edges *BC, DF, AG, CD, CE* as their weights are in ascending order and they are not forming loop.

Edge	Weight (cost)	Add
BC	5	Yes
DF	6	Yes
AG	7	Yes
CD	7	Yes
CE	7	Yes
AB	17	Yes
AD	17	No
FH	17	Yes
GH	17	No
EF	17	No
FG	20	No

(iii) Now we will have two edges AB and AD. But selection of both of them creates a circuit. Let us choose one of them say AB.

(iv) Similarly, we can choose one of GH and FH.

(v) Now we got seven edges ($n - 1$ edges).

Finally, we will be having spanning tree shown in Fig.3.84.

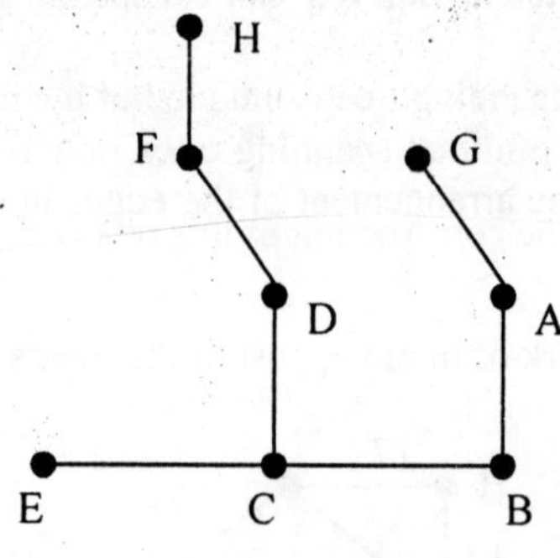

Fig. 3.84

∴ the minimal cost for construction of this road network (shortest spanning tree) is

$$5 + 6 + 7 + 7 + 7 + 17 + 17 = 66.$$

Example 3.43

Find the minimal spanning tree of the graph given in the Fig. 3.85.

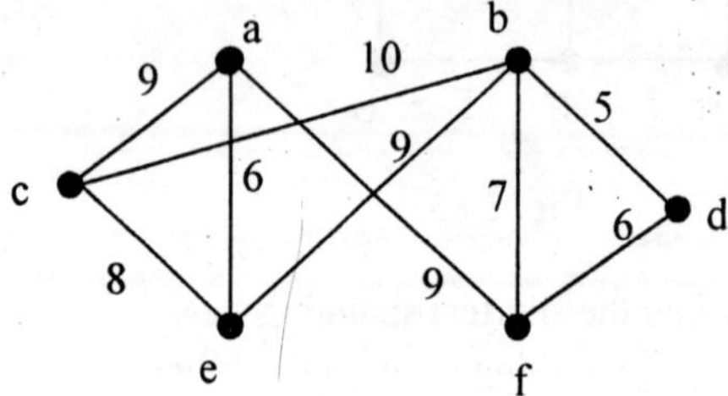

Fig. 3.85

Solution

As the given figure has six vertices, hence spanning tree will have five edges. Let us first collect the weights in nondecreasing order in the following table:

Edges	Weight	Add
bd	5	Yes
ae	6	Yes
df	6	Yes
bf	7	No
ce	8	Yes
ac	9	No
af	9	Yes
be	9	No
bc	10	No

Applying Kruskal's algorithm, we have following minimal spanning tree.

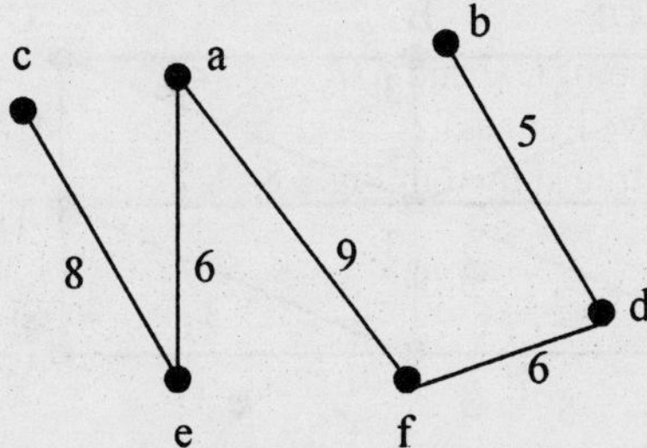

Therefore, the minimal weight for the construction of this tree is 5 + 6 + 6 + 8 + 9 = 34.

Example 3.44

Find the shortest spanning tree using Kruskal's algorithm in the graph shown in the Fig. 3.86.

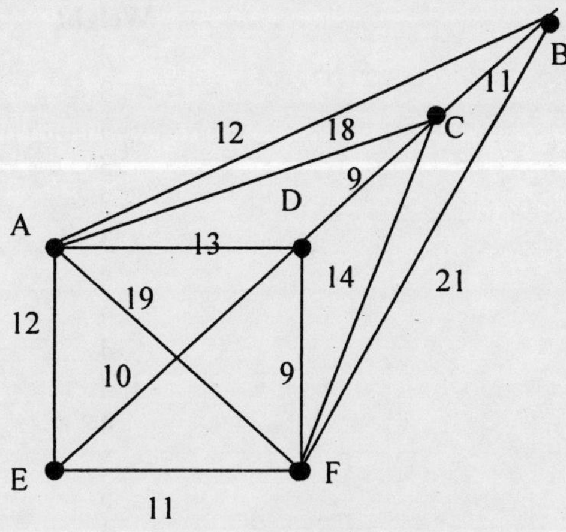

Fig. 3.86

Solution

Applying Kruskal's algorithm, we have following minimal spanning tree. Minimal spanning tree will contain the edges *DF, DC, DE, CB, DA.*

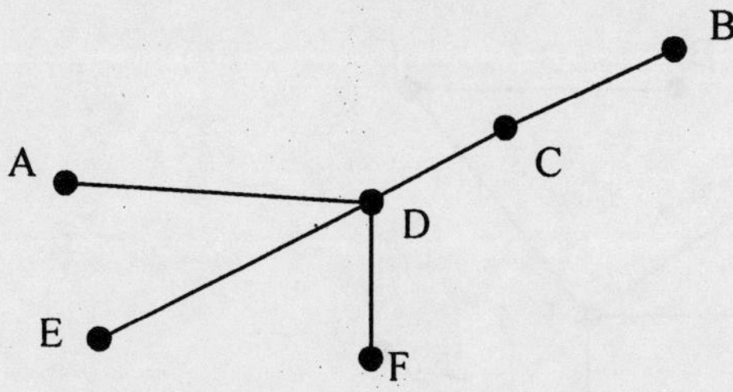

Example 3.45

Find a minimal spanning tree *T* for the weighted graph *G* shown in Fig.3.87.

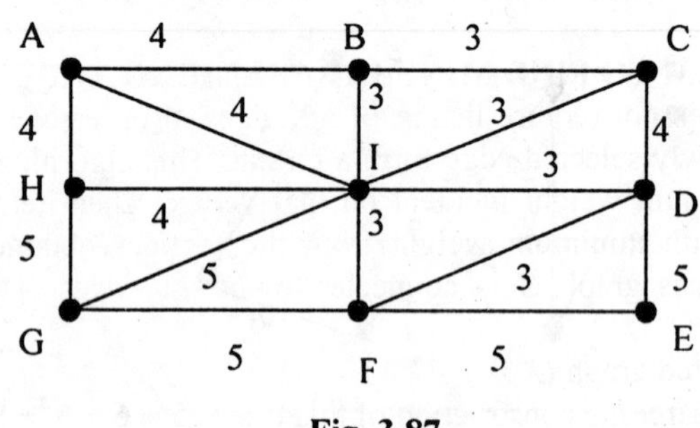

Fig. 3.87

Solution

As total number of vertices is 9, minimal spanning tree T must have n -1 = 8 edges. Collecting the weights in nondecreasing order in a table, we get

Edges	Weight
BC	3
IC	3
ID	3
BI	3
IF	3
FD	3
AB	4
AI	4
AH	4
CD	4
HI	4
HG	5
G F	5
FE	5
DE	5

Applying Kruskal's algorithm by keep on adding edges with minimum weight and without forming any circuit until 8 edges are added. We get the minimal spanning tree as shown in the Fig.3.88.

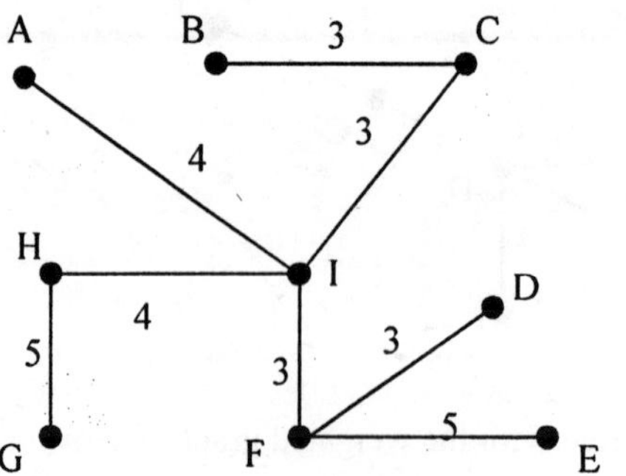

Fig. 3.88

Minimum Weight $= 3 + 3 + 3 + 3 + 4 + 4 + 5 + 5 = 30.$

3.24 PRIM'S ALGORITHEM TO FIND MINIMAL SPANNING TREE

Prim's algorithm does not require listing of all edges in order of non decreasing weight or checking at each step if a newly selected edge form a circuit. This algorithm firstly chooses a vertex and then an edge with minimum weight incident on that vertex. Then iterate by choosing an edge which is not yet selected with minimum weight from the vertices obtained so far. Continue this process until all the n vertices graph G is connected by n - 1 edges. The algorithm consists of following steps:

Input: A connected weighted graph G.

Output: A minimal spanning tree.

Step 1 : Select any vertex v_1 of G and choose an edge which has minimum weight among the edges of G incident with the selected vertex v_1.

Step 2 : Now we have two vertices v_1 and v_2 (say). From these two vertices select the edge which is incident on these vertices and has minimum weight. In case of tie of weights select any one of them.

Step 3 : Continue step 2 with latest vertices in such a way that precisely one end of the new edge (to be selected) lie on the vertices already selected and other end should not lie on any of the selected vertex. Stop when all the n vertices are included.

Example 3.46

Find the minimal spanning tree of the following graph by Prim's algorithm.

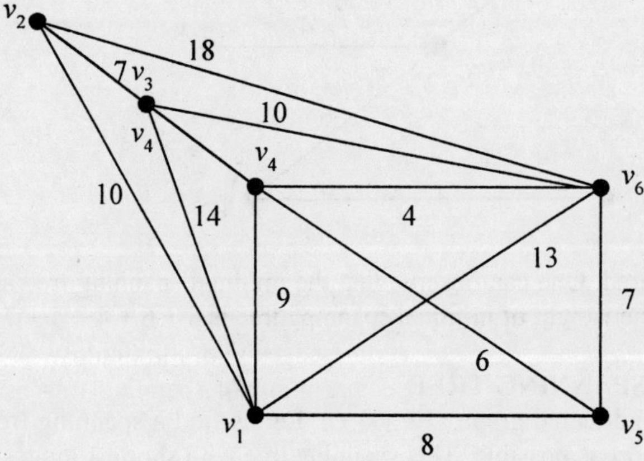

Solution

Step 1 : Let us choose the vertex v_1. Here edge with smallest weight incident on v_1 is (v_1, v_5).

v_1 ——— v_5

8

Step 2 : Now weights of different edges incident on v_1 and v_5 is given by

$w(v_1, v_6) = 13$, $w(v_1, v_4) = 9$, $w(v_1, v_3) = 14$, $w(v_1, v_2) = 10$,

$w(v_1, v_5) = 7$, $w(v_5, v_4) = 6$. Out of these weights of different edge we choose edge (v_5, v_4) since it is minimum.

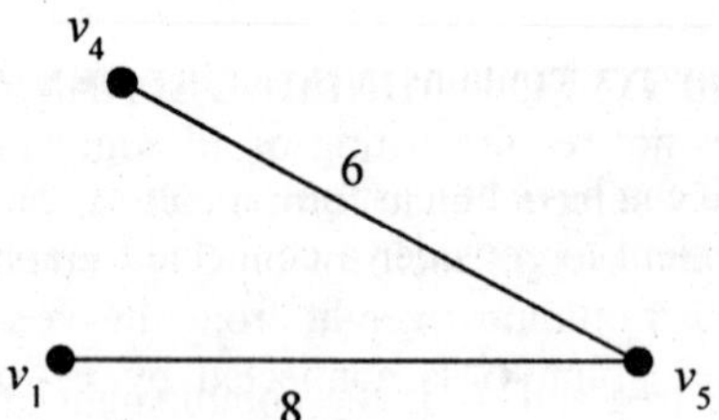

Step 3 : Again take weights of the all edges incident on v_1, v_4 and v_5 (except those whose both end lie on v_1, v_4 and v_5):

$w(v_1, v_2) = 10$, $w(v_1, v_3) = 14$, $w(v_1, v_6) = 13$, $w(v_4, v_3) = 4$, $w(v_4, v_6) = 4$, $w(v_5, v_6) = 7$.

Out of these weights of different edge we choose edge (v_4, v_6) since it is minimum.

It is to be noted here that in place of edge (v_4, v_6) we could also have selected edge (v_4, v_3) since its weight is also 4. Selection of either of edge will give same minimal spanning tree. Proceeding similarly next sequences of edges of the resulting minimal tree will be (v_4, v_3) and (v_3, v_2).

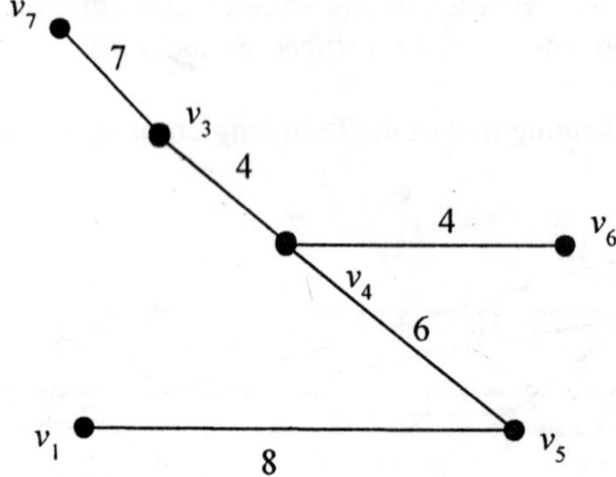

Now all the (6) vertices are covered. One can observe that the minimal spanning tree as shown in the above figure contains 5 edges. The weight of minimal spanning tree is $8 + 6 + 4 + 4 + 7 = 29$.

3.25 MINIMUM DIAMETER SPANNING TREE

Consider a connected undirected graph $G = (V, E)$. Let us find a spanning tree T of G such that the longest path in T is as short as possible. This spanning tree with shortest longest path is called **minimum diameter spanning tree**.

3.26 BRANCH

An edge in a spanning tree T is called a branch of T.

3.26-1 CHORD

An edge of G that is not in a given spanning tree T is called **chord**.

In Fig.3.71 b_1, b_2, b_3 and b_4 are branches while c_1, c_2, c_3, c_4 and c_5 are chords. Here, it is to be noted that the branches and chords are defined with respect to a given spanning tree. Branches and chords with respect to different spanning trees will be different.

NOTE

(i) A connected graph always contains a spanning tree. A graph can have more than one spanning tree.

(ii) Every connected graph can have at least one spanning tree.

(iii) It is sometimes convenient to consider a connected graph G as a union of two subgraphs, T and $\overline{T}$, i.e. $G = T \cup \overline{T}$,

where T is a spanning tree and $\overline{T}$ is the complement of T in G. $\overline{T}$ is collection of chords, hence sometimes referred to as the chord set (or **tie set** or **cotree**) of T.

(iv) A connected graph $G = (V, E)$ of n vertices and e edges with respect to any of its spanning tree will have $n - 1$ tree branches and $e - n + 1$ chords, where $e = |E|$ = number of edges and $n = |V|$ = number of vertices.

3.27 RANK AND NULLITY

Consider a graph G. If n is the number of vertices, e the number of edges, and k the number of components of G then **rank** and **nullity** is denoted by r and μ respectively and defined as

$$\text{rank } r = n - k,$$

$$\text{nullity } \mu = e - n + k.$$

When $k = 1$, the graph is connected, then the rank and nullity of G will be $n - 1$ and $e - n + 1$ respectively.

REMARK

It can be observed that rank of graph G = number of branches in any spanning tree (or forest) of G. Nullity of G = number of chords in G, and rank + nullity = number of edges in G. The nullity of a graph is also referred to as its **cyclomatic number** or **first betti number**.

3.28 CUT SET

A **cut set** is a set of edges in a graph such that the deletion of the set will increase the number of connected components in the remaining subgraph, whereas the deletion of any proper subset of it will not. In other words it is the minimal set of edges in a connected graph whose removal leaves the graph disconnected. Consider the graph shown in Fig.3.89 (a), the set of edges $\{a, e, f, g, d\}$ is a cut set, as its deletion will leave an unconnected subgraph as shown in the Fig.3.89 (b), whereas deletion of any of the proper subset will not. This is also the set of edges that join the vertices into two subsets $\{v_1, v_5\}$ and $\{v_2, v_3, v_4\}$.

OBSERVATION

From the above discussion one can conclude that in a connected graph, deletion of cut set will separate the graph into two parts. Some authors refer to a cut-set as **minimal cut set, a proper cut set** or **a simple cut set**.

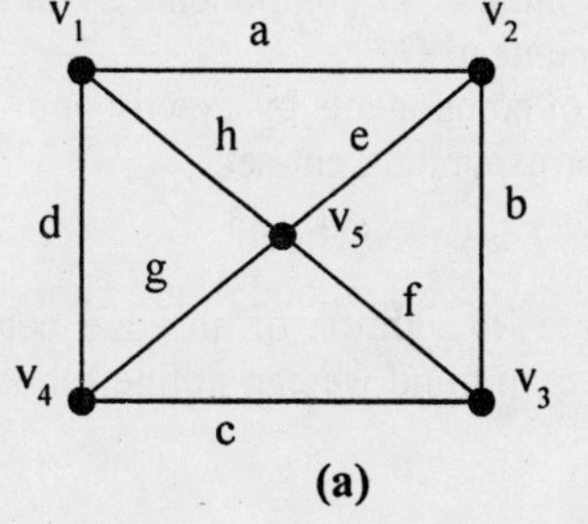

(a)

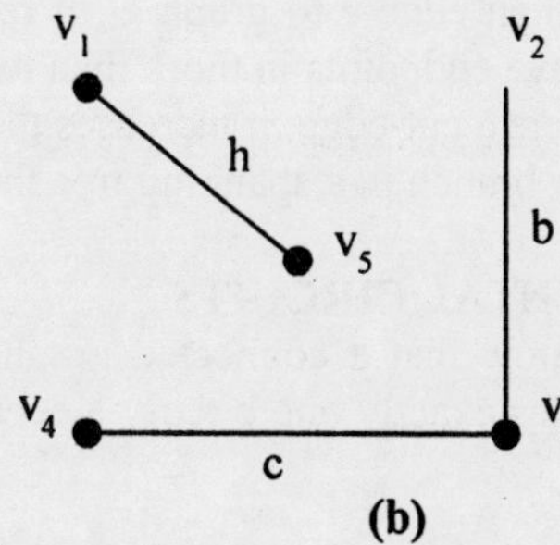

(b)

Fig. 3.89

NOTE

1. A cut set always "cuts" a graph into two.
2. Removal of the cut-set from the graph reduces the rank of the graph by one.
3. Every edge of a tree is a cut-set.
4. Consider a connected graph G, a spanning tree T in it and an arbitrary cut-set S in G. Then it is not possible for S to have any edge common with a spanning tree T of G. Otherwise, deletion of the cut-set S from G would not disconnect the graph.

THEOREM 3.15

In a connected graph G, any minimal set of edges containing at least one branch of every spanning tree of G is a cut-set.

PROOF

Consider a graph G and let S be the sub set of edges of G such that S contains at least one branch of every spanning tree having minimum number of edges. If G-S is connected it means G-S has spanning tree and no edge of spanning tree present in S. Therefore, it must contain an edge of spanning tree, however it may contain more than one branch of spanning tree. Here S is a minimal cut set containing branch of the tree.

3.28-1 CUT VERTEX AND CUT EDGE

A cut vertex of a graph $G = (V, E)$ is a vertex whose removal increases the number of components. Analogously an edge whose deletion from graph G produces a graph with more connected components than the edges is called **cut set** or **bridge**. The notation for the subgraph obtained by removal of a vertex v and an edge e is $G - v$ and $G - e$ respectively.

The graph shown in the Fig. 3.90 has three components.

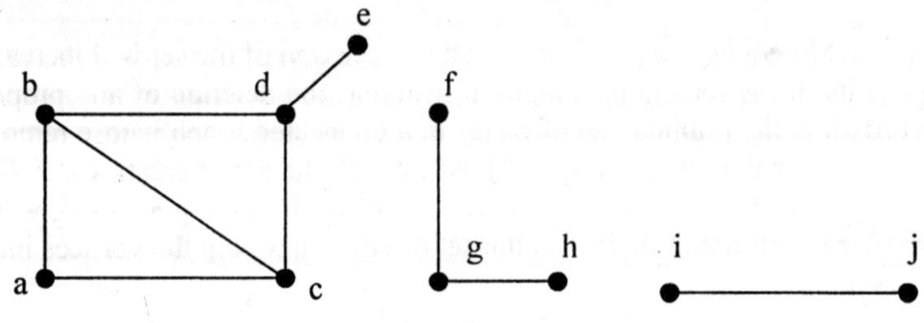

Fig. 3.90

The cut vertices of the graph are d and g. Its cut-edges are de, fg, gh and ij.

NOTE

1. If we add an edge e to graph G, it reduces the number of components by at most one, as e cannot have endpoints in more than two components of G.
2. If we delete a cut-edge, it increases the number of components by exactly one.
3. For every branch in a spanning tree there is a corresponding cut-set.

3.29 FUNDAMENTAL CIRCUITS

As we know that a connected graph G is a tree iff addition of an edge between any two vertices in G creates exactly one circuit. Keeping this fact in mind we can define fundamental circuit as follows:

Consider a spanning tree T in a connected graph G. Addition of a chord to T will create exactly one circuit. Formation of such a circuit due to addition of a chord to a spanning tree is called a **fundamental circuit**. Since a fundamental circuit contains exactly one chord of the spanning tree, it is referred to as the fundamental circuit corresponding to the chord.

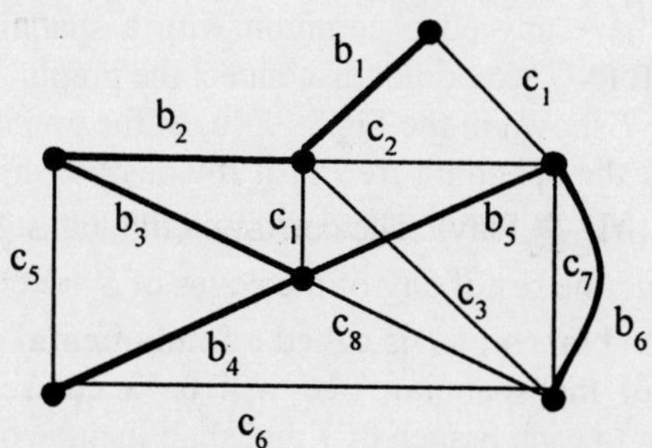

Fig. 3.91

Consider the tree $\{b_1, b_2, b_3, b_4, b_5, b_6\}$ shown in the Fig.3.91. If we add c_l to it, we get a subgraph $\{b_1, b_2, b_3, b_4, b_5, b_6, c_1\}$, which has one circuit $\{b_1, b_2, b_3, b_4, b_5, c_1\}$, that is also a fundamental circuit. It is to be noted that if we add chord c_2 instead of c_l to the tree, we will get a different fundament circuit $\{b_2\ b_3, b_5, c_2\}$. If we add both chords c_1 and c_2 to the tree, the subgraph $\{b_1, b_2, b_3, b_4, b_5, b_6, c_1, c_2\}$ will be fundamental circuit (just mentioned) and a third circuit $\{b_1, c_1, c_2\}$ is not a fundamental circuit.

REMARK

The number of fundamental circuit and the total number of fundamental circuits in a graph is fixed but the circuit that become fundamental, changes with respect to the spanning trees.

THEOREM 3.17

A circuit and the complement of any spanning tree must have at least one edge in common.

PROOF

Had there been a circuit without common edge with the complement of a spanning tree, the circuit is contained in the spanning tree. But, this is not possible, as a tree doesn't contain any circuit.

THEOREM 3.18

A cut-set and any spanning tree must have at least one edge in common.

PROOF

Had there been a cut-set without having a common edge with a spanning tree, the deletion of cut-set would have left the spanning tree intact.

$\Rightarrow$ the deletion of the cut-set will not separate the graph into two components, but this contradicts the definition of a cut-set.

THEOREM 3.19

Every circuit has an even number of edges in common with every cut-set.

PROOF

As we know corresponding to a cut set, the vertices of the graph will be divided into two subsets. These two subsets are the sets of vertices in the two components of the graph when the edges in the cut set are removed. This implies that a path connecting two vertices in one subset must traverse the edges in the cut-set an even number of times.

3.30 FUNDAMENTAL CUT-SETS

Consider a spanning tree T shown in the Fig.3.92 (b) of the graph G shown in the Fig.3.92(a). If we remove the branch e_1 from the spanning tree T, it divides the graph into two disjoint sets of vertices $\{b, c, e, d\}=V_1$ (say) and $\{a\}= V_2$ (say). The corresponding cut set is $\{e_1, e_4, e_5\}$. Cut set S will contain only one branch e_1 of T and the rest if any of the edges of S are chords with respect to T. Such a cut-set S will contain exactly one branch of T is called a **fundamental cut-set** or **basic cut-set** with respect to T. As every branch of the spanning tree will be a cut set, therefore there will be a fundamental cut set corresponding to each branch of T and shall include only that branch of T.
We can formally define fundamental cut set as:

Any cut set S of G that also partitions all the vertices of G into the same two disjoint subsets V_1 and V_2 and will contain exactly one branch of spanning tree T is called a fundamental cut set.

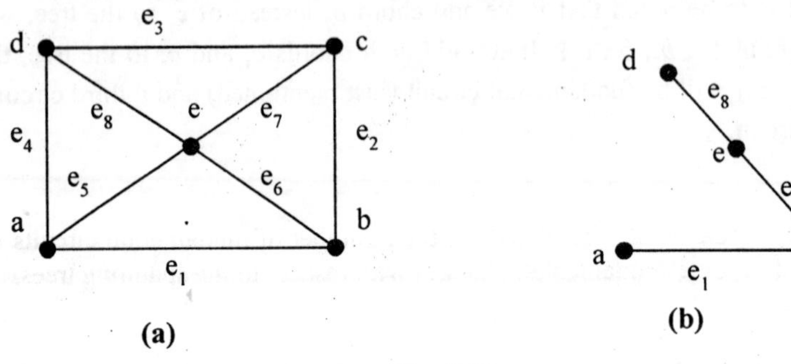

Fig. 3.92

NOTE

1. Every chord of a spanning tree defines a unique fundamental circuit.
2. Every branch of a spanning tree defines a unique fundamental cut-set.

THEOREM 3.20

With respect to a given spanning tree T, a chord c_i that determines a fundamental circuit C occurs in every fundamental cut-set associated with the branches in C and in no other fundamental cut set.

PROOF

Let T is a spanning tree of a given connected graph. Further assume that C is a fundamental circuit formed by a chord a_i with respect to T. Let $e_1, e_2, e_3, \ldots, e_k$ are the branches of fundamental circuit C, then $C = \{ a_i, e_1, e_2, e_3, \ldots, e_k\}$ will be fundamental circuit with respect to T.

We know that every branch of spanning tree T has a fundamental cut set associated with it. Let us suppose that C_l be a fundamental cut set associated with branch e_l of T. Let fundamental cut set C_l contains m chords other then branch e_l. So, $S_l = \{e_l, a_l, a_2, a_3, \ldots, a_m\}$ is a fundamental cut set with respect to T. But we know that number of edges common to fundamental circuit C and any fundamental cut set S_l (say) must be even. It is to be noted here that both C and $S_l$contain e_l. Therefore, there is only one edge e_l in C that is common to both c and S_l. We can observe here that a_i is the only other edge that can possibly be in both C and S_l. This implies both e_l and a_i are common, hence a_i is contained in the set of chords $\{a_l, a_2, a_3, \ldots, a_m\}$. Similar arguments can be applied to the fundamental circuits corresponding to the branches $e_2, e_3, e_4, e_5, \ldots, e_k$. Therefore, the chord c_i is contained in every fundamental cut set associated with branches in C.

Further a_i is not contained in any other fundamental cut set, say S' with respect to T which is associated with branches other then $\{e_1, e_2, e_3, \ldots, e_k\}$. This is due to the reason that none of the branches in C are in S'. If a_i is contained in any other fundamental cut set S', then there would be only one edge a_i contained in both C and S'. But this contradicts the theorem that every circuit has an even number of edges in common with any cut set.

THEOREM 3.21 (CONVERSE OF THEOREM 3.20)

With respect to a given spanning tree T, a branch e_i that determines a fundamental cut set S occurs in every fundamental circuit associated with the chords in S and is not contained in any other fundamental circuit.

PROOF

Let T be a spanning tree of a given connected graph. Further assume that $S = \{e_i, a_1, a_2, a_3, \ldots, a_n\}$ is a fundamental cut set formed by a branch e_i with respect to T consisting of n chords . Let us suppose that $C_l = \{a_1, e_1, e_2, e_3, \ldots, e_k\}$ is a fundamental circuit determined by chord a_l.

One chord e_i is contained both in S and C_l. We also know that the number of chords common to fundamental circuit S and fundamental cut set C_l (say) must be even. Thus e_i must exist in C_l (e_l is one of the branches $e_1, e_2, e_3, \ldots, e_k$). Similar arguments can be applied to the fundamental circuits corresponding to the chords $a_2, a_3, \ldots, a_n$.

Further a_i is not contained in any other fundamental cut set, say S' with respect to T which is associated with branches other then $\{e_1, e_2, e_3, \ldots, e_k\}$. This is due to the reason that none of the branches in C are in S'.

If we assume e_i to be contained in any other fundamental circuit C' corresponding to the chord other then $a_1, a_2, a_3, \ldots, a_n$ then there would be only one edge e_i contained in both S and C' (since none of the chords $a_1, a_2, a_3, \ldots, a_n$ is contained in C'). But this contradicts the theorem that every circuit has an even number of edges in common with any cut set. Thus e_i is not contained in any other fundamental circuit C' corresponding to any chord other then those contained in S.

Example 3.47

In the graph shown in Fig. 3.93(A) consider branch e_8 of spanning tree $\{e_3, e_4, e_6, e_7, e_8\}$ shown in Fig. 3.93(B). The fundamental cut-set corresponding to e_8 is $\{e_2, e_8, e_5\}$. The fundamental circuits corresponding to the chord e_2 is $\{e_2, e_7, e_8\}$ and fundamental circuit corresponding to the chord e_5 is $\{e_3, e_4, e_5, e_8\}$. It is to be noted here that e_8 is contained in both these circuits, but none of the remaining three fundamental circuits contain branch e_8.

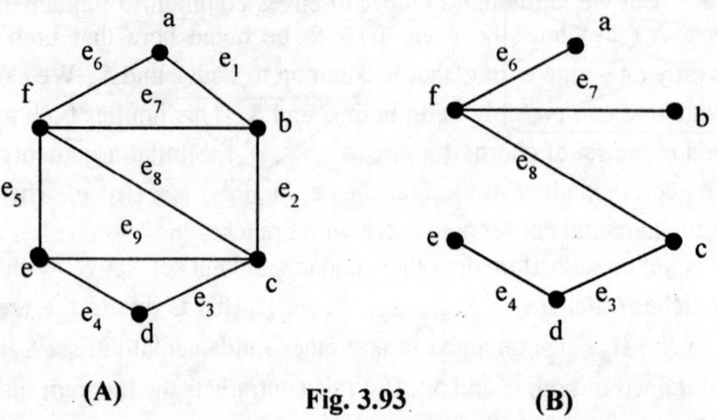

Fig. 3.93

Example 3.48

Consider the spanning tree $\{e_1, e_2, e_3, e_4, e_5\}$. The fundamental circuit corresponding to chord c_3 is $\{c_4, e_3, e_4, e_5\}$. The three fundamental cut set formed by the branch, b_3 is $\{c_2, b_3, c_4\}$; the branch e_4 is $\{c_4, c_3, e_4\}$; the branch e_5 is $\{c_4, c_3, e_5\}$. One can observe here that c_4 is contained in each of these fundamental cut sets but none of the remaining fundamental cut sets contain the chord c_4.

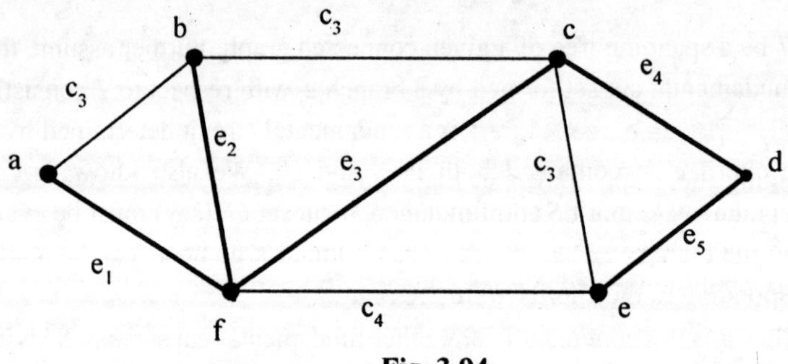

Fig. 3.94

3.31 CONNECTIVITY AND SEPARABILITY

3.31-1 Edge Connectivity

If G is a connected graph, then its **edge connectivity** is the size of the smallest cut set in G. Thus, we can say edge connectivity is the minimum number of edges that we need to delete in order to disconnect G.

Example 3.49

The edge connectivity of graphs shown in Fig.3.95 (a), b) and (c) is one, two and three respectively. Edge connectivity of a tree is always one.

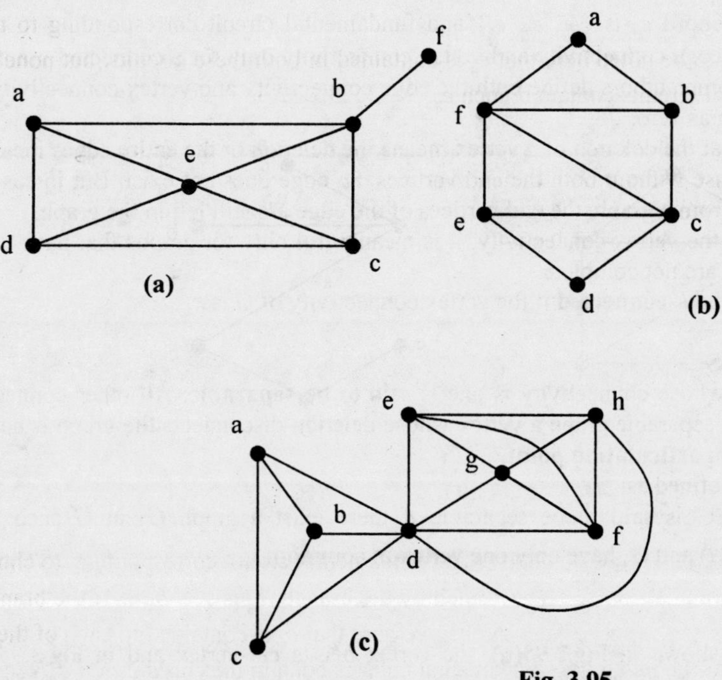

Fig. 3.95

NOTE

Edge connectivity of a connected graph is also said to be the minimum number of edges whose deletion reduces the rank by one.

3.31-2 VERTEX CONNECTIVITY

Examining the graph shown in Fig.3.95(c), one can observe that the deletion of no single edge disconnects the graph but deletion of single vertex *d* disconnects it. Now, we define the **vertex connectivity** of a connected graph as the smallest number of vertices whose deletion from *G* leaves the remaining graph disconnected. Vertex connectivity is sometimes also called as **connectivity**.

Example 3.50

Vertex connectivity of a tree is one. Vertex connectivity of the graph shown in Fig.3.95 *(a)*, *(b)* and *(c)* are one, two and one, respectively.

Example 3.51

Find the edge connectivity and vertex connectivity of the graph given below.

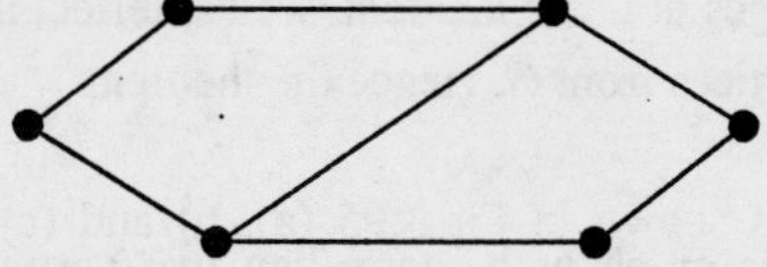

In this graph edge connectivity as well as vertex connectivity is 2.

REMARK

1. Here, we are taking edge connectivity and vertex connectivity only for a connected graph. It is to be noted that some authors define both the edge connectivity and vertex connectivity of a disconnected graph as zero.
2. It is to be recalled that the deletion of a vertex means the deletion of the entire edges incident on that vertex, because without both the end vertices, an edge does not exist. But in case of removal of an edge from a graph, the end vertices of the edge are still left in the graph.
3. As per definition of the vertex connectivity, it is meaningful only for graphs that have three or more vertices and are not complete.
4. A graph G is said to be ***k*-connected** if the vertex connectivity of G is k.

3.32 SEPARABLE GRAPHS

A connected graph whose connectivity is one is said to be **separable**. All other connected graphs are nonseparable. In a separable graph a vertex whose deletion disconnects the graph is called a **cut-vertex**, a **cut node**, or an **articulation point**.

Separable graph can also be defined as:

A connected graph G is said to be separable if there exist a graph G_1 in G such that complement of G_1 in G (i.e. $\bar{G}$) and G_1 have only one vertex in common.

Example 3.52

Consider the graph shown in Fig.3.95(a), the vertex b is a cut-vertex and in Fig.3.95(c), vertex d is a cut vertex.

NOTE

1. It is to be noted that in a tree every vertex with degree greater than one is a cut-vertex.
2. A **1-connected graph** and a separable graph are same.

THEOREM 3.22

The edge connectivity of a graph will never be more than the degree of the vertex with the smallest degree in G.

PROOF

Consider a vertex v with the smallest degree say $d(v)$ in graph G. Vertex v can be separated from G by deleting $d(v)$ edges incident on vertex v. Hence the theorem.

THEOREM 3.23

The vertex connectivity of any graph G will never be more than the edge connectivity of G.

PROOF

Let edge connectivity of graph G be k. This implies, there exists a cut-set C in G with k-edges. Let cut set C divides the vertices of G into subsets V_1 and V_2. By deleting maximum of k-vertices from V_1 (or V_2) on which the edges in C are incident, we can effect the deletion of C together with all other edges incident on these vertices from G. Hence the theorem.

COROLLARY

Every cut-set in a nonseparable graph with more than two vertices contains at least two edges.

REMARK

1. A vertex v in a connected graph G is a cut-vertex iff there exist two vertices v_i and v_j in G such that every path between v_i and v_j passes through v.
2. Vertex connectivity of a connected graph G is α (i.e. α connected) iff every pair of vertices in G is joined by α or more path that do not intersect, and at the minimum one pair of vertices is joined by exactly α-non intersecting paths (Here, it is to be noted that paths with no common vertices, except the two terminal vertices, are called **non-intersecting paths** or **vertex-disjoint path**).
3. A graph G is α-edge-connected iff every pair of vertices in G is joined by α or more disjoint paths*, and at the minimum one pair of vertices is joined by exactly k-edge disjoint paths.

THEOREM 3.24

Consider a graph G with v number of vertices and e number of edges ($e \geq v - 1$). The maximum vertex connectivity obtained in G is the integral part of the number $2\,e/n$; i.e. $\lfloor 2e/n \rfloor$.

PROOF

As we know all edges of graph G contributes two degrees. Since we have here v vertices and total $2e$ degrees is divided among v vertices. This implies, there must exist at least one vertex in G whose degree $\leq 2e/v$. From the theorems 3.23 and 3.24, we can conclude that vertex connectivity of G cannot exceed this number.

To show that one can get maximum connectivity $= \lfloor 2e/n \rfloor$, first construct an n-vertex regular graph of degree $\lfloor 2e/n \rfloor$ and then add the remaining $e-(n/2).\lfloor 2e/n \rfloor$ edges arbitrarily. Rest do yourself.

NOTE

1. Vertex connectivity $\leq$ edge connectivity $\leq 2e/n$
2. Maximum vertex connectivity possible $= \lfloor 2e/n \rfloor$

Example 3.53

What is the maximum vertex connectivity and edge connectivity for the graph shown in the Fig.3.96.

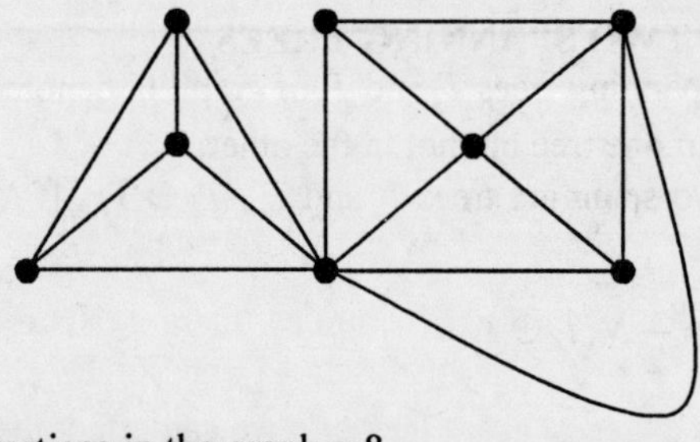

Fig. 3.96

Solution

Total number of vertices in the graph = 8

Total number of edges in the graph = 16

∴ Maximum edge connectivity and hence vertex connectivity

* Edge-disjoint path means the paths that may intersect, but have no edge in common.

$$= \frac{2e}{v} = \frac{2 \cdot 16}{8} = 4$$

3.33 CYCLIC INTERCHANGE

In many applications of graph theory, we require all spanning trees. Consider the graph G shown in Fig. 3.97. For the generation of spanning trees, we can start with any spanning tree say $T_1(e_1, e_2, e_3, e_4)$ in the Fig.3.97.

Addition of a chord say e_8 to tree T_1 forms a fundamental circuit. If we remove a branch say e_3 from fundamental circuit e_2, e_3, e_8, e_4 just formed, will form a new spanning tree T_2. This generation of one spanning tree from another, through addition of a chord and deletion of an appropriate branch is called a **cyclic Interchange or elementary tree transformation.**

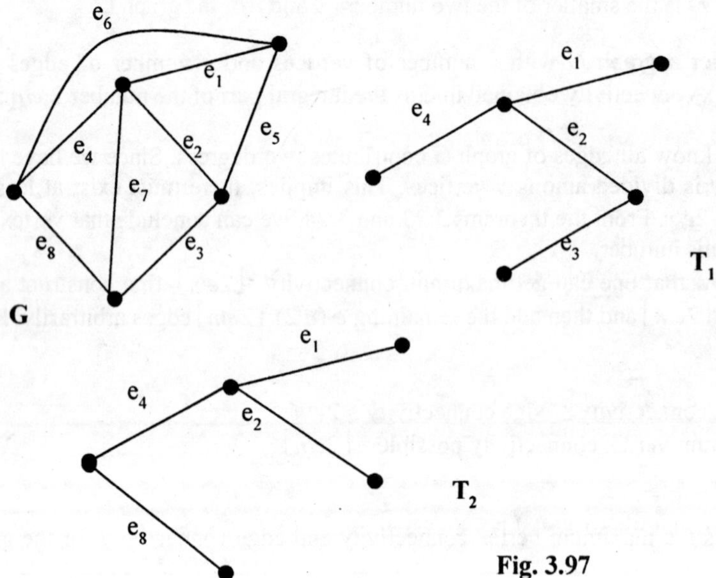

Fig. 3.97

3.34 THE DISTANCE BETWEEN TWO SPANNING TREES

The distance between two spanning trees T_i and T_j of graph G denoted by $d(T_i, T_j)$ is defined as the number of edges of G present in one tree but not in the other.

Consider the ring sum of two spanning trees T_i and T_j, $T_i \oplus T_j$. If $N(G)$ denote the number of edges in a graph G. Then

$$d(T_i, T_j) = \frac{1}{2} N(T_i \oplus T_j)$$

NOTE

The number $d(T_i, T_j)$ is the minimum number of cyclic interchanges involved in going from T_i to T_j.

THEOREM 3.25

Starting from any spanning tree of a graph G, we can obtain every spanning tree of G by successive cyclic exchanges.

REMEMBER

(i) The maximum distance between any two spanning trees in a connected graph G of rank r having r edges is given by

$$\max \; d(T_i, T_j) = \frac{1}{2} \max \quad N(T_i \oplus T_j)$$
$$\leq r, \quad \text{which is the rank of } G$$

(ii) If μ is the nullity of G, then there is no more than μ edges of a tree T_i that can be replaced to get another tree T_j

$\Rightarrow$ $\max d(T_i, T_j) \leq \mu$

$\Rightarrow$ $\max d(T_i, T_j) \leq \min (\mu, r)$,

where min (μ, r) is the smaller of the two numbers μ and r of the graph G.

3.35 CENTRAL TREE

Consider a graph G and a spanning tree T_0 of it and let the maximal distance between T_0 and any other spanning tree T_i of G denoted by $\max_i d(T_0, T_i)$, then T_0 is called a **central Tree** of G if

$$\max_i d(T_0, T_i) \leq \max_j d(T, T_j) \text{ for every tree of } G.$$

NOTE

In general a central tree in a graph is not unique.

3.36 TREE GRAPH

Consider a graph G. The **tree graph** of graph G is defined as a graph in which each vertex corresponds to a spanning tree of G, and each edge corresponds to a cyclic interchange between the spanning trees of G represented by the two end vertices of the edge.

NOTE

Tree graph of any graph is connected.

EXERCISE 3.4

1. (i) Determine the different nonisomorphic spanning trees for the graph.

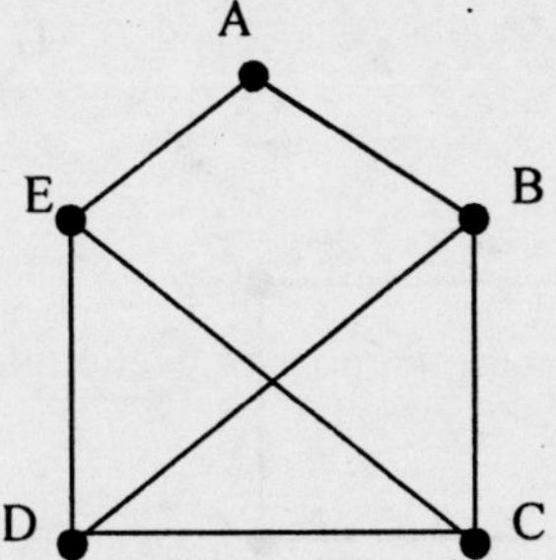

(ii) How many different spanning trees does each of the following simple graphs have

(a) K_3 (b) K_4 (c) $K_{2,2}$

2. Show that a path is its own spanning tree.
3. Find all spanning trees of the graph shown in the following figure.

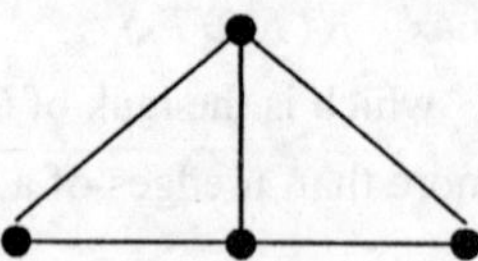

4. Show that the distance between two spanning trees is a metric.
5. Define spanning tree of a graph G and prove that any given edge of a connected graph G is a branch of some spanning tree.
6. Find all spanning trees of the graphs shown in Fig.3.98.
7. Show that a Hamiltonian path is a spanning tree.
8. Prove that any edge of a connected simple graph G is an edge of some spanning tree of G.
9. Determine a minimum spanning tree for the graph shown in Fig.3.98.

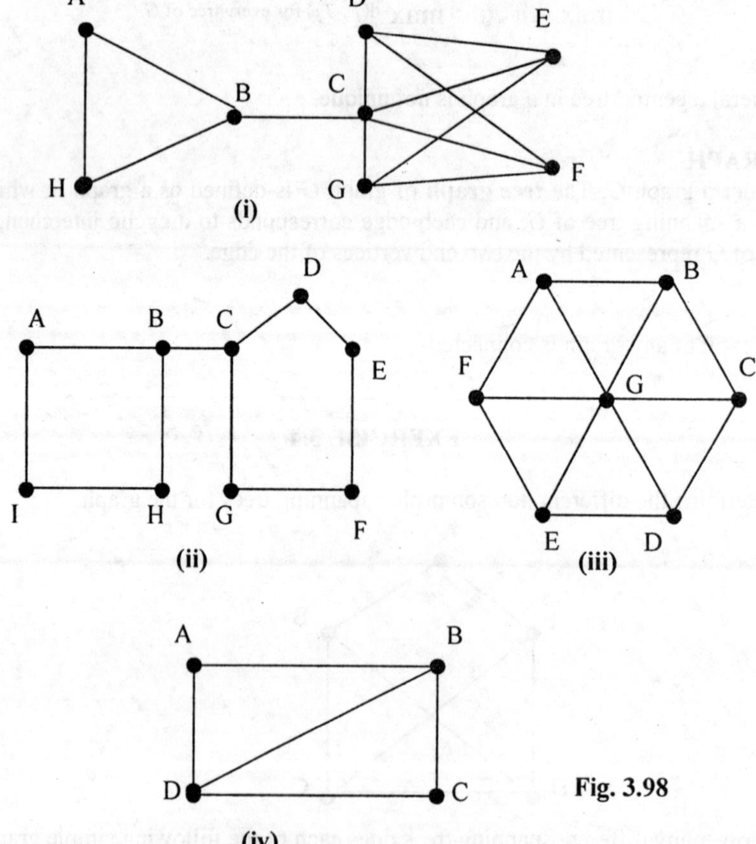

Fig. 3.98

(v) The star graph $K_{1,n}$

(vi) C_n, the n-cycle graph.

10. Consider a connected weighted graph in which every edge belongs to some circuit. If e_m is the edge with weight greater than that of any other edge in G, show that no spanning tree in G will contain e_m.

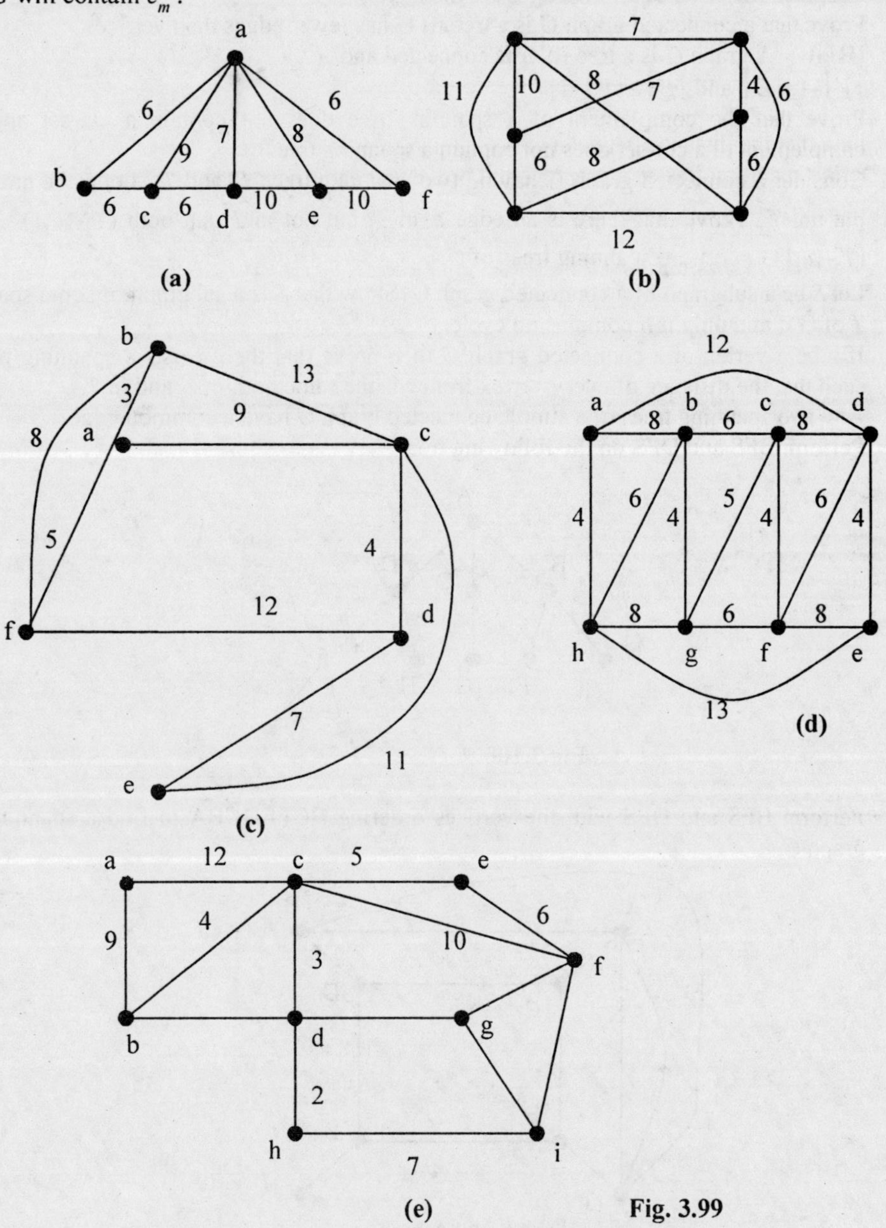

Fig. 3.99

11. Prove that if a simple connected graph G has only one spanning tree, then G itself is a tree.
12. Show that a tree is a bipartite graph.
13. Prove that a pendant edge in a connected graph G is contained in every spanning tree of G.
14. Disprove that a connected graph G with more than two edges has a unique minimal spanning tree, then the costs of all the edges are distinct.
15. Prove that a connected graph G is a tree iff G has fewer edges than vertices.
 [Hint A graph G is a tree iff G is connected and $|V|-1=|E|$ and $|E|\geq|V|-1$]
16. Prove that the complement of a spanning tree does not contain a cut-set and that the complement of a cut-set does not contain a spanning tree.
17. Consider a connected graph G having two spanning trees T_1 and T_2. Let e_1 be an edge in T_1 but not T_2. Prove that there is an edge e_2 in T_2 but not in T_1 s.t. both $(T_1\text{-}\{e_1\}) \cup \{e_2\}$ and $(T_2\text{-}\{e_1\}) \cup \{e_2\}$ are spanning trees of G.
18. Let S be a subgraph of a connected graph G. Show that S is a subgraph of some spanning tree T of G if and only if S contains no cycle.
19. If v be a vertex of a connected graph G then prove that there exists a spanning tree T in G such that the distance of every vertex from v is the same both in G and in T.
20. Any two spanning trees of a simple connected graph G have a common edge.
21. Perform BFS and DFS on the following tree.

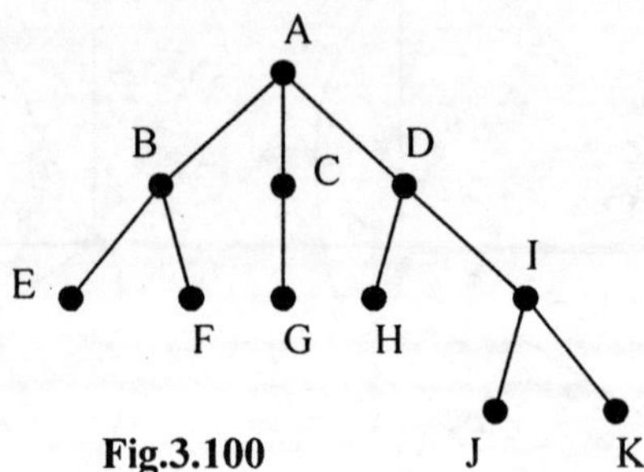

Fig.3.100

22. Perform BFS and DFS with the vertices ordering HGFEDCBA to find a spanning tree for graph G.

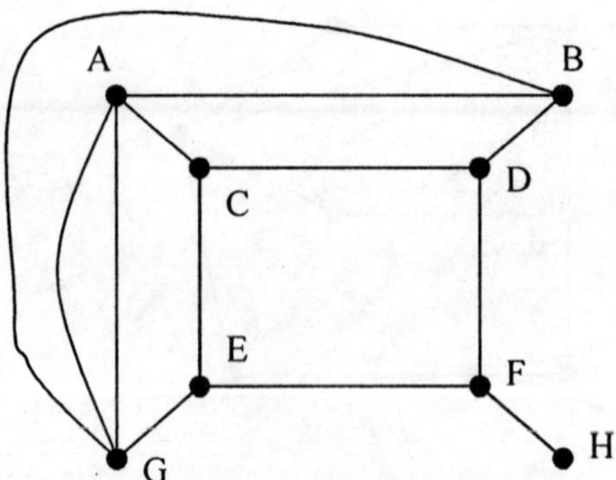

Fig. 3.101

23. Perform DFS to produce a spanning tree for the following graph

(i)

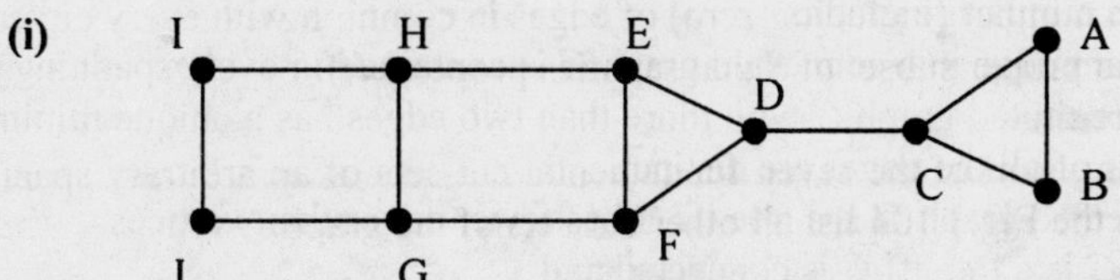

(ii)

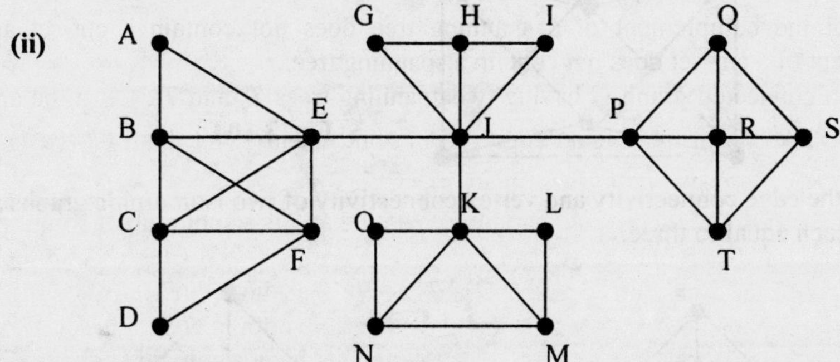

Also perform BFS to find the spanning tree.

23. Take any spanning tree in the following Fig.3.102. List all the seven (n -1 = 7) fundamental cut-sets w.r.t to this tree.

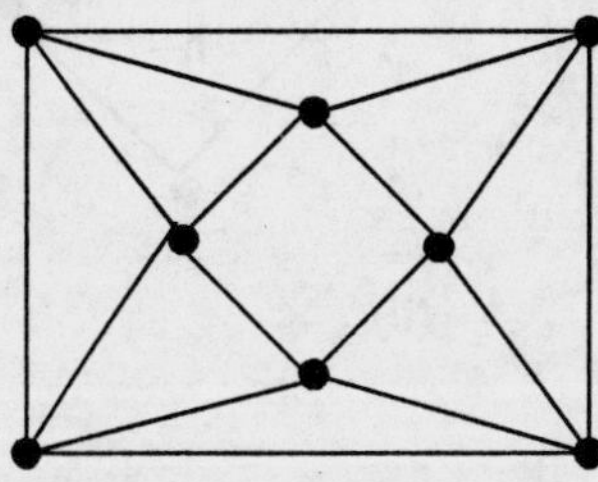

Fig. 3.102

25. Prove that in a tree every vertex of degree greater than one is a cut vertex.
26. Prove that every connected graph with three or more vertices has at least two vertices which are not cut vertices.
27. Prove that an Euler graph cannot have a cut- set with an odd number of edges. [**Hint** The number of vertices of odd degree in a graph is always even]
28. Find all cut-sets w.r.t the vertex pair a, b in the graph shown below in the Fig.3.103.

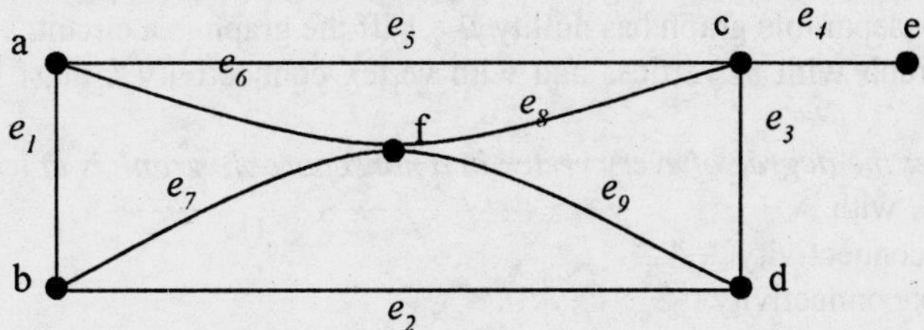

Fig. 3.103

29. Consider a connected graph G, and set of edges S with following properties
    (i) S has even number (including zero) of edges in common with every cut-set of G.
    (ii) There is no proper subset of S that satisfies property (*i*).
    Prove that S is a circuit.
30. With the help of ring sum of the seven fundamental cut-sets of an arbitrary spanning tree in the graph shown in the Fig.3.104 list all other cut-sets of the graph.

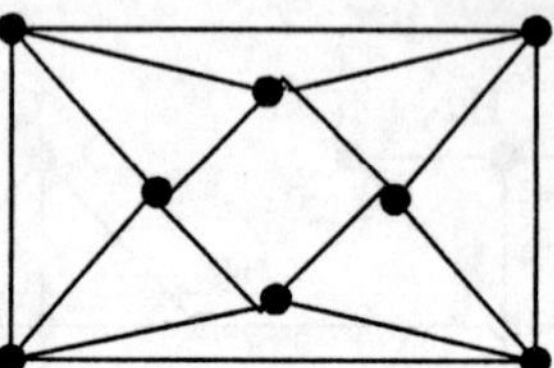

Fig. 3.104

31. Show that the edge connectivity and vertex connectivity of two isomorphic graphs shown below are each equal to three.

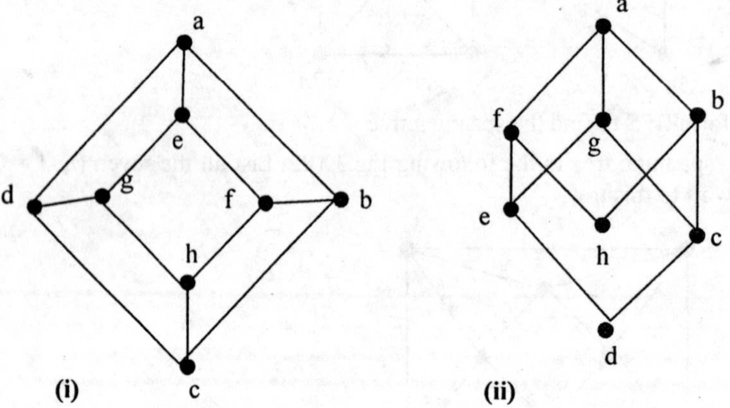

Fig. 3.105

32. What is the edge connectivity of the complete graph of n-vertices?
33. (i) Show that a graph G is nonseparable iff every vertex pair in G can be placed in some circuit in G.
    (ii) How can you use above result to obtain an algorithm for identifying every block of a large separable graph?
34. Prove that in a nonseparable graph G the set of edges incident on each vertex of G is a cut-set.
35. Prove that a nonseparable graph has nullity $\mu = 1$ iff the graph is a circuit.
36. Show that a graph with m vertices and with vertex connectivity k must have at least $kn/2$ edges.
    Particular case: the degree of every vertex in a nonseparable graph is at least two.
37. Draw a graph G with
    (i) Edge connectivity $= 4$
    (ii) Vertex connectivity $= 3$

(iii) Degree of every vertex of $G \geq 5$.

38. Show that a simple graph is nonseparable iff for any two given arbitrary edges a circuit can always be found that will include these two edges.

39. Is every regular graph of degree $d \geq 3$ nonseparable? If not, give a simple regular graph of degree that is separable.

ANSWERS

1(ii). (a) 3 (b) 16 (c) 4

3.

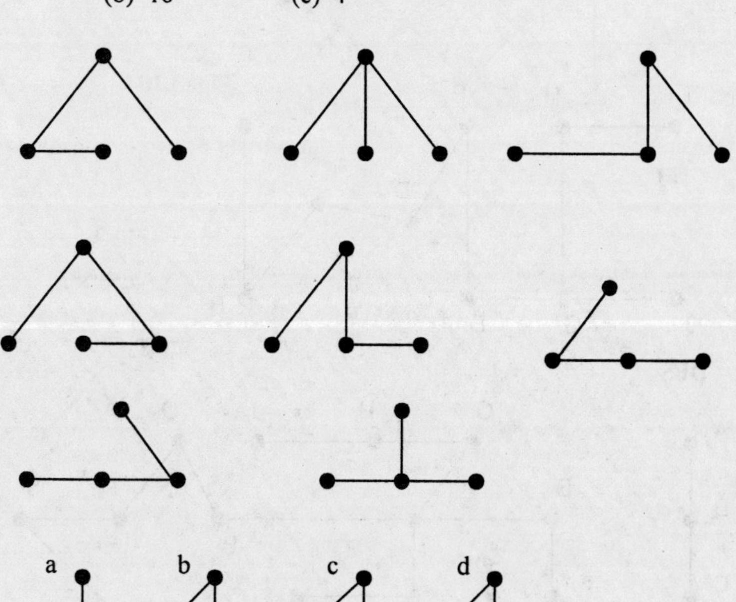

10. (d)

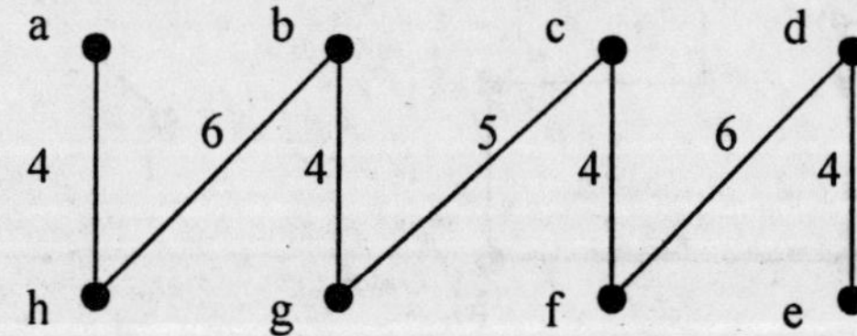

21. We obtain the following labelled trees, where label corresponds to the order in which the vertices are visited.

BFS DFS

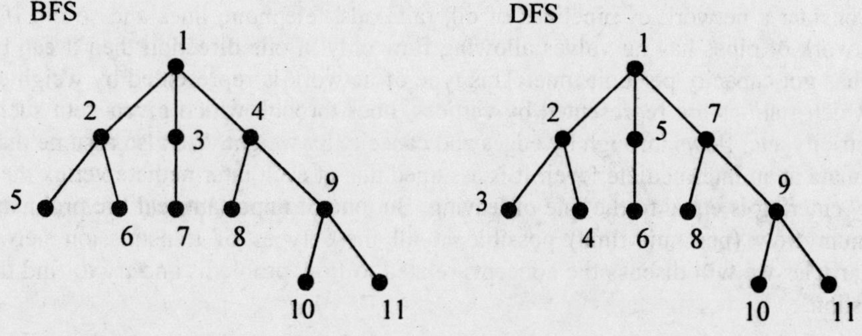

22. **DFS**: Path (*H, F, E, G, B, D, C, A*)

BFS

23. **DFS**

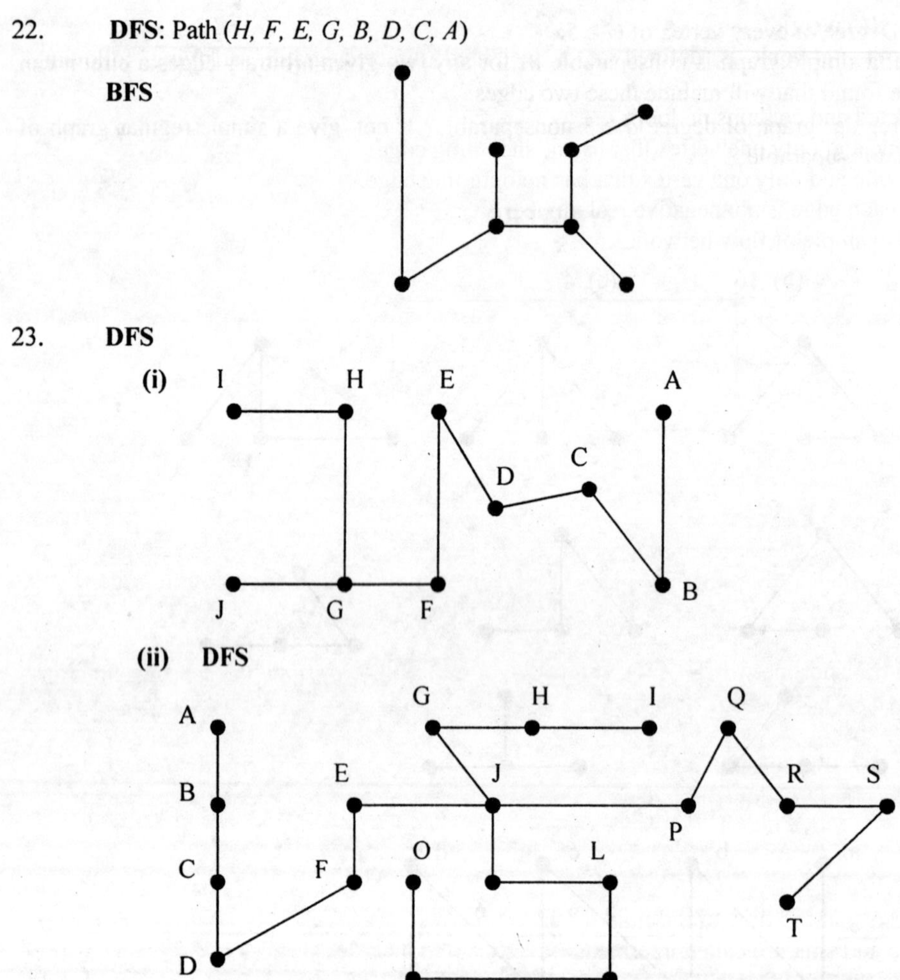

3.37 NET WORK FLOW PROBLEMS (TRANSPORT NETWORKS)

Consider a network of pipelines of oil, railroads, telephone lines and so on. If we take the case of network of pipes having valves allowing flow only in one direction then it can be noted that each pipe has got capacity per unit time. This type of network is represented by weighted connected graph in which stations are represented by vertices, lines through which given item such as oil, gas, water, electricity, etc. flows through by edges and capacity by weight. We also assume that flow could not accumulate at an intermediate level. It is assumed that at each intermediate vertex the total rate of commodity entering is equal to the rate of leaving. But one of important real life problems is: what is the maximum flow (per unit time) possible in all these types of transmission network? In the following articles we will discuss the concepts related to flow problems and try to find the answer of above question.

3.37-1 TRANSPORT NETWORK

A **transport network** is defined as a weighted directed graph satisfying following conditions:

(a) It is connected and contains no loops.
(b) Contains one and only one vertex that has no incoming edge.
(c) It contains one and only one vertex that has no outgoing edge.
(d) Weight of each edge is nonnegative real number.

Fig.3.106 shows an example of flow network.

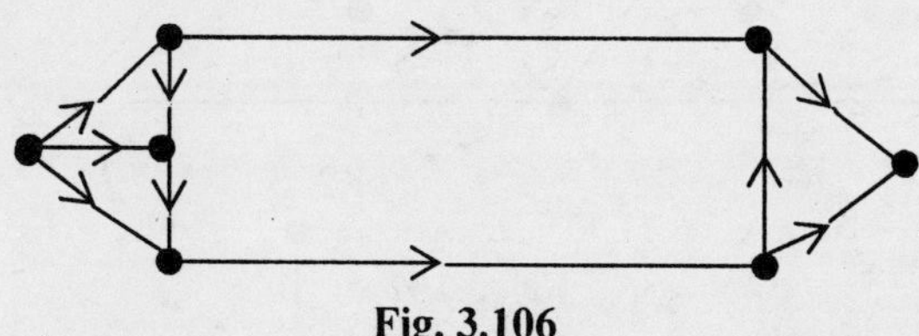

Fig. 3.106

3.37-2 SINK, SOURCE AND CAPACITY

In a network flow, the vertex that has no incoming edge is called the **source**, the vertex that has no outgoing edge is called the **sink**. The weight of an edge is called the **capacity** of the edge.

3.37-3 FLOW

Consider a transport network (G, w) with source A and sink Z. Assume the capacity function w is defined on the edge (i, j) of G. A **flow** in G is a nonnegative real valued function $f(i, j)$ defined on each edge (i, j) such that

1. $0 \le f(i, j) \le w(i, j)$ for each directed edge (i, j).
   This suggests that the flow through any edge (i, j) does not exceed its capacity.
2. $\sum_{\text{all } j} f(i, j) = \sum_{\text{all } j} f(j, i)$ for each vertex j except the source and sink.
   or $\sum_{\text{all } j} f(i, j) - \sum_{\text{all } j} f(j, i) = 0$.
   This suggests that the flow of the fluid from any intermediary vertex, say j (all vertices other then source and sink are called intermediary vertices) to all other vertices in G is same as the total flow of the fluid from all other intermediary in G to the intermediate vertex j.
3. $f(i, j) = 0$ for any edge (i, j) incident to the source or incident from the sink.

It is to be noted here that the condition (1) ensures that amount of flow along an edge does not exceed capacity for that edge, and condition (2) ensures that except at the source and at the sink, the amount of item flowing into vertex must equal to the fluid flowing out of the intermediate vertices. Condition (1) is called the **capacity constraint** and condition (2) is called the **conservation equation**. Condition (2) ensures that the flow is conserved at each intermediary vertex, i.e. the total flow through all the edges incident into any intermediate vertex j in G is equal to the total flow through all the edges incident out of the same intermediary vertex j. Condition (3) ensures that the flow moves from source to sink and not in the opposite direction.

NOTE

Every transport network has at least one flow, namely the zero flow where $f(i, j) = 0$ for each edge (i, j).

We follow the convention of labeling each edge in such a way that the first number represents the capacity of the edge and the second number represents the flow in the edge. Fig.3.107 shows the flow in the transport network.

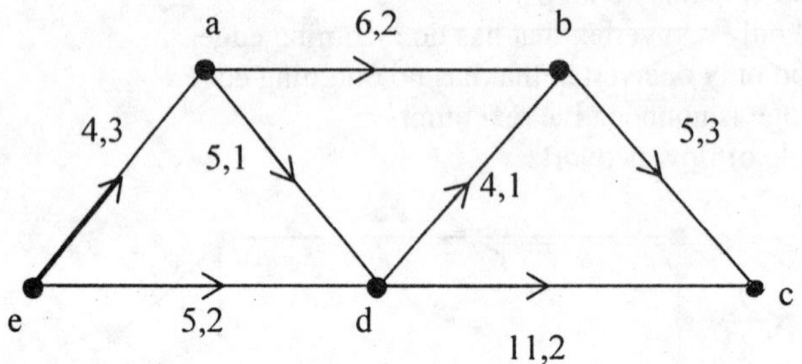

Fig. 3.107

3.37-4 VALUE OF THE FLOW

The quantity $\Sigma_{\text{all } i} f(A, a_i)$ is said to be the **value of the flow** f, denoted by f_v.

$$f_v = \Sigma_{\text{all } i} f(A, a_i) = \Sigma_{\text{all } k} f(b_k, Z) \quad \text{...(1)}$$

where A and Z are source and sink of the network respectively and a_i denote all the vertices that are incident from the source and b_k denote all the vertices incident to the sink.

Equation (1) says the total outgoing flow at the source is equal to the total incoming flow at the sink. Mathematically we write

$$\begin{aligned} f_v &= f(A, a_1) + f(A, a_2) + f(A, a_3) + \cdots f(A, a_i) \ldots + f(A, a_m) \\ &= f(A, V(G)) \quad \forall i \\ &= f(b_1, Z) + f(b_2, Z) + \cdots f(b_k, Z) \ldots + f(b_n, Z) \\ &= f(V(G), Z) \quad \forall k \end{aligned}$$

OBSERVATION

From the definition of flow we can have following observations:

(a) Flow through any edge does not exceed its capacity.

(b) There is a specific vertex, A (called source) in every flow network for which

$$\sum_i f_{Ai} - \sum_i f_{iA} = f_v$$

where summation is taken over all the vertices in f_v and f_v is the value of the function.
One can see here that the total flow of fluid through all the edges incident out of source A exceeds the total flow through all the edges that are incident into the source A. In other words we can say the total flow of the fluid from the source to all the vertices in G exceeds the total flow of fluid from all the vertices in G to the source A by a real positive quantity f_v.

(c) Like source there exist another specified vertex Z (called sink) for which

$$\sum_i f_{Zi} - \sum_i f_{iZ} = -f_v \quad \text{... (1)}$$

Here we can see reverse to that of source, (1) indicates that total flow through the edges that are incident out of sink Z fall short of all the edges that are incident into the sink A. In other words the total flow of fluid from all the vertices in G falls short of the flow of fluid from the vertices in G to the sink by a real positive quantity f_v. It states that the net flow of fluid into the sink is f_v.

Example 3.54

The flow in Fig. 3.108 has value 9.

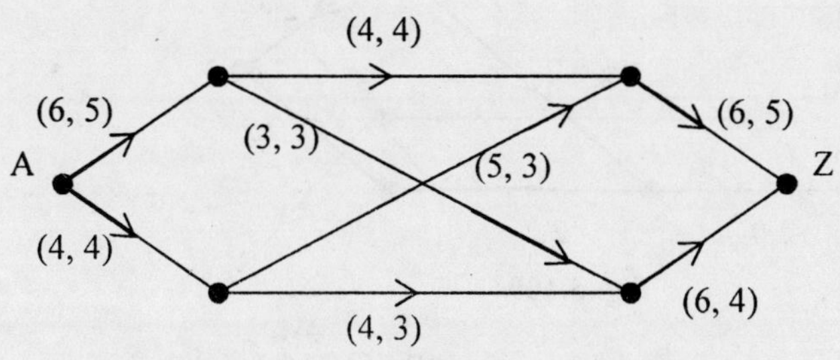

Fig. 3.108

THEOREM 3.26

Let A and Z be the source and sink respectively of a network G. Let f be a flow in G. Then the flow out of $A = f(A, V(G)) = f(V(G), Z)$ = the flow into Z, where $V(G)$ is the set of vertices of G.

PROOF

We can see that

$$f(V(G), V(G)) = \sum_{x \in V(G)} f(x, V(G)) = \sum_{y \in V(G)} f(V(G), y)$$

[But we know that, if a is any vertex, different from the source or the sink, then the sum of all values $f(a, b)$ such that $b \in A(a)$ must equal to the sum of all values $f(c, a)$ such that $c \in B(a)$.]

$$\therefore \quad f(y, V(G)) = f(V(G), y)) \text{ for each } y \in V(G) \mid \{A, Z\}.$$

$\therefore$ Thus, the above sum becomes

$$f(A, V(G) + f(Z, V(G)) = f(V(G), A)) + f(V(G), Z))$$

But, we also know that

$$f(V(G), A)) = 0 = f(Z, V(G)).$$

Since there is no flow into A and no flow out of Z.

$\therefore \quad f(A, V(G)) = f(V(G), Z)$. Hence the theorem.

3.37-5 SATURATED AND UNSATURATED FLOW

The flow along an edge (i, j) is said to be **saturated** if $f(i, j) = w(i, j)$, and if $f(i, j) < w(i, j)$ then the flow is said to be **unsaturated**.

A flow is said to be **feasible** if it satisfies the capacity constraint for each edge and conservation constraint for each vertex.

A **maximum flow** or **maximal flow** in a transport network is a flow that achieves the largest possible value. *It is to be noted that in a transport network there may be more than one maximum flow.*

Example 3.55

Consider the network shown in the Fig.3.109. The **zero flow** assigns flow 0 to each edge (i.e. it is feasible). In the network shown in Fig.3.109 each edge is having capacity 2. Here $f(s,c) = f(b,t) = 0$ and $f(i, j) = 2$ for every other edge (i, j). This is a feasible flow of value 2.

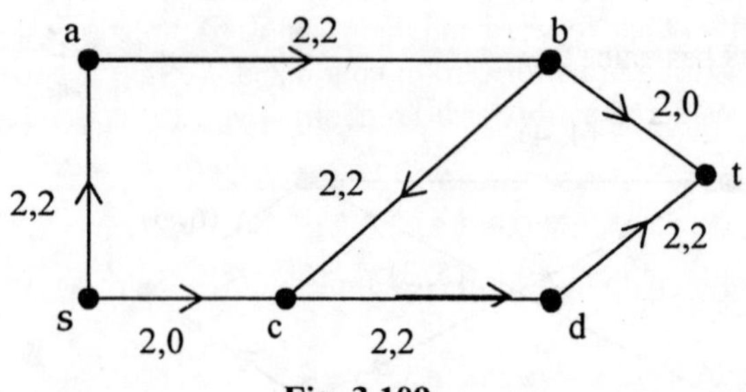

Fig. 3.109

Example 3.56

Given the following network, find the maximal flow by trial and error.

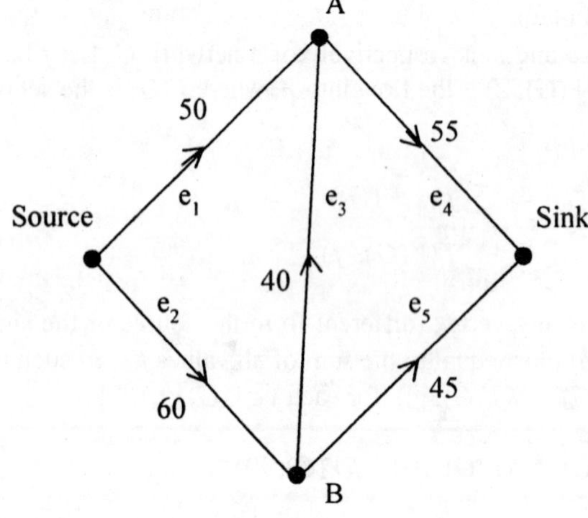

Fig.3.110

Solution

One flow says f_1 may be defined as:

$f_1(e_1) = 50, f_1(e_2) = 45, f_1(e_3) = 0,$

$f_1(e_4) = 50, f_1(e_5) = 45$, and the value of the flow $f_{1v} = 95$.

The total flow into the sink cannot be more than 100; f_1 is not very far from maximal flow.

Another flow f_2 may be defined as follows:

$f_2(e_1) = 50, f_2(e_2) = 50, f_2(e_3) = 10, f_2(e_4) = 55$ and $f_2(e_5) = 45$ with value of the flow $f_{2v} = 100$.

This is clearly the maximal flow and this solution is obviously not unique.

REMARK

1. The sum of capacities into the sink cannot always be obtained by a flow.
2. Sum of the capacities of the source cannot always be realized.

3.37-6 FLOW AUGMENTING PATH

Let f be a flow in a network (V, E). A **flow augmenting** path with respect to f is a path from source to sink using edges both in their forward and their backward directions such that for each edge in the path $w(e) - f(e) > 0$ if e is used in its forward direction and, $f(e) > 0$ if e is used in the backward direction [Source to sink is considered as forward direction].Here $w(e)$ is the capacity and $f(e)$ is the value of the edge e.

Example 3.57

Find the flow augmenting path to f_1 for the example 3.56.

Solution

A flow augmenting path would be (e_2, e_3, e_4).

$$w(e_2) - f_1(e_2) = 60 - 45 = 15$$
$$w(e_3) - f_1(e_3) = 40 - 0 = 40$$
$$w(e_4) - f_1(e_4) = 55 - 50 = 5$$

Three positive differences represent unused capacities, and the smallest value represents the amount of flow that can be added to each edge in the path. By adding 5 to each edge in one path we obtain f_2, maximal flow.

NOTE

If an edge with a positive flow is used in the backward direction, it is contributing a movement of material that is counter productive to the objective of maximizing the flow. This is the reason why the algorithm decreases the flow through edge.

3.37-7 FORD AND FULKERSON ALGORITHM

Ford and Fulkerson have given the algorithm to find the maximal flow, which is as follows:

1. Define the flow function f_0 by $f_0(e) = 0$ for each $e \in E$
2. $i = 0$
3. (i) If possible, find a flow augmenting path with respect to f
   (ii) If a flow augmenting path exists, then determine
   (*a*) $m = \min \{\{w(e) - f_i(e)$: e is used in the forward direction$\}$, $\{f_i(e)$: e is used in the backward direction $\}\}$.
   (*b*) Define f_{i+1} by
   $f_{i+1}(e) = f_i(e)$ if e is not used in the path
   $= f_i(e) + m$ if e is not used in the forward direction
   $= f_i(e) - m$ if e is used in the backward direction.
   (c) i: $i + 1$ until no flow augmenting path exists.
4. Terminate with a maximal flow f_i.

REMARK

1. Every flow augmenting path leads to a flow of increased value.
2. None of the capacities can be violated.

Example 3.58

Find the maximal flow for the network shown in Fig.3.111.

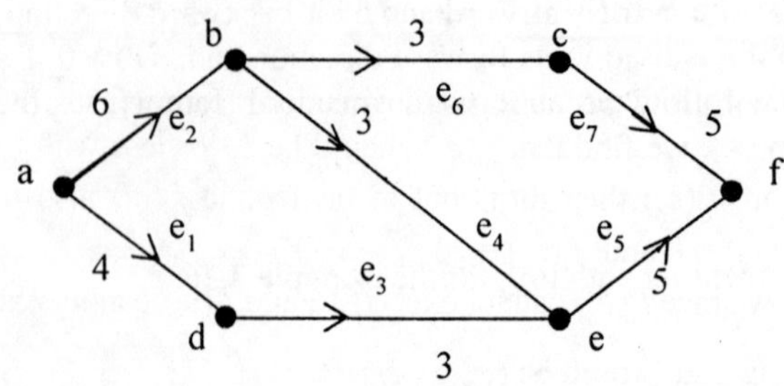

Fig. 3.111

Solution

We can observe that $f_1(e_i)$ is a flow, where

$f_1(e_1) = 3$ since $w(e_3) = 3$,

$f_1(e_2) = 6$ since $w(e_4) = 3$ and $w(e_6) = 3$,

$f(e_3) = 3, f(e_4) = 2$ since $w(e_5) = 5$, $f(e_6) = 3$,

$f(e_7) = 3$ since $f(e_6) = 3$ and $f(e_5) = 5$.

Note that f is obviously maximal, since no flow augmenting path exists as indicated by the Fig.3.112.

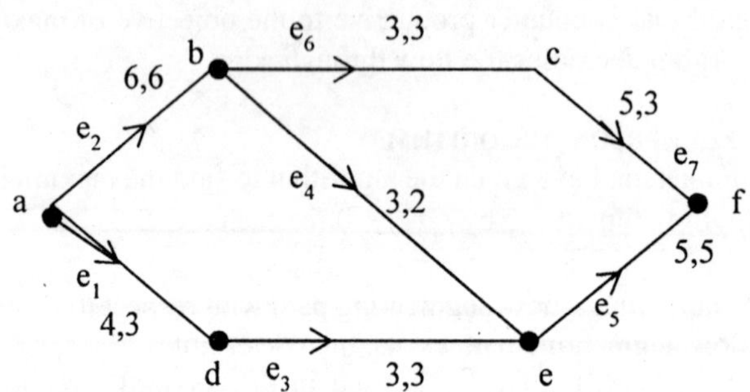

Fig. 3.112

Example 3.59

Use Ford and Fulkerson Algorithm to find the maximal flow of the network.

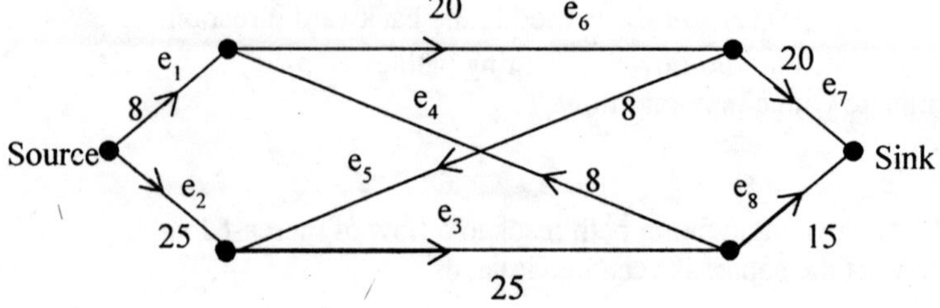

Fig.3.113

Solution

Let the flow f_1 is defined as follows:

$f_1(e_1) = 8$, $f_1(e_2) = 23$, $f_1(e_3) = 23$, $f_1(e_4) = 8$, $f_1(e_5) = 0$, $f_1(e_6) = 16$, $f_1(e_7) = 16$, $f_1(e_8) = 15$.

This is a maximal flow because no augmenting path exists using Ford and Fulkerson algorithm. If we see e_4 and e_6, we find that $f_1(e_3)$ should be 23 or less with $f_1(e_5)$ as zero to satisfy the definition of flow function. $f_1(e_6)$ then turns out to be 16 and $f_1(e_7)$ also turns out to be 16. Value cannot exceed 31 since $f_1(e_1) = 8$ and $f_1(e_2) = 23$.

f_1 is maximal flow since $f_1(e_6)$ cannot exceed 16 and $f_1(e_8)$ cannot exceed 15.

3.37-8 Cut and its Capacity

A **cut** in a transport network is a cut-set of the undirected graph, obtained from the transport network by avoiding the direction of the edges that separates the source from the sink.

The notation $(S, \bar{S})$ denotes the cut-set that divides the network into two portions, one containing source and the other containing sink (i.e. S contains the source and $\bar{S}$ contains the sink).

The **capacity of a cut** denoted by $W(S, \bar{S})$ is defined as the sum of the capacities of those edges incident from the vertices in S to vertices in $\bar{S}$; i.e.

$$W(S,\bar{S}) = \sum_{i \in S, j \in \bar{S}} W(i, j)$$

Specifically, we call a cut $(X, \bar{X})$ as a **minimum cut** if there is no $(Y, \bar{Y})$ cut such that sum of all capacities of edges from Y to $\bar{Y}$ is less than the sum of capacities of all edges from X to $\bar{X}$ i.e. $W(Y, \bar{Y}) < W(X, \bar{X})$.

For instance, in Fig. 3.114 the dashed line identifies a cut that separates the vertex subset $S = \{v_1, v_4\}$ from the vertex subset $\bar{S} = \{v_2, v_3, v_5\}$ in such a way that vertex subset S contains the source v_1 and vertex subset $\bar{S}$ contains sink v_5. The capacity of this cut is equal to the sum of capacities of edges $\{v_1, v_2\} + \{v_1, v_4\} + \{v_4, v_5\} = 10 + 9 + 12 = 31$.

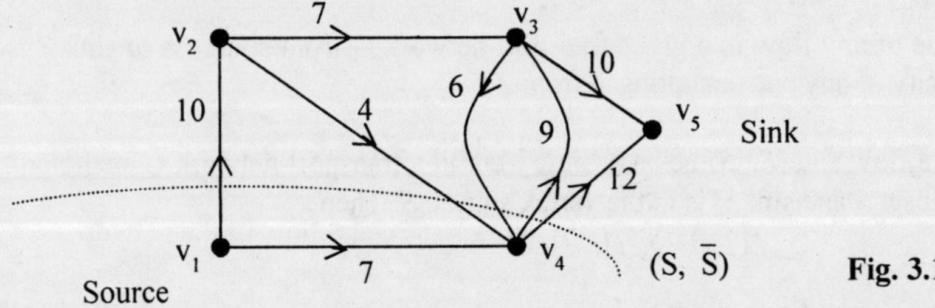

Fig. 3.114

Example 3.60

List all $(S, \bar{S})$ cuts in the transport network shown in the following figure.

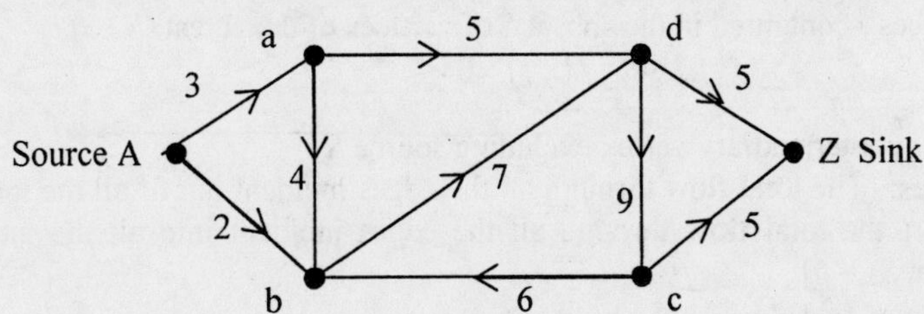

Solution

Possible $(S, \bar{S})$ cuts

S	$\bar{S}$	Capacity of $(S, \bar{S})$ $W(S, \bar{S})$
$\{A\}$	$\{a,b,c,d,Z\}$	5
$\{A,a\}$	$\{b,c,d,Z\}$	11
$\{A,b\}$	$\{a,c,d,Z\}$	10
$\{A,c\}$	$\{a,b,d,Z\}$	16
$\{A,d\}$	$\{a,b,c,Z\}$	19
$\{A,a,b\}$	$\{c,d,Z\}$	12
$\{A,a,c\}$	$\{b,d,Z\}$	22
$\{A,a,d\}$	$\{b,c,Z\}$	20
$\{A,b,c\}$	$\{a,d,Z\}$	15
$\{A,b,d\}$	$\{a,c,Z\}$	17
$\{A,c,d\}$	$\{a,b,Z\}$	21
$\{A,a,b,c\}$	$\{d,Z\}$	17
$\{A,a,b,d\}$	$\{c,Z\}$	14
$\{A,a,c,d\}$	$\{b,Z\}$	22
$\{A,b,c,d\}$	$\{a,Z\}$	13
$\{A,a,b,c,d\}$	$\{Z\}$	10

NOTE

In the above example when $S = \{A, a, b\}$, $f(S, \bar{S}) = 12 = f(a, d) + f(b,d) = 5 + 7$. The capacity of edge (c, b) does not enter into the computation of the capacity of S because the edge (c, b) is in the wrong direction, that is (c, b) is an edge from $\bar{S}$ to S and not from S to $\bar{S}$. The same is the case for (A, a), where edge (c, b) and (a, b) don't contribute any value of the capacity.

THEOREM 3.27

The value of any flow in a given transport network G from source A to sink Z is less than or equal to the capacity of any cut separating A from Z.

PROOF

Let f be a flow and $(S, \bar{S})$ be an arbitrary cut in a transport network G such that the source A is in the vertex subset S and sink Z is in the vertex subset $\bar{S}$. Then

$$\sum_i f(j,i) - \sum_i f(i,j) = 0 \qquad \ldots(1)$$

[Equation (1) states: Total flow of edges incident out of the vertex j - total flow of edges into vertex j = 0.]

Since equation (1) is true for all intermediary vertices in G, hence summing equation (1) for all the intermediary vertices j contained in the subset S of vertices of the cutest $(S, \bar{S})$.

$$\sum_i f(j,i) - \sum_i f(i,j) = 0 \qquad \ldots(2)$$

where j denotes any intermediary vertex excluding source S.

[Equation (2) states: (The total flow through all the edges incident out of all the intermediary vertices contained in S) - (the total flow through all the edges incident into all the intermediary vertices contained in subset S) = 0]

But we know that

$$\sum_{i} f(A,i) - \sum_{i} f(i,A) = f_v \qquad \ldots(3)$$

[Equation (3) states: The total flow through all the edges incident out of the source A - the total flow through all the edges incident into the source A = f_v.]

One can note at this point that in the equation (2) all the vertices of the set S is included except the source vertex A, whereas the equation (3) includes only the source vertex A.

Adding equations (2) and (3), we get

$$\sum_{j\in S}\left[\sum_{i\in V} f(j,i) - \sum_{i\in V} f(i,j)\right] = f_v, \text{ where } V \text{ is the set of all vertices.}$$

or

$$\sum_{j\in S}\sum_{i\in V} f(j,i) - \sum_{j\in S}\sum_{i\in V} f(i,j) = f_v \qquad \ldots(4)$$

[Equation (4) states: Total flow through all the edges incident out of all the vertices contained in the set S - total flow through all the edges incident into all the vertices contained in S = f_v.]

Equation (4) can also be written as

$$\sum_{\substack{j\in S\\ i\in V}} f(j,i) - \sum_{\substack{j\in S\\ i\in V}} f(i,j) = f_v$$

[Summation over j in equation (2) does not include the source vertex A whereas in equation (4) it includes.]

Breaking both the terms of equation (4) by taking summation over $i \in V$ = summation over $i \in S$ + summation over $i \in \bar{S}$ (since $V = S \cup \bar{S}$), we get

$$\left(\sum_{j\in S}\sum_{i\in S} f(j,i) + \sum_{j\in S}\sum_{i\in \bar{S}} f(j,i)\right) - \left(\sum_{j\in S}\sum_{i\in S} f(i,j) + \sum_{j\in S}\sum_{i\in \bar{S}} f(i,j)\right) = f_v$$

or

$$\sum_{\substack{j\in S\\ i\in S}} f(j,i) + \sum_{\substack{j\in S\\ i\in \bar{S}}} f(j,i) - \sum_{\substack{j\in S\\ i\in S}} f(i,j) - \sum_{\substack{j\in S\\ i\in \bar{S}}} f(i,j) = f_v \qquad \ldots(5)$$

Readers can note here that all the edges taken out (into) all the vertices contained in equation (4) have been taken as equal to all the edges incident out of (into) all the vertices contained in S and incident into (out of) vertices contained in S + all the edges incident out (into) all the vertices contained in S and into (out of) all the vertices contained in $\bar{S}$ in equation (5).

But the total flow through all the edges is given by

$$\sum_{\substack{j\in S\\ i\in S}} f(j,i) - \sum_{\substack{j\in S\\ i\in S}} f(i,j) = 0 \qquad \ldots(6)$$

Since both sums run through all the vertices in S. Therefore, substituting equation (6) into equation (5), we have

$$\sum_{\substack{j\in S\\ i\in \bar{S}}} f(j,i) - \sum_{\substack{j\in S\\ i\in \bar{S}}} f(i,j) = f_v \qquad \ldots(7)$$

Since $\sum_{\substack{j\in S\\ i\in \bar{S}}} f(i,j)$ is always a nonnegative quantity, hence equation (7) becomes

$$f_v \le \sum_{\substack{j \in S \\ i \in \bar{S}}} f(j,i) \le \sum_{\substack{j \in S \\ i \in \bar{S}}} W(j,i) = W(S,\bar{S})$$

3.37-9 MAX-FLOW MIN-CUT THEOREM

In a given transport network G, the maximum value of a flow from source A to sink Z is equal to the minimum value of the capacities all the cuts in G that separates A from G.

PROOF

We have seen in above theorem that the value of the flow f_v from source A to sink Z is less than or equal to the capacity of any cut separating A from Z. Hence here we need only to prove that there exist a flow in a network G such that the flow w_0 from source A to sink Z is equal to the capacity $W(S, \bar{S})$ of some cut $(S, \bar{S})$ separating A from Z.

Consider a flow pattern in G such that the value of the flow from A to Z is at maximum possible value w_0.

Let us define a vertex set S in G recursively in following way:

(i) $s \in S$

(ii) If vertex $i \in S$ and either $f(j, i) > 0$ or $f(i, j) < W(i, j)$, then $j \in S$. Any vertex not in S belongs to $\bar{S}$.

The vertices that are not in S will be in $\bar{S}$.

Further, it can be observed that the vertex Z cannot be in S. Because if sink Z will be in S then there would be a path, say Γ from source A to sink Z, say $A, v_1, v_2, \ldots, v_j, v_{j+1}, \ldots, v_k, Z$ for which in every edge either flow $f(v_{j+1}, v_j) > 0$ or $f(v_j, v_{j+1}) < W(v_j, v_{j+1})$ [by assumption (ii) above]. In the path Γ if an edge (v_j, v_{j+1}) is directed from v_j to v_{j+1} then it is called a *forward edge* and an edge (v_{j+1}, v_j) directed from v_{j+1} to v_j is called a *backward edge*. See following figure:

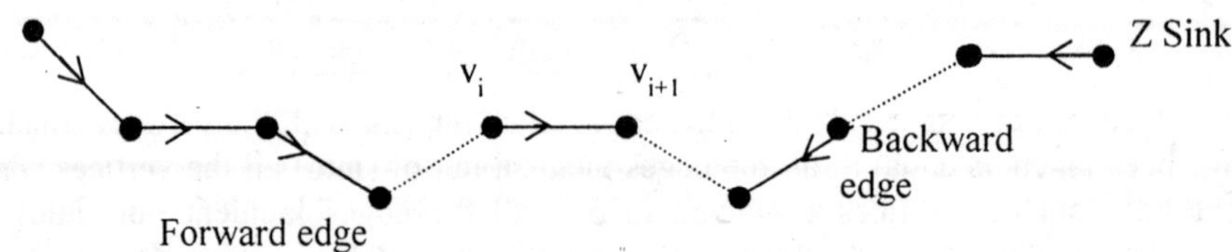

Let e_1 be the least of all differences of flow $W(v_j, v_{j+1}) - f(v_j, v_{j+1})$ in forward edges and e_2 be the least of all such flows in backward edges. e_1 and e_2 are positive quantities. Let $e = \min(e_1, e_2)$. Now to increase the flow of the network G, we have to increase the flow in each forward edge and decrease the flow in each backward edge by an amount e. But this contradict the assumption that w_0 is the maximum flow. This implies that Z should be there in the vertex set $\bar{S}$ not in S. In other words, we can say the cut $(S, \bar{S})$ separates A from Z.

Now, for the condition (ii), each vertex j in S and i in $\bar{S}$, we have

$f(j, i) = W(j, i)$ and $f(i, j) = 0$

$\therefore$ from above theorem, we get the value of the flow:

$$W_0 = \sum_{\substack{j \in S \\ i \in \bar{S}}} f(j,i) - \sum_{\substack{j \in S \\ i \in \bar{S}}} f(i,j)$$
$$= \sum_{\substack{j \in S \\ i \in \bar{S}}} W(j,i) = W(S,\bar{S})$$

Hence the theorem.

REMARK

1. If a transport network (G, W) has a flow f with value f_v, then network either contains an f-augmenting path or a cut $(S, \bar{S})$ such that capacity $W(S, \bar{S}) = f_v$.
2. A flow f is maximal iff there is no f-augmenting path from A to Z.
3. A cut $(S, \bar{S})$ is a minimal cut iff every maximal flow f is such that
   (a) $f(e) = W(e)$ for each edge $e \in (S, \bar{S})$ and
   (b) $f(e) = 0$ for each edge $e \in (S, \bar{S})$.

Example 3.61

In the Fig. 3.115 the value of the flow and the capacity of the cut are both 8, therefore, the flow is maximal and the cut is minimal.

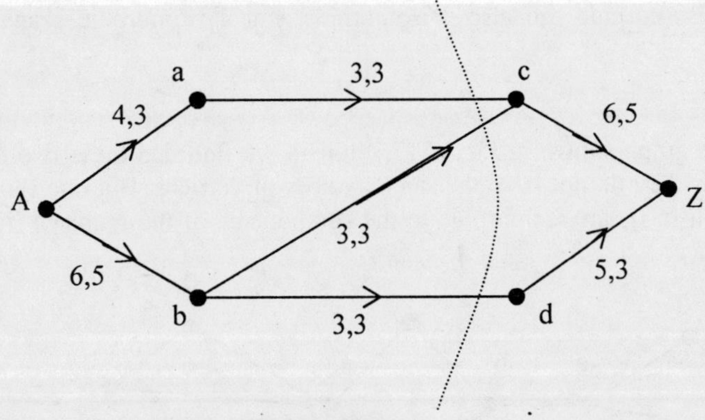

Fig.3. 115

3.38 BLOCKS

A separable graph can be decomposed into two or more than two subgraphs. A **block** of graph G is a maximal connected nonseparable subgraph of G. Block of a graph G can also be defined as a maximal connected subgraph of G that has no cut vertex i.e. if G itself is connected and has no cut-vertex (nonseparable) then G is a block. Graph shown in the Fig.3.116 have three blocks and two cut vertices (a and b).

NOTE

1. Some authors use the term component for block.
2. An edge of a cycle cannot itself be a block because it belongs to a larger subgraph having no cut-vertex. Therefore, an edge is a block of G iff it is a cut-edge of G. The blocks of a tree are its edge.

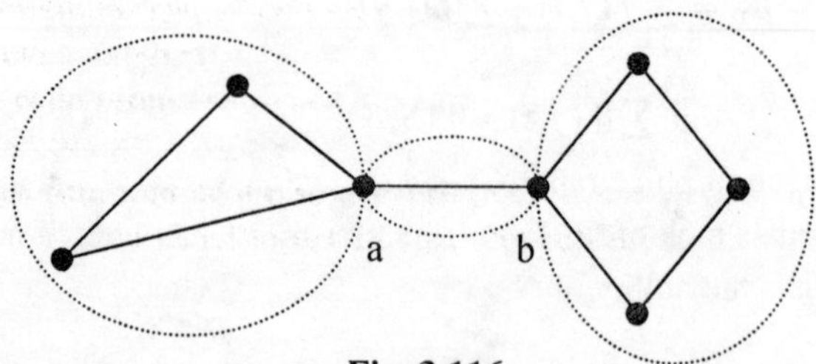

Fig. 3.116

3. Two blocks in a graph share at most one vertex.

3.38-1 1-ISOMORPHISM

Two graphs G_1 and G_2 are said to be **1-isomorphic** if they become isomorphic to each other when we repeatedly "decompose", a cut-vertex into two vertices to produce two disjoint subgraphs.

NOTE

1. Two nonseparable graphs are 1-isomorphic iff they are isomorphic to each other.
2. A nonseparable graph is one block; 1-isomorphism for nonseparable graphs is the same as isomorphism.
3. For separable graph 1-isomorphism is different from isomorphism.
4. Graphs that are isomorphic are also 1-isomorphic, but 1-isomorphic graph may not be isomorphic.

Example 3.62

If we compare the graph shown in Fig.3.117(i) to (ii), we find that these two graphs are not isomorphic to each other as they do not have the same number of vertices. But one can observe here that the blocks of the graph in (i) are isomorphic to the components of the graph in (ii) hence these two graphs are 1-isomorphic.

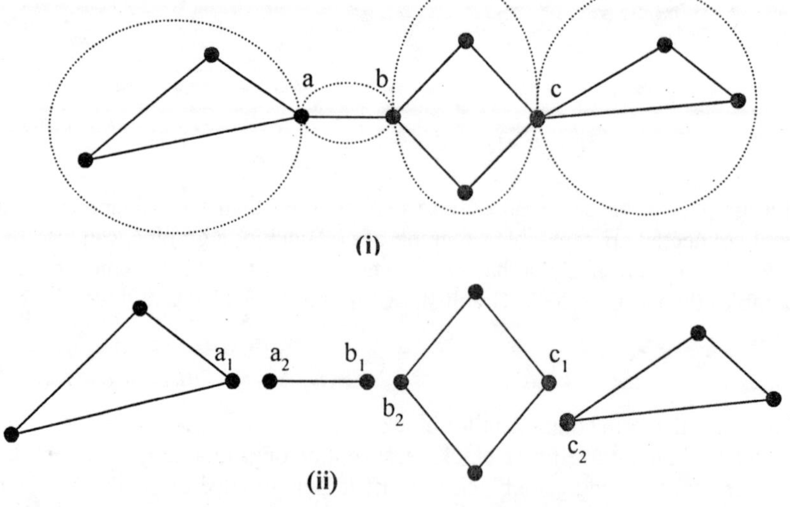

Fig. 3.117

Example 3.63

The graph shown in Fig.3.118 has $\{v_1, v_2, v_3, v_4\}$, $\{v_5, v_6, v_7, v_8\}$, $\{v_{10}, v_{11}\}$, $\{v_1, v_9\}$, $\{v_1, v_5, v_{11}\}$ blocks.

THEOREM 3.28

Consider two 1-isomiorphic graphs G_1 and G_2,

1. the rank of G_1 equals rank of G_2.
2. nullity of G_1 equals the nullity of G_2.

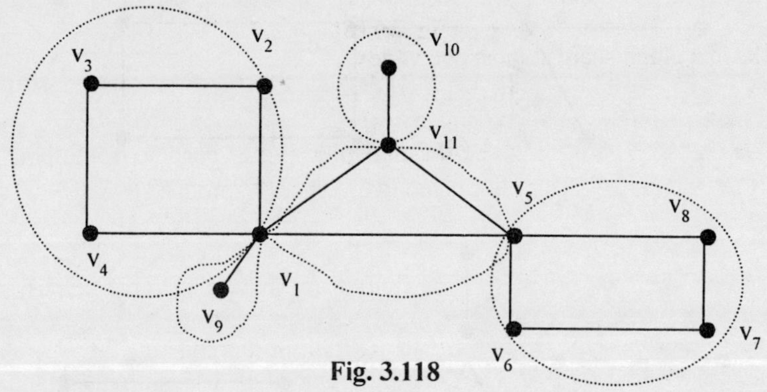

Fig. 3.118

PROOF

When we decompose a cut-vertex into two vertices, the number of components of G increases by 1. As we know,

rank of G = number of vertices in G - number of components in G.

And after performing the operation, the rank will remain same. Further, since no edges are destroyed or no new edges created by the operation, two 1-isomorphic graphs have equal number of edges. Since

Nullity = number of edges – rank

and two graphs with same number of edges and ranks will have the same nullity.

Example 3.64

The graphs shown in Fig.3.119 (i), (ii) and (iii) are 1-isomorphic to one another.

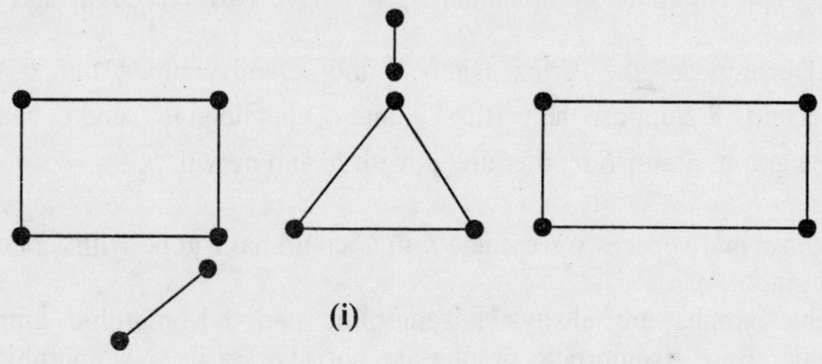

(i)

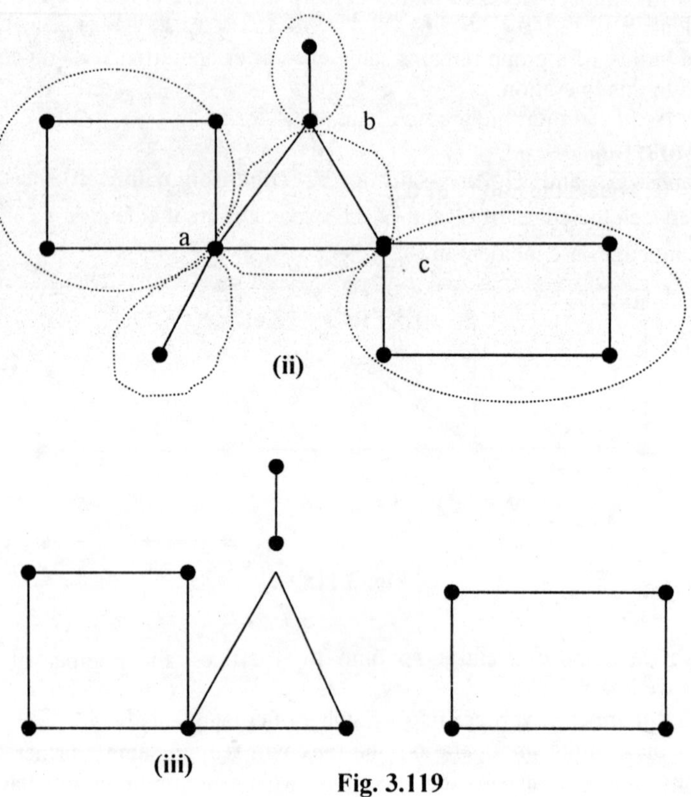

Fig. 3.119

3.38-2 2-ISOMORPHISM

Two graphs G_1 and G_2 are said to be **2-isomorphic** if they become isomorphic after undergoing operation 1 or operation 2 or both any number of times.

Operation 1 "Decompose" a cut-vertex into two vertices to produce two disjoint subgraphs.

Consider a 2-connected graph G and two vertices a and b. Let the deletion of two vertices a and b from G leave the remaining graph disconnected. The same can also be said in other words, G consists of a subgraph S and its complement $\bar{S}$, and have two vertices a and b in common. The operation 2 on G is

Operation 2 " Decompose" the vertex a into a_1 and a_2 and vertex b into b_1 and b_2 such that G decomposes into S and $\bar{S}$. Suppose the vertices a_1 and a_2 contained in S and b_1 and b_2 contained in $\bar{S}$. Now reconnect the graphs S and $\bar{S}$ by merging a_1 with b_2 and a_2 with b_1.

NOTE

1. Edges whose end vertices were a and b in G could have gone with S or $\bar{S}$, without affecting the final graph.
2. Isomorphic graphs are always 1-isomorphic and 1-isomorphic graphs are always 2-isomorphic. But 2-isomorphic graphs are not necessarily 1-isomorphic and 1-isomorphic graphs are not necessarily isomorphic.

3. A graph with connectivity 3 or more; isomorphism, 1-isomorphism, and 2-isomorphism are synonymous.
4. Rank and nullity of a graph remains same even after operation 2, as no edges or vertices are destroyed in this operation.

3.38-3 Edge Isomorphic

Two graphs G_1 and G_2 are said to be **edge-isomorphic** if there is a one-to-one correspondence between the edges of G_1 and G_2 s.t. two edges are incident (at a common vertex) in G_2 iff the corresponding edges are incident in G_2.

EXERCISE 3.6

1. Determine which of the following graph is transport network. Explain your answer.

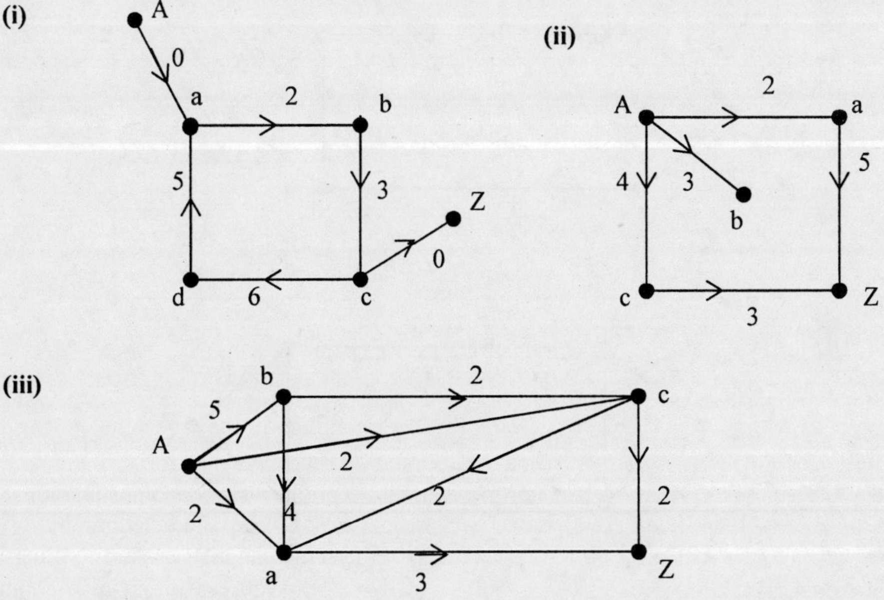

2. Determine the value of the flow and way to increase the value of the flow in the following networks with indicated flows. If it is not possible to increase the value, explain why not.

(i)

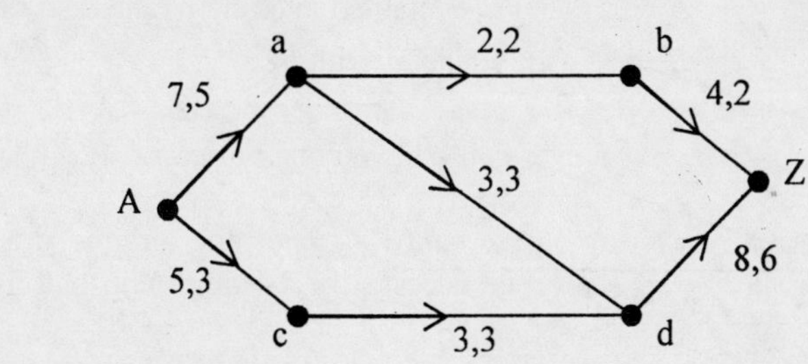

(ii)

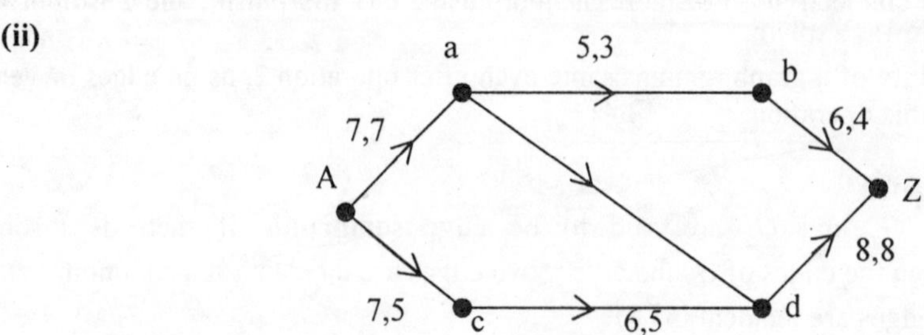

3. Fill in the missing edge flows so that the result is a flow in the given network. Also determine the values of the flow.

(i)

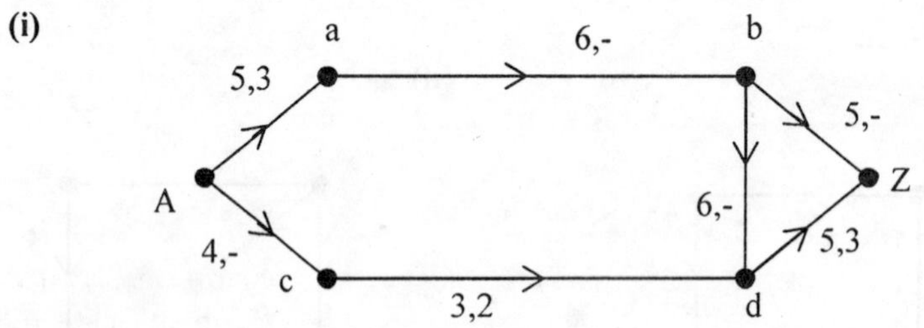

(ii)

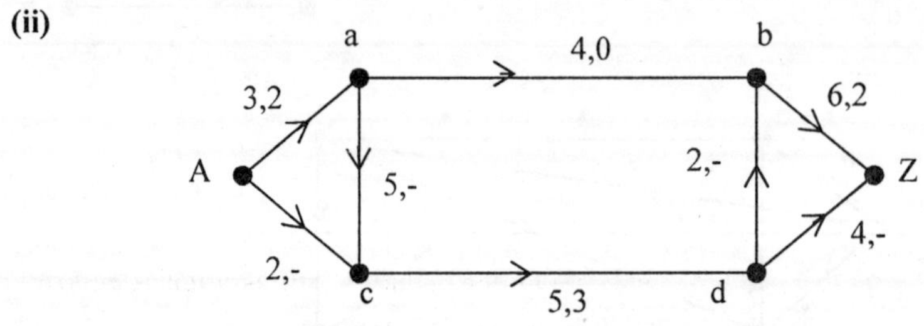

(iii)

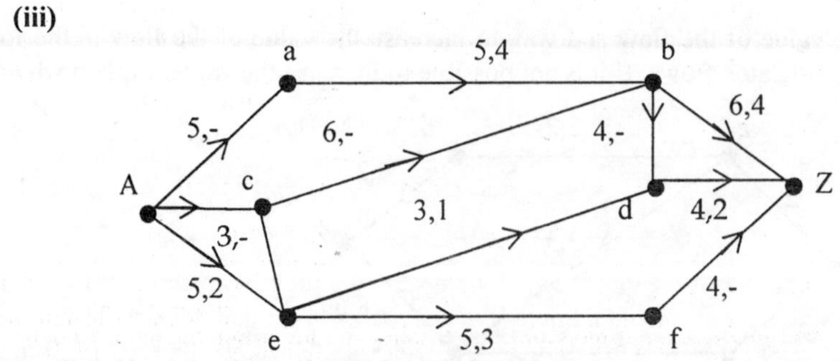

4. Consider the Fig. 3.120. Using the capacities network, determine which of the following Fig.121 are flows. Explain.

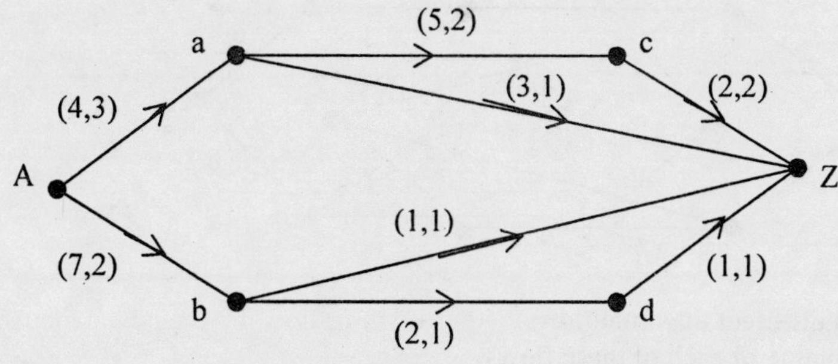

Fig.3.120

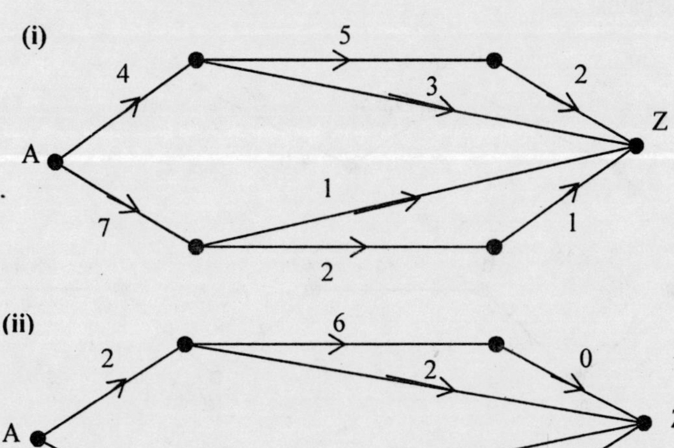

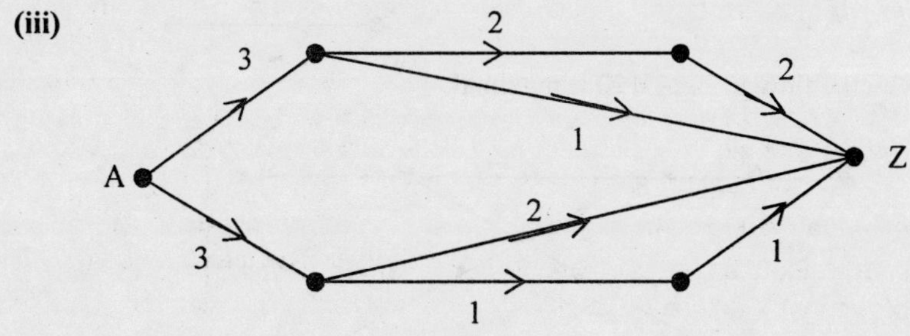

Fig. 3.121

5. Consider the network with the capacities shown below in figure.

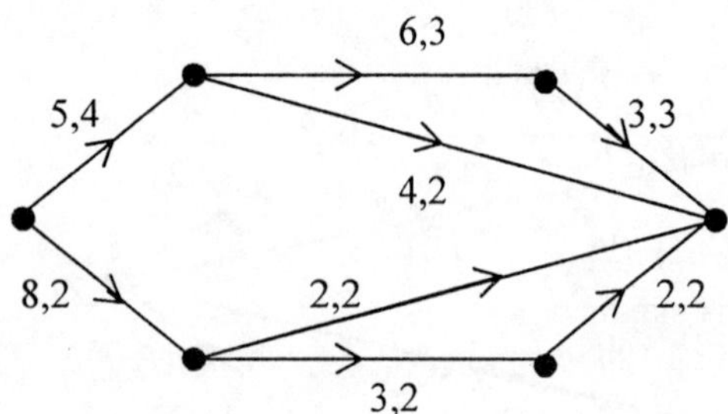

(i) find two different maximal flows.
(ii) find the value of each of these flows.

6. Find the maximum possible increase in the flow obtainable by altering the flow in the edges in the following paths.

(i)

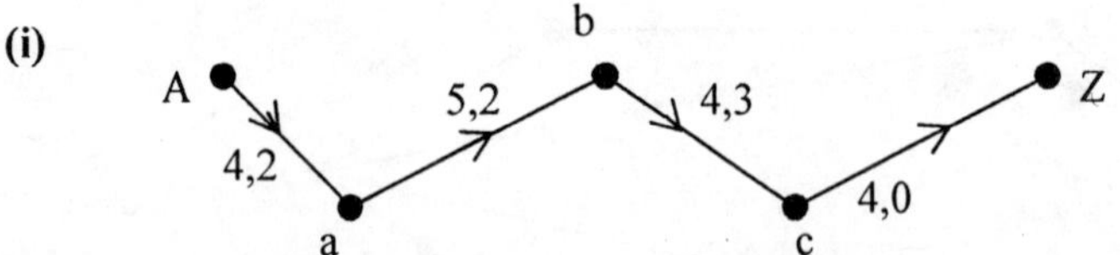

(ii)

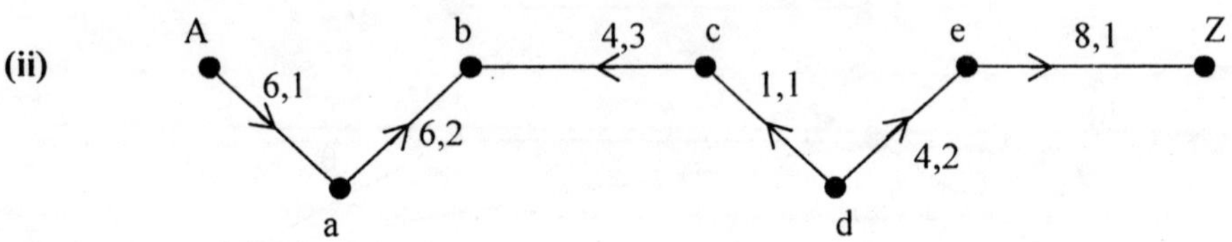

(iii)

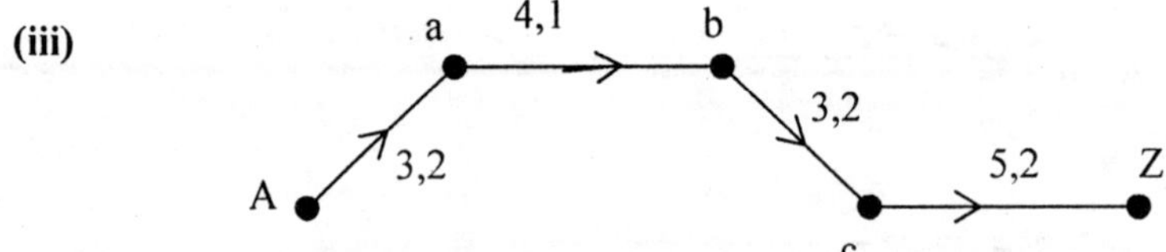

7. Explain why the flow in Fig.3.122 is maximal.

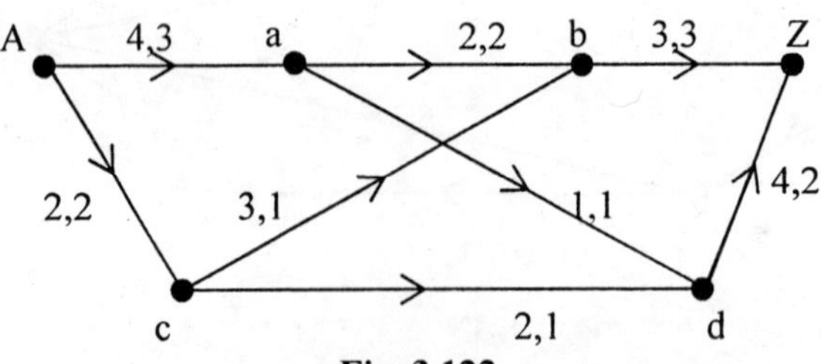

Fig. 3.122

[**Hint** As the value of the flow defined on the network is such that

$f = f(A, a) + f(A, c)$

$f(A,c) \le W(A,c) \le 2$

and $f = f(A,a) = f(a,d) + f(a,b)$

$\le W(a,d) + W(a,b)$

$= 1 + 2 = 3$

$\therefore \quad f_v = f(A,a) + f(A,c) \le 5$

$\Rightarrow$ the indicáted flow is maximal .]

8. For the network flow f in Fig.3.124, compute $f(P, Q)$, where $f(P,Q)$

(i)	$P = \{a,b,c\}$	$Q = \{d,e\}$
(ii)	$P = \{A,c\}$	$Q = \{a,b\}$
(iii)	$P = \{A,c\}$	$Q = \{b\}$
(iv)	$P = \{b\}$	$Q = \{A,c\}$
(v)	$P = \{a\}$	$Q = \{b\}$
(vi)	$P = \{A,a,b,c\}$	$Q = \{Z\}$
(vii)	$P = \{A,b,c\}$	$Q = \{a,d,e,Z\}$

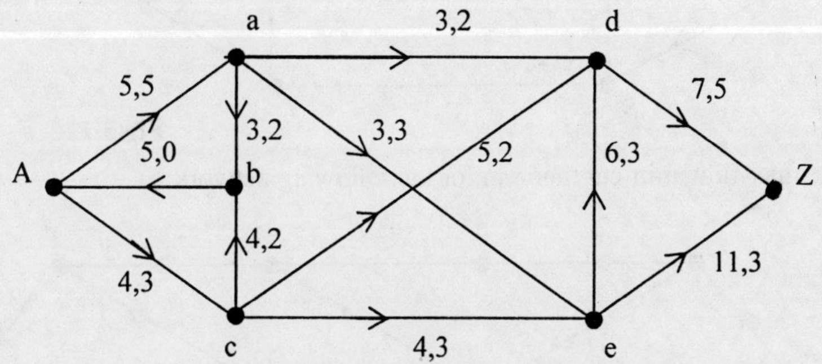

Fig. 3.123

9. Find the capacity for the cut $(S, \bar{S})$. Also determine whether the cut is minimal.

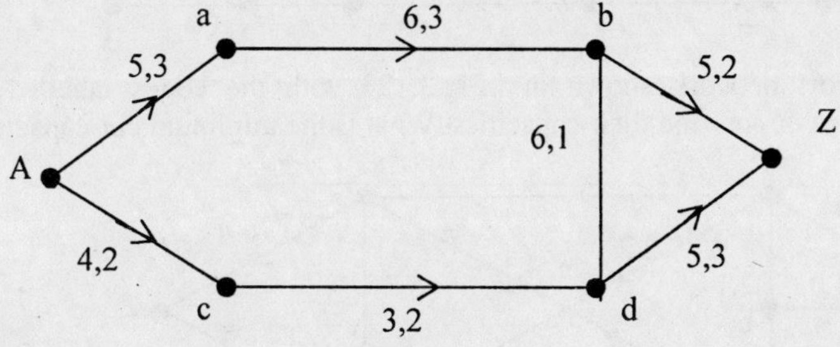

Fig.3.124

10. (i) Find a minimal cut of the following network.

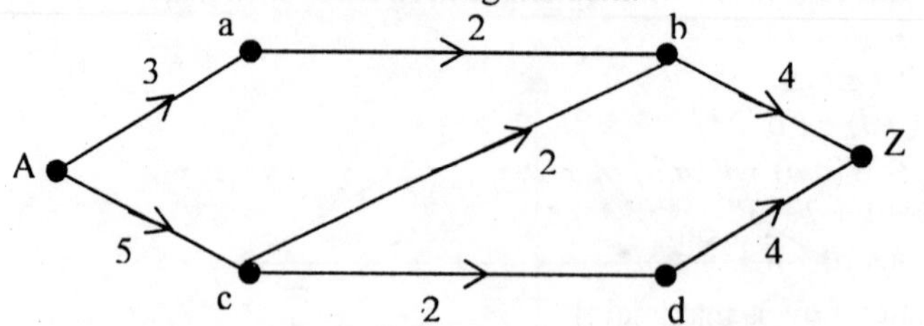

Fig. 3.125

(ii) Fig. 3.125

11. Consider the network of the Fig 3.125.

(i) List all the cuts in the network, and find a minimum cut.

(ii) Find a maximum flow, and verify the max-flow min-cut theorem.

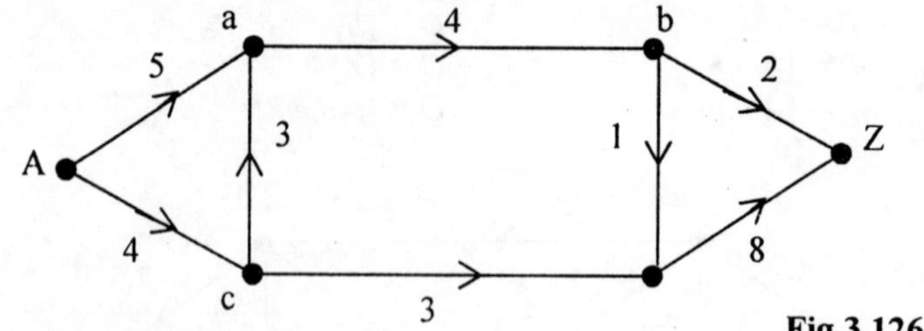

Fig.3.126

12. Verify the max-flow min-cut theorem for the following network.

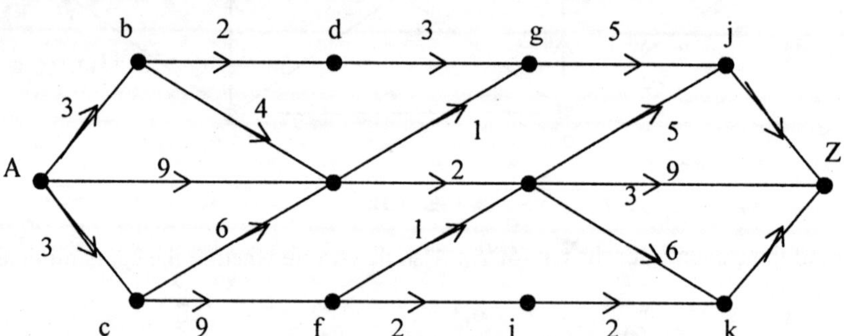

13. Consider a transport network shown in Fig.3.127 with the edges labelled with their capacities. Find all A-Z cuts and their capacities. What is the minimum cut-capacity?

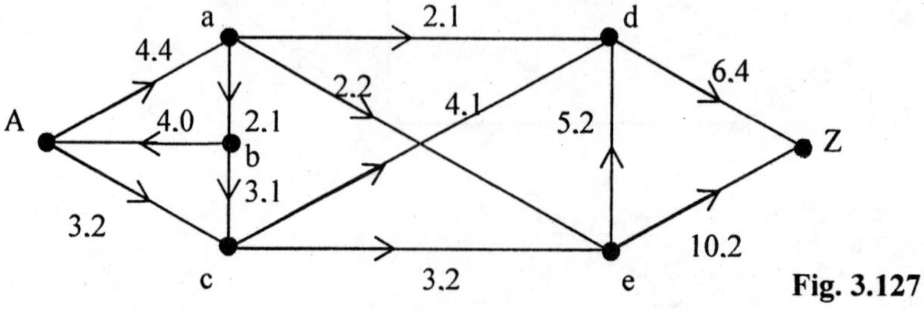

Fig. 3.127

14. Find the maximal flows for the following networks.

(i)

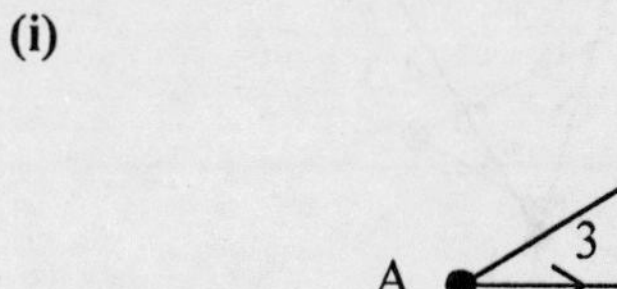
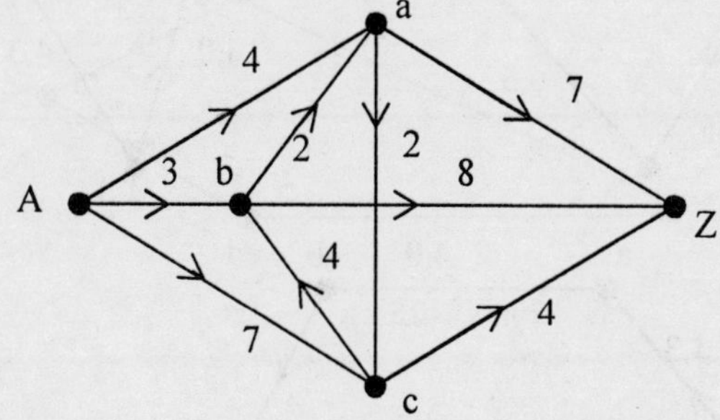

(ii)

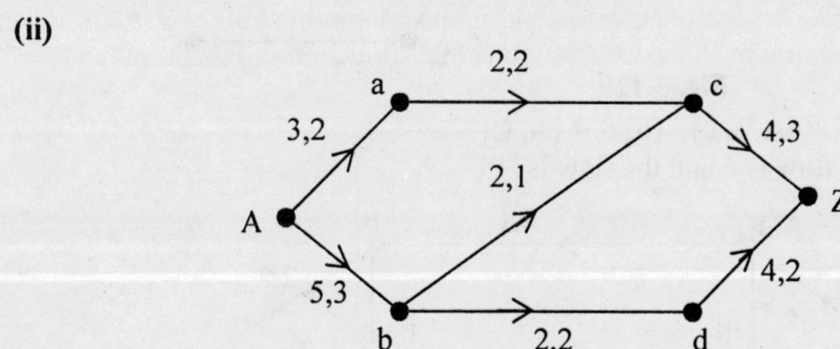

(iii)

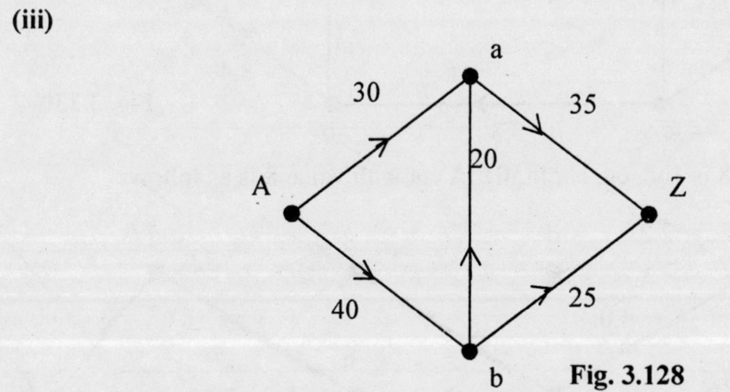

Fig. 3.128

5. Construct an example to show that edge isomorphic graphs may not be isomorphic.

ANSWERS

2. (i) 8, maximal flow

3. (i) $a, b : 6,3$; $A, e : 4,2$; $f_v = 5$
$b, d : 6,1$; $b, z : 5,2$
(ii) $a, c : 5,2$; $d, b : 2,2$; $A, c : 2,1$; $d, z : 4,1$; $f_v = 3$
(iii) $A, a, : 5, 3$; $A, c : 3,1$; $c, b : 6, 0$; $c,e : 2,1$; $e, d : 3,1$; b, Z: 5,2; $b,d : 4,1$; f, Z = 4,2

6. (ii)

(iii)

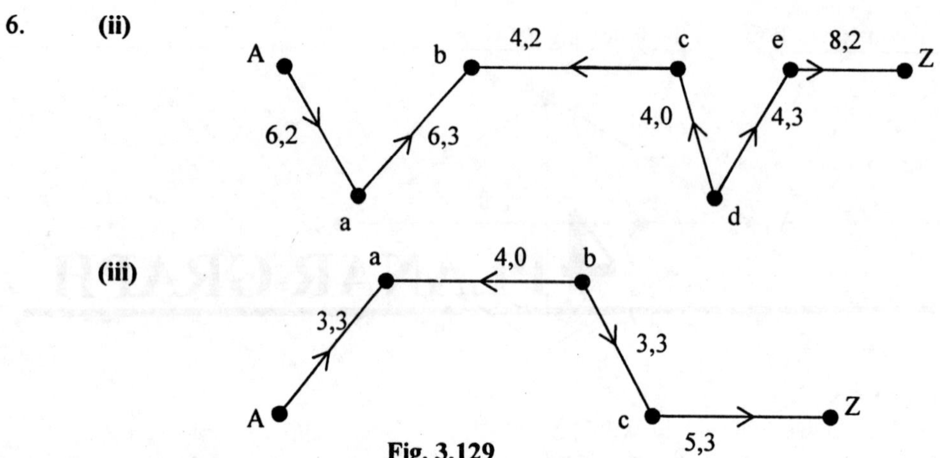

Fig. 3.129

9. 8, Minimal 9 (i). $S = \{A, a, c\}$ (ii) S = {A, C}.

10. Value of maximum flow is 6 and the flow is

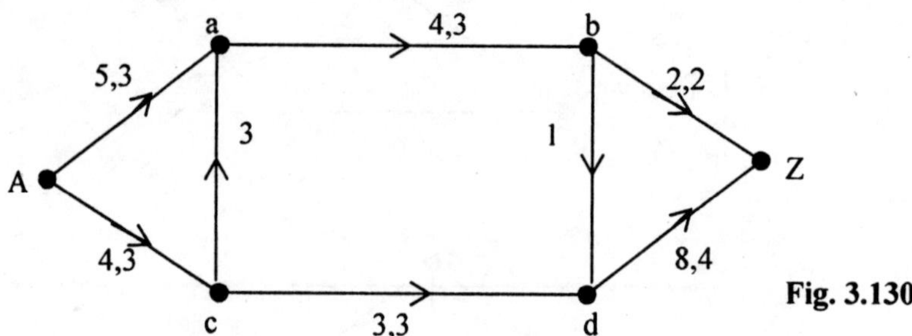

Fig. 3.130

12. A cut with capacity 8 is {bd, eg, eh, fh, fi}. A cut with value 8 is as follows

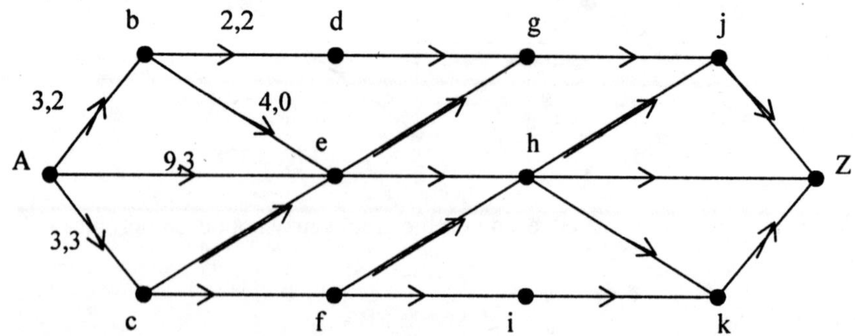

13. The 16 cuts have capacities 8, 8, 8, 9, 9,10, 11, 11, 12, 13, 14, 16, 17, 18, 19, 21.

14. (i) 14, (ii) 6 (iii) 60, many maximal flows.

4 PLANAR GRAPH

4.0 INTRODUCTION

In this chapter we will mainly study how to draw a graph without its edges crossing over. The planarity and related concepts are very useful in many practical situations. The knowledge of planarity is essential mainly to the electrical engineers when it is required to make a connection without an extra layer of insulation.

4.1 PLANAR GRAPH

A graph G is said to be planar if it can be drawn on a plane in such a way that no two of its edges intersect (do not cross); otherwise G is said to be nonplanar. It is to be noted that if a graph G has been drawn with crossing edges, this doesn't imply that G is nonplanar, as there are different ways to draw the graph without crossovers*.

For example, the graph shown in the Fig.4.1 (a) seems to be nonplanar, but it can be redrawn in Fig.4.1 (b) without crossovers.

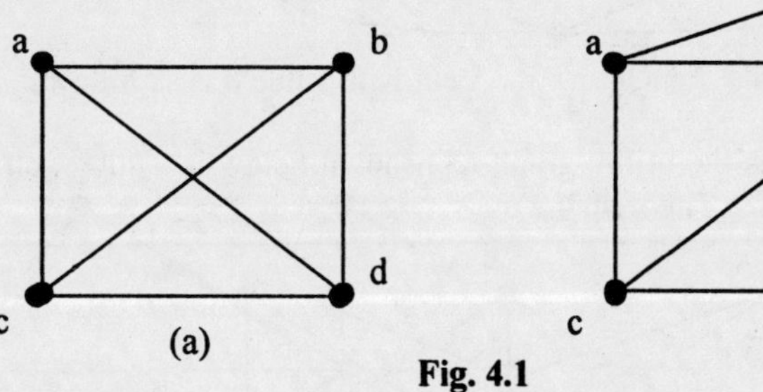

Fig. 4.1

NOTE

The meeting of edges at a vertex is not considered an intersection.

4.1-1 EMBEDDING

Embedding is a drawing of a geometric representation of graph on any surface. An embedding of a planar graph G on a plane such that not two edges intersect is called a **plane**

---

* The point of intersection of edges are called crossovers and the crossing edges are called cross-over each other.

representation of G. We say that a planar graph is a plane graph if it is drawn in a plane so that no two edges crossover.

Example 4.1

Draw a planar representation of both the graphs shown in Fig. 4.2, if possible.

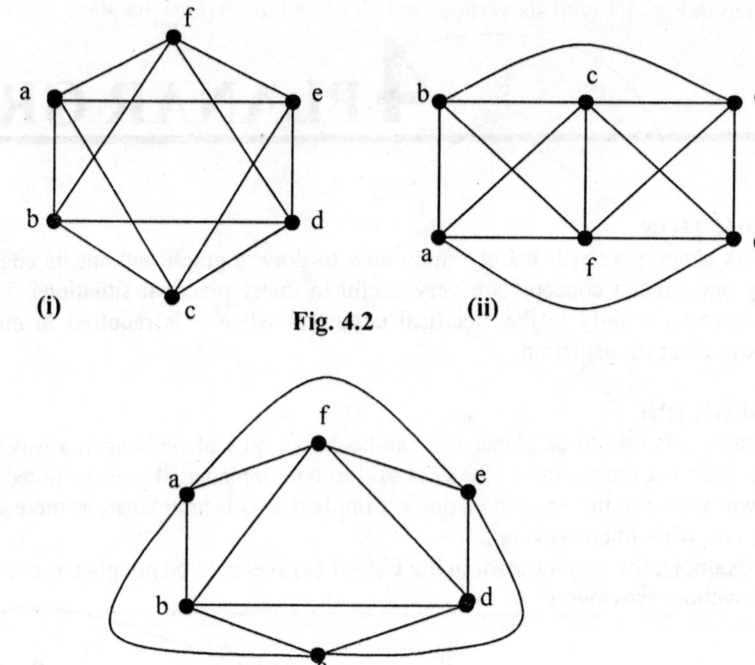

Fig. 4.2

Solution(i)

Fig. 4.3

(ii) This graph is nonplanar as the utility graph $K_{3,3}$ shown below is a subgraph where we have redrawn the position of d and e. (See Theorem 4.2)

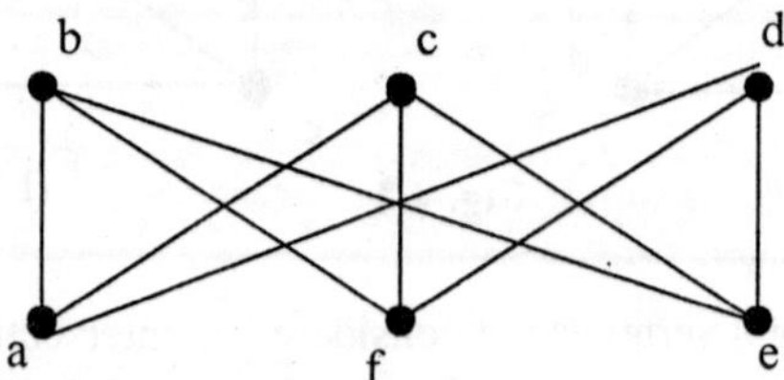

Utility graph

The utility problem consists of three houses and three utility companies - say, gas, electric, and water - and asks if each utility can be connected to each house without having any of the gas/ water/ electric lines/pipes pass over any other. H. E. Dudeney first posed the problem in this form in 1917. The utility graph UG is the graph showing the relationships described above. It is known in the more formal parlance of graph theory as the complete bipartite graph $K_{3,3}$.

4.1-2 MAPS AND REGIONS

A particular planar representation of a finite planar multigraph is called a **map**. If the underlying multigraph is connected then we say that the map is **connected**.

A plane representation of a graph divides the plane into **regions** or in other words we can say a given map divides the plane into various **regions** (also called **windows**, **faces**, or **meshes**). For example, the map in Fig. 4.4 with six vertices and eleven edges divides the plane into six regions.

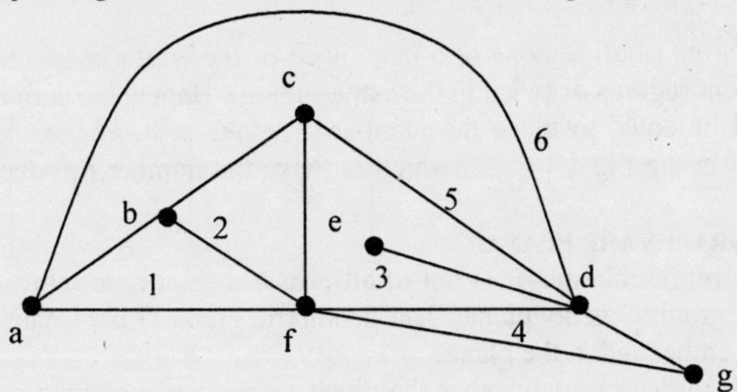

Fig. 4.4

Here, it is to be noted that five regions (1, 2, 3, 4, 5) are bounded while region 6 is not bounded.

NOTE

(i) A region is known by the set of edges or the set of vertices forming its boundary.

(ii) Region is not defined for a nonplanar graph.

(iii) Region is not defined for a planar graph which is not embedded in a plane. For example graph shown in Fig. 4.5 has no region.

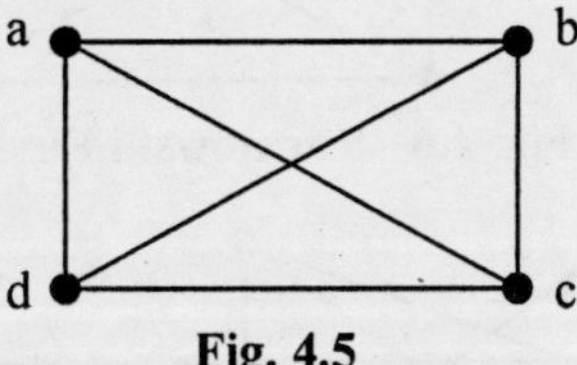

Fig. 4.5

(iv) Border of each region of a map consists of edges. Edges not always form a cycle.

(v) Region is a property of the specific plane representation of a graph.

(vi) The portion of the plane lying outside a graph embedded in a plane is infinite in extent. Such a region is said to be **infinite**, **outer**, **unbounded** or **exterior** region for that particular plane representation. The region 6 shown in the Fig.4.4 is unbounded. Any bounded region is called **interior region**.

(vii) A planar graph has exactly one infinite region.

(viii) By changing the embedding of a given planar graph, the infinite region can be changed.

4.1-3 DEGREE OF A REGION

The **degree** of any region r, represented as deg(r), means the length of the cycle or closed walk which borders region r. We can observe that each edge either borders two regions or is contained in a region and will occur twice in any walk along the border of the region.

In the graph shown in the Fig.4.4 boundary of all the regions are cycles except for region 3. But, in case we move anticlockwise in region 3, starting at the vertex say *c*, then the closed path obtained is (*c, d, e, d, f, c*), where the edge (*e, d*) occurs twice. The degrees of region 1, 2, 3, 4, 5, and 6 are respectively 3, 3, 5, 3, 4, 4.

THEOREM 4.1

The sum of the degrees of the regions of a map is equal to twice the number of edges.

PROOF

While calculating degree of all regions of a map, each of the edges is counted twice (either once in each of the two adjacent regions or twice in the same region). Hence, the sum of the degree of all the regions of a map will be equal to twice the number of edges in it. For example the sum of degrees of the regions shown in the Fig.4.4 is 22, which is twice the number of edges (see example 4.16).

4.1-4 DETERMINING WHEN GRAPHS ARE PLANAR

If none of geometric representation of *G* out of all possible geometric representation can be embedded in a plane then the graph *G* is nonplanar. But geometric graph *G* is planar if one can draw isomorphic graph of *G* that is embedded in the plane.

But the question how to determine whether the given graph is planar/nonplanar is answered by Kuratowski (Polish Mathematician) in the two graphs known by his name. A complete graph with five vertices (K_5) is the first graph of Kuratowski, and regular connected graph with six vertices and nine edges ($K_{3,3}$) is second graph of Kuratowski.

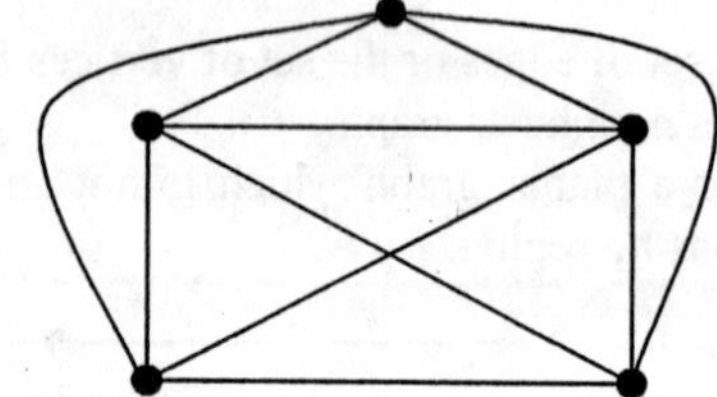

Fig. 4.6 Kuratowski First Graph K_5

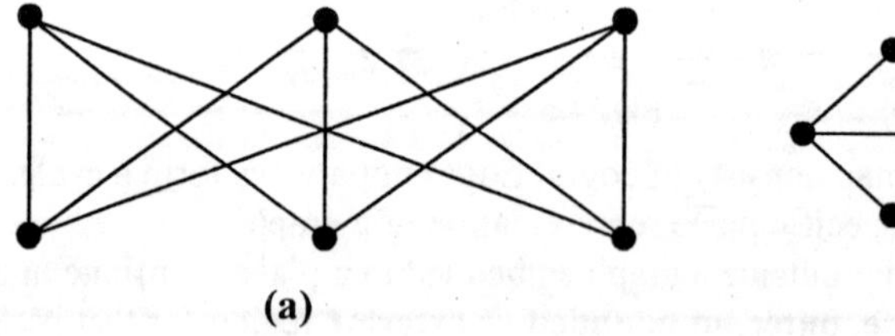

(a)

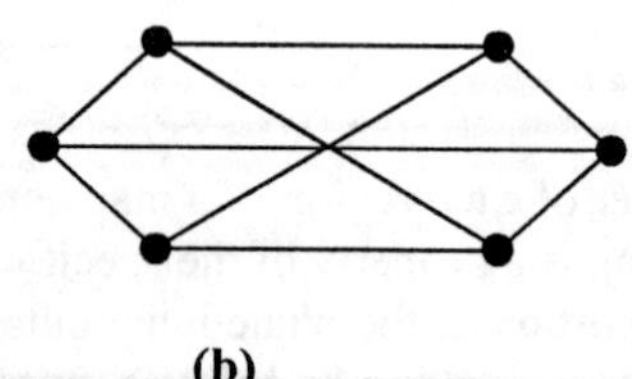

(b)

Fig. 4.7 Kuratowski Second Graph $K_{3,3}$

Example 4.2

Is $K_{3,3}$ (Kuratowski's second graph) a planar graph?

Solution

Let us first try to find a systematic way to draw a graph in the plane without intersection of edges. The method we will adopt involves following two ideas:

1. In geometric representation of a cycle in the plane, any edge not on the cycle must lie either inside the cycle or outside the cycle or the edge must intersect one of the edges of the cycle.

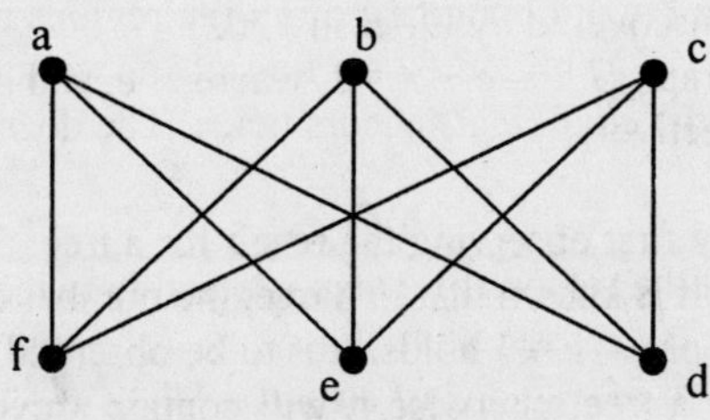

Fig. 4.8

2. The graph can be redrawn so that edges and vertices formerly outside the cycle are now inside the cycle and vice-versa i.e. the roles of being inside or outside the cycle are symmetric (equivalent). Consider the graph $K_{3,3}$ shown in Fig. 4.8.

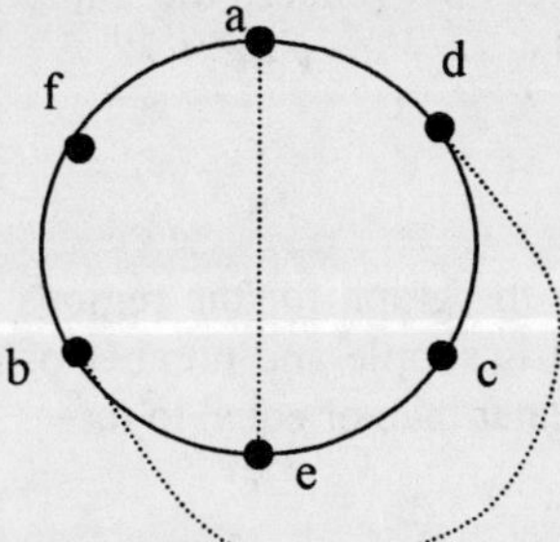

Fig. 4.9

Draw the cycle *a-d-c-e-b-f-a*. As the role of inside and outside the cycle is symmetric (equivalent). Let us take the edge (*a e*) inside the cycle then the edge (*b d*) must be drawn outside the cycle otherwise they will intersect. But we can observe that there is no way to draw the edge (*c f*) either inside or outside the cycle without crossing over either (*b d*) or (*a e*).

Hence $K_{3,3}$ is not planar.

Example 4.3

The complete graph of five vertices K_5 (Kuratowski first graph) is nonplanar. See theorem 4.4.

THEOREM 4.2 (Kuratowski's Theorem)

A necessary and sufficient condition for a graph G to be planar is that G does not contain either of Kuratowski's two graphs or any graph homeomorphic to either of them i.e. K_5 and $K_{3,3}$.

4.1-5 CRITICAL PLANAR

A graph G is **critical planar** if G is nonplanar but any subgraph obtained by removing a vertex is planar. For example $K_{n,3}$, $n \leq 3$ is critical planar.

4.2 EULER'S FORMULA

A planar graph may be represented differently but the question is whether the number of regions obtained from each embedding is same or not? The answer is same, which can be obtained with the help of Euler's formula taken below.

THEOREM 4.3 (Euler's Formula. Discovered by Euler in 1752)

For any connected planar graph G, $v - e + r = 2$, where v, e, and r are the number of vertices, edges, and regions of the graph respectively.

PROOF

Let us prove this formula by first observing the result for a tree. Since a tree determines only one region (by convention). Further, it is known that in a tree, the number of edges is one less than the number of vertices, i.e. $v - e + 1 = 2$ or $v - e = 1$ holds. It is to be observed here that a connected plane graph with only one region must be a tree otherwise it will contain a cycle in G. And if it contains cycle in G, there must exist an internal region and an external region.

The general result can be proved by induction on m, the number of regions determined by G. If $r = 1$, the result is already proved. Let us assume that the result holds for $r \geq 1$ and show that the result holds for $r + 1$ regions. Since G is connected and suppose it has $r + 1$ regions. Deleting an edge common to the boundary of two separate regions we obtain a plane graph G' with v vertices, e' edges and r' region having same number of vertices but reduces the number of edges and vertices by one. Applying the induction hypothesis, we find $v - e + r = v - e' + r'$

$$\Rightarrow \qquad v - e + r = 2.$$

REMARK

Above theorem actually holds for the graph for the regions of degree one; but through out this section we will assume that the graph is simple and number of edges is greater than one. This implies that the degree of each region is greater than or equal to three.

OBSERVATION

In a plane graph G, if the degree of each region $\geq k$ for $k \geq 1$,then $r \leq 2e$, here r is the number of regions and e is the number of edges.

COROLLARY

In a simple, connected planar graph G, with r regions, v vertices ($v \geq 3$) and e edges ($e > 1$), then

(i) $e \geq (3/2)\, r$ (ii) $e \leq 3v-6$

(iii) There is a vertex v of G such that $\deg(v) \leq 5$.

PROOF

By Euler's Formula

$$v - e + r = 2.$$

As the sum of the degrees of the region equals $2e$. But each region has degree ≥ 3, hence

$$2e \geq 3r$$

$$\Rightarrow \qquad e \geq \frac{3}{2} r$$

Substituting this in Euler's formula, we get

$$e \geq \frac{3}{2}(e - v + 2)$$

As for (iii) let us assume that each vertex has degree ≥ 6, then since degree of vertices of G is twice the number of edges, i.e. $\deg(v) = 2e$,

$$6v \leq 2e$$

$$\text{or} \quad v \le \frac{1}{3}e$$

and as we know from (i)

$$r \le \frac{2}{3}e$$

But, from Euler's formula $r + v = e + 2$, hence we have

$$\frac{2}{3}e + \frac{1}{3}e \ge r + v = 2 + e$$

$$\text{or} \quad e \ge e + 2$$

$$\text{or} \quad 0 \ge 2$$

which is a contradiction. Hence deg(v) is less than 5.

THEOREM 4.4

A complete graph K_n is planar if $n \le 4$.

PROOF

It can be seen that K_n is planar for $n = 1, 2, 3, 4$ (By corollary (iii) of Euler's formula). Thus, we have only to show that K_n is nonplanar if $n > 5$. As K_5 is a complete graph of $v = 5$ vertices and has $e = 10$ edges. If the graph is planar then by $e \le 3v - 6$, we have

$$0 = e \le 3v - 6 = 3\times5 - 6 = 9$$

which is impossible. Thus K_5 (Kuratowski's first graph) is nonplanar.

THEOREM 4.5

A complete bipartite graph $K_{m,n}$ is planar iff $m \le 2$ or $n \le 2$.

PROOF

It can be seen easily that $K_{m,n}$ is planar if $m \le 2$ or $n \le 2$(in all these cases Euler's formula is satisfied). To prove for $m \ge 3$ and $n \ge 3$, it is sufficient to prove that $K_{3,3}$ is nonplanar. As we know that bipartite graph $K_{3,3}$ has six vertices and nine edges. If $K_{3,3}$ were planar by Euler's formula $r = e - v + 2 = 9 - 6 + 2 = 5$. But $K_{3,3}$ is a bipartite graph, it cannot have a cycle of odd length. This implies that each cycle has length ≥ 4. Therefore, degree of each region would have to be greater than or equal to 4. But for this we would have $4r \le 2e$ or $20 = 4\cdot5 \le 2\cdot9 = 18$, which is a contradiction. Hence $K_{m,n}$ is planar iff $n \le 2$ or $n \le 2$. Therefore, Kuratowski's second graph $K_{3,3}$ is nonplanar.

4.3 SOME MORE PLANAR GRAPHS

4.3-1 MAXIMAL PLANAR GRAPH

A simple planar graph to which no edge can be added without disturbing its planarity while keeping the graph simple is called a **maximal planar graph**.

4.3-2 COMPLETELY REGULAR PLANAR GRAPH

A planar graph G is said to be **completely regular** if the degree of all vertices of G is equal and every region is bounded by the same number of edges. Graph shown in Fig. 4.11 is completely regular planar graph.

4.3-3 POLYHEDRAL

A connected planar graph is said to be **polyhedral** if degree(r) ≥ 3 for each region r and if in addition, degree(v) ≥ 3 for each vertex in G. Graph shown in Fig. 4.12 is polyhedral.

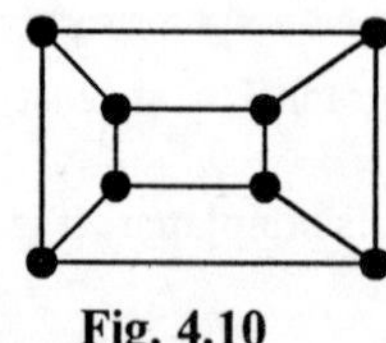
Fig. 4.10

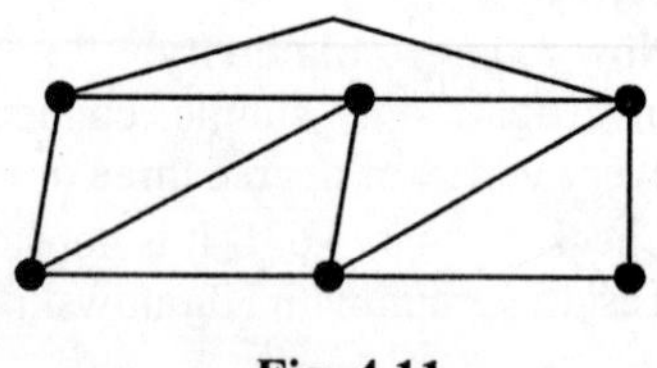
Fig. 4.11

4.3-4 MINIMAL NONPLANAR GRAPH

A nonplanar graph s.t. every proper sub graph is planar is called **minimal nonplanar graph**.

4.4 DETECTION OF PLANARITY

For the deletion of planarity let us discuss the following simple and efficient criterion.

4.4-1 ELEMENTARY REDUCTION

Step1: As we know that a separable graph is planar iff each of its block is planar and a disconnected graph is planar iff each of its component is planar. We need consider only one component at a time. Now, for the given arbitrary graph G, determine the set

$$G = \{G_1, G_2 ...., G_k\}$$

where each G_i ($i = 1,2,\ldots k$)is a nonseparable block of G. Then check each G_i for planarity.

Step2: Remove all self loops, as it does not affect the planarity.

Step3: Eliminate edges in parallel by deleting all but one edge between every pair of vertices, as parallel edges do not affect planarity.

Step4: Eliminate all edges in series, as deletion of vertex of degree 2 does not affect planarity. [In a graph two edges are said to be in series if they have exactly one vertex in common and if this vertex is of degree two. In the Fig.4.12 edges e_1 and e_2 , and e_4 and e_6 are in series]

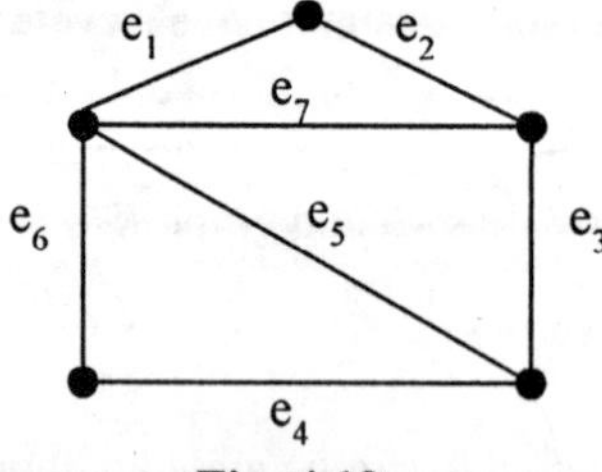

Fig. 4.12

Suppose the nonseparable graph G_i reduced to a new graph S_i after applications (repeated) of steps 3 and 4.

The look of the graph S_i is followed by the Theorem 4.6.

THEOREM 4.6

The graph S_i obtained by repeated applications of step 3 and 4 above may be:

(a) A single edge or

(b) A complete graph of four vertices

(c) A nonseparable, simple graph with $n \geq 5$ and $e \geq 7$.

OBSERVATION

Now to check planarity:

(i) Investigate only simple, connected, nonseparable graphs of at least five vertices and with every vertex of degree three or more.

(ii) Check if $e \leq 3v - 6$. If it is not satisfied, the graph S_i is nonplanar. If it is satisfied then

(iii) Test the graph with Kuratowski's theorem.

Example 4.4

Is it necessary for a nonplanar graph to have either of Kuratowski graphs as a subgraph?

It is not necessary for a nonplanar graph to have either of Kuratowski graphs as a subgraph as this theorem is sometimes misstated. The nonplanar graph may have subgraph homeomorphic to a Kuratowski's graph.

One can see the graph shown in the Fig. 4.13 (i) is nonplanar but it does not have either Kuratowski's graph as subgraph. If we delete edges (v_2, v_6) and (v_1, v_3) from this graph, we get a subgraph as shown in the Fig. 4.13 (ii). Observe that this subgraph is homeomorphic to the graph as shown in Fig. 4.13 (iii). The graph as shown in Fig. 4.13 (iii) is isomorphic to Kuratowski's second graph $K_{3,3}$.

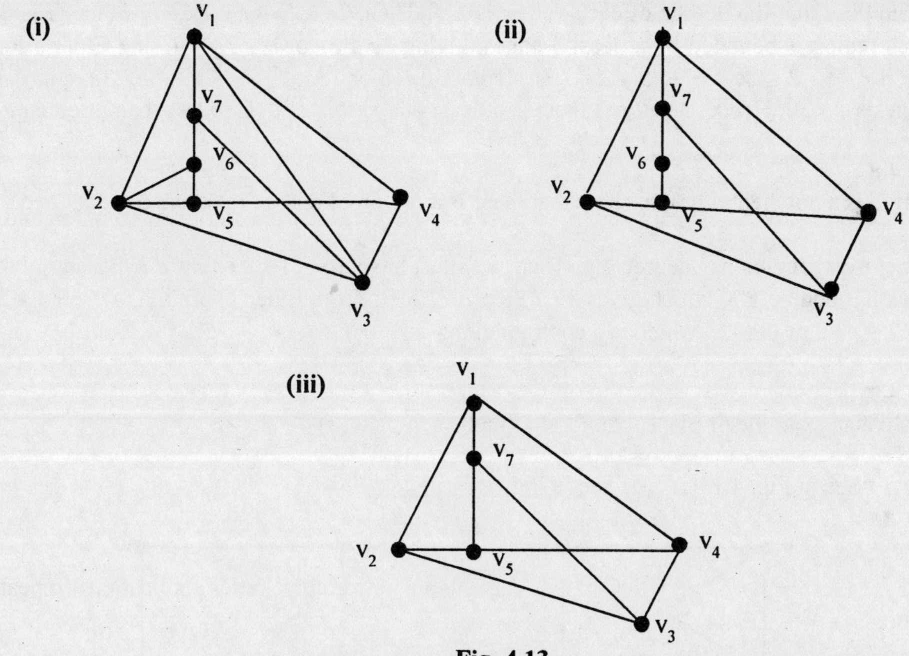

Fig. 4.13
Nonplanar graph with a subgraph homeomorphic to $K_{3,3}$

Example 4.5

A polyhedral graph is a connected plane graph in which each region has degree equal to or greater than 3 and each vertex has degree equal to or greater than 3. Prove that there exists no polyhedral planar graph with exactly 7 edges.

Solution

Since each region has 3 edges or more. $3r \leq 2e = 14$. Since r must be an integer, it follows that $r \leq 14/3$, which implies $r \leq 4$. As the degree of each vertex is greater than or equal to 3, $3r \leq 2e = 14$ or $r \leq 14/3$; and since v must be an integer $v \leq 4$. From Euler's formula, $r + v = e + 2 = 9$. But since both r and v are less than or equal to 4, this is a contradiction. Hence there doesn't exist any planar graph with exactly 7 edges.

Example 4.6

Prove that no polyhedral graph with exactly 30 edges and 11 regions exist.

Solution

By Euler's formula, we have $e = v + r - 2$ or $30 = v + 11 - 2$, giving $v = 21$. Also, since each vertex has degree equal to or greater than v, we must have $3v \leq 2e$ or $3\times21 \leq 2\times21$ or $63 \leq 60$, which is a contradiction. Hence there doesn't exist any polyhedral graph with exact 30 edges and 11 regions.

Example 4.7

Show that a plane connected graph with less than 30 edges has a vertex of degree ≤ 4.

Solution

Suppose that the degree of each vertex is ≥ 5. Then, $2e \geq 5v$ or $v \geq 2e/5$. Since G is a plane graph, each region is bounded by at least 3 edges, giving $2e \geq 3r$ or $r \leq 2e/3$. Using Euler's formula, we have $e = r + v - 2 \leq 2e/3 + 2e/5 - 2 = (6e + 10e - 30)/15$ or $15e \leq 16e - 30$ or $e \geq 30$, contradicting the hypothesis $e < 30$. Hence, it follows there is at least one vertex of degree less than or equal to 4.

Example 4.8

If G is a polyhedral graph, show that there is a region of degree less than or equal to 5.

Solution

Let every region has degree ≥ 6. Then, we must have, $(6r/2) \leq e$ or $r \leq e/3$. Being polyhedral, each vertex has degree ≥ 3, which gives $3v \leq 2e$ or $v \leq 2e/3$. From Euler's formula, $e = r + v - 2 \leq e/3 + (2e/3) - 2 = e - 2$ or $0 \leq -2$, which is a contradiction.

Example 4.9

Give an example of plane, connected graph such that (a) $e = 3v - 6$ (b) $e < 3v - 6$.

Solution

(a) See in Fig. 4.14(i) (b) See in Fig. 4.14(ii).

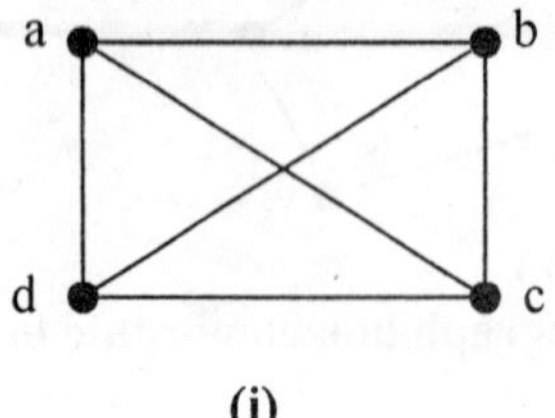

(i)

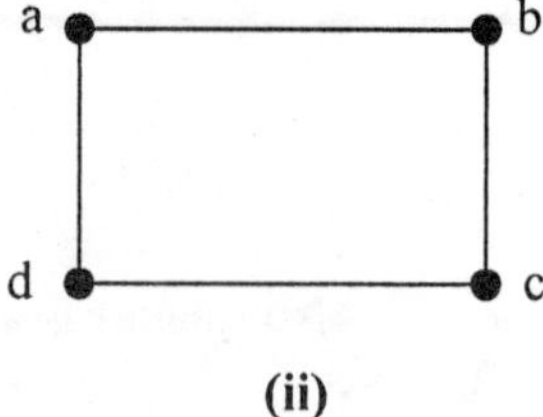

(ii)

Fig. 4.14

Example 4.10

Show that for a connected planar graph with $v \geq 3$, $5V_1 + 4V_2 + 3V_3 + 2V_2 \geq V_7 + 2V_8 + \ldots$

$(v - 6)\ V_v + 12$, where V_i is vertices with degree i and v is the number of vertices.

Solution

Since in a planar graph $e \le 3v - 6$ or $2e \le 6v - 12$.

Hence, $\quad 6v - 2e \ge 12.$

Put $\quad v = V_1 + V_2 + \ldots + V_v$

and $\quad 2e = V_1 + 2V_2 + 3V_3 + 4V_4 + 5V_5 + 6V_6 + \ldots + vV_v.$

Substituting in the above relation,

$$6(V_1 + V_2 + \ldots\ V_6 + \ldots\ V_v) - (V_1 + 2V_2 + \ldots + vV_v) \ge 12$$

or

$$5V_1 + 4V_2 + 3V_3 + 2V_2 \ge V_7 + 2V_8 + \ldots + (v - 6)\ V_v + 12.$$

Example 4.11

If G consists of all cycles of length greater than or equal to k, then prove that

$$e \le \frac{k(v-2)}{(k-2)}.$$

Solution

We have $kr \le 2e$ or $r \le 2e/k$. Also,

$$e\left(1 - \frac{2}{k}\right) \le (v - 2)$$

we have $\quad r = e - v + 2$ or $e = r + v - 2 \le (2e/k) + v - 2$

$$e \le \frac{k(v-2)}{(k-2)}.$$

Example 4.12

To show that Grotzsch graph is non-Planar.

Solution

Grotzsch graph G is given in Fig.4.15 (a). By removing its vertex v_{11} we obtain its sub-graph G_1 contained in G shown in Fig.4.15 (b).

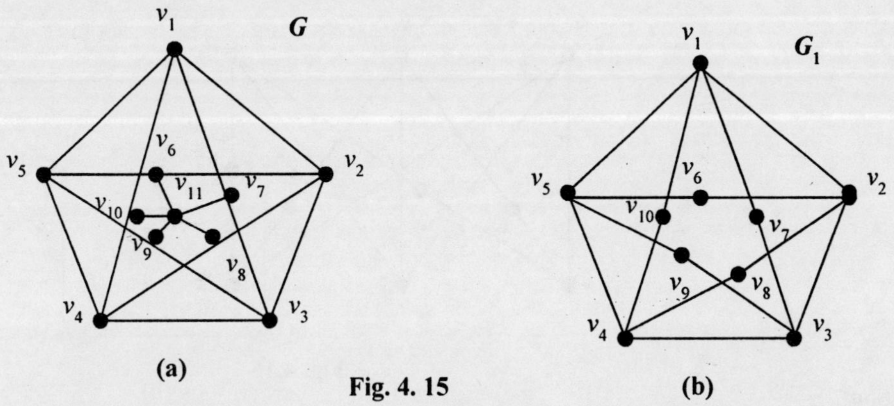

Fig. 4. 15

This graph is homeomorphic with Kuratowski's first graph shown in Fig.4.15 (c) as the latter graph is obtained by merging three pairs of edges in series.

Hence the graph G shown in Fig. 4.15 (a) is not planar.

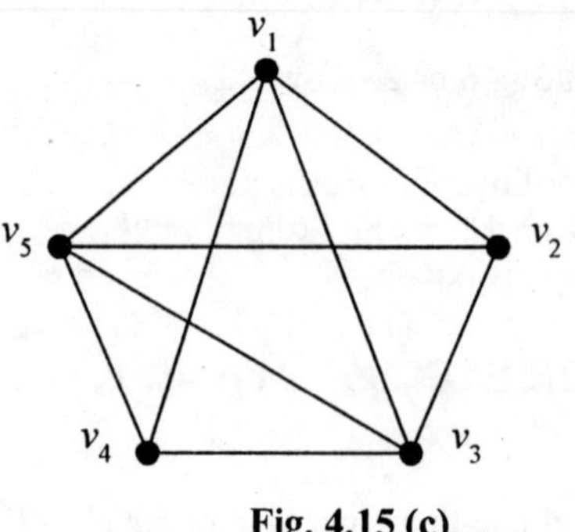

Fig. 4.15 (c)

Example 4.13

Show that Petersen graph is non planar.

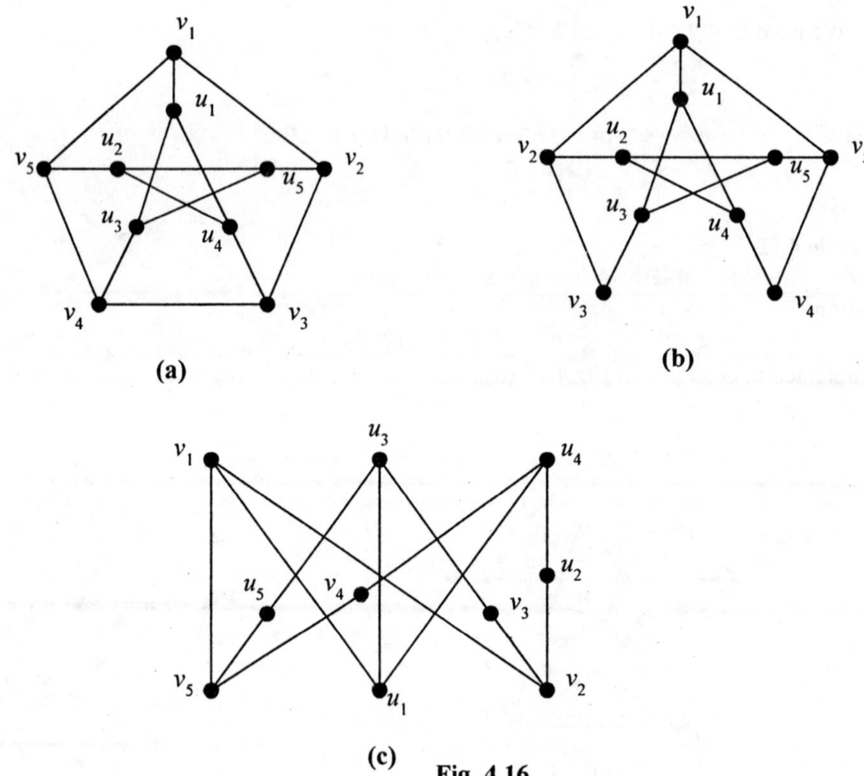

Fig. 4.16

Solution

Petersen graph G is shown in Fig. 4.16 (a). From this graph we obtain its sub-graph by removing its two edges and (v_2, v_4) shown in Fig. 4.16 (b). This graph which is contained in *G* is homeomorphic with Kuratowski's second graph given in Fig.4.16 (c).

EXERCISE 4.1

1. Which of the following graphs shown in the Fig. 4.17 are planar? If planar draw.
2. Show that $K_5 - e$ and $K_{3,3} - e$ are planar for any edge e of K_5 and $K_{3,3}$ respectively.
3. Show that any graph with 4 or fewer vertices is planar.
4. Show that there are only five possible simple completely regular planar graphs, excluding the trivial graphs with degree ≤ 2. Sketch them.

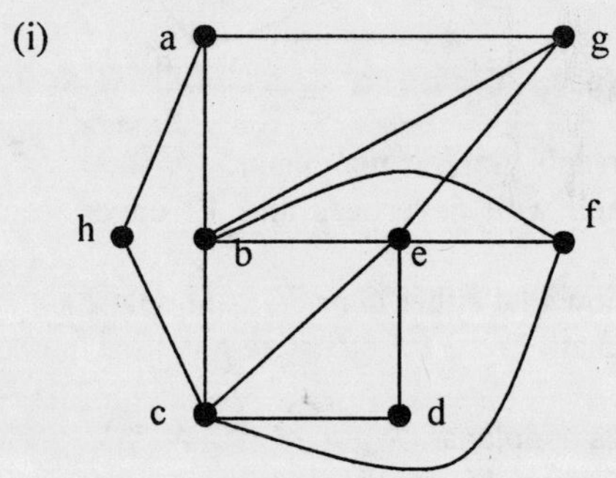

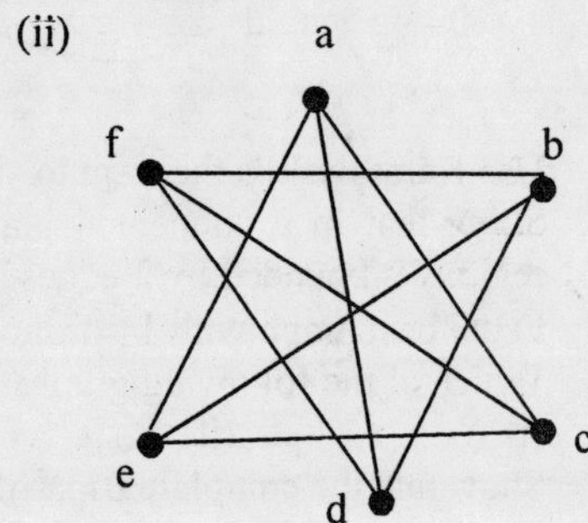

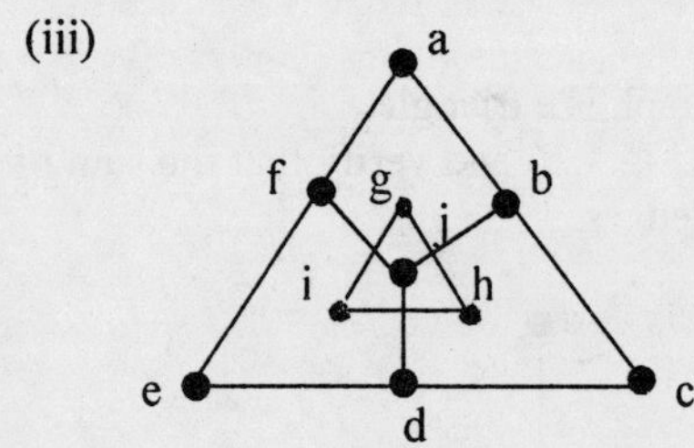

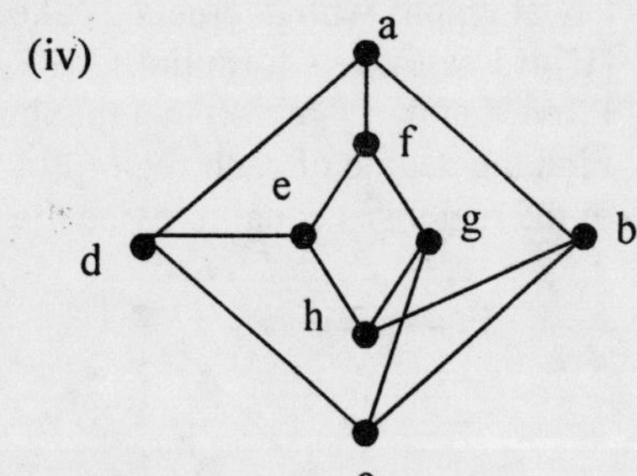

Fig. 4.17

5. Show that each of the graph shown in Fig. 4.18 is not planar by finding a subgraph homeomorphic to either K_5 or $K_{3,3}$.

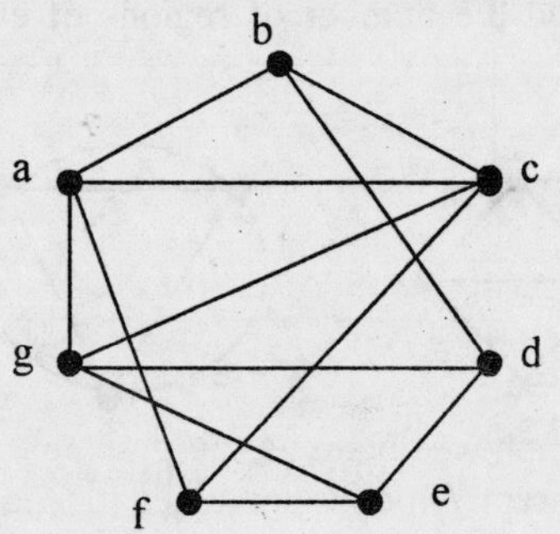

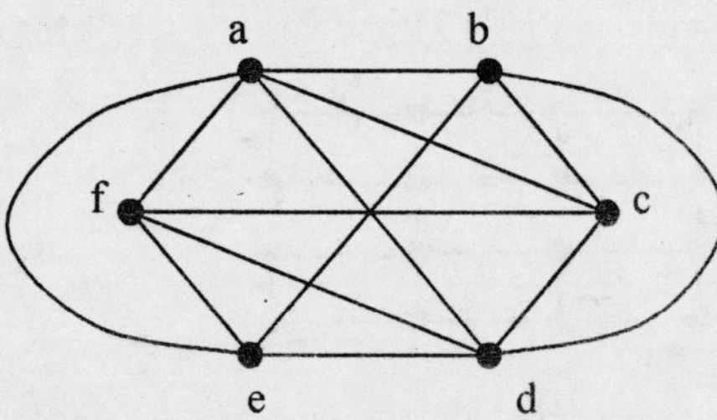

Fig. 4.18

6. Check whether the following graphs are planar. If it is, redraw it so that no edges cross, otherwise find a subgraph homeomorphic to either K_5 or $K_{3,3}$.

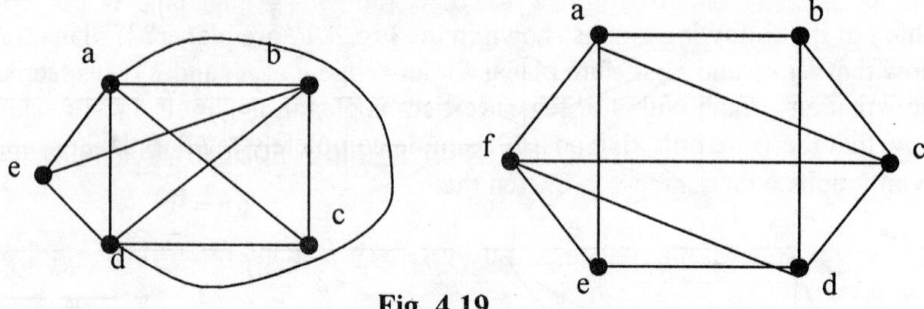

Fig. 4.19

7. Use Kuratowski's theorem to show that Petersen's graph is nonplanar.
8. Show that in a connected planar linear graph with 6 vertices and 12 edges, each of the regions is bounded by 3 edges.
9. Let G be a graph with 11 or more vertices. Show that either G or $\bar{G}$ is nonplanar.
10. Which of the following graphs are critical planar
(i) K_6 (ii) $K_{4,3}$ (iii) $K_{3,3}$.
11. Show that the complete tripartite graph $K_{1,2,3}$ is nonplanar.
12. How many edges, vertices and regions do the wheel W_n have?
13. Show that there are only five possible simple completely regular planar graphs, excluding the trivial graphs with degree ≤ 2. Sketch them.
[**Hint** Use Euler's formula].
14. Prove that every region in a maximal planar graph is a triangle.
15. Find the degree of each region for the map in Fig.4.20 and verify that the sum of the degrees of the regions is equal to twice the number of edges.

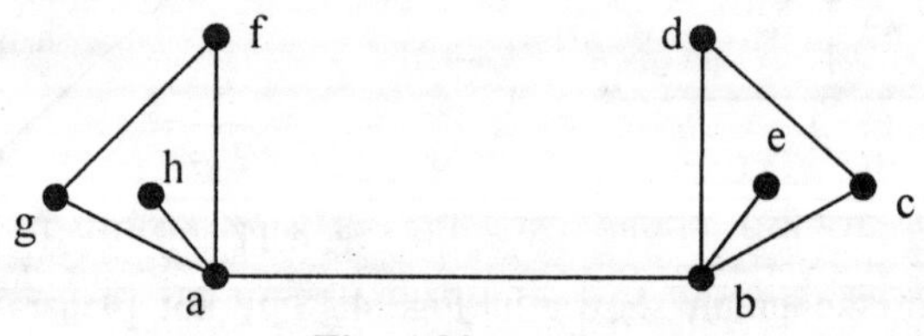

Fig. 4.20

16. Count the number of vertices, number of edges, and the number of regions of each map shown in Fig. 4.21 and verify Euler's formula.

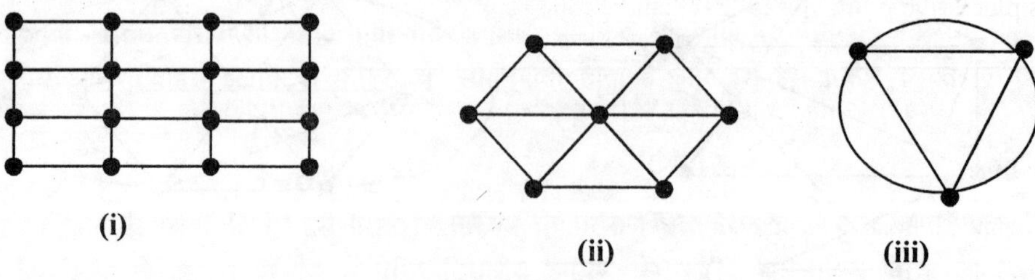

Fig. 4.21

ANSWERS

5(i).

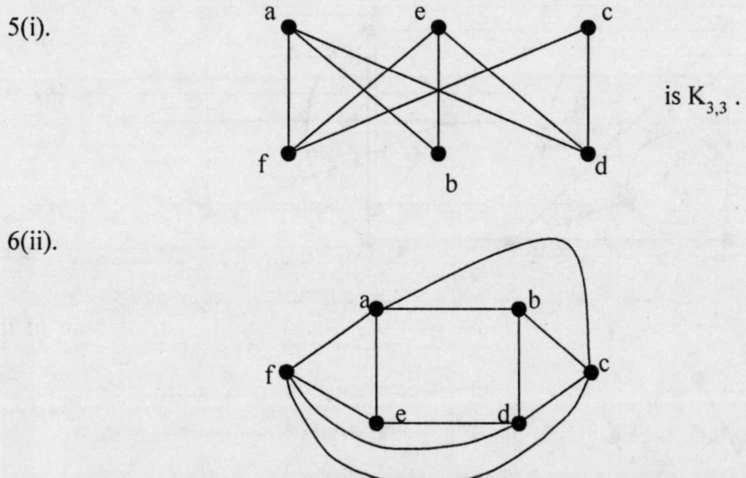

6(ii).

15. The outside region has degree 8, and other two regions have degree 5.

16. (i) 12, 18,10 (ii) 7, 12, 7 (iii) 3, 6, 5.

4.5 DUAL GRAPH

We can take a geographic map on the plane or on the sphere as a plane graph in which regions or faces are the territories of the map, the nodes are the places where several boundaries meet, and edges are the portions of the boundaries that join two vertices. We allow full generality of loops and multiple edges. We can form a plane graph from any plane graph called its **dual**.

Suppose G is a plane graph. We define another multigraph G^* of G as follows:

Corresponding to each region (face) r of G there is a vertex v^* of G^*, and if e is an edge of G that has region r_1 on one side and region r_2 on the other side, then the corresponding dual edge e^* of G^* is an edge joining the vertices a, b of G^* that correspond to the regions r_1 and r_2 of G. In particular, a loop is added at a vertex v^* of G^* for each cut-edge of G that belongs to the boundary of the region r. The multigraph G^* is called the **dual** (or a geometric dual) of G.

Example 4.14

Consider a plane graph G shown in the Fig.4.22(a) with five regions R_1, R_2, R_3, R_4 and R_5. Let us place five points v_1, v_2, v_3, v_4 and v_5 one each in five regions as shown in Fig. 4.22(*b*).

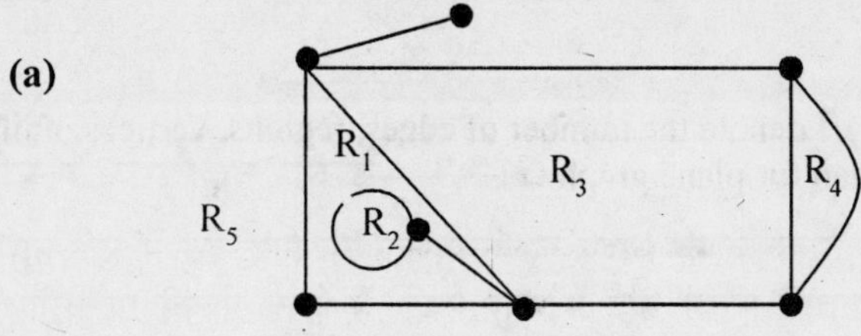

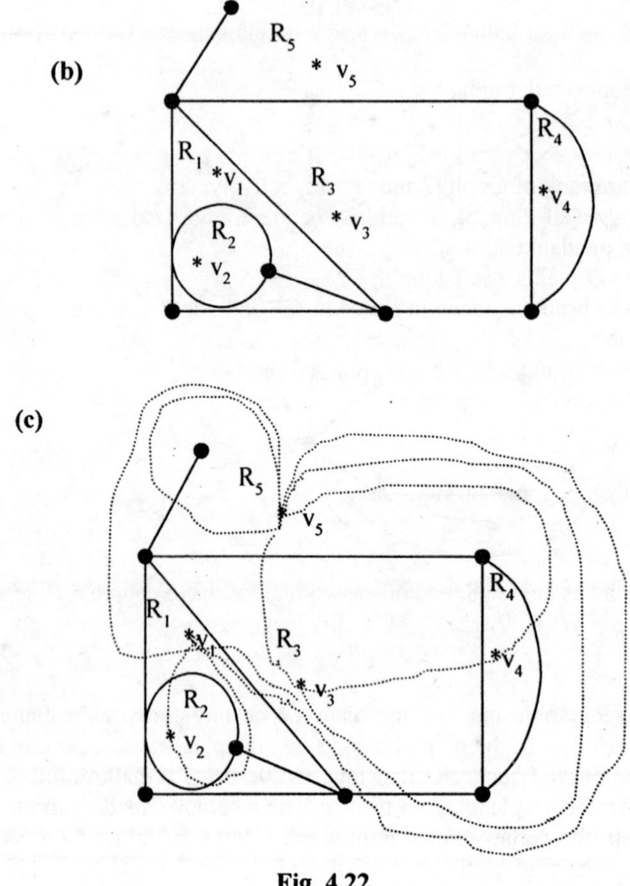

Fig. 4.22

If two regions r_i and r_j are adjacent (i.e. have a common edge) draw a line joining points v_i and v_j that crosses the common edge r_i and r_j exactly once. In case there exist more than one edge common between r_i and r_j, draw one edge between points vi and v_j for each of the common edges of G. For an edge e lying entirely in one region, say r_k, draw a self loop at point v_k crossing e exactly once. Note that if e is a loop of G, the e^* is a cut edge of G^* and conversely. This procedure is illustrated in Fig.4.22(c), where the dual edges are indicated by dashed lines and dual vertices by asterisks.

OBSERVATION

1. If e, r, v μ and e^*, r^*, v^*, μ^* denote the number of edges, regions, vertices, nullity of graph G and dual graph G* respectively. Then for plane graph G :

(i) $e^* = e$

(ii) $v^* = r$

(iii) the degree of the vertex v^* in G^* is the same as the degree of the corresponding region R determined by G.

(iv) If G is connected graph $r^* = v$

(v) $n^* = \mu$

(vi) $\mu^* = n$

where n and n^* are rank of graph G and G^* respectively.

2. An edge forming a self-loop in G yields a pendant edge (An edge incident on a pendant vertex is called a pendant edge) in G^*.
3. A pendant edge in G yields a self-loop in G^*.
4. Parallel edges in G produce edges in series in G^* and edges that are in series in G produce parallel edges in G^*.
5. Graph G^* is embedded in the plane and is therefore planar.
6. G and G^* graphs are dual to each other.

Example 4.15

Different embedding of a planar graph may have nonisomorphic duals. As one can see in the Fig. 4.23 (a) has regions of lengths 6, 4, 3, 3 while the plane graph shown in the Fig. 4.23 (b) has regions of lengths 5, 5, 3, 3. Here, one can observe that both these graphs are isomorphic but their duals have degree sequences 6, 4, 3, 3 and 5, 5, 3, 3 respectively, and consequently the duals of these isomorphic graphs are nonisomorphic. Every 3-connected planar graph has essentially one embedding and hence all duals of a 3-connected graph are isomorphic.

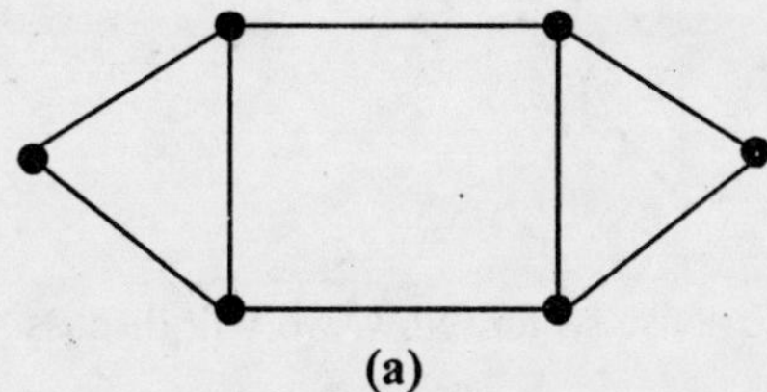

(a) **Fig. 4.23** (b)

It is to be noted that all duals of a planar graph are 2-isomorphic, and every graph 2-isomorphic to a dual of G is also a dual of G.

Example 4.16

Prove that the sum of degrees of the regions of a map is equal to twice the number of edges.

Proof

Let $\sum$degree(r) and $\sum$degree(r^*) denote the sum of degrees of regions of graph G and its dual graph G* respectively. Then

$$\sum \deg(r) = \sum \deg(r^*) \quad \text{from (iii) above}$$

But we know that the sum of the degrees of all the vertices in any graph is twice the number of edges, i.e., in particular

$\sum$degree(r^*) = 2 e^* , where e^* is the number of edges of G*

But $e = e^*$ by(i) above. Hence the theorem.

4.6 THE PLANE VERSUS THE SPHERE

We haven't differentiated between graphs in the plane and maps on the sphere because a graph embeds in the plane iff it embeds on the sphere. Given an embedding on the sphere, we can puncture the graph within any face and project from there on to a plane tangent to the antipodal point (place on opposite sides of the earth to each other) to obtain a planar embedding. The punctured region on the sphere becomes the unbounded region in the plane, and the process is reversible.

REMARK

1. A statement about a connected graph becomes a statement about the dual graph when we interchange the roles of vertices and regions.
2. Edges incident to a vertex become edge bounding a region, and vice-versa.
3. The length of a region in a plane graph G is the length of the walk that bounds it.
4. A cut-edge belongs to the boundary of only one region, and it contributes twice to the boundary of that region as we traverse the boundary.
5. The graph shown in Fig. 4.24 has three regions, with lengths 3, 6, 7.

The sum of the lengths is 16, which is twice the number of edges.

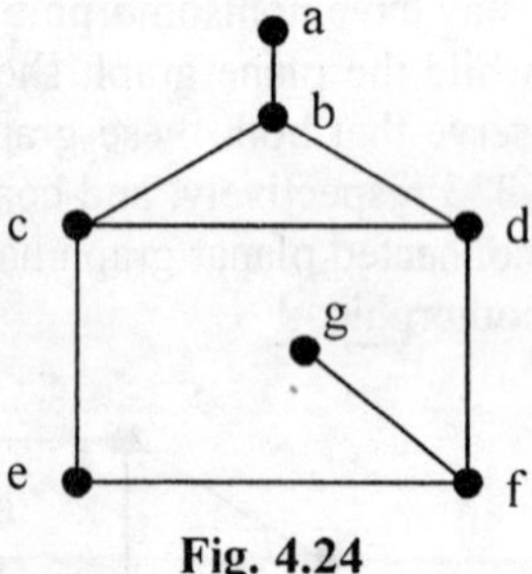

Fig. 4.24

4.7 COMBINATORIAL DUAL

Duality of a graph has been discussed in geometric sense so far. Now, we will discuss some equivalent definition of duality other than geometric notions.

Fig. 4.25

4.7-1 COMBINATORIAL DUAL

Two planar graphs G and G^* are said to be **combinatorial dual** (or duals) of each other if there exist a one-to-one correspondence between edges of G and G^* such that if S is a subgraph of G and s is a corresponding subgraph of G^*, then

Rank of (G^*-s) = rank of G^* - nullity of S

This relationship can be shown diagrammatically as shown in the Fig.4.26.

Example 4.17

Consider two graphs shown in Fig.4.26 (A) and (B), where graph (*B*) is a dual of graph (*A*). Now, consider subgraph $\{d^*, e^*, f^*, g^*,\}$ of Fig.4.26 (B) and corresponding subgraph $\{ d, e, f, g \}$ of Fig. 4.26 (A)

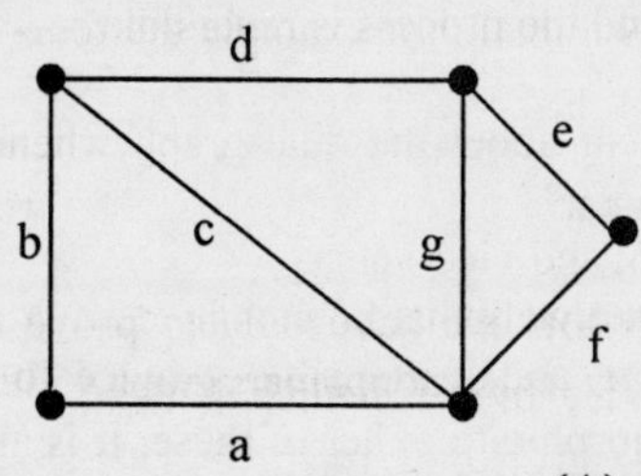

f*
a*
e*
d*
g*
b*
c*

(A) Fig. 4.26 (B)

Rank of $(G^* - \{d^*, e^*, f^*, g^*\})$ = rank of $\{a^*, b^*, c^*,\} = 2$

Rank of $G^* = 3$

Nullity of $\{d, e, f, g\} = 1.$

and $2 = 3 - 1$

Note that this definition is independent from geometric connotation.

Example 4.18

Let *G* be a planar graph with each vertex of degree 3. Show that the dual of *G* will have an odd number of regions of finite degree.

Solution

Since each vertex has degree equal to 3, we have $3v = 2e$ or $v = 2e/3 \Rightarrow v$ must be multiple of 3, say $v = 3k$. Then,.

$$v = \frac{2 \times 3k}{3} = 2k .$$

Hence (v-1) is odd. But the number of regions in the dual excluding the infinite region is v-*1*, which is odd.

4.8 DUAL OF A SUBGRAPH

Consider a planar graph *G* and its dual G^* and an edge *e* in *G* and the corresponding edge e^* in G^*. If we want to find the dual of *G* after deleting an edge *e* from *G*, i.e. of $G - e$, then there exist following two possibilities:

(i) *e* may be lying on the boundary of two regions. In this case, deletion of an edge will merge these two regions. And to obtain the dual of $(G - e)^*$, remove the corresponding edge *e* from G^* and fuse the two end vertices of e^* in $G^* - e^*$.

(ii) If *e* is not lying on the boundary, e^* form a self-loop. But, in this situation $G^* - e^* = (G - a)^*$. Thus, if a graph *G* has a dual G^*, then the dual of any subgraph of *G* can be obtained by successive application of this method.

4.9 DUAL OF A HOMEOMORPHIC GRAPH

Consider a planar graph *G* and dual G^* of it. Let *e* be an edge in *G* and e^* be corresponding additional edge in G^*. Let us introduce an additional vertex of degree two in edge *e* in *G*. By doing so, it will add an edge parallel to e^* in G^*.

Similarly, the reverse process of merging two edges in series will eliminate one of the corresponding parallel edges in G^*.

Therefore, if a graph G has a dual G^*, the dual of a graph homeomorphic to G can be obtained from G^* by the above method.

4.9-1 SELF-DUAL GRAPHS

A plane graph G is **self-dual** if it is isomorphic to its own dual. For example the four-vertex complete graph is a self-dual graph.

Example 4.19

Prove that a graph G has a dual G^* if and only if it is planar.

Proof

We need to prove just the only if part or we can say that we have only to prove that a nonplanar graph does not have a dual. Let us assume that G is a nonplanar graph. Then by Kuratowski's theorem, G contains K_5 or $K_{3,3}$ or a graph homeomorphic to either of these. It is already shown earlier that a graph G can have a dual only if every subgraph S of G and every graph homeomorphic to S has a dual. Thus if it has been proved that neither K_5 or $K_{3,3}$ has a dual, we have proved the theorem.

4.10 THICKNESS AND CROSSINGS

4.10-1 THICKNESS

The **thickness** of graph G is the minimum number of planar subgraphs whose union is G. In other words the thickness of graph G is defined to be the minimum number of (edge-disjoint) planar subgraphs into which G can be decomposed.

Example 4.20

(i) Thickness of a planar graph is one.

(ii) Thickness of each Kuratowski's graph is two.

(iii) Thickness of a complete graph of 9 vertices is three (show it).

4.10-2 CROSSING NUMBER

There are many parameters, which measure a graph's deviation from planarity, one comes directly from the definition of planarity. The **crossing number** of a graph G is the minimum number of crossing (intersections) in drawing of G in the plane.

Example 4.21

By definition, the crossing number of a planar graph is zero. Crossing number of both of Kuratowski's graph is one.

EXERCISE 4.2

1. Draw the graph which is dual to the graph shown in the Fig. 4.27

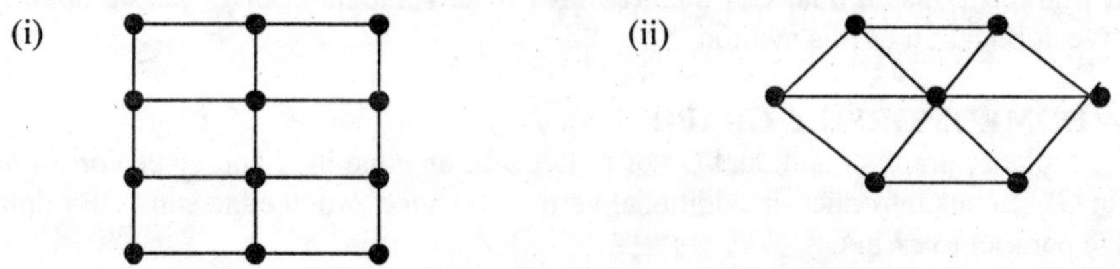

(iii)

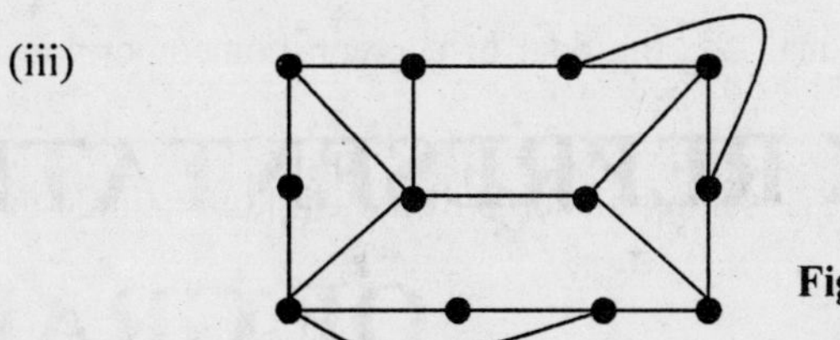

Fig. 4.27

2. Prove that a self-loop-free planar graph is 2-connected iff its dual is also 2-connected.
3. Show that the wheel graph W_n of n vertices is isomorphic to its dual.
4. Compute the crossing number of $K_{4,4}$ and the Petersen graph.
5. Show that the complete graph of four vertices is self-dual.
6. Find a self-complimentary planar graph with 8 vertices. What is the thickness of K_8?
7. Prove that the crossing number of the graph drawn in Fig. 4.28 is at most 5.

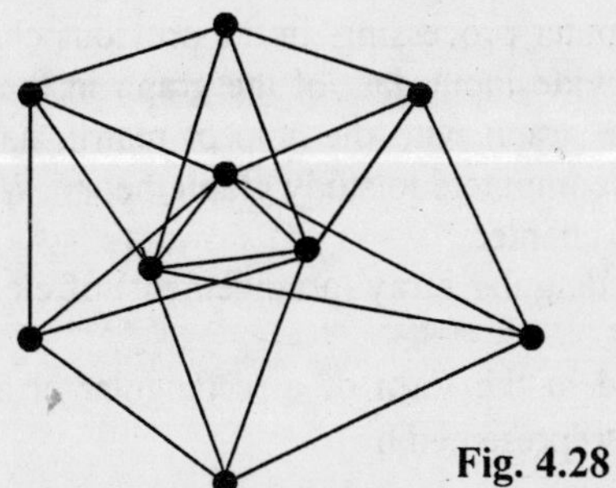

Fig. 4.28

8. Show by construction that the dual of two (2-isomorphic graphs in Fig. 4.29(i) and (ii) are isomorphic

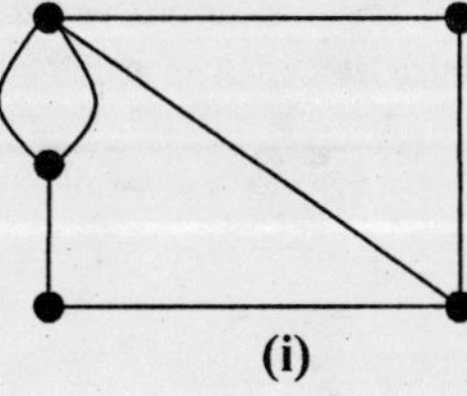

(i)

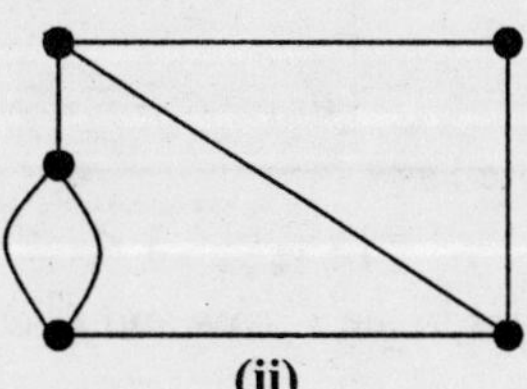

(ii)

Fig. 4.29

9. Give an example of a graph in which a finite region becomes infinite region when it is redrawn.
10. Demonstrate by an example that G, the dual of graph G, may not be isomorphic to G, but is 2-isomorphic to it.
11. Show that the dual of a wheel is a wheel.

5 MATRIX REPRESENTATION OF GRAPHS

5.0 INTRODUCTION

We have discussed the graph terminology in chapter 2 where terms such as graph, node, edge, path, cycle, circuits, etc. were introduced. A diagrammatic representation of a graph may have a limited usefulness. We can represent the graphs by another important mathematical structure, i.e. matrices which is very useful in computer processing. In the previous chapters we have seen that the graphical representation does not provide input data of the graph in the form of numerals 0 and 1. Here in this chapter we represent the graph with the help of matrix having elements 0 and 1.This matrices can be used as input data by computers to study graph theory. We will study several types of matrix representation of graphs in this chapter.

A **matrix** is defined as a rectangular array (arrangement) of elements and subject to certain rules of operations.

If *m.n* numbers are arranged in the form of a rectangular array *A*, having *m* rows and *n* columns, then *A* is called $m \times n$ matrix, represented by

$$A = \begin{bmatrix} a_{11} & a_{12} & a_{13} & \cdots & a_{1n} \\ a_{21} & a_{22} & a_{23} & \cdots & a_{2n} \\ \vdots & \vdots & \vdots & & \vdots \\ a_{m1} & a_{m2} & a_{m3} & \cdots & a_{mn} \end{bmatrix}$$

or it may be expressed as

$$A = [a_{ij}]_{m \times n} \qquad 1 \le i \le m,\ 1 \le j \le n,$$

where a_{ij} lies in the ith row and j^{th} column.

5.1 ADJACENCY MATRIX

Suppose $G = (V, E)$ is a simple graph with no parallel edges and *m* vertices labelled $v_1, v_2, \ldots v_m$. Then the **adjacency matrix** $A(G) = (a_{ij})$ of the graph *G* is the $m \times m$ matrix defined as follows:

$$a_{ij} = \begin{cases} 1 & \text{If } v_i \text{ is adjacent ot } v_j\text{, i.e. if there is an edge } (v_i, v_j) \\ 0 & \text{If there is no edge between } v_i \text{ and } v_j. \end{cases}$$

Element of the adjacency matrix is either 0 or 1. Such a matrix sometimes also called as a **bit matrix** or a **Boolean matrix**.

DEFINITION

Alternatively adjacency matrix for a graph G which contains parallel edges as well as self loop is defined as:

Let G be a graph with vertices $v_1, v_2, v_3, \ldots, v_n$ then the adjacency matrix $A(G)$ is a $n \times n$ matrix such that

$$A(G) = [a_{ij}],$$

where the matrix element a_{ij} denotes the number of edges joining the vertex v_i to v_j.

Example 5.1

Find the adjacency matrix of the following graph using ordering of vertices

(a) v_1, v_2, v_3, v_4 (b) v_2, v_3, v_1, v_4.

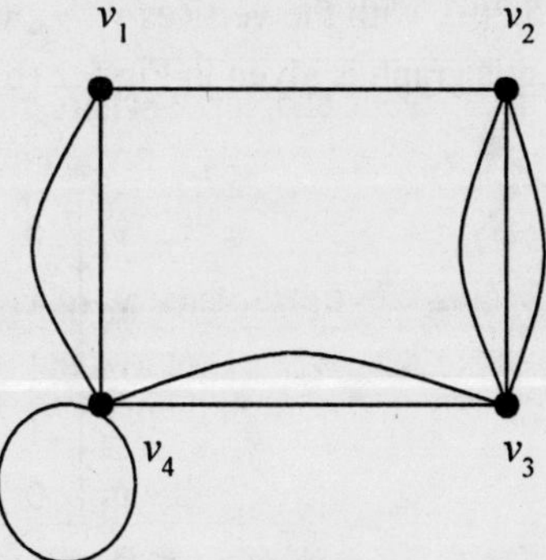

Solution

(a)
$$\begin{array}{c} \\ v_1 \\ v_2 \\ v_3 \\ v_4 \end{array} \begin{array}{c} \begin{array}{cccc} v_1 & v_2 & v_3 & v_4 \end{array} \\ \begin{bmatrix} 0 & 1 & 0 & 2 \\ 1 & 0 & 3 & 0 \\ 0 & 3 & 0 & 2 \\ 2 & 0 & 2 & 1 \end{bmatrix} \end{array}$$

(b)
$$\begin{array}{c} \\ v_2 \\ v_3 \\ v_1 \\ v_4 \end{array} \begin{array}{c} \begin{array}{cccc} v_2 & v_3 & v_1 & v_4 \end{array} \\ \begin{bmatrix} 0 & 3 & 1 & 0 \\ 3 & 0 & 0 & 2 \\ 1 & 0 & 0 & 2 \\ 0 & 2 & 2 & 1 \end{bmatrix} \end{array}$$

Example 5.2

Find the graph corresponding to following adjacency matrix.

$$\begin{array}{c} \\ v_1 \\ v_2 \\ v_3 \\ v_4 \\ v_5 \\ v_6 \\ v_7 \\ v_8 \end{array} \begin{array}{c} \begin{array}{cccccccc} v_1 & v_2 & v_3 & v_4 & v_5 & v_6 & v_7 & v_8 \end{array} \\ \begin{bmatrix} 0 & 2 & 0 & 0 & 0 & 0 & 0 & 0 \\ 2 & 0 & 1 & 2 & 0 & 0 & 0 & 0 \\ 0 & 1 & 0 & 1 & 0 & 0 & 0 & 0 \\ 0 & 2 & 1 & 0 & 1 & 0 & 0 & 0 \\ 0 & 0 & 0 & 1 & 0 & 0 & 0 & 0 \\ 0 & 0 & 0 & 0 & 0 & 1 & 1 & 0 \\ 0 & 0 & 0 & 0 & 0 & 1 & 0 & 3 \\ 0 & 0 & 0 & 0 & 0 & 0 & 3 & 0 \end{bmatrix} \end{array}$$

Solution

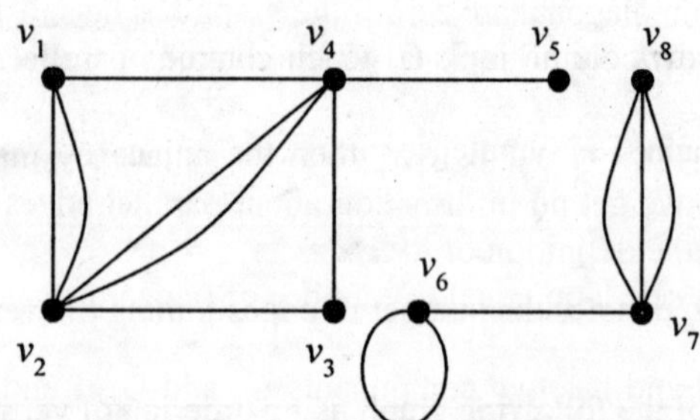

Fig. 5.1

Example 5.3

Consider the graph given in Fig.5.2(a), with the vertices v_1, v_2, v_3, v_4 and v_5 and edges e_1, e_2, e_3, e_4, e_5 and e_6. The adjacency matrix for this graph is given in Fig.5.2 (b).

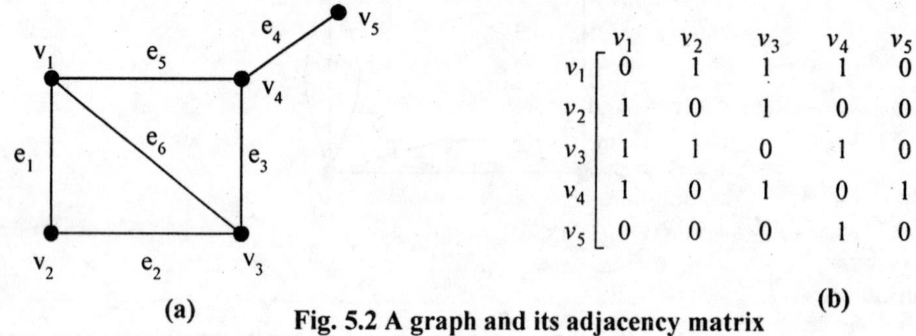

$$\begin{array}{c|ccccc} & v_1 & v_2 & v_3 & v_4 & v_5 \\ \hline v_1 & 0 & 1 & 1 & 1 & 0 \\ v_2 & 1 & 0 & 1 & 0 & 0 \\ v_3 & 1 & 1 & 0 & 1 & 0 \\ v_4 & 1 & 0 & 1 & 0 & 1 \\ v_5 & 0 & 0 & 0 & 1 & 0 \end{array}$$

(b)

Fig. 5.2 A graph and its adjacency matrix

Example 5.4

Draw a graph whose adjacency matrix is given below

$$A(G) = \begin{array}{c} \\ a \\ b \\ c \\ d \\ e \\ f \end{array} \begin{array}{c} \begin{array}{cccccc} a & b & c & d & e & f \end{array} \\ \begin{bmatrix} 0 & 1 & 0 & 1 & 0 & 0 \\ 1 & 1 & 1 & 0 & 1 & 0 \\ 0 & 1 & 0 & 0 & 1 & 0 \\ 1 & 0 & 0 & 0 & 1 & 0 \\ 0 & 1 & 1 & 1 & 0 & 1 \\ 0 & 0 & 0 & 0 & 1 & 1 \end{bmatrix} \end{array}$$

Solution

Observing the adjacency matrix we get the graph as

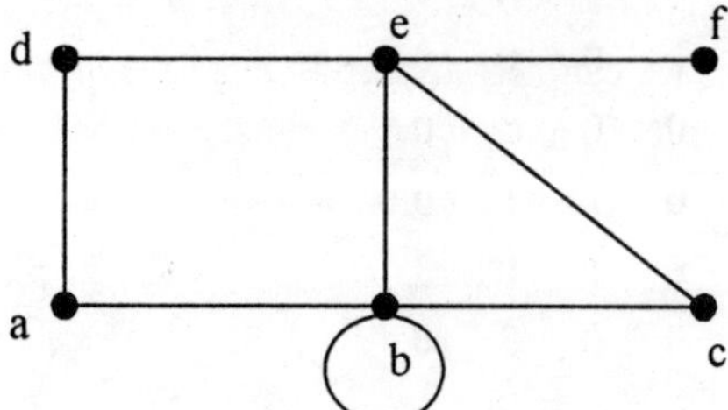

Fig. 5.3

NOTE

1. All the entries along the main diagonal are zero if the graph has no self-loop.
2. Sum of all entries in any row/column is equal to the degree of the vertex corresponding to that row /column.
3. The adjacency matrix of graph G is symmetric matrix i.e. $a_{ij} = a_{ji}$ for every i and j.
4. From the adjacency matrix we get no information about parallel edges or multiple loops, so we ignore parallel edges in the definition of $A(G)$.
5. Number of vertices having self loop equal to number of nonzero diagonal entries in matrix $A(G)$.
6. A graph G is disconnected and has two components G_1 and G_2 if and only if its adjacency matrix $A(G)$ can be partitioned as

$$A(G) = \begin{bmatrix} A(G_1) & 0 \\ 0 & A(G_2) \end{bmatrix}$$

where $A(G_1)$ is the adjacency matrix of the component G_1 and $A(G_2)$ is that of component G_2. The partitioning clearly indicates that there exists no edge joining any vertex in subgraph G_1 to any vertex in subgraph G_2.
7. The adjacency matrix of the graph G depends on the ordering of the vertices of G, i.e. a different ordering of the vertices may result in a different adjacency matrix. However, the resultant matrices from two different ordering are closely related and can be obtained from the other by simply interchanging rows and columns. Hence, two graphs G_1 and G_2 (with no parallel edges) are isomorphic if and only if their adjacency matrices $A(G_1)$ and $A(G_2)$ are related as:

$$A\,(G_2) = P^{-1}.A\,(G_1)\,.P,$$

where P is a permutation matrix.
8. If the corresponding rows and corresponding columns are interchanged (rows into rows and columns into columns) in the same order and the graph obtained after this is not altered then the graph of this changed matrix and original matrix shall be isomorphic.
9. If the adjacency matrices of two graphs are identical then the two graphs are isomorphic.

5.2 APPLICATIONS OF ADJACENCY MATRIX

Determination of isomorphism of two graphs

We have seen in the earlier chapter that the necessary condition for determining isomorphism of two graphs G_1 and G_2 are:

1. Number of vertices and edges in G_1 is equal to that of G_2 respectively.
2. The degree of sequence of vertices in graph G_1 is the same as the degree sequence of vertices in graph G_2.

To get the sufficient condition i.e. one to one correspondence between the vertex set of G_1 and vertex set of G_2 for the isomorphism we can use adjacency matrices of graph G_1 and G_2. If these two adjacency matrices are same then graphs G_1 and G_2 are isomorphic otherwise not.

Example 5.5

Show that the following two graphs G_1 and G_2 are isomorphic.

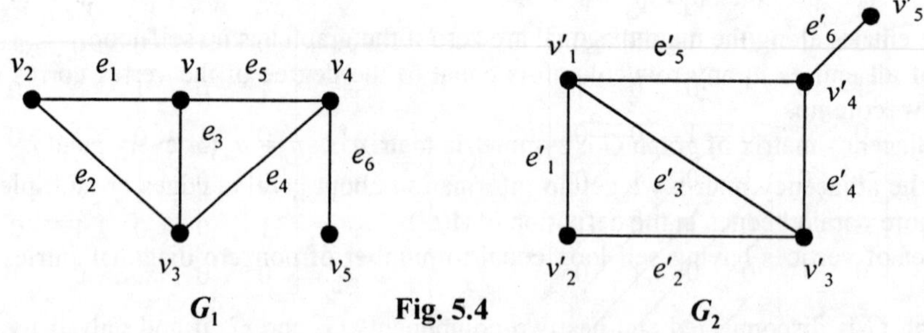

G_1 **Fig. 5.4** G_2

Solution

We can observe that both the graphs G_1 and G_2 are having same number of vertices and edges. Both the graphs G_1 and G_2 are having same degree sequences of vertices 1, 2, 3, 3, 3. Now to check the one-one correspondence let us form the identity matrices for both the graphs G_1 and G_2.

$$A(G_1) = \begin{array}{c} \\ v_1 \\ v_2 \\ v_3 \\ v_4 \\ v_5 \end{array} \begin{array}{c} \begin{array}{ccccc} v_1 & v_2 & v_3 & v_4 & v_5 \end{array} \\ \begin{bmatrix} 0 & 1 & 1 & 1 & 0 \\ 1 & 0 & 1 & 0 & 0 \\ 1 & 1 & 0 & 1 & 0 \\ 1 & 0 & 0 & 0 & 1 \\ 0 & 0 & 0 & 1 & 0 \end{bmatrix} \end{array}, \quad A(G_2) = \begin{array}{c} \\ v'_1 \\ v'_2 \\ v'_3 \\ v'_4 \\ v'_5 \end{array} \begin{array}{c} \begin{array}{ccccc} v'_1 & v'_2 & v'_3 & v'_4 & v'_5 \end{array} \\ \begin{bmatrix} 0 & 1 & 1 & 1 & 0 \\ 1 & 0 & 1 & 0 & 0 \\ 1 & 1 & 0 & 1 & 0 \\ 1 & 0 & 1 & 0 & 1 \\ 0 & 0 & 0 & 1 & 0 \end{bmatrix} \end{array}$$

Since $A(G_1) = A(G_2)$, hence two graphs are isomorphic.

Example 5.6

Check the isomorphism of the following two graphs.

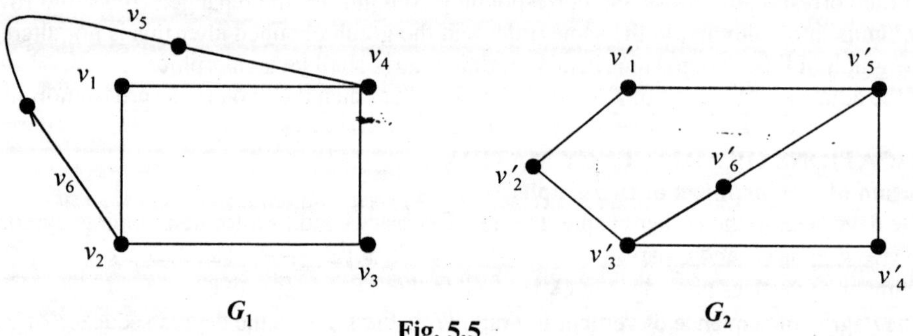

G_1 **Fig. 5.5** G_2

Solution

We can observe from the graphs G_1 and G_2 that both are having same, i.e. 5 number of vertices and 7 number of edges. The degree sequence of the graph G_1 is 2, 2, 2, 2, 3, 3 and that of G_2 is 2, 2, 2, 2, 3, 3 which are also same.

Now to check the one to one correspondence of the vertices of the given graphs form the adjacency matrix.

$$A(G_1) = \begin{array}{c} \\ v_1 \\ v_2 \\ v_3 \\ v_4 \\ v_5 \\ v_6 \end{array} \begin{array}{c} \begin{array}{cccccc} v_1 & v_2 & v_3 & v_4 & v_5 & v_6 \end{array} \\ \begin{bmatrix} 0 & 1 & 0 & 1 & 0 & 0 \\ 1 & 0 & 1 & 0 & 0 & 1 \\ 0 & 1 & 0 & 1 & 0 & 0 \\ 1 & 0 & 1 & 0 & 1 & 0 \\ 0 & 0 & 0 & 1 & 0 & 1 \\ 0 & 1 & 0 & 0 & 1 & 0 \end{bmatrix} \end{array}, \quad A(G_2) = \begin{array}{c} \\ v'_1 \\ v'_2 \\ v'_3 \\ v'_4 \\ v'_5 \\ v'_6 \end{array} \begin{array}{c} \begin{array}{cccccc} v'_1 & v'_2 & v'_3 & v'_4 & v'_5 & v'_6 \end{array} \\ \begin{bmatrix} 0 & 1 & 0 & 0 & 1 & 0 \\ 1 & 0 & 1 & 0 & 0 & 0 \\ 0 & 1 & 0 & 1 & 0 & 1 \\ 0 & 0 & 1 & 0 & 1 & 0 \\ 1 & 0 & 0 & 1 & 0 & 1 \\ 0 & 0 & 1 & 0 & 1 & 0 \end{bmatrix} \end{array}$$

We can observe that by permutations of rows and columns this matrix can be changed to that of $A(G_1)$. Hence graphs G_1 and G_2, are isomorphic.

Alternative: We find the correspondence of vertices of graph G_1 and G_2, in such a way that v_1, v_2, v_3, v_4, v_5 and v_6 of G_1 corresponds to v'_6, v'_3, v'_4, v'_5, v'_1, v'_2 respectively of G_2. If we form the adjacency matrix of G_2 in this order of vertices, we get the same adjacency matrix as $A(G_1)$.

$$A(G_2) = \begin{array}{c} \\ v'_6 \\ v'_3 \\ v'_4 \\ v'_5 \\ v'_1 \\ v'_2 \end{array} \begin{array}{c} \begin{array}{cccccc} v'_6 & v'_3 & v'_4 & v'_5 & v'_1 & v'_2 \end{array} \\ \begin{bmatrix} 0 & 1 & 0 & 1 & 0 & 0 \\ 1 & 0 & 1 & 0 & 0 & 1 \\ 0 & 1 & 0 & 1 & 0 & 0 \\ 1 & 0 & 1 & 0 & 1 & 0 \\ 0 & 0 & 0 & 1 & 0 & 1 \\ 0 & 1 & 0 & 0 & 1 & 0 \end{bmatrix} \end{array},$$

5.3 POWERS OF ADJACENCY MATRIX

Consider the powers A, A^2, A^3... of an adjacency matrix $A=[a_{ij}]$ of the graph G. The element in the ith row and j^{th} column of matrix A^n may be expressed as

$$a_n(i, j) = \text{the } ij^{\text{th}} \text{ entry in the matrix } A^n.$$

THEOREM 5.1

If A be an adjacency matrix of a graph G then $a_n(i,j)$ gives the number of paths of the length n from v_i to v_j (i.e. number of different edge sequences[1] of n edges) whenever $i \neq j$.

PROOF

The theorem can be proved with the help of following example. Consider the graph given in Fig. 5.2(a) whose adjacency matrix is given in Fig. 5.2(b). The matrices A^2, A^3 and A^4 for the same graph is as follow:

---

[1] An **edge sequence** is a sequence of edges in which each edge has one vertex in common with the edge preceding it and one vertex in common with the edge following it. An edge may appear more than once in an edge sequence.

Observe that $a_2(2,4) = 2$, so there are two paths of length 2 from v_2 to v_4, i.e. $\{e_2, e_3\}$ and $\{e_1, e_4\}$. Also $a_3(3, 5) = 1$, so there is one path of length 3 from v_3 to v_5, i.e. $\{e_6, e_4, e_5\}$; and $a_4(5, 2) = 2$, so there are two paths of length 4 from v_5 to v_2.

$$A^2 = \begin{bmatrix} 3 & 1 & 2 & 1 & 1 \\ 1 & 2 & 1 & 2 & 0 \\ 2 & 1 & 3 & 1 & 1 \\ 1 & 2 & 1 & 3 & 0 \\ 1 & 0 & 1 & 0 & 1 \end{bmatrix} \qquad A^3 = \begin{bmatrix} 4 & 5 & 4 & 5 & 1 \\ 5 & 2 & 5 & 2 & 2 \\ 5 & 5 & 4 & 6 & 1 \\ 6 & 2 & 6 & 2 & 3 \\ 1 & 2 & 1 & 2 & 0 \end{bmatrix}$$

$$A^4 = \begin{bmatrix} 14 & 8 & 14 & 9 & 5 \\ 9 & 10 & 9 & 12 & 2 \\ 15 & 9 & 16 & 10 & 6 \\ 12 & 12 & 10 & 15 & 2 \\ 5 & 2 & 5 & 2 & 2 \end{bmatrix}$$

OBSERVATION

1. A^2 is symmetric.
2. The diagonal elements in A^2 equals the degree of corresponding vertex ,for example $a_2(i, i)$ = degree of v_i.

5.4 CALCULATING A PATH BETWEEN VERTICES

In the previous example we have seen the existence of an edge between two vertices v_i and v_i is shown by an entry of 1 in the ith row and j^{th} column of the adjacency matrix. This entry represents a path of length 1 from v_i to v_j. To compute a path of length 2, the matrix of length 1 must be multiplied by itself and the product matrix is the matrix representation of path of length 2. Similarly to get a path of length 3, the matrix of length 2 should be multiplied by matrix of length 1. In general, to generate the matrix of path of length n, take the matrix of path length n - 1 and multiply it with the matrix of path length 1.

Example 5.7

Determine the number of paths of length 4 from a to d in the graph shown in Fig.5.6.

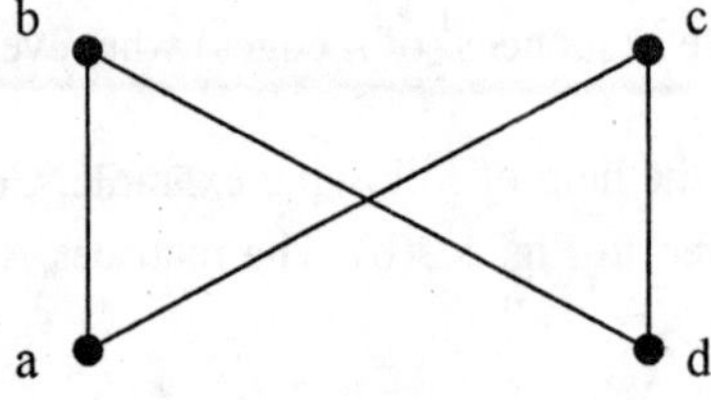

Fig. 5.6

Solution

The adjacency matrix of graph G is

$$A(G) = \begin{array}{c|cccc} & a & b & c & d \\ \hline a & 0 & 1 & 1 & 0 \\ b & 1 & 0 & 0 & 1 \\ c & 1 & 0 & 0 & 1 \\ d & 0 & 1 & 1 & 0 \end{array}$$

The number of paths of length 4 from a to d is obtained by a_4 (1, 4) in $A^4(G)$.

$$A^4(G) = \begin{bmatrix} 8 & 0 & 0 & 8 \\ 0 & 8 & 8 & 0 \\ 0 & 8 & 8 & 0 \\ 8 & 0 & 0 & 8 \end{bmatrix}$$

We can see from the matrix $A^4(G)$ that $a_4(1, 4) = 8$, hence there are exactly eight paths of length 4 from a to d.

Example 5.8

(i) The adjacency matrix of a complete graph is all 1's except for 0's on the diagonal.

(ii) The adjacency matrix of a complete bipartite graph $K_{m,n}$ has the form

$$\begin{bmatrix} 0 & J \\ J^T & 0 \end{bmatrix}$$

Where J is an $m \times n$ matrix of all ones and 0 denotes zero matrix.

THEOREM 5.2

Each diagonal entry in A^3 equals twice the number of different circuits (triangles) of length 3 passing through the corresponding vertex.

PROOF

The theorem can be proved with the help of an example. Examine the matrix corresponding to the graph shown in Fig. 5.2 (a).

$$A^3 = \begin{bmatrix} 4 & 5 & 4 & 5 & 1 \\ 5 & 2 & 5 & 2 & 2 \\ 5 & 5 & 4 & 6 & 1 \\ 6 & 2 & 6 & 2 & 3 \\ 1 & 2 & 1 & 2 & 0 \end{bmatrix}$$

For the graph G in Fig.5.2 (a) we see that $a_3(2, 2) = 2$ so there is a circuit of length 3 passing through the vertex v_2 i.e. $c_1 = \{e_1, e_6, e_2\}$. Similarly, $a_3(3, 3) = 4$, i.e. there are two circuit passing through the vertex v_3, namely $c_1 = \{e_2, e_1, e_6\}$ and $c_2 = \{e_6, e_4, e_3\}$.

REMARK

Suppose v_i and v_j be two vertices of graph $G = (V, E)$ which is having n vertices, and A is the adjacency matrix. If we define the matrix B_n as $B_n = A + A^2 + A^3 + ... + A^{n-1}$.

Then from the matrix B_n we can determine the number of paths of length n or less from v_i to v_j (for $i \neq j$).

THEOREM 5.3

Let G be a graph with m vertices $v_1, v_2, \ldots, v_m$ and suppose A be the adjacency matrix of G.

And $$B_m = A + A^2 + A^3 + \ldots + A^{m-1}$$

then G is a connected graph iff for every pair of distinct indices i and j, we have $b_{ij} \neq 0$, i.e. iff B has no zero entries off the main diagonal.

PROOF

This result can be proved with the help of following example.

Adjacency matrix of graph G shown in Fig. 5.7 is given below by A. Check whether it is connected. Since, here $m = 5$, we look at

$$B_5 = A + A^2 + A^3 + A^4.$$

$$A = \begin{bmatrix} 0 & 0 & 1 & 0 & 0 \\ 0 & 0 & 0 & 1 & 0 \\ 1 & 0 & 0 & 0 & 1 \\ 0 & 1 & 0 & 0 & 1 \\ 0 & 0 & 1 & 1 & 0 \end{bmatrix}$$

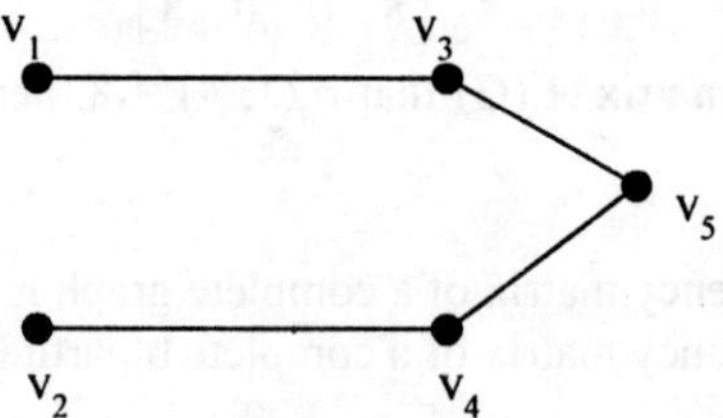

Fig. 5.7

We find A^2, A^3, A^4 and A^5 by matrix multiplication.

$$A^2 = \begin{bmatrix} 1 & 0 & 0 & 0 & 1 \\ 0 & 1 & 0 & 0 & 1 \\ 0 & 0 & 2 & 1 & 0 \\ 0 & 0 & 1 & 2 & 0 \\ 1 & 1 & 0 & 0 & 2 \end{bmatrix}; \quad A^3 = \begin{bmatrix} 0 & 0 & 2 & 1 & 0 \\ 0 & 0 & 1 & 2 & 0 \\ 2 & 1 & 0 & 0 & 3 \\ 1 & 2 & 0 & 0 & 3 \\ 0 & 0 & 3 & 3 & 0 \end{bmatrix}; \quad A^4 = \begin{bmatrix} 2 & 1 & 0 & 0 & 3 \\ 1 & 2 & 0 & 0 & 3 \\ 0 & 0 & 5 & 4 & 0 \\ 0 & 0 & 4 & 5 & 0 \\ 3 & 3 & 0 & 0 & 6 \end{bmatrix}$$

Now $B_5 = A + A^2 + A^3 + A^4$.

$$= \begin{bmatrix} 3 & 1 & 3 & 1 & 4 \\ 1 & 3 & 1 & 3 & 4 \\ 3 & 1 & 7 & 5 & 4 \\ 1 & 3 & 5 & 7 & 4 \\ 4 & 4 & 4 & 4 & 8 \end{bmatrix}$$

Since B_5 has all non-zero entries. The graph is connected.

NOTE

This theorem can also be used to determine the shortest path between any two vertices. In a connected graph the distance between two vertices v_i and v_j $(i \neq j)$ is k, iff k is the least integer for which the $(i, j)^{th}$ element in A^k is non zero.

For example, in the graph shown in Fig. 5.7 if we want to determine the distance between (i) v_1 and v_2 (ii) v_1 and v_4, then

(i) (1, 2)th elements in A, A^2, A^3 and A^4 are 0, 1, 0 and 2 respectively which is positive (for least value of k in A^k)in A^2. Hence distance $(v_1, v_2) = 1$.

(ii) (1, 4)th element in A, A^2, A^3 and A^4 are 0, 0, 1 and 0 respectively which is positive (for least value of k in A^k)in A^3. Hence distance $(v_1, v_4) = 1$.

5.5 PATH MATRIX

Suppose $G(V, E)$ is a simple graph with no parallel edges and m vertices labelled $v_1, v_2 \ldots v_m$. Then the **path matrix** or **reachability matrix** for (v_i, v_j) vertices is denoted by $P(v_i, v_j)$ or $[P_{ij}]$, where

$$P_{ij} = \begin{cases} 1, & \text{If there is an edge between } v_i \text{ and } v_j \\ 0, & \text{Otherewise} \end{cases}$$

The rows in by $P_{ij} = P(v_i, v_j)$ corresponds to different paths between vertices v_i and v_j and column corresponds to the edge in graph G.

Example 5.9

Consider the graph G in Fig.5.8 with vertices v_1, v_2, v_3, v_4 and v_5. There are three different paths from v_1 to v_3 i.e. $p_1 = \{e_6\}$, $p_2 = \{e_4, e_3\}$, $p_3 = \{e_1, e_2\}$. And seven edges in G as e_1, e_2, e_3, e_4, e_5, e_6 and e_7.

The path matrix $P(v_1, v_3)$ is of order 3×7 is given in Fig.5.8 (b).

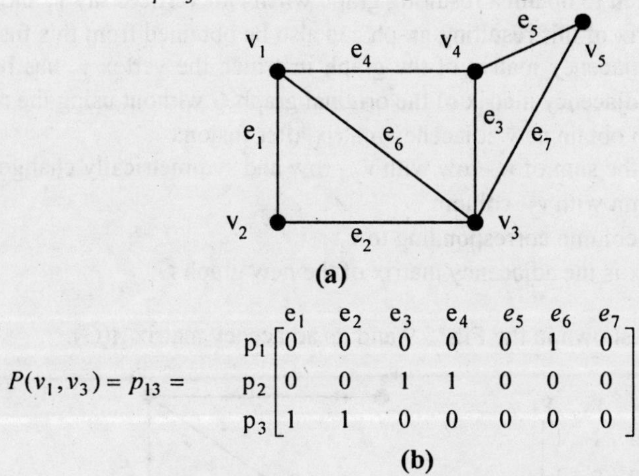

$$P(v_1, v_3) = p_{13} = \begin{array}{c} \\ p_1 \\ p_2 \\ p_3 \end{array} \begin{array}{c} \begin{array}{ccccccc} e_1 & e_2 & e_3 & e_4 & e_5 & e_6 & e_7 \end{array} \\ \begin{bmatrix} 0 & 0 & 0 & 0 & 0 & 1 & 0 \\ 0 & 0 & 1 & 1 & 0 & 0 & 0 \\ 1 & 1 & 0 & 0 & 0 & 0 & 0 \end{bmatrix} \end{array}$$

(b)

Fig. 5.8 A graph and its path matrix

NOTE

1. All 0's entry of a column correspond to an edge that does not lie on any path between v_i and v_j.
2. All 1's entry of a column correspond to an edge that lie in every path between v_i and v_j.
3. There is no information about the loop of a graph by a path matrix $P(v_i, v_j)$.
4. If two vertices lie in different components then we can not define the path matrix. It implies that path matrix is possible only for connected graph.
5. In a path matrix there exist at least one 1's in it.
6. Number of 1's in a row is equal to the length of path matrix corresponding to the row.

7. The ring sum of any two rows in $P(v_i, v_j)$ corresponds to a circuit or an edge-disjoint union of circuits.
8. The path matrix of a tree is a row matrix.
9. If two rows are identical, there are multiple edges between the vertices which lie in the path corresponds to row.

Example 5.10

Consider the graph G in Fig. 5.2(a). The ring sum of any two rows in $P(v_1, v_3)$ corresponds to a circuit or an edge disjoint union of circuits.

$$p_1 \oplus p_2 = \begin{bmatrix} \overset{e_1}{0} & \overset{e_2}{0} & \overset{e_3}{1} & \overset{e_4}{1} & \overset{e_5}{0} & \overset{e_6}{1} & \overset{e_7}{0} \end{bmatrix}$$

We see that there is a circuit of length 3, i.e. $\{e_3, e_4, e_6\}$. Similarly,

$$p_2 \oplus p_3 = \begin{bmatrix} 1 & 1 & 1 & 1 & 0 & 0 & 0 \end{bmatrix}$$

We see that there is a circuit of length 4, i.e. $\{e_1, e_2, e_3, e_4\}$. The same can be generalized for any two rows in $P(v_i, v_j)$.

5. 6 FUSION OF VERTICES IN AN ADJACENCY MATRIX

Consider a graph G with n vertices $v_1, v_2, v_3, \ldots, v_n$. Let $A(G)$ be the adjacency matrix of the graph G. We have already seen to obtain a resulting graph when two vertices say v_1 and v_2 are fused .

The adjacency matrix of this resulting graph can also be obtained from this fused graph. Here we see how to obtain the adjacency matrix of the graph in which the vertex v_2 has been fused with vertex v_1, directly from the adjacency matrix of the original graph G without using the resulting graph. Following is the algorithm to obtain new adjacency matrix after fusion:

Step 1: Change v_1 - row to the sum of v_1- row with v_2 - row and symmetrically change v_1 - column to the sum of v_1- column with v_2- column.

Step 2: Delete the row and column corresponding to v_2.

The resulting matrix is the adjacency matrix of the new graph G.

Example 5.11

Consider a graph G shown in the Fig. 5.9 and its adjacency matrix $A(G)$.

$$A(G) = \begin{array}{c} \\ v_1 \\ v_2 \\ v_3 \\ v_4 \end{array} \begin{array}{c} \begin{array}{cccc} v_1 & v_2 & v_3 & v_4 \end{array} \\ \begin{bmatrix} 0 & 0 & 1 & 1 \\ 0 & 0 & 0 & 1 \\ 1 & 0 & 0 & 1 \\ 1 & 1 & 1 & 0 \end{bmatrix} \end{array}$$

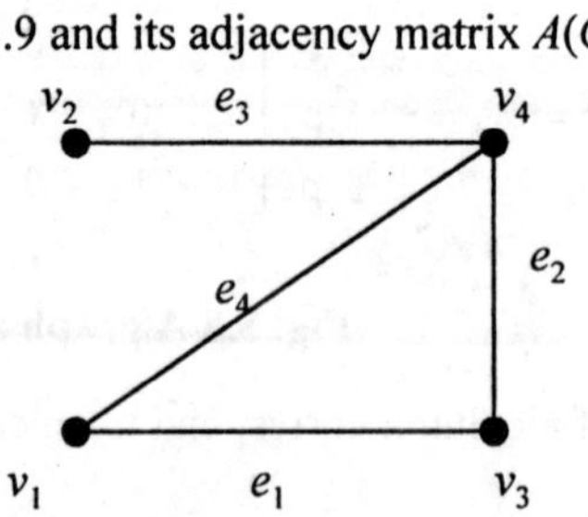

Fig. 5.9

If we fuse vertex v_1 with v_3 then the resulting graph is obtained by replacing v_1 and v_3 with a single vertex say v in such a way that every edge that is incident with either v_1 or v_3 in G is now incident with the new vertex.

We can observe that the resultant graph G_1 shown in Fig.5.10 has one less vertex than G but same number of edges as G. The degree of the new vertex v is the sum of the degree of vertices v_1 and v_3.

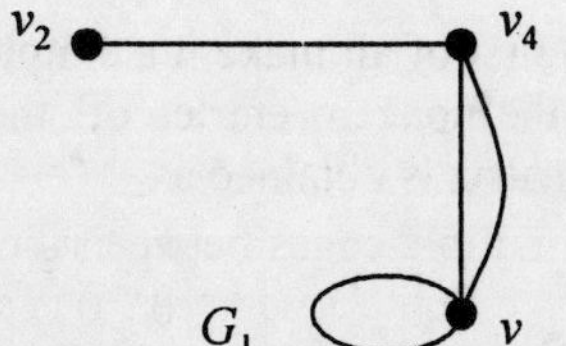

Fig. 5.10

Now we can obtain the adjacency matrix $A(G_1)$ of graph G_1 by following the algorithm discussed above. In the adjacency matrix $A(G)$ replacing the v_1- row by the sum of v_1 and v_3 - rows and name as v-row. Similarly replacing v_1-column by the sum of v_1 and v_3 -columns and name as v-column. Delete the rows and columns corresponding to v_3. Finally we get the following adjacency matrix.

$$A(G_1) = \begin{array}{c} \\ v \\ v_2 \\ v_4 \end{array}\begin{array}{c} \begin{array}{ccc} v & v_2 & v_4 \end{array} \\ \begin{bmatrix} 1 & 0 & 2 \\ 0 & 0 & 1 \\ 2 & 1 & 0 \end{bmatrix} \end{array}$$

5.7 FUSION ALGORITHM FOR CONNECTEDNESS

We have seen previously how to check connectedness of a graph using adjacency matrix. Here we give fusion algorithm for connectedness:

Step 1: Replace graph G by its corresponding simple graph (if given graph G is not simple graph). To get the adjacency of new graph replaces all the non-zero entries off the diagonal by 1 and make all entries on the diagonal 0. Denote this simple graph by G_1.

Step 2: Fuse vertex v_1 to the first of the vertices $v_2, v_3, \ldots, v_n$ with which it is adjacent to get a new graph say G_2, where new vertex is also denoted by v_1.

Step 3: Obtain adjacency matrix of G_1 which will also be adjacency matrix of G.

Step 4: Repeat step 1 and 2 with vertex v_1 until v_1 is not adjacent to any of the other vertices.

Step 5: Repeat step 2 and 4 on the vertex v_2 of the last graph obtained in step 4. Continue this process on all other remaining vertices of the resulting graphs.

The final graph obtained in this way is a null graph. The number of isolated vertices in the null graph obtained shall give the number of connected components of the original graphs G.

Example 5.12

Consider the following graph and its adjacency matrix. Use fusion algorithm to check the connectedness of the graph.

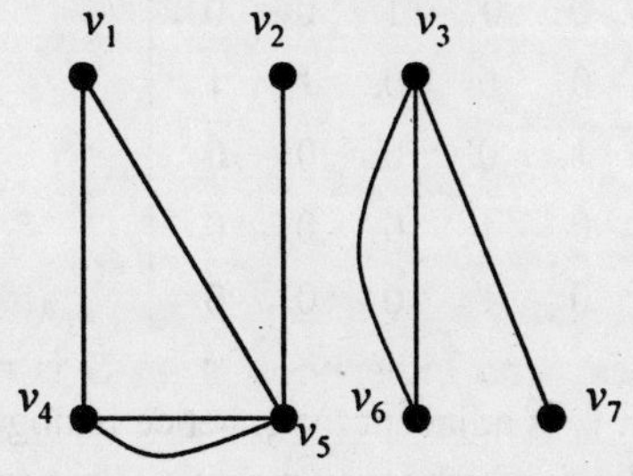

$$A(G) = \begin{array}{c} \\ v_1 \\ v_2 \\ v_3 \\ v_4 \\ v_5 \\ v_6 \\ v_7 \end{array}\begin{array}{c} \begin{array}{ccccccc} v_1 & v_2 & v_3 & v_4 & v_5 & v_6 & v_7 \end{array} \\ \begin{bmatrix} 0 & 0 & 0 & 1 & 1 & 0 & 0 \\ 0 & 0 & 0 & 0 & 1 & 0 & 0 \\ 0 & 0 & 0 & 0 & 0 & 2 & 1 \\ 1 & 0 & 0 & 0 & 2 & 0 & 0 \\ 1 & 1 & 0 & 2 & 0 & 0 & 0 \\ 0 & 0 & 2 & 0 & 0 & 0 & 0 \\ 0 & 0 & 1 & 0 & 0 & 0 & 0 \end{bmatrix} \end{array}$$

Step1: Since the graph is not simple hence first of all make it a simple graph. The adjacency of this graph is obtained by replacing all the nonzero entries off the diagonal by 1 and make all entries on the diagonal 0. The new matrix is obtained as

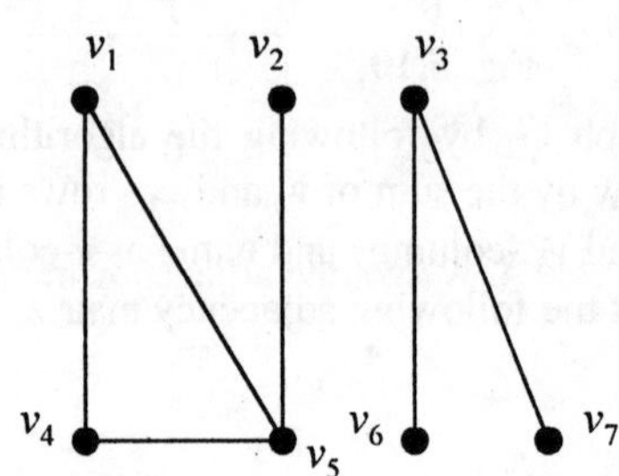

	v_1	v_2	v_3	v_4	v_5	v_6	v_7
v_1	0	0	0	1	1	0	0
v_2	0	0	0	0	1	0	0
v_3	0	0	0	0	0	1	1
v_4	1	0	0	0	1	0	0
v_5	1	1	0	1	0	0	0
v_6	0	0	1	0	0	0	0
v_7	0	0	1	0	0	0	0

Step 2: Now we observe that v_1 is adjacent to vertex v_4 hence we fuse vertex v_1 with v_4 and name the resulting new vertex as v_1. Now to obtain adjacency matrix from the matrix obtained in step 1 change v_1 -row to the sum of v_1 and v_4 rows and v_1 column to the sum of v_1 and v_4 columns and then delete the rows and columns corresponding to v_4. The new adjacency matrix is given by

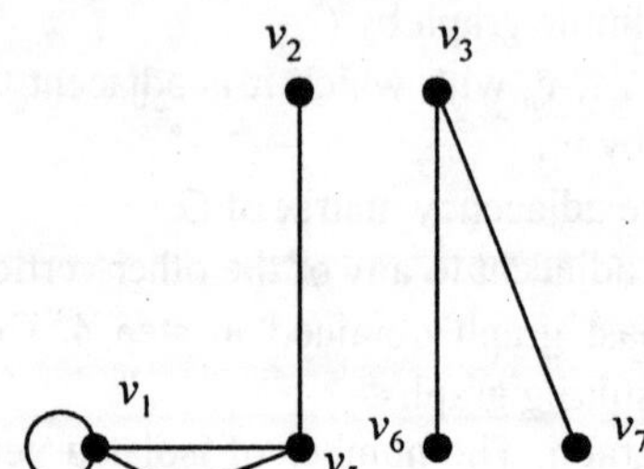

	v_1	v_2	v_3	v_5	v_6	v_7
v_1	1	0	0	2	0	0
v_2	0	0	0	1	0	0
v_3	0	0	0	0	1	1
v_5	2	1	0	0	0	0
v_6	0	0	1	0	0	0
v_7	0	0	1	0	0	0

Step 3: Now repeating step 1 on the graph obtained in step 2, we get the following matrix corresponding to simple graph.

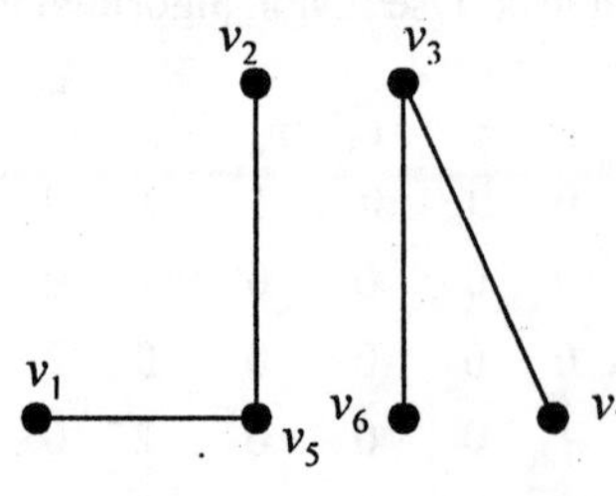

	v_1	v_2	v_3	v_5	v_6	v_7
v_1	0	0	0	1	0	0
v_2	0	0	0	1	0	0
v_3	0	0	0	0	1	1
v_5	1	1	0	0	0	0
v_6	0	0	1	0	0	0
v_7	0	0	1	0	0	0

Fusing v_1 with v_5

Step 4: From the adjacency matrix obtained in step 4 we see that v_1 is adjacent to v_5, hence fusing v_1 to v_5 and naming the new vertex again v_1, we get the following adjacency matrix.

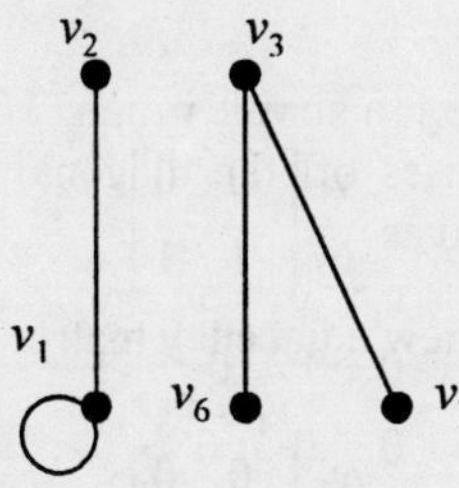

$$\begin{array}{c} \\ v_1 \\ v_2 \\ v_3 \\ v_6 \\ v_7 \end{array}\begin{array}{c} \begin{array}{ccccc} v_1 & v_2 & v_3 & v_6 & v_7 \end{array} \\ \begin{bmatrix} 1 & 1 & 0 & 0 & 0 \\ 1 & 0 & 0 & 0 & 0 \\ 0 & 0 & 0 & 1 & 1 \\ 0 & 0 & 1 & 0 & 0 \\ 0 & 0 & 1 & 0 & 0 \end{bmatrix} \end{array}$$

Step 5: Again applying step 1 on the matrix obtained in step 4, we get the adjacency matrix of the corresponding simple graph as follows:

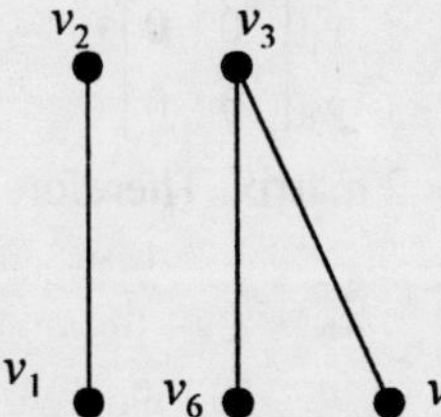

$$\begin{array}{c} \\ v_1 \\ v_2 \\ v_3 \\ v_6 \\ v_7 \end{array}\begin{array}{c} \begin{array}{ccccc} v_1 & v_2 & v_3 & v_6 & v_7 \end{array} \\ \begin{bmatrix} 0 & 1 & 0 & 0 & 0 \\ 1 & 0 & 0 & 0 & 0 \\ 0 & 0 & 0 & 1 & 1 \\ 0 & 0 & 1 & 0 & 0 \\ 0 & 0 & 1 & 0 & 0 \end{bmatrix} \end{array}$$

Step 6: Here, we observe that v_1 is adjacent to v_2, hence fusing v_1 to v_2 naming the new vertex as v_1,, we get the following adjacency matrix.

$$\begin{array}{c} \\ v_1 \\ v_3 \\ v_6 \\ v_7 \end{array}\begin{array}{c} \begin{array}{cccc} v_1 & v_3 & v_6 & v_7 \end{array} \\ \begin{bmatrix} 1 & 0 & 0 & 0 \\ 0 & 0 & 1 & 1 \\ 0 & 1 & 0 & 0 \\ 0 & 1 & 0 & 0 \end{bmatrix} \end{array}$$

Step 7: Applying step 1, we get the adjacency matrix of the resulting simple graph as follows:

$$\begin{array}{c} \\ v_1 \\ v_3 \\ v_6 \\ v_7 \end{array}\begin{array}{c} \begin{array}{cccc} v_1 & v_3 & v_6 & v_7 \end{array} \\ \begin{bmatrix} 0 & 0 & 0 & 0 \\ 0 & 0 & 1 & 1 \\ 0 & 1 & 0 & 0 \\ 0 & 1 & 0 & 0 \end{bmatrix} \end{array}$$

Step 8: Here we observe that there is no vertex adjacent to v_1. Therefore, we move to the next vertex v_3. Since v_3 is adjacent to v_6, we fuse them. Name the new vertex as v_3. The new adjacency matrix is given by

$$\begin{array}{c} \\ v_1 \\ v_3 \\ v_7 \end{array}\begin{array}{c} \begin{array}{ccc} v_1 & v_3 & v_7 \end{array} \\ \begin{bmatrix} 0 & 0 & 0 \\ 0 & 1 & 1 \\ 0 & 1 & 0 \end{bmatrix} \end{array}$$

Step 9: Applying step 1, the adjacency matrix of the corresponding simple graph is obtained as

$$\begin{array}{c} \\ v_1 \\ v_3 \end{array}\begin{array}{c} \begin{array}{cc} v_1 & v_3 \end{array} \\ \begin{bmatrix} 0 & 0 \\ 1 & 1 \end{bmatrix} \end{array}$$

Step 10: Now fusing vertex v_3 with its adjacent vertex v_7, the new adjacency matrix is

$$\begin{array}{c} \\ v_1 \\ v_3 \end{array}\begin{array}{c} \begin{array}{cc} v_1 & v_3 \end{array} \\ \begin{bmatrix} 0 & 0 \\ 0 & 1 \end{bmatrix} \end{array}$$

Step 11: Applying steps 1, the adjacency matrix of the corresponding simple graph is obtained as

$$\begin{array}{c} \\ v_1 \\ v_3 \end{array}\begin{array}{c} \begin{array}{cc} v_1 & v_3 \end{array} \\ \begin{bmatrix} 0 & 0 \\ 0 & 0 \end{bmatrix} \end{array}$$

This is a null matrix. The final adjacency matrix is 2×2 matrix. Therefore, the original graph is disconnected and has two connected components.

5.8 INCIDENCE MATRIX

Suppose $G = (V, E)$ is a simple graph (with no parallel edges), m vertices listed v_1, v_2, v_3,...,v_m and t edges listed $e_1, e_2, \ldots, e_t$. The **incidence matrix** of $m \times t$ order is denoted by $[m_{ij}]$, where

$$M(G) = m_{ij} = \begin{cases} 1, & \text{If } e_i \text{ edge is incident to } v_i \text{ vertex} \\ 0, & \text{otherwise} \end{cases}$$

Example 5.13

Construct the incidence matrix M for the graph shown in Fig.5.11.

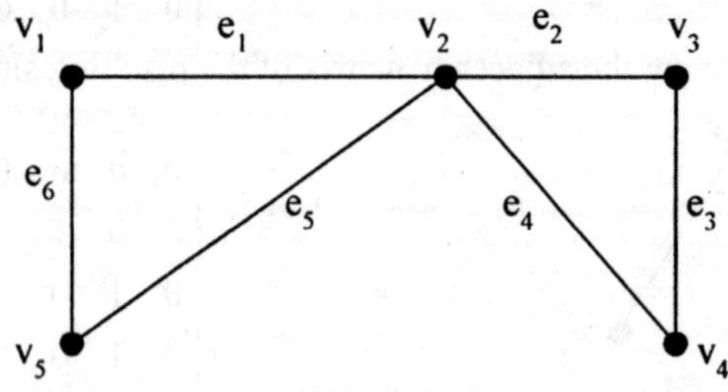

Fig. 5.11

Solution

$$M(G) = \begin{array}{c} \\ v_1 \\ v_2 \\ v_3 \\ v_4 \\ v_5 \end{array}\begin{array}{c} \begin{array}{cccccc} e_1 & e_2 & e_3 & e_4 & e_5 & e_6 \end{array} \\ \begin{bmatrix} 1 & 0 & 0 & 0 & 0 & 1 \\ 1 & 1 & 0 & 1 & 1 & 0 \\ 0 & 1 & 1 & 0 & 0 & 0 \\ 0 & 0 & 1 & 1 & 0 & 0 \\ 0 & 0 & 0 & 0 & 1 & 1 \end{bmatrix} \end{array}$$

NOTE

1. Sum of all entries in i^{th} row is equal to the degree of the vertex v_i.
2. Every edge is incident on exactly two vertices hence each column correspond to an edge has exactly two one's.
3. All 0's entries of a row represent an isolated vertex.
4. Two graphs G_1 and G_2 are isomorphic iff their incidence matrices $M(G_1)$ and $M(G_2)$ differ only by permutations of rows and columns.
5. The incidence matrix of a disconnected graph can be partitioned in a form as

$$M(G) = \begin{bmatrix} M(g_1) & \vdots & 0 \\ \cdots & \vdots & \cdots \\ 0 & \vdots & M(g_2) \end{bmatrix}$$

   where $M(g_1)$ and $M(g_2)$ are incidence matrix of components of g_1 and g_2. This can also be true for any number of components of disconnected graph.
6. Permutation of any two columns represents the relabeling of the edges of graph G.
7. Permutation of any two rows corresponds to the rearrangement of the vertices of graph G.

THEOREM 5.4

Two graphs G_1 and G_2 are isomorphic if and only if their incidence matrices $M(G_1)$ and $M(G_2)$ differ only by permutations of rows and columns.

5.9 RANK OF THE INCIDENCE MATRIX

Consider a graph with m vertices and n edges. The incidence matrix contains only two entries 0 and 1. Therefore, this matrix can be treated as a matrix over any commutative ring with identity. But in certain cases it will be convenient to take this ring as Z_2 the field with two elements. Then each column of $M(G)$ has exactly two 1's. Hence as a matrix over Z_2 the sum of all the rows of $M(G)$ is zero (one can observe that this is a modulo 2 sum of the corresponding entries). This imply that over Z_2 the rows of $M(G)$ are linearly dependent. Consequently, the rank of $M(G)$ over Z_2 can never exceed n - 1.

THEOREM 5.5

Let $M(G)$ is an incidence matrix of a connected graph with m vertices; then the rank of $M(G)$ is m - 1.

PROOF

Suppose $M(G)$ is an incidence matrix with m vertices and n edges then it may be represented as

$$M(G) = \begin{bmatrix} m_{11} & m_{12} & m_{13} & \cdots & m_{1n} \\ m_{21} & m_{22} & m_{23} & \cdots & m_{2n} \\ \vdots & \vdots & \vdots & & \vdots \\ m_{m1} & m_{m2} & m_{m3} & \cdots & m_{mn} \end{bmatrix}$$

Let the vector in each row is called M_1, M_2 ...., M_m respectively. Thus

$$M(G) = \begin{bmatrix} M_1 \\ M_2 \\ \vdots \\ \vdots \\ M_m \end{bmatrix}$$

Because the graph is connected, hence there are exactly two 1's in every column of M. In other words we can say the sum (mod 2) of all these vectors are zero. Thus vectors $M_1, M_2, \ldots M_m$ are not linearly independent. So the rank of M is less than m; i.e rank of $A \leq m - 1$.

Now, consider the sum of any k rows of these m vectors ($k \leq m$-1). We see that if the graph is connected, $M(G)$ cannot be partitioned. In other words we can say no $k \times k$ submatrix of M can be found for $k \leq m - 1$. Hence, the addition of all vectors taken k at a time exhausts all possible linear combinations of m-1 row vectors, i.e. there is no linear combination of k row vectors of $M(G)$ equal to zero. So the rank of $M(G)$ must be at least $m - 1$.

Hence the rank of incidence matrix of connected graph is $m - 1$.

5.10 REDUCED INCIDENCE MATRIX

Consider connected graph G with m vertices and t edges. The **reduced incidence matrix** is a submatrix of order (m - 1) by t obtained on removing any one row from the incidence matrix.

Thus (m -1) rows of an incidence matrix are sufficient to specify the information of corresponding graph completely. Also, these (m - 1) row vectors are linearly independent. The vertex corresponding to the deleted row in M_f is called the **reference vertex.**

THEOREM 5.6

Let the columns of incidence matrix $M(G)$ and path matrix $P(v_i,v_j)$ of a connected graph G are arranged in same order, then the product (mod 2) is given by

$$M \cdot P^T(v_i, v_j) = X,$$

where X is a matrix with 1's in two rows v_i and v_j and the rest of the m - 2 rows have all 0's.

PROOF

Let us prove the theorem with the help of following example. Consider the graph given in Fig. 5.8(a) and path matrix for vertices v_1 and v_3. The incidence matrix of graph is as follows:

$$M = \begin{bmatrix} 1 & 0 & 0 & 1 & 0 & 1 & 0 \\ 1 & 1 & 0 & 0 & 0 & 0 & 0 \\ 0 & 1 & 1 & 0 & 0 & 1 & 1 \\ 0 & 0 & 1 & 1 & 1 & 0 & 0 \\ 0 & 0 & 0 & 0 & 1 & 0 & 1 \end{bmatrix}$$

Then

$$M \cdot P^T(v_1, v_3) = \begin{bmatrix} 1 & 0 & 0 & 1 & 0 & 1 & 0 \\ 1 & 1 & 0 & 0 & 0 & 0 & 0 \\ 0 & 1 & 1 & 0 & 0 & 1 & 1 \\ 0 & 0 & 1 & 1 & 1 & 0 & 0 \\ 0 & 0 & 0 & 0 & 1 & 0 & 1 \end{bmatrix} \begin{bmatrix} 0 & 0 & 1 \\ 0 & 0 & 1 \\ 0 & 1 & 0 \\ 0 & 1 & 0 \\ 0 & 0 & 0 \\ 1 & 0 & 0 \\ 0 & 0 & 0 \end{bmatrix} \pmod 2$$

$$= \begin{bmatrix} 1 & 1 & 1 \\ 0 & 0 & 0 \\ 1 & 1 & 1 \\ 0 & 0 & 0 \\ 0 & 0 & 0 \end{bmatrix}$$

5.11 CIRCUIT MATRIX

Consider a graph $G = (V, E)$ with n different circuits and t edges. Then the **circuit matrix** $C = [c_{ij}]$ of graph G is a $n \times t$ binary matrix defined as follows:

$$C(G) = c_{ij} = \begin{cases} 1, & \text{if } i^{th} \text{ circuit consists of } j^{th} \text{ edge} \\ 0, & \text{otherwise} \end{cases}$$

Example 5.14

Construct the circuit matrix of the graph given in Fig.5.12.

Solution

The six different circuits of the graph are given by
$c_1 = \{e_1, e_6, e_8\}$, $c_2 = \{e_3, e_4, e_9\}$, $c_3 = \{e_2, e_8, e_5, e_9\}$, $c_4 = \{e_1, e_2, e_3, e_7\}$, $c_5 = \{e_6, e_5, e_4, e_7\}$, $c_6 = \{e_1, e_6, e_5, e_4, e_3, e_2\}$ and ten edges. The circuit matrix of graph shown in Fig.5/12 is as follows:

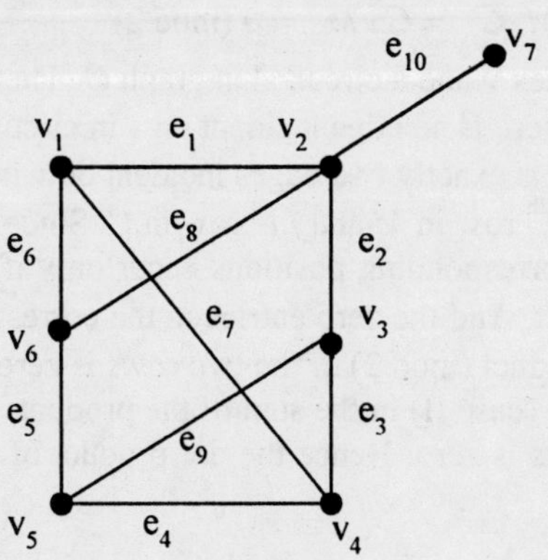

Fig. 5.12

$$C = \begin{array}{c} \\ c_1 \\ c_2 \\ c_3 \\ c_4 \\ c_5 \\ c_6 \end{array} \begin{array}{c} \begin{array}{cccccccccc} e_1 & e_2 & e_3 & e_4 & e_5 & e_6 & e_7 & e_8 & e_9 & e_{10} \end{array} \\ \begin{bmatrix} 1 & 0 & 0 & 0 & 0 & 1 & 0 & 1 & 0 & 0 \\ 0 & 0 & 1 & 1 & 0 & 0 & 0 & 0 & 1 & 0 \\ 0 & 1 & 0 & 0 & 1 & 0 & 0 & 1 & 1 & 0 \\ 1 & 1 & 1 & 0 & 0 & 0 & 1 & 0 & 0 & 0 \\ 0 & 0 & 0 & 1 & 1 & 1 & 0 & 1 & 0 & 0 \\ 1 & 1 & 1 & 1 & 1 & 1 & 0 & 0 & 0 & 0 \end{bmatrix} \end{array}$$

NOTE

1. All 0's entry of a column corresponds to an edge that does not belong to any circuit.
2. The circuit matrix of a disconnected graph can be written in a block diagonal form as

$$C(G) = \begin{bmatrix} C(g_1) & \vdots & 0 \\ \ldots & \vdots & \ldots \\ 0 & \vdots & C(g_2) \end{bmatrix}$$

where $C(g_1)$ and $C(g_2)$ are circuit matrices of two components g_1 and g_2. It implies that there is no common edge between the circuit g_1 and g_2.

3. The number of 1's in a row is equal to the number of edges in the corresponding circuit.
4. Two graphs G_1 and G_2 are 2-isomorphic iff they have same circuit matrix.
5. A row with exactly one 1's in it corresponds to a self-loop.
6. Permutation matrix of circuit matrix corresponds to relabeling the circuits and edges.

THEOREM 5.7

Every row of circuit matrix C is orthogonal to every row of incidence matrix M in same graph G, i.e.

$$M \cdot C^T = C \cdot M^T = 0 \text{ (mod 2)}$$

PROOF

If M and C are incidence matrix and circuit matrix of graph G respectively then

$$M \cdot C^T = C \cdot M^T = 0 \text{ (mod 2)}$$

To prove this consider a vertex v and a circuit Δ in graph G. The following two cases arise:

Case I: v is not in Δ, i.e. there is no edge incident on v in circuit Δ.

Case II v is in Δ, i.e. there is exactly two edges incident on v in the circuit Δ.

Now, consider the ith row in M and j i^{th} row in C. Since the edges are arranged in same order, the nonzero entries in the corresponding positions occur only if the particular edge is incident on the ith vertex and in the jth circuit. And the zero entries in the corresponding positions occur if i^{th} vertex is not in the jth circuit, and product (mod 2) of the two rows is zero. Also, when i^{th} vertex is in j^{th} circuit, there will be exactly two 1's (case II) in the sum of the products of individual entries. Since, again the product (mod 2) of two rows is zero. Hence the dot product of the two arbitrary rows of M and C is zero. Hence the theorem.

Example 5.15

Verify $M{\cdot}C^T = C{\cdot}M^T = 0$ (mod 2) where M and C are the incidence and circuit matrix respectively of the following graph G:

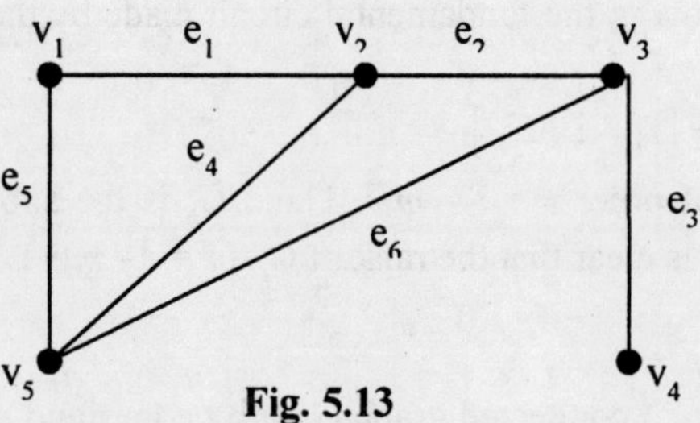

Fig. 5.13

Solution

The graph shown in Fig 5.6 has circuits: $c_1 = \{e_1, e_4, e_5\}$, $c_2 = \{e_1, e_2, e_6, e_5\}$ and $c_3 = \{e_4, e_2, e_6\}$. The circuit matrix and incidence matrix of graph with five vertices and six edges are given by

$$C = \begin{array}{c} \\ c_1 \\ c_2 \\ c_3 \end{array}\begin{array}{c} \begin{matrix} e_1 & e_2 & e_3 & e_4 & e_5 & e_6 \end{matrix} \\ \begin{bmatrix} 1 & 0 & 0 & 1 & 1 & 0 \\ 1 & 1 & 0 & 0 & 1 & 1 \\ 0 & 1 & 0 & 1 & 0 & 1 \end{bmatrix} \end{array}; \qquad M = \begin{array}{c} \\ v_1 \\ v_2 \\ v_3 \\ v_4 \\ v_5 \end{array}\begin{array}{c} \begin{matrix} e_1 & e_2 & e_3 & e_4 & e_5 & e_6 \end{matrix} \\ \begin{bmatrix} 1 & 0 & 0 & 0 & 1 & 0 \\ 1 & 1 & 0 & 1 & 0 & 0 \\ 0 & 1 & 1 & 0 & 0 & 1 \\ 0 & 0 & 1 & 0 & 0 & 0 \\ 0 & 0 & 0 & 1 & 1 & 1 \end{bmatrix} \end{array}$$

$$M \cdot C^T = \begin{bmatrix} 1 & 0 & 0 & 0 & 1 & 0 \\ 1 & 1 & 0 & 1 & 0 & 0 \\ 0 & 1 & 1 & 0 & 0 & 1 \\ 0 & 0 & 1 & 0 & 0 & 0 \\ 0 & 0 & 0 & 1 & 1 & 1 \end{bmatrix} \begin{bmatrix} 1 & 1 & 0 \\ 0 & 1 & 1 \\ 0 & 0 & 0 \\ 1 & 0 & 1 \\ 1 & 1 & 0 \\ 0 & 1 & 1 \end{bmatrix} = \begin{bmatrix} 2 & 2 & 0 \\ 2 & 2 & 2 \\ 0 & 2 & 2 \\ 0 & 0 & 0 \\ 2 & 2 & 2 \end{bmatrix} = 0 \pmod{2}$$

and

$$C \cdot M^T = \begin{bmatrix} 1 & 0 & 0 & 1 & 1 & 0 \\ 1 & 1 & 0 & 0 & 1 & 1 \\ 0 & 1 & 0 & 1 & 0 & 1 \end{bmatrix} \begin{bmatrix} 1 & 1 & 0 & 0 & 0 \\ 0 & 1 & 1 & 0 & 0 \\ 0 & 0 & 1 & 1 & 0 \\ 0 & 1 & 0 & 0 & 1 \\ 1 & 0 & 0 & 0 & 1 \\ 0 & 0 & 1 & 0 & 1 \end{bmatrix} = \begin{bmatrix} 2 & 2 & 0 & 0 & 2 \\ 2 & 2 & 2 & 0 & 2 \\ 0 & 2 & 2 & 0 & 2 \end{bmatrix} = 0 \pmod{2}$$

Hence the theorem.

5.11-1 FUNDAMENTAL CIRCUIT MATRIX

Consider a connected graph G with t edges and m vertices. The **fundamental circuit matrix** C_f of order $t - m + 1$ by t is a submatrix (of a circuit matrix) in which all rows correspond to a set of fundamental circuits with respect to some spanning tree.

In matrix C_f all the $t - m + 1$ chords correspond to the first $t - m + 1$ columns and are rearranged in such a way that the first row corresponds to the fundamental circuit made by the chord in the first column, the second row to the fundamental circuit made by the second and so on. Thus, C_f can be written as

$$C_f = [I_r \ \vdots \ C_b]$$

where I_r is an identity matrix of order $r = t - m + 1$ and C_b is the submatrix corresponding to the branches of the spanning tree. It is clear that the rank of $C_f = r = t - m + 1$.

Hence, rank of $C \geq t - m + 1$.

THEOREM 5.8

If C is a circuit matrix of a connected graph G with t edges and m vertices, then rank of $C = t - m + 1$.

PROOF

Suppose M is an incidence matrix of graph G, then from theorem 5.7, we have

$$M{\cdot}C^T = 0 \pmod 2$$

Therefore, according to Sylvester's theorem*

$$\text{rank of } M + \text{rank of } C \leq t.$$

i.e. rank of $C \leq t$ - rank of M

But rank of $M = m - 1$ (by theorem 5.6)

$\therefore$ rank of $C \leq t - m + 1$.

But rank of $C \geq t - m + 1$

So, we must have rank of $C = t - m + 1$.

5.12 CUT-SET MATRIX

Consider a graph $G = (V, E)$ with n different circuits, t edges and p cut-sets. Then the cut-set matrix $C(G) = [e_{ij}]$ of graph G is a $p{\times}t$ binary matrix defined as follows:

$$C(G) = c_{ij} = \begin{cases} 1, & \text{if } i^{th} \text{ cut set contains } j^{th} \text{ edge} \\ 0, & \text{otherwise} \end{cases}$$

Example 5.16

Consider the graph G in Fig.5.14(a) and its cut sets in fig 5.14(b).Construct the cut set matrix.

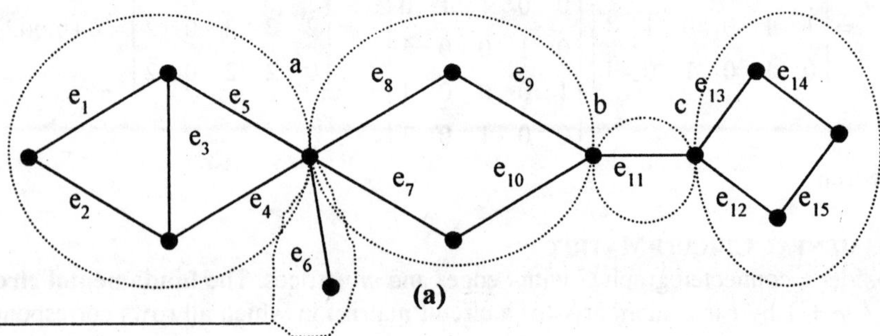

(a)

---

* If P is $k \times n$ matrix and Q is an $k \times n$ matrix, then the nullity of the product cannot exceed the sum of the nullities of the factors, i.e.

Nullity of $P{\cdot}Q \leq$ nullity of P+ nullity of Q.

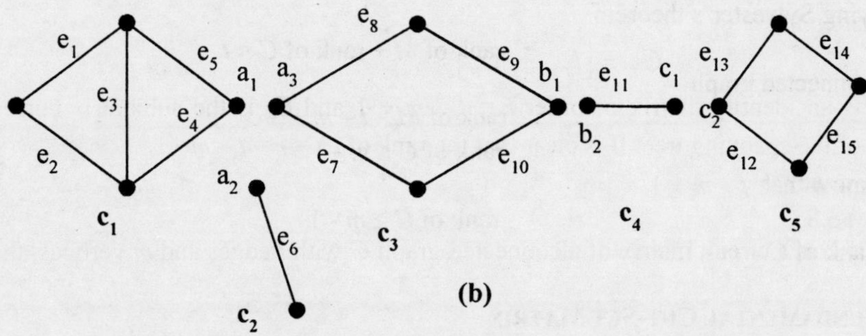

Fig. 5.14

Solution

The graph has five different cut sets: c_1, c_2, c_3, c_4 and c_5. The cut set vectors are represented as:

$c_1 = \{e_1, e_2, e_3, e_4, e_5\}$; $c_2 = \{e_6\}$; $c_3 = \{e_7, e_8, e_9, e_{10}\}$;

$c_4 = \{e_{11}\}$; $c_5 = \{e_{12}, e_{13}, e_{14}, e_{15}\}$.

Total number of edges in the graph shown in Fig. 5.14(a) are fifteen, then the cut set matrix of order 5×15 is as follows:

$$C = \begin{array}{c} \\ c_1 \\ c_2 \\ c_3 \\ c_4 \\ c_5 \end{array} \begin{array}{c} \begin{array}{ccccccccccccccc} e_1 & e_2 & e_3 & e_4 & e_5 & e_6 & e_7 & e_8 & e_9 & e_{10} & e_{11} & e_{12} & e_{13} & e_{14} & e_{15} \end{array} \\ \begin{bmatrix} 1 & 1 & 1 & 1 & 1 & 0 & 0 & 0 & 0 & 0 & 0 & 0 & 0 & 0 & 0 \\ 0 & 0 & 0 & 0 & 0 & 1 & 0 & 0 & 0 & 0 & 0 & 0 & 0 & 0 & 0 \\ 0 & 0 & 0 & 0 & 0 & 0 & 1 & 1 & 1 & 1 & 0 & 0 & 0 & 0 & 0 \\ 0 & 0 & 0 & 0 & 0 & 0 & 0 & 0 & 0 & 0 & 1 & 0 & 0 & 0 & 0 \\ 0 & 0 & 0 & 0 & 0 & 0 & 0 & 0 & 0 & 0 & 0 & 1 & 1 & 1 & 1 \end{bmatrix} \end{array}$$

NOTE

1. All 0's entries of a column correspond to an edge forming a self-loop.
2. Permutation matrix of cut set matrix corresponds relabeling the cut sets and edges.
3. For a separable graph, the cut set matrix contains the incidence matrix of block.
4. For nonseparable graphs, every set of edges incident on a vertex is a cut set.

   ⇒ every row of incidence matrix $M(G)$ will be there as a row in the $C(G)$. In other words we can say $C(G)$ contains $M(G)$.

   Hence rank of $C(G) \geq$ rank of $M(G)$

   ⇒ rank of $C(G) \geq m - 1$ [theorem 5.6]
5. The number of edges common to a cut set matrix and circuit matrix is always even.

THEOREM 5.9

The rank of cut set matrix $C(G)$ is equal to the rank of graph G with m vertices and t edges and rank of incidence matrix $M(G)$, i.e. $(m - 1)$.

PROOF

Consider a graph G with m vertices and t edges. If the edges of circuit matrix C and cut set matrix C are arrange in the same order then every row in C^T is orthogonal to every row in C.

$$C \cdot C^T = C \cdot C^T = 0 \pmod 2$$

Because there are even number of edges common to a cut-set and a circuit matrix.
On applying Sylvester's theorem

$$\text{rank of } M + \text{rank of } C \leq t$$

But for connected graph

$$\text{rank of } M = t - m + 1$$
$$\text{rank of } C \leq m - 1$$

But we know that

$$\text{rank of } C \geq m - 1$$

Hence, rank of $C = m - 1$.

5.12-1 Fundamental Cut-Set Matrix

Consider the connected graph G with m vertices and t edges. The fundamental cut set matrix C_f is a sub matrix of order $(m - 1)$ by t of C such that the set of fundamental cut sets with respect to same spanning tree are represented by rows.

A C_f can also be partitioned into two submatrices and represented as

$$C_f = [C_c \;\vdots\; I_{m-1}]$$

where I_{m-1} is an identity matrix of order m-1 and C_c is a cut set matrix corresponds to first $t - m + 1$ columns (chords).

Example 5.17

Consider the connected graph shown in Fig.5.15. Construct the fundamental cut-set matrix with respect to spanning tree shown by heavy lines.

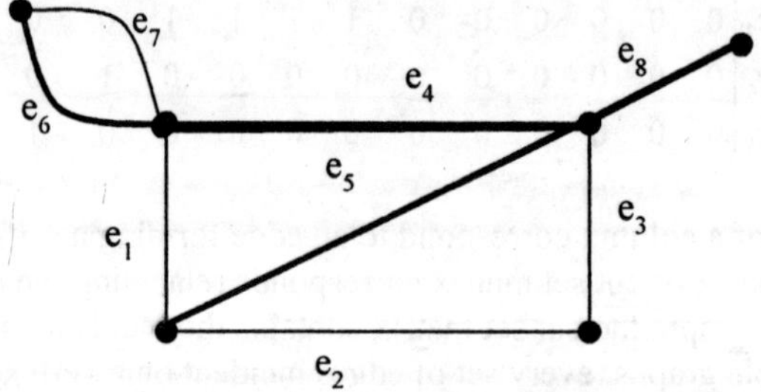

Fig. 5.15

Solution

The fundamental cut-set matrix of given connected graph with respect to spanning tree (shown by heavy lines) is

$$C_f = \begin{array}{c} \\ c_1 \\ c_2 \\ c_3 \\ c_4 \end{array} \begin{array}{c} \begin{array}{cccccccc} e_1 & e_2 & e_3 & e_7 & e_6 & e_8 & e_5 & e_4 \end{array} \\ \left[\begin{array}{cccc:cccc} 0 & 0 & 0 & 1 & 1 & 0 & 0 & 0 \\ 0 & 0 & 0 & 0 & 0 & 1 & 0 & 0 \\ 1 & 1 & 0 & 0 & 0 & 0 & 1 & 0 \\ 0 & 0 & 1 & 0 & 0 & 0 & 0 & 1 \end{array}\right] \end{array}$$

$$= [C_c \;\vdots\; I_4]$$

where $c_1=\{e_6, e_7\}$, $c_2=\{e_8\}$, $c_3=\{e_1, e_2, e_5\}$ and $c_4=\{e_3, e_4\}$ are cut set vectors.

5.13 RELATIONS AMONG REDUCED INCIDENCE MATRIX, FUNDAMENTAL CIRCUIT MATRIX AND FUNDAMENTAL CUT-SET MATRIX

Consider reduced incidence matrix M_f, the fundamental circuit matrix C_f and the fundamental cut-set matrix C_f of a connected graph $G = (V, E)$. We know that

$$C_f = [I_r \;\vdots\; C_b] \quad \text{... (1)}$$

$$C_f = [C_c \;\vdots\; I_{m-1}] \quad \text{... (2)}$$

where C_b denotes the submatrix corresponding to the branches of a spanning tree, and C_c denotes the submatrix corresponding to the chords.

Consider the graph $G = (V, E)$ and spanning tree T for both equations (1) and (2). Let the order of edges in equations (1) and (2) be the same. Now, consider the incidence matrix M_f of order $(m - 1)$ by t (number of edges) arranged in the same order as in C_f and C_f. Now, the partition of M_f into two submatrices can be represented as

$$M_f = [M_c \;\vdots\; M_b] \quad \text{... (3)}$$

where M_c and M_b are the submatrices corresponding to $t - m + 1$ chords and branches of the spanning tree T respectively.

But the columns in M_f and C_f are arranged in the same order, then from theorem 5.7, we have

$$M_f \cdot C_f' = 0$$

$$or \quad [M_c \;\vdots\; M_b]\begin{bmatrix} I_r \\ \cdots \\ C_b' \end{bmatrix} = 0$$

$$i.e. \quad M_c + M_b \cdot C_b' = 0 \quad \text{... (4)}$$

Here M_b is a nonsingular matrix. Multiplying both sides of equation (4) by M_b^{-1}, we get

$$M_b^{-1} . M_c = -M_b^{-1} \cdot M_b \cdot C_b'$$

$$\text{or} \quad M_b^{-1} \cdot M_c = -C_b'$$

$$\text{or} \quad M_b^{-1} . M_c = C_b' \quad [\because \text{ in mod 2 Arithmetic } -1 = 1] \quad \text{... (5)}$$

Similarly, the columns in C_f and C_f are arranged in the same order, then according to theorem 5.7, we have

$$C_f \cdot C_f' = 0$$

$$\text{or} \quad [C_c \;\vdots\; I_{m-1}]\begin{bmatrix} I_r \\ \cdots \\ C_b' \end{bmatrix} = 0$$

or

$$C_c + C_b' = 0 \quad \text{... (6)}$$

$$C_c = C_b^{-1} \cdot M_c \quad \text{[from (5)]}$$

EXERCISE 5.1

1. Find the following from the graph shown in Fig. 5.16

(a) G-U where $U = \{x_1, x_3, x_5\}$.

(b) G-F where $F = \{e_2, e_4, e_6, e_{10}, e_{12}\}$.

(c) Find a subgraph isomorphic to K_3.

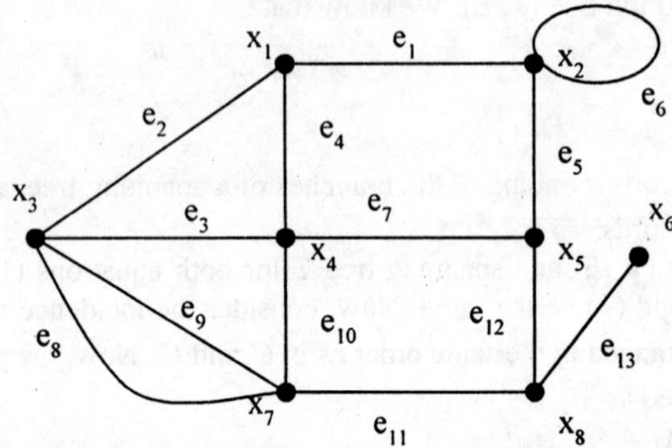

Fig. 5.16

2. Draw the graph having the following matrix as its adjacency matrix.

$$\begin{bmatrix} 0 & 1 & 2 & 3 \\ 1 & 0 & 3 & 2 \\ 2 & 3 & 0 & 1 \\ 3 & 2 & 1 & 0 \end{bmatrix}$$

3. A graph G has the following adjacency matrices. Check whether it is connected:

$$(a) \quad A(G) = \begin{bmatrix} 0 & 0 & 1 & 0 \\ 0 & 0 & 0 & 1 \\ 0 & 1 & 0 & 0 \\ 0 & 0 & 1 & 1 \end{bmatrix} \qquad (b) \quad A(G) = \begin{bmatrix} 0 & 1 & 0 \\ 1 & 0 & 0 \\ 0 & 0 & 1 \end{bmatrix}$$

$$(c) \quad A(G) = \begin{bmatrix} 0 & 0 & 0 & 1 & 0 \\ 0 & 0 & 0 & 0 & 1 \\ 0 & 1 & 0 & 0 & 1 \\ 0 & 0 & 1 & 1 & 0 \end{bmatrix} \qquad (d) \quad A(G) = \begin{bmatrix} 1 & 1 & 0 & 1 & 1 \\ 1 & 1 & 1 & 0 & 1 \\ 1 & 1 & 1 & 1 & 0 \\ 1 & 0 & 1 & 1 & 0 \\ 1 & 1 & 0 & 0 & 1 \end{bmatrix}$$

4. Find the adjacency matrix A for the following graph G:

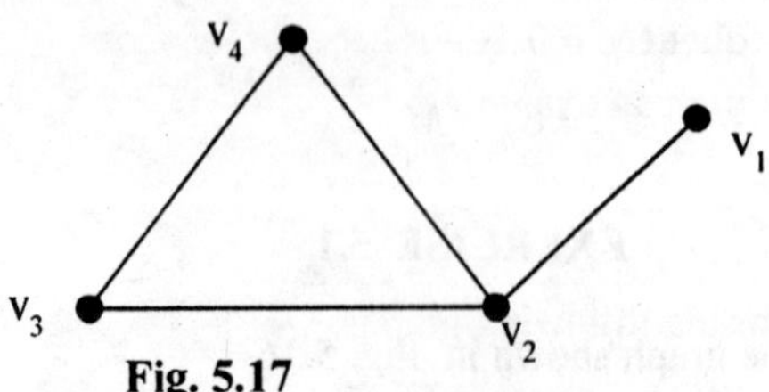

Fig. 5.17

Also find A^2, A^3.

5. Find the adjacency matrix A of the graph G given below:

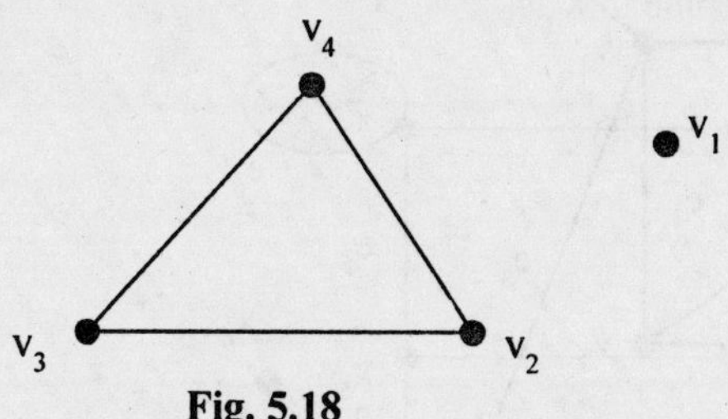

Fig. 5.18

and deduce from the matrix $X = A + A^2 + A^3$ that G is disconnected.

6. Find the adjacency matrix $A = [a_{ij}]$ of each graph G shown in Fig. 5.19.

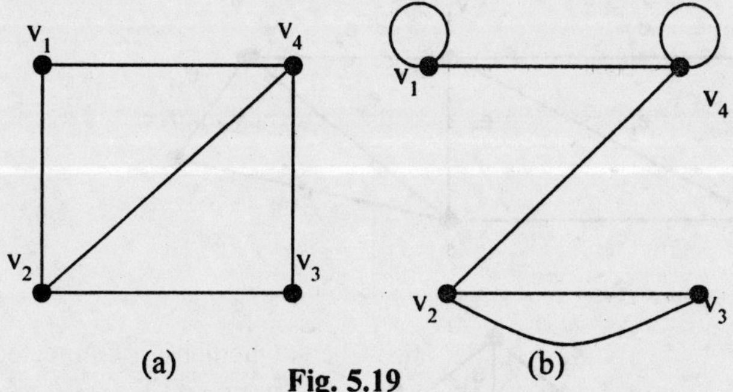

Fig. 5.19

7. Draw the graph G corresponding to each adjacency matrix:

$$(a)\ A = \begin{bmatrix} 0 & 1 & 0 & 1 & 0 \\ 1 & 0 & 0 & 1 & 1 \\ 0 & 0 & 0 & 1 & 1 \\ 1 & 1 & 1 & 0 & 1 \\ 0 & 1 & 1 & 1 & 0 \end{bmatrix} \qquad (b)\ A = \begin{bmatrix} 1 & 3 & 0 & 0 \\ 3 & 0 & 1 & 1 \\ 0 & 1 & 2 & 2 \\ 0 & 1 & 2 & 0 \end{bmatrix}$$

8. Show that each diagonal entry in A^3 equals twice the number of triangles passing through the corresponding vertex.

9. Let A be the adjacency matrix of a simple graph G. Then prove that ij^{th} entry in A^r is the number of different edge sequences of r edges between vertices v_i and v_j.

10. Define the following terms with an example:
    (a) Adjacency Matrix
    (b) Path Matrix
    (c) Incidence Matrix

11. Find the incidence matrix for the following graphs:

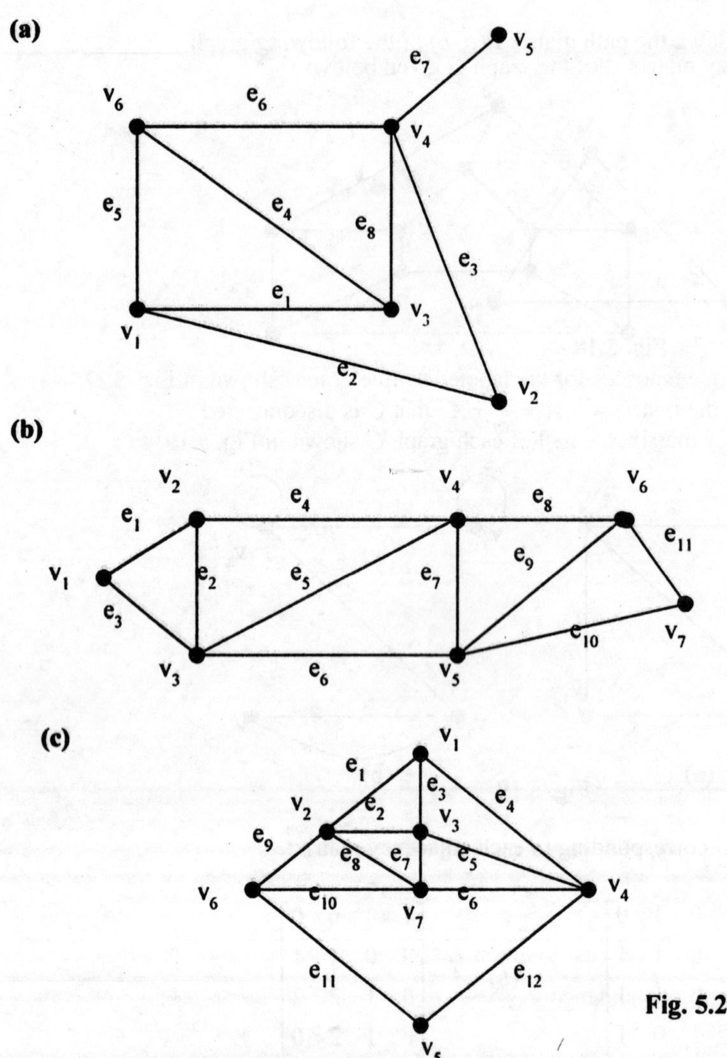

Fig. 5.20

12 **(a). Obtain the adjacency and incidence matrix of the graph given below. Find the elementary paths of length 1 and 2 from v_3 to v_1.**

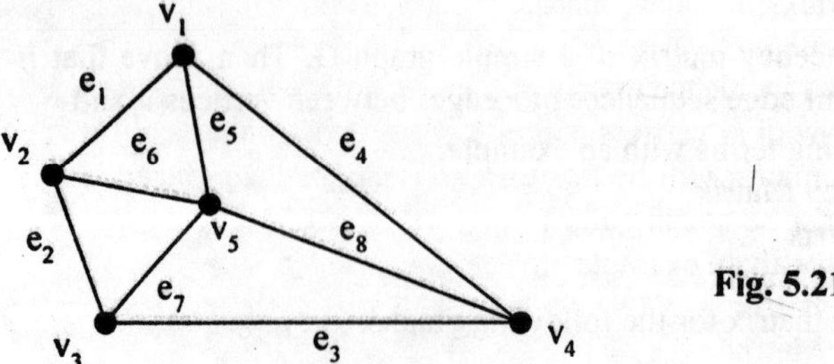

Fig. 5.21

(b). Write down the path matrix $P(a, b)$ of the following graph.

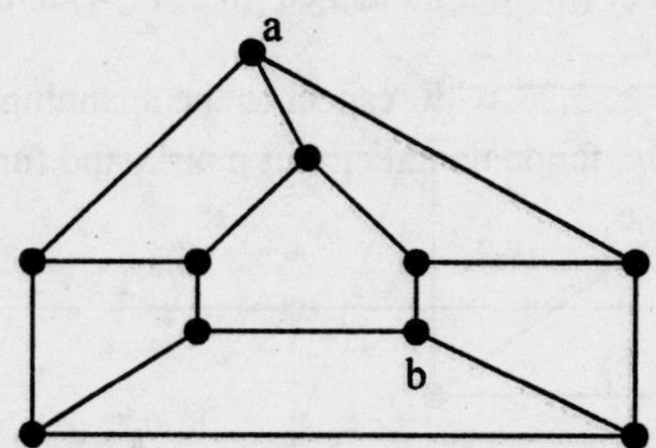

13. Write the incidence matrices for the labeled simple graphs shown in Fig. 5.22.

(a)

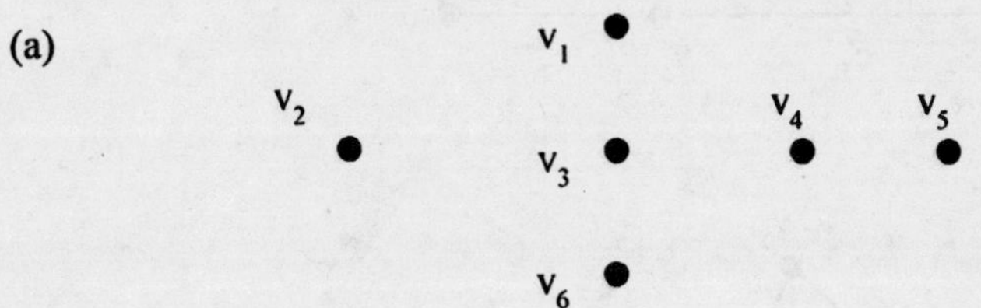

(b)

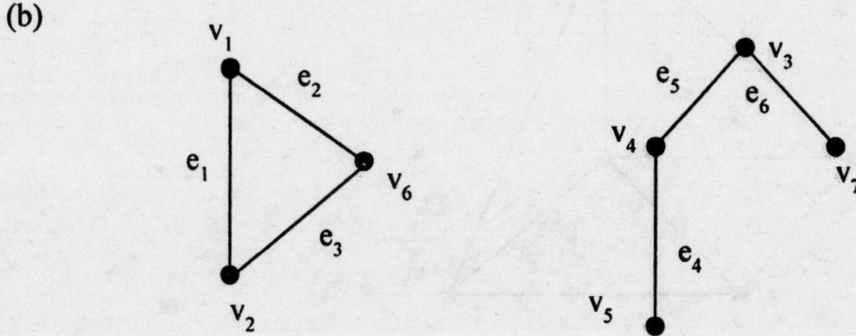

Fig. 5.22

14. Prove that the rank of incidence matrix is m-k, if G is a disconnected graph with m vertices and k components.

15. Prove that (m-1) by (m-1) submatrix of $M(G)$ is nonsingular iff the m-1 edges corresponding to the m-1 columns of this matrix constitute a spanning tree in G.
[Hint The reduced incidence matrix of the subgraph G with m-1 edges is a square matrix of order m-1 in $M(G)$. But square matrix of $M(G)$ is nonsingular iff corresponding subgraph is a spanning tree with m-1 vertex.]

16. Prove that reduced incidence matrix of a tree is nonsingular.
[Hint: A graph with m vertices and m-1 edges is disconnected if it is not a tree. So the rank of this incidence matrix will be less than m-1. Hence the reduced incidence square matrix of order (m-1) will not be nonsingular.]

17. Define the terms with an example
(a) Circuit Matrix
(b) Cut-Set Matrix

(c) Fundamental Cut-Set Matrix

18. Find the adjacency matrix (A), the incidence matrix (M) and circuit matrix (C) of the graphs shown in Fig. 5.22.

19. Consider the graph in Fig. 5.23 with respect to the spanning tree $\{e_2, e_3, e_5, e_7, e_8, e_9\}$. Write reduced incidence matrix, fundamental circuit matrix and fundamental cut set matrix.

(a)

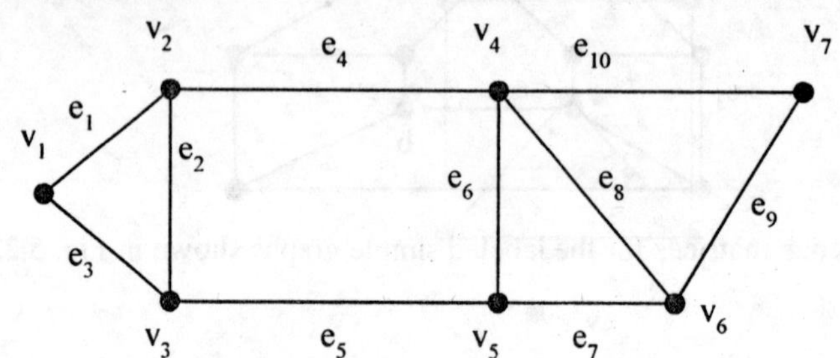

(b)

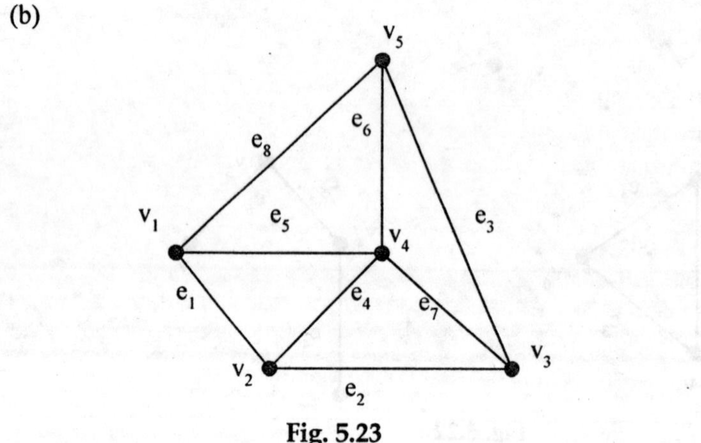

Fig. 5.23

20. Show that for a simple disconnected graph of k components, m vertices and t edges, the rank of matrices M, C and C are $m - k$, $t - m + k$ and $m - k$ respectively.

21. Define the terms:
(a) Isomorphic Graph
(b) Separable Graph
(c) Non-separable Graph

22. Determine G for the given list of all spanning trees.
$\{e_1, e_3, e_4, e_5\}$, $\{e_1, e_3, e_4, e_6\}$, $\{e_2, e_3, e_4, e_5\}$,
$\{e_2, e_3, e_4, e_6\}$, $\{e_1, e_3, e_5, e_6\}$, $\{e_2, e_3, e_5, e_6\}$,
$\{e_1, e_4, e_5, e_6\}$, $\{e_2, e_4, e_5, e_6\}$, $\{e_1, e_2, e_4, e_5\}$,
$\{e_1, e_2, e_4, e_6\}$, $\{e_1, e_2, e_5, e_6\}$.

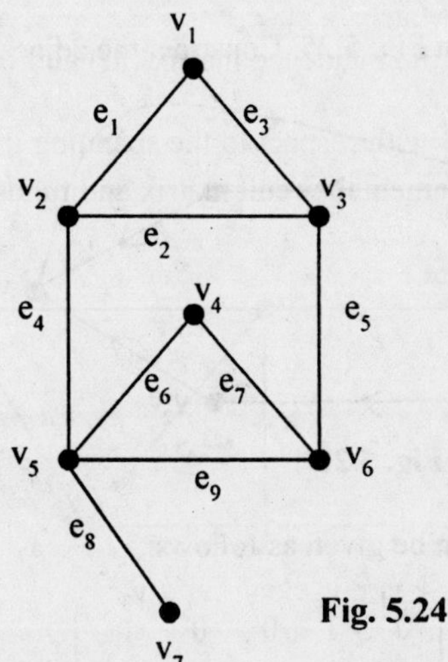

Fig. 5.24

23. Obtain the circuit matrix and cut-set matrix of graph shown in Fig. 5.14(a).

24. Show that every circuit in simple non separable graph G is obtained as a *mod 2* sum of two rows of corresponding path matrix $P(v_i, v_j)$.

[Hint Every circuit Δ in G falls in one of following three categories

(i) Δ passes through both v_i and v_j.

(ii) Δ passes through neither v_i nor v_j.

(iii) Δ passes through either v_i or v_j.]

25. Characterize simple, self-dual graphs in terms of their circuit and cut-set matrices.

26. Show that the submatrix $C(v_i, v_j)$ of cut-set matrix C of a nonseparable graph G containing only those rows of C that represent cut-sets with respect to vertices v_i and v_j contains a fundamental cut-set matrix C_f of G.

27. The diameter of a connected graph is defined as the largest distance between two vertices in the graph. How will you determine the diameter of the corresponding graph of given adjacency matrix A.

5.14 MATRICES OF DIRECTED GRAPHS

5.15 ADJACENCY MATRIX

Suppose $G = (V, E)$ be a simple directed graph with m vertices. Then the **adjacency matrix** $A=(a_{ij})$ of the digraph is a $m \times m$ matrix defined as follows:

$$A = a_{ij} = \begin{cases} 1 & \text{if there is an edge directed from vertex } v_i \text{ to } v_j \\ 0 & \text{otherwise} \end{cases}$$

This matrix is also called as **transition matrix, connection matrix, precedence matrix** or **prefrence matrix, predecessor matrix** in different discipline where it is used.

Example 5.18

Consider the directed graph G in Fig. 5.25. Construct the adjacency matrix for the digraph.

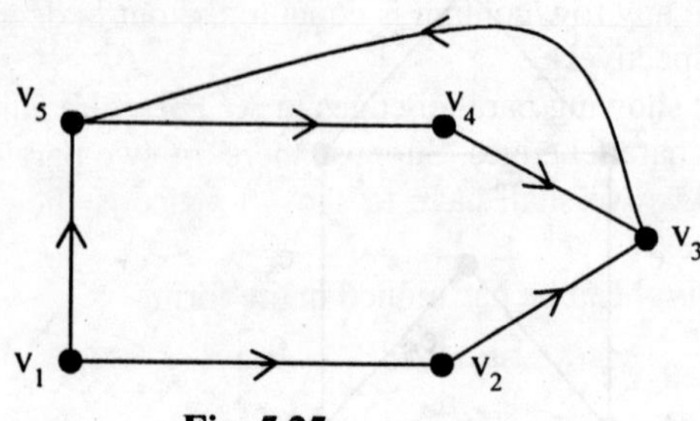

Fig. 5.25

Solution

The adjacency matrix A of G can be given as follows:

$$A = \begin{array}{c} \\ v_1 \\ v_2 \\ v_3 \\ v_4 \\ v_5 \end{array} \begin{array}{c} \begin{array}{ccccc} v_1 & v_2 & v_3 & v_4 & v_5 \end{array} \\ \begin{bmatrix} 0 & 1 & 0 & 0 & 1 \\ 0 & 0 & 1 & 0 & 0 \\ 0 & 0 & 0 & 0 & 1 \\ 0 & 0 & 1 & 0 & 0 \\ 0 & 0 & 0 & 1 & 0 \end{bmatrix} \end{array}$$

Example 5.19

Write down the adjacency matrix of the graph.

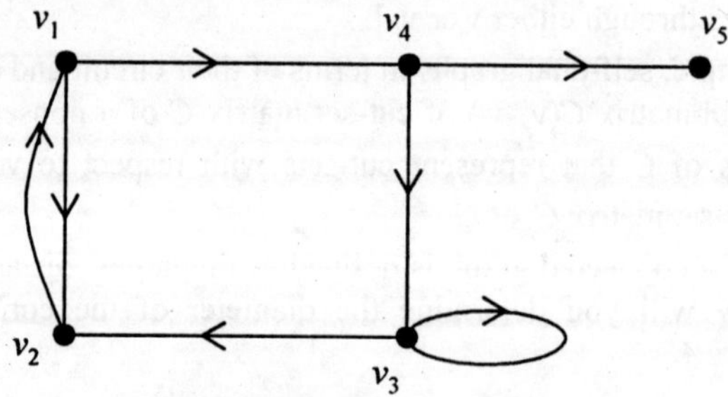

Fig. 5.26

Solution

The adjacency matrix of the above graph is

$$A = \begin{array}{c} \\ v_1 \\ v_2 \\ v_3 \\ v_4 \\ v_5 \end{array} \begin{array}{c} \begin{array}{ccccc} v_1 & v_2 & v_3 & v_4 & v_5 \end{array} \\ \begin{bmatrix} 0 & 1 & 0 & 1 & 0 \\ 1 & 0 & 0 & 0 & 0 \\ 0 & 1 & 1 & 1 & 0 \\ 0 & 0 & 1 & 0 & 1 \\ 0 & 0 & 0 & 0 & 0 \end{bmatrix} \end{array}$$

We can observe in this adjacency matrix that all the main diagonal elements are not zero because there is a self loop at vertex v_3.

NOTE

1. All the entries along the main diagonal of a digraph are zero iff the graph has no self-loop.
2. Sum of all entries in any row/column is equal to the out/in degree of vertex corresponding to that row/column respectively.
3. There is no element showing parallel edges in A. Therefore adjacency matrix is defined only for digraph without parallel edges. Suppose there are two parallel edges e_1 and e_2 connecting vertices $v_1$and v_2, then we shall have to show 1 twice in the row of v_1 in the column of v_2 which is not possible.
4. The adjacency matrix A can be partitioned in the form

$$\begin{bmatrix} A_1 & \vdots & 0 \\ \cdots & \vdots & \cdots \\ 0 & \vdots & A_2 \end{bmatrix}$$

where A_1 and A_2 are square sub matrices of A if and only if digraph is disconnected.

5. If the adjacency matrix A can be partitioned in more than one way then the digraph is weakly connected and vice versa.
6. The number of non-zero entries in A gives the number of edges.
7. Permutation of any two rows accompanied by permutation of corresponding columns does not alter the graph. The permutation merely corresponds to reordering of vertices. This implies two graphs are isomorphic if and only if their permutation matrix differs only by such permutations.

5.16 INCIDENCE MATRIX OF A DIGRAPH

Incidence matrix of a diagraph with n vertices, e edges, and no self-loops is an n by e matrix $A = [a_{ij}]$ whose row correspond to vertices and columns correspond to edges, such that

$$\begin{aligned} a_{ij} &= 1 \text{ if } j^{th} \text{ edge is incident out of } i^{th} \text{ vertex.} \\ &= 1 \text{ if } j^{th} \text{ edge is incident into } i^{th} \text{ vertex.} \\ &= -1 \text{ if } j^{th} \text{ edge is incident on } i^{th} \text{ vertex.} \end{aligned}$$

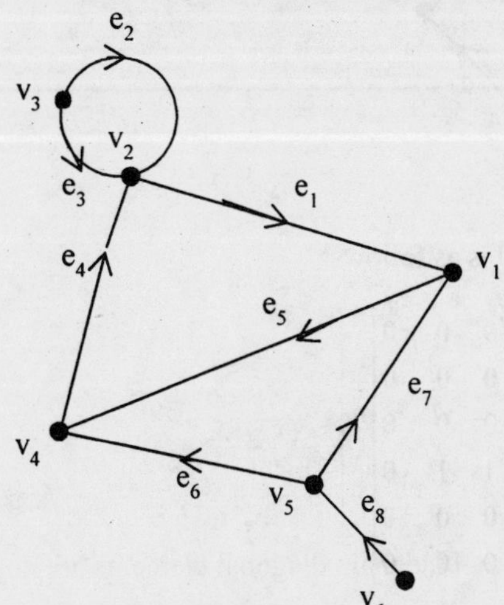

$$\begin{array}{c} \\ v_1 \\ v_2 \\ v_3 \\ v_4 \\ v_5 \\ v_6 \end{array} \begin{array}{c} \begin{array}{cccccccc} e_1 & e_2 & e_3 & e_4 & e_5 & e_6 & e_7 & e_8 \end{array} \\ \begin{bmatrix} -1 & 0 & 0 & 0 & 1 & 0 & -1 & 0 \\ 1 & -1 & -1 & -1 & 0 & 0 & 0 & 0 \\ 0 & 1 & 1 & 0 & 0 & 0 & 0 & 0 \\ 0 & 0 & 0 & 1 & -1 & -1 & 0 & 0 \\ 0 & 0 & 0 & 0 & 0 & 1 & 1 & -1 \\ 0 & 0 & 0 & 0 & 0 & 0 & 0 & 1 \end{bmatrix} \end{array}$$

Properties of incidence matrix of a digraph is similar to an undirected graph with minor changes.

5.17 CIRCUIT MATRIX OF A DIGRAPH

Consider a graph $G = (V, E)$ with n vertices and t edges. Then a circuit matrix $C = (C_{ij})$ of graph G is a $n \times t$ matrix defined as follows:

$$C_{ij} = \begin{cases} -1 & \text{if } i^{th} \text{ circuit consists of } j^{th} \text{ edge but the orientation of the two are opposite.} \\ 1 & \text{if } i^{th} \text{ circuit consists of } j^{th} \text{ edge but the orientation of the two coincide.} \\ 0 & \text{if } i^{th} \text{ circuit does not consists of } j^{th} \text{ edges.} \end{cases}$$

Example 5.20

Construct the circuit matrix of the graph given in Fig.5.27.

Solution

The graph has six different circuits such as

$C_1 = \{e_1, e_7, e_6\}$, $C_2 = \{e_2, e_5, e_6\}$, $C_3 = \{e_3, e_4, e_5\}$,

$C_4 = \{e_9, e_8\}$, $C_5 = \{e_1, e_2, e_5, e_7\}$,

$C_6 = \{e_1, e_2, e_3, e_4, e_7\}$.

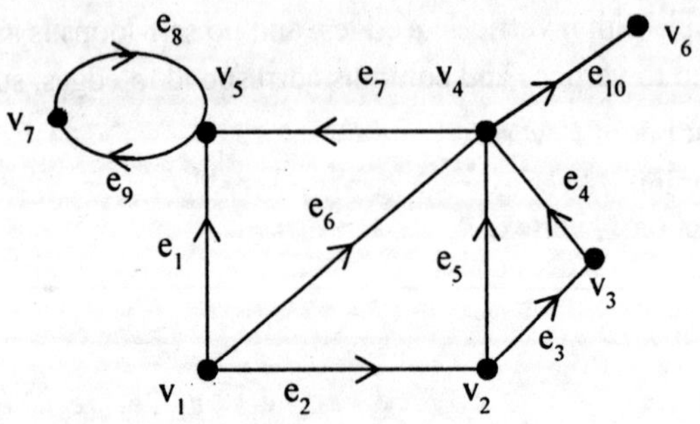

Fig. 5.27

and ten edges. Then the circuit matrix of order 6×10 is as follows:

$$C = \begin{array}{c} \\ C_1 \\ C_2 \\ C_3 \\ C_4 \\ C_5 \\ C_6 \end{array} \begin{array}{c} \begin{matrix} e_1 & e_2 & e_3 & e_4 & e_5 & e_6 & e_7 & e_8 & e_9 & e_{10} \end{matrix} \\ \begin{bmatrix} -1 & 0 & 0 & 0 & 0 & 1 & 1 & 0 & 0 & 0 \\ 0 & 1 & 0 & 0 & 1 & -1 & 0 & 0 & 0 & 0 \\ 0 & 0 & 1 & 1 & -1 & 0 & 0 & 0 & 0 & 0 \\ 0 & 0 & 0 & 0 & 0 & 0 & 0 & 1 & 1 & 0 \\ -1 & 1 & 0 & 0 & 1 & 0 & 1 & 0 & 0 & 0 \\ -1 & 1 & 1 & 1 & 0 & 0 & 1 & 0 & 0 & 0 \end{bmatrix} \end{array}$$

NOTE

Theorem 5.7, i.e. $M \cdot C^T = C \cdot M^T = 0$ (*mod 2*) for undirected graph also holds in digraph.

5.18 PATH MATRIX

The **path matrix** or **reachability matrix** of digraph $G(V,E)$ with no parallel edge and m vertices is defined as follows:

$$P(v_i, v_j) = p_{ij} = \begin{cases} 1, & \text{if there is an edge (path) directed from } v_i \text{ to } v_j \\ 0, & \text{otherwise} \end{cases}$$

If there is a path from vertex v_i to v_j in G then there is a simple path from v_i to v_j when $v_i \neq v_j$ and a cycle from v_i to v_j where $v_i = v_j$.

Since the number of vertices of graph G is m. We can observe here that

(i) The length of such a simple path will be either less than or equal to m - 1.

(ii) The length of such a cycle will be either less than or equal to m.

Hence if we consider the powers $A, A^2, A^3, \ldots$ of an adjacency matrix $A = [a_{ij}]$ of digraph G. The element in the ith and j^{th} column is given by

$$B_m = A + A^2 + A^3 + \ldots + A^{m-1} \text{ which is nonzero.}$$

The $(i, j)^{th}$ entry of the matrix B_m gives the number of paths of length r from its vertices v_i to v_j.

Here it is to be noted that the path matrix P and matrix B_n have the same non-zero entries.

A directed graph G is said to be strongly connected (Refer chapter 7) if for any pair of vertices v_i and v_j of G there exist an edge between v_i and v_j both ways. The same can be said in terms of path matrix as it has nonzero entries (or matrix B_m has nonzero entries). It can be stated in terms of the following theorem.

THEOREM 5.10

A graph of m vertices is strongly connected if and only if the matrix.

$B_m = A + A^2 + A^3 + \ldots + A^{m-1}$ has no zero entry, where A is the adjacency matrix of a graph having m vertices.

Example 5.21

Construct the path matrices of the following digraph G shown in Fig. 5.28.

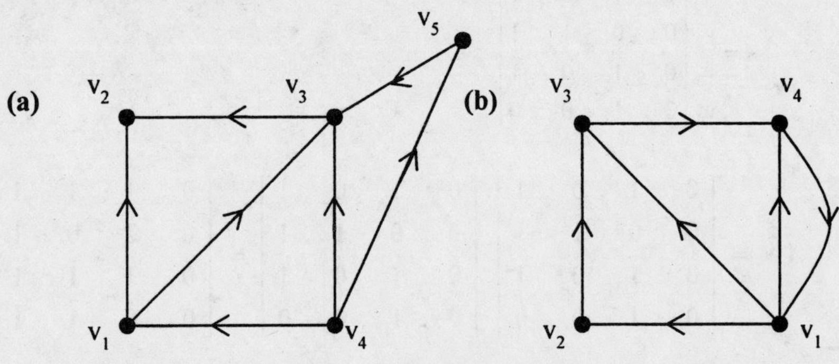

Fig.5.28

Solution

(a)

$$P = \begin{array}{c} \\ v_1 \\ v_2 \\ v_3 \\ v_4 \\ v_5 \end{array} \begin{array}{c} \begin{array}{ccccc} v_1 & v_2 & v_3 & v_4 & v_5 \end{array} \\ \begin{bmatrix} 0 & 1 & 1 & 0 & 0 \\ 0 & 0 & 0 & 0 & 0 \\ 0 & 1 & 0 & 0 & 0 \\ 1 & 1 & 1 & 0 & 1 \\ 0 & 1 & 1 & 0 & 0 \end{bmatrix} \end{array}$$

(b)

$$P = \begin{array}{c} \\ v_1 \\ v_2 \\ v_3 \\ v_4 \end{array} \begin{array}{c} \begin{array}{cccc} v_1 & v_2 & v_3 & v_4 \end{array} \\ \begin{bmatrix} 1 & 1 & 1 & 1 \\ 1 & 1 & 1 & 1 \\ 1 & 1 & 1 & 1 \\ 1 & 1 & 1 & 1 \end{bmatrix} \end{array}$$

Example 5.22

Consider the digraph G with four vertices v_1, v_2, v_3 and v_4 in Fig. 5.29.

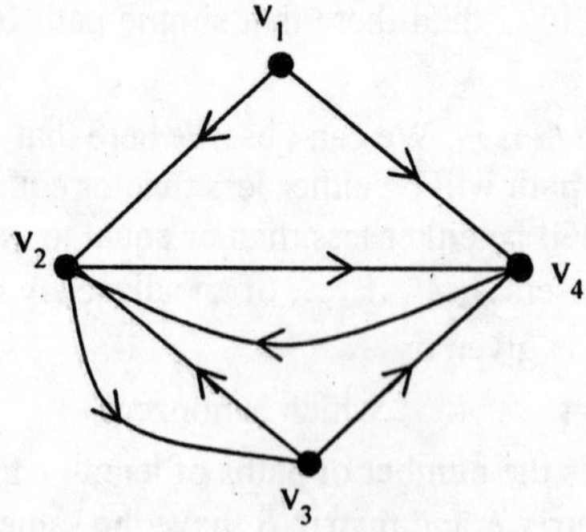

Fig. 5.29

(a) Obtain the adjacency matrix A of the graph G.

(b) Calculate A^2, A^3, A^4.

(c) Find AA^T and A^TA.

(d) Calculate B_4 i.e. $B_4 = A + A^2 + A^3$.

Solution

(a) The adjacency matrix A of G is given by

$$A = \begin{array}{c} \\ v_1 \\ v_2 \\ v_3 \\ v_4 \end{array} \begin{array}{c} \begin{array}{cccc} v_1 & v_2 & v_3 & v_4 \end{array} \\ \begin{bmatrix} 0 & 1 & 0 & 1 \\ 0 & 0 & 1 & 1 \\ 0 & 1 & 0 & 1 \\ 0 & 1 & 0 & 0 \end{bmatrix} \end{array}$$

(b)

$$A^2 = \begin{bmatrix} 0 & 1 & 0 & 1 \\ 0 & 0 & 1 & 1 \\ 0 & 1 & 0 & 1 \\ 0 & 1 & 0 & 0 \end{bmatrix} \begin{bmatrix} 0 & 1 & 0 & 1 \\ 0 & 0 & 1 & 1 \\ 0 & 1 & 0 & 1 \\ 0 & 1 & 0 & 0 \end{bmatrix} = \begin{bmatrix} 0 & 1 & 1 & 1 \\ 0 & 2 & 0 & 1 \\ 0 & 1 & 1 & 1 \\ 0 & 0 & 1 & 1 \end{bmatrix}$$

$$A^3 = \begin{bmatrix} 0 & 1 & 1 & 1 \\ 0 & 2 & 0 & 1 \\ 0 & 1 & 1 & 1 \\ 0 & 0 & 1 & 1 \end{bmatrix} \begin{bmatrix} 0 & 1 & 0 & 1 \\ 0 & 0 & 1 & 1 \\ 0 & 1 & 0 & 1 \\ 0 & 1 & 0 & 0 \end{bmatrix} = \begin{bmatrix} 0 & 2 & 1 & 2 \\ 0 & 1 & 2 & 2 \\ 0 & 2 & 1 & 2 \\ 0 & 2 & 0 & 1 \end{bmatrix}$$

(c)

$$A^T = \begin{bmatrix} 0 & 0 & 0 & 0 \\ 1 & 0 & 1 & 1 \\ 0 & 1 & 0 & 0 \\ 1 & 1 & 1 & 0 \end{bmatrix}$$

$$X = A \cdot A^T = \begin{bmatrix} 0 & 1 & 0 & 1 \\ 0 & 0 & 1 & 1 \\ 0 & 1 & 0 & 1 \\ 0 & 1 & 0 & 0 \end{bmatrix} \begin{bmatrix} 0 & 0 & 0 & 0 \\ 1 & 0 & 1 & 1 \\ 0 & 1 & 0 & 0 \\ 1 & 1 & 1 & 0 \end{bmatrix} = \begin{bmatrix} 2 & 1 & 2 & 1 \\ 1 & 2 & 1 & 0 \\ 2 & 1 & 2 & 1 \\ 1 & 0 & 1 & 1 \end{bmatrix}$$

$$Y = A^T \cdot A = \begin{bmatrix} 0 & 0 & 0 & 0 \\ 1 & 0 & 1 & 1 \\ 0 & 1 & 0 & 0 \\ 1 & 1 & 1 & 0 \end{bmatrix} \begin{bmatrix} 0 & 1 & 0 & 1 \\ 0 & 0 & 1 & 1 \\ 0 & 1 & 0 & 1 \\ 0 & 1 & 0 & 0 \end{bmatrix} = \begin{bmatrix} 0 & 0 & 0 & 0 \\ 0 & 3 & 0 & 2 \\ 0 & 0 & 1 & 1 \\ 0 & 2 & 1 & 3 \end{bmatrix}$$

(d) $B_4 = A + A^2 + A^3$

$$= \begin{bmatrix} 0 & 1 & 0 & 1 \\ 0 & 0 & 1 & 1 \\ 0 & 1 & 0 & 1 \\ 0 & 1 & 0 & 0 \end{bmatrix} + \begin{bmatrix} 0 & 1 & 1 & 1 \\ 0 & 2 & 0 & 1 \\ 0 & 1 & 1 & 1 \\ 0 & 0 & 1 & 1 \end{bmatrix} + \begin{bmatrix} 0 & 2 & 1 & 2 \\ 0 & 1 & 2 & 2 \\ 0 & 2 & 1 & 2 \\ 0 & 2 & 0 & 1 \end{bmatrix}$$

$$= \begin{bmatrix} 0 & 4 & 2 & 2 \\ 0 & 3 & 3 & 4 \\ 0 & 4 & 2 & 4 \\ 0 & 3 & 1 & 2 \end{bmatrix}$$

OBSERVATION

The following observations can be made from the above results (see Fig 5.29):

1. $a_n(i, j)$ gives the number of paths of length n from v_i to v_j in A^n matrix. For example $a_2(2, 4) = 1$, implies there is one path of length 2 from v_2 to v_4 i.e. $v_2\, v_3\, v_4$ and in $a_4(2, 4) = 3$, there exist three path of length 4 from v_2 to v_4 i.e. $v_2\, v_3\, v_4\, v_2\, v_4$, $v_2\, v_4\, v_2\, v_3\, v_4$, $v_2\, v_3\, v_2\, v_3\, v_4$.

2. All the entries along the main diagonal of A^n gives the number of cycles of length n at that vertex.
3. The graph whose adjacency matrix is A^T will be given by:

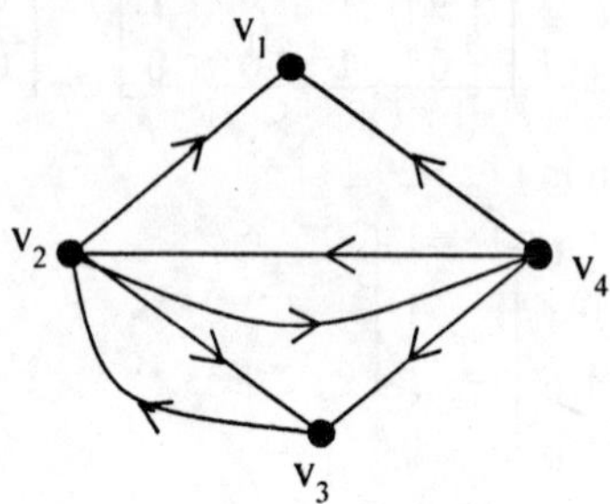

If we compare it from the graph given in Fig. 5.29, we find that it is nothing but the converse of the graph G.
4. All the entries along the main diagonal of AA^T give the out degrees of the vertices of graph.
5. All the entries along the main diagonal of $A'A$ give the in-degrees of the vertices.
6. The x_{ij} $(i \neq j)$ entry in AA' gives the number of vertices which are terminal vertices of edges from both v_i and v_j, e.g. $x_{14} = 2$ means the number of vertices which are terminal vertices of edges from v_1 and v_4 is 2 i.e. v_2 and v_4.
7. The y_{ij} $(i \neq j)$ entry in $A'A$ gives the number of vertices of the graph which are such that the edges initiating from these vertices terminate in both v_i and v_j. For example $y_{24} = 2$ means there are 2 nodes which terminate in both v_2 and v_4, i.e v_1 and v_3.
8. The b_{ij} entry of B_m gives the number of paths of length less than or equal to r from v_i to v_j. For example in B_4, $b_{23} = 4$ means there are four paths, from v_2 to v_3 of length less than or equal to 4 namely

$$v_2 v_3,\ v_2 v_4 v_2 v_3,\ v_2 v_3 v_2 v_3,\ v_2 v_3 v_4 v_2 v_3$$

9. On replacing the non-zero entries in B_4 by 1, we obtain the path matrix P of the graph G. Here

$$P(G) = \begin{bmatrix} 0 & 1 & 1 & 1 \\ 0 & 1 & 1 & 1 \\ 0 & 1 & 1 & 1 \\ 0 & 1 & 1 & 1 \end{bmatrix}$$

Hence, according to the definition of path matrix and Theorem 5.3, G is not strongly connected. In particular, vertex v_1 is not reachable from any of the other vertices.

REMARK

All entries in adjacency matrix A and path matrix $P(G)$of a graph G are 0 and 1. Hence they may be viewed as logical matrices where 0 represents "false" or "OFF" and 1 represents "True" or "ON". Thus the logical operator AND ($\wedge$) and OR ($\vee$) may be applied to the entries of A and P. The operations $\wedge$ and $\vee$ are given by:

$\wedge$	0 1
0	0 0
1	0 1

(a) AND

$\vee$	0 1
0	0 1
1	1 1

(b) OR

Now consider the adjacency matrix of graph in Fig.5.29.

$$A = \begin{bmatrix} 0 & 1 & 0 & 1 \\ 0 & 0 & 1 & 1 \\ 0 & 1 & 0 & 1 \\ 0 & 1 & 0 & 0 \end{bmatrix}$$

Then

$$A \wedge A = A^{(2)} = \begin{bmatrix} 0 & 1 & 0 & 1 \\ 0 & 0 & 1 & 1 \\ 0 & 1 & 0 & 1 \\ 0 & 1 & 0 & 0 \end{bmatrix} \wedge \begin{bmatrix} 0 & 1 & 0 & 1 \\ 0 & 0 & 1 & 1 \\ 0 & 1 & 0 & 1 \\ 0 & 1 & 0 & 0 \end{bmatrix}$$

$$= \begin{bmatrix} 0 & 1 & 1 & 1 \\ 0 & 1 & 0 & 1 \\ 0 & 1 & 1 & 1 \\ 0 & 0 & 1 & 1 \end{bmatrix}$$

$$A^{(3)} = A^{(2)} \wedge A = \begin{bmatrix} 0 & 1 & 1 & 1 \\ 0 & 1 & 0 & 1 \\ 0 & 1 & 1 & 1 \\ 0 & 0 & 1 & 1 \end{bmatrix} \wedge \begin{bmatrix} 0 & 1 & 0 & 1 \\ 0 & 0 & 1 & 1 \\ 0 & 1 & 0 & 1 \\ 0 & 1 & 0 & 0 \end{bmatrix}$$

$$= \begin{bmatrix} 0 & 1 & 1 & 1 \\ 0 & 1 & 1 & 1 \\ 0 & 1 & 1 & 1 \\ 0 & 1 & 0 & 1 \end{bmatrix}$$

$$A^{(4)} = A^{(3)} \wedge A = \begin{bmatrix} 0 & 1 & 1 & 1 \\ 0 & 1 & 1 & 1 \\ 0 & 1 & 1 & 1 \\ 0 & 1 & 0 & 1 \end{bmatrix} \wedge \begin{bmatrix} 0 & 1 & 0 & 1 \\ 0 & 0 & 1 & 1 \\ 0 & 1 & 0 & 1 \\ 0 & 1 & 0 & 0 \end{bmatrix}$$

$$= \begin{bmatrix} 0 & 1 & 1 & 1 \\ 0 & 1 & 1 & 1 \\ 0 & 1 & 1 & 1 \\ 0 & 1 & 1 & 1 \end{bmatrix}$$

NOTE

1. The main difference between A^m and $A^{(m)}$ is that in A^m the $a_m(i,j)$ entry gives the number of paths of length m from v_i to v_j whereas in Boolean matrix $A^{(m)}$, the $a^{(m)}$ (i, j) entry is 1 if there is at least one path of length m from v_i to v_j.
2. If the number of vertices of a digraph G is m then

$$P = A \vee A^{(2)} \vee A^{(3)} \vee \ldots \vee A^{(m)}$$
$$= \bigvee_{r=1}^{m} A^{(r)}$$

In above graph G,

$$P = A \vee A^{(2)} \vee A^{(3)} \vee A^{(4)}$$
$$= \begin{bmatrix} 0 & 1 & 0 & 1 \\ 0 & 0 & 1 & 1 \\ 0 & 1 & 0 & 1 \\ 0 & 1 & 0 & 0 \end{bmatrix} \vee \begin{bmatrix} 0 & 1 & 1 & 1 \\ 0 & 1 & 0 & 1 \\ 0 & 1 & 1 & 1 \\ 0 & 0 & 1 & 1 \end{bmatrix} \vee \begin{bmatrix} 0 & 1 & 1 & 1 \\ 0 & 1 & 1 & 1 \\ 0 & 1 & 1 & 1 \\ 0 & 1 & 0 & 1 \end{bmatrix} \vee \begin{bmatrix} 0 & 1 & 1 & 1 \\ 0 & 1 & 1 & 1 \\ 0 & 1 & 1 & 1 \\ 0 & 1 & 1 & 1 \end{bmatrix}$$

$$= \begin{bmatrix} 0 & 1 & 1 & 1 \\ 0 & 1 & 1 & 1 \\ 0 & 1 & 1 & 1 \\ 0 & 1 & 1 & 1 \end{bmatrix}$$

Now, consider

$$P^T = \begin{bmatrix} 0 & 0 & 0 & 0 \\ 1 & 1 & 1 & 1 \\ 1 & 1 & 1 & 1 \\ 1 & 1 & 1 & 1 \end{bmatrix}$$

And

$$P \wedge P^T = \begin{bmatrix} 0 & 1 & 1 & 1 \\ 0 & 1 & 1 & 1 \\ 0 & 1 & 1 & 1 \\ 0 & 1 & 1 & 1 \end{bmatrix} \wedge \begin{bmatrix} 0 & 0 & 0 & 0 \\ 1 & 1 & 1 & 1 \\ 1 & 1 & 1 & 1 \\ 1 & 1 & 1 & 1 \end{bmatrix} = \begin{bmatrix} 0 & 0 & 0 & 0 \\ 0 & 1 & 1 & 1 \\ 0 & 1 & 1 & 1 \\ 0 & 1 & 1 & 1 \end{bmatrix}$$

which is obtained by element-wise product of the elements.

EXERCISE 5.2

1. **Obtain the adjacency matrix A of the digraph shown in Fig. 5.30. Find the elementary paths of lengths 1 and 2 from v_1 to v_3.**

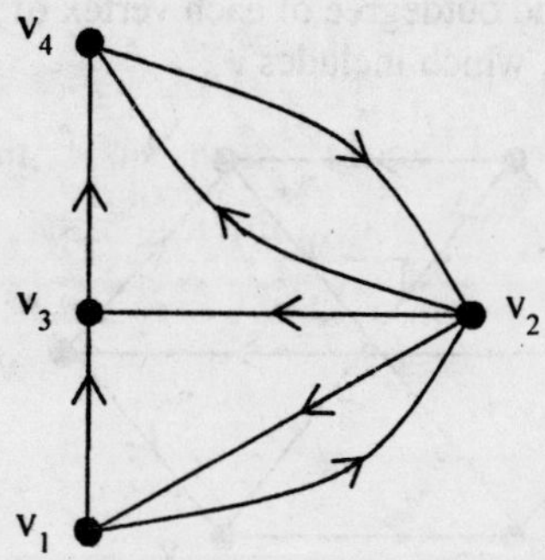

Fig. 5.30

2. Define the following terms
   (a) Directed Graph
   (b) Adjacency Matrix of a digraph
   (c) Path matrix of a digraph
   (d) Circuit matrix of a digraph
   (e) Cut set matrix of a digraph.
3. Obtain the B_6 matrix of graph G in Fig. 5.31. Also find the path matrix.

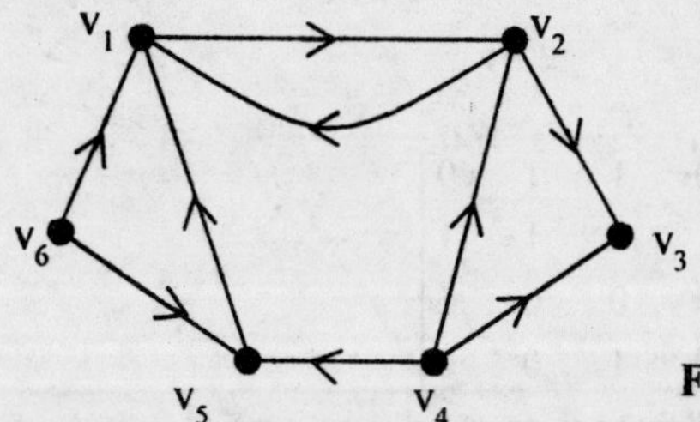

Fig. 5.31

4. For the digraph given in Fig. 5.32 write down the incidence matrix and adjacency matrix.

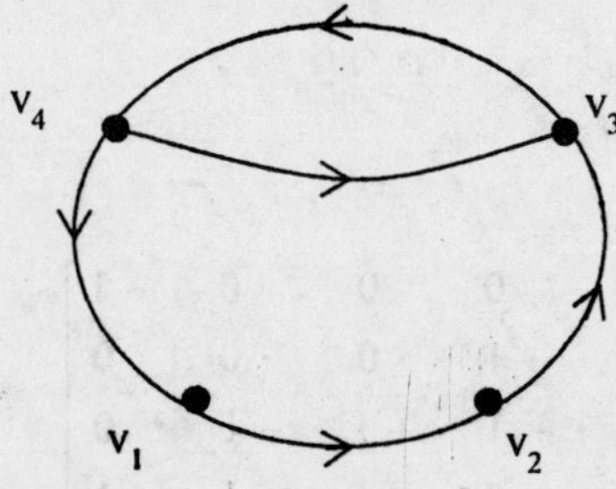

Fig. 5.32

Using its adjacency matrix find how many paths of length 3 exist from v_4 to v_3

5. Consider the graph G in Fig. 5.33. Suppose the vertices are v_1, v_2, v_3, v_4, v_5 and v_6.

(a) Find the adjacency matrix A of the graph G.
(b) Find the path matrix P of G using power of A.
(c) Is G strongly connected?
(d) Find the indegree and outdegree of each vertex of G.
(e) Find all cycles in G, which includes v_4.

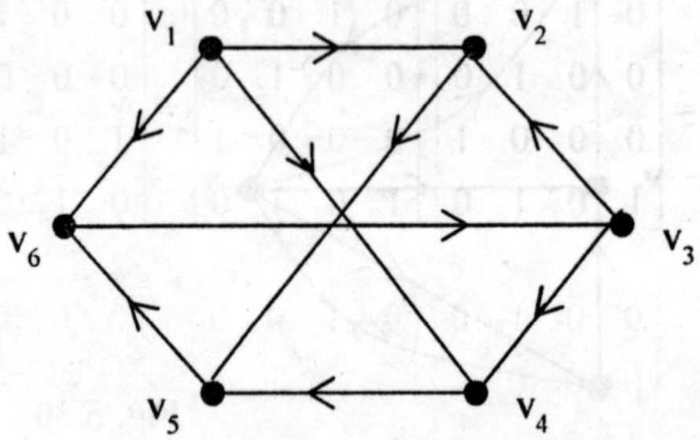

Fig. 5.33

6. Sketch all different simple digraph with 1, 2 and 3 vertices.
7. Prove that in any digraph the sum of the indegree of all vertices is equal to the sum of their degrees; and this sum is equal to the number of edges in the digraph.

ANSWERS

1.

$$A = \begin{array}{c} \\ v_1 \\ v_2 \\ v_3 \\ v_4 \end{array} \begin{array}{c} \begin{array}{cccc} v_1 & v_2 & v_3 & v_4 \end{array} \\ \begin{bmatrix} 0 & 1 & 1 & 0 \\ i & 0 & 1 & 1 \\ 0 & 0 & 0 & 1 \\ 0 & 1 & 0 & 0 \end{bmatrix} \end{array}$$

Elementary path of length 1 from v_1 to v_3 : $v_1\ v_3$

Elementary path of length 2 from v_1 to v_3 : $v_1\ v_2\ v_3$

4. If we consider the edges

$$e_1 = \{v_1, v_2\};\quad e_2 = \{v_2, v_3\};\ e_3 = \{v_3, v_4\};$$
$$e_4 = \{v_4, v_3\};\ e_5 = \{v_4, v_1\}$$

Incidence matrix is

$$M = \begin{bmatrix} +1 & 0 & 0 & 0 & -1 \\ -1 & +1 & 0 & 0 & 0 \\ 0 & -1 & +1 & -1 & 0 \\ 0 & 0 & -1 & +1 & 1 \end{bmatrix}$$

and Adjacency matrix is

$$A = \begin{array}{c} \\ v_1 \\ v_2 \\ v_3 \\ v_4 \end{array} \begin{array}{c} \begin{array}{cccc} v_1 & v_2 & v_3 & v_4 \end{array} \\ \begin{bmatrix} 0 & 1 & 0 & 0 \\ 0 & 0 & 1 & 0 \\ 0 & 0 & 1 & 1 \\ 1 & 0 & 1 & 0 \end{bmatrix} \end{array}$$

Now,

$$A^2 = \begin{bmatrix} 0 & 1 & 0 & 0 \\ 0 & 0 & 1 & 0 \\ 0 & 0 & 0 & 1 \\ 1 & 0 & 1 & 0 \end{bmatrix} \begin{bmatrix} 0 & 1 & 0 & 0 \\ 0 & 0 & 1 & 0 \\ 0 & 0 & 0 & 1 \\ 1 & 0 & 1 & 0 \end{bmatrix} = \begin{bmatrix} 0 & 0 & 1 & 0 \\ 0 & 0 & 0 & 1 \\ 1 & 0 & 1 & 0 \\ 0 & 1 & 0 & 1 \end{bmatrix}$$

$$A^3 = \begin{bmatrix} 0 & 0 & 1 & 0 \\ 0 & 0 & 0 & 1 \\ 0 & 1 & 0 & 0 \\ 0 & 1 & 0 & 1 \end{bmatrix} \begin{bmatrix} 0 & 1 & 0 & 0 \\ 0 & 0 & 1 & 0 \\ 0 & 0 & 0 & 1 \\ 1 & 0 & 1 & 0 \end{bmatrix} = \begin{bmatrix} 0 & 0 & 0 & 1 \\ 1 & 0 & 1 & 0 \\ 0 & 1 & 0 & 1 \\ 1 & 0 & 2 & 0 \end{bmatrix}$$

There are two path of length 3 from v_4 to v_3 since $(4, 3)^{th}$ element of $A^3 = 2$

6 COLOURINGS

6.0 INTRODUCTION

We have already discussed problems such as shortest path algorithms, Travelling salesman problem and Chinese postman problem in chapter 2. These problems are solved with the help of graph theory. Similarly, simple decision making problems such as scheduling of n jobs on m different machines, scheduling of committee meeting times without overlapping, networking of pipes characterized by flow of information from source to sink, waiting line problem which is defined by arrival of customers randomly at one or more service facility etc. may also be represented by graph and solved by colouring. Colouring has also got applications in compilers frequency assignment in mobiles, radios and pattern matching. Graph colouring leads to most effective utilisation of resources.

6.1COLOURING

In graph theory, colouring is an assignment of colours to certain object (vertex, edge, region etc.) in a graph. Vertex colouring is most important kind out of these different colourings. It is the starting point of the entire subject and other colouring problem can be transformed into a vertex version. **Vertex colouring** means colouring the vertices of a graph. Colouring the vertices of a graph in such a ways so that no two adjacent vertices have the same colour is called **proper colouring**. When used without any qualification, a colouring is nearly always assumed to be proper coluring. Further, when used without any qualification a colouring always assumed to be vertex colouring. If we consider a graph G with m vertices coloured with n colours then G is said to be ***n*-colourable**.

Further for colouring of map, any two regions having a common border are assigned different colours. When the number of regions are large, our interest will be in finding the least number of colours required to colour the map in such a way that the adjacent regions do not have the same colour. In the following articles we will discuss all these problems in detail.

Example 6.1

(a) Consider a complete graph G as shown in Fig. 6.1(a) with six vertices. Since every vertex is adjacent to each other vertex, therefore, G requires six different colours for colouring. Hence G is 6-colourable.

(b) Consider the graph G in Fig. 6.1(b). One colour p_1 is assigned to vertices v_1, v_3 and v_5. The second colour p_2 is assigned to vertices v_2, v_4 and v_6 as they are adjacent to v_1, v_3 and v_5. All the vertices have been assigned with two colours in such a way that no two adjacent vertices have same colour. Hence G is 2-colourable.

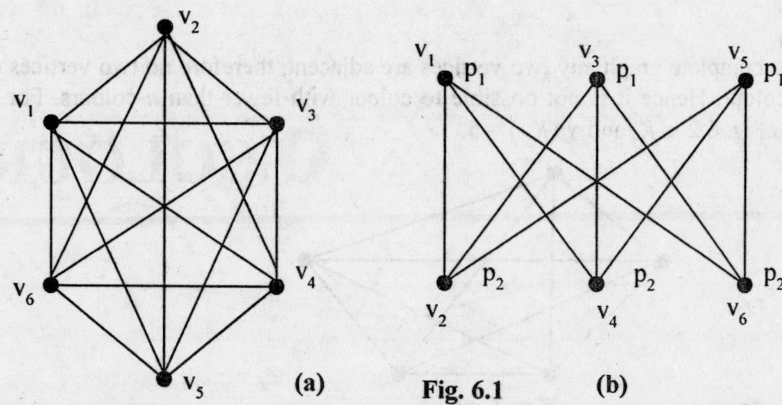

(a) **Fig. 6.1** (b)

6.2 CHROMATIC NUMBER

Now the question arises, how many minimum numbers of colours are required for proper colouring? The answer of this question can be obtained by chromatic number.

The **chromatic number** of a graph G is the minimum number of colours n needed for colouring of the graph G. It is denoted by $\chi(G)$ and $\chi(G) = n$ means G is n-chromatic. A graph that can be assigned n-colouring is n-colourable, and it is n- chromatic if its chromatic number is exactly n. For colouring problems, we need only simple connected graph because discarding loops and replacing all parallel edges with single edge without affecting adjacency does not change chromatic number.

One must note here a proper colouring of a graph with k-colours is called k-colouring of a graph while chromatic number is the minimum number of such colours.

6.3 SOME PROPERTIES OF CHROMATIC NUMBER χ

1. $\chi(G) = 1$ if G is a null graph. A graph with only isolated vertices is 1-chromatic.
2. A graph is 1-chromatic if it is totally disconnected.
3. $\chi(G) \leq |V|$, where $|V|$ in the number of vertices.
4. If subgraph of G is n-chromatic then $\chi(G) \leq n$.
5. If degree of vertex is d then at most d colours are required to colour the vertices adjacent to it.
6. A complete graph of n vertices K_n requires n colours for proper colouring i.e. $\chi(K_n) = n$. In particular a triangle always requires three colours.
7. Every n-chromatic graph has at least n vertices v such that deg $(v) \geq n - 1$.
8. Chromatic number of every bipartite graph $K_{m,\,n}$ is 2.
9. A cycle graph C_n has chromatic number 2 or 3 according as it has even or odd number of vertices.
10. A wheel graph W_n has chromatic number 3 or 4 according as it has odd or even number of vertices.
11. $\chi(G) \leq \Delta(G)$, where $\Delta(G)$ is the largest degree of any vertex of G.
12. For any graph G, $\chi(G) \leq \Delta(G) + 1$.

13. Every tree is 2- colourable.

Proof of (6)

In a complete graph any two vertices are adjacent, therefore no two vertices can be coloured with same colour. Hence it is not possible to colour with fewer than n-colours. For example, graph shown in the Fig. 6.2 is K_5 and $\chi(K_5) = 5$.

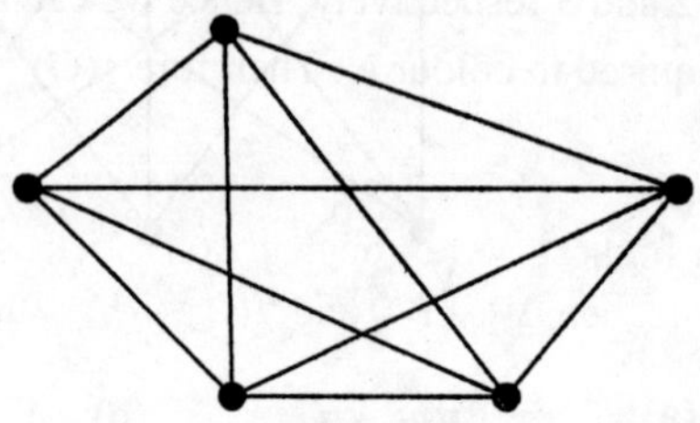

Fig. 6.2

Proof of (8)

The vertex set of a bipartite graph can be selected into two partitions M and N such that no two vertices in M or N are adjacent. Hence one can give one colour to vertices of M and another colour to vertices of N. For example, consider the graph $K_{3,4}$. Here vertices a, b, c can be coloured with one colour whereas vertices d, e, f and g can be coloured with another colour.

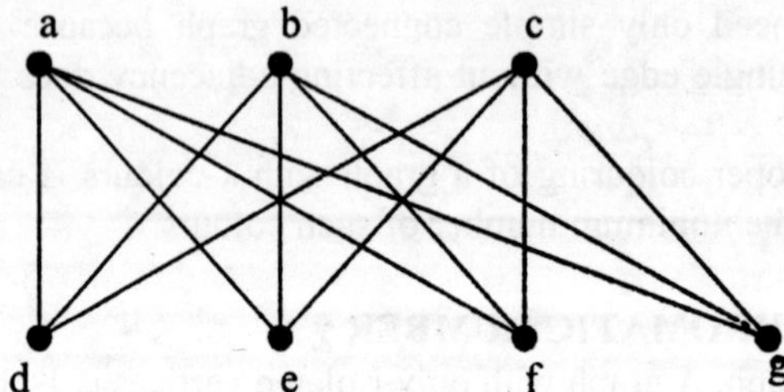

Proof of (10)

Consider wheel graph W_5 shown in the Fig. 6.3.

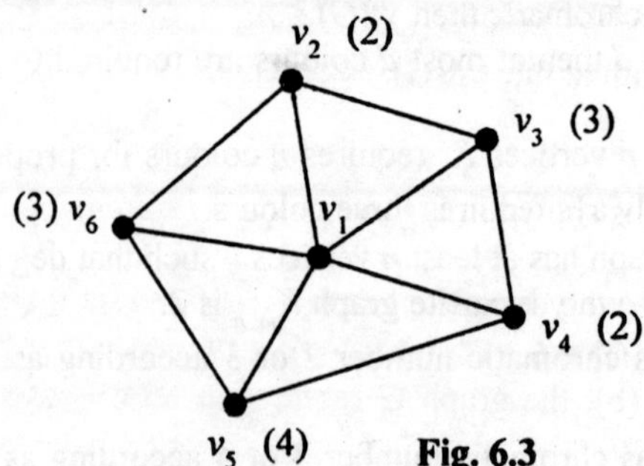

Fig. 6.3

We can observe here deg $(v_1) = 5$ and all other vertices are of degree 3, therefore $\Delta(G) = 5$ and hence $\chi(G) = 1 + \Delta(G) = 6$. But graph G has a triangle subgraph, therefore $\chi(G) \geq 3$. If $\chi(G) \geq 5$, the property (7) would imply that there would be at least 5 vertices of degree greater than or equal to 4. But for graph G it is not true, hence $\chi(G) \leq 4$. Now we conclude $3 \leq \chi(G) \leq 4$.

Let us first choose the triangle v_1, v_2, v_3 and colour them with colours 1, 2 and 3 respectively. Here we can observe that v_4 is adjacent to vertex $v_1$and v_3. Since v_1 and v_3 being coloured with 1 and 3 respectively, hence it is being forced to colour 2 to vertex v_4. Again the adjacent vertices v_2 and v_6 can not have same colour, hence it is coloured with colour 3. Now we are left with vertex v_5, which is adjacent to v_1, v_4 and v_6 having colours 1, 2 and 3 respectively. Hence we can not colour v_5 with either of these colours. Hence fourth colour is required to colour v_5. Therefore $\chi(G) = 4$.

Proof of (12)

Consider a simple graph G with highest vertex degree $\Delta(G)$. Let us prove this theorem by induction on number of vertices $|V(G)| = 1$, then $\Delta(G) = 0$ and $\chi(G)=1$. Assume that $\chi(G) \leq \Delta(G) + 1$ for $|V(G)| \leq n$.

Now consider graph G with $|V(G)| = n + 1$. If we delete any vertex v and its incident edges from the graph G then let the new graph obtained is G_1. See Fig. 6.4.

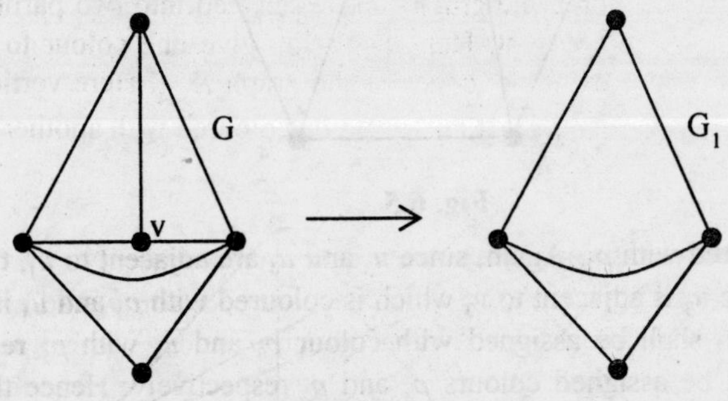

Fig. 6.4

It is clear $\Delta(G) \geq \Delta(G_1)$. But by induction hypothesis

$$\Delta(G) \leq \Delta(G_1) + 1.$$

Therefore, the graph G_1 can be coloured with at most $\Delta(G) + 1$ colours. Since deg $(v) = \Delta(G)$, the graph G can be coloured with at most $\Delta(G) + 1$ colours.

Example 6.2

Consider the graphs shown in Fig. 6.1. Find the chromatic number.

Solution

(a) In Fig. 6.1(a) the complete graph G requires six different colours for proper colouring, so the chromatic number $\chi(G) = 6$.

(b) In Fig. 6.1 (b), the graph G requires at least two colours for proper colouring. Thus $\chi(G) = 2$.

Example 6.3

Find the chromatic number of Petersen graph and Grotzsch graph.

Solution

(a) Petersen Graph

Consider the Petersen graph shown in the Fig. 6.5. The outer edges of the graph consist of vertices v_1, v_2, v_3, v_4 and v_5. We can observe that number of these vertices are odd (5 in number), hence it requires 3 colours to paint. Let the vertices v_1, v_2, v_3, v_4 and v_5 are coloured with P_1, P_2, P_1, P_2 and P_3 respectively. Now take the inner vertices: u_1 being adjacent to v_1 cannot be coloured with P_1.

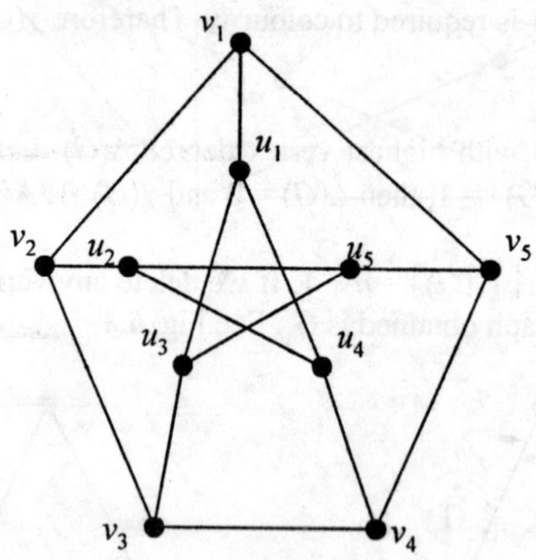

Fig. 6.5

Let it is coloured with p_2. Again, since u_3 and u_4 are adjacent to u_1, they can be painted with colours p_1 and p_3. Since u_3 is adjacent to v_3 which is coloured with p_1 and u_4 is adjacent to v_4 which is coloured with p_2, so u_3 shall be assigned with colour p_3 and u_4 with p_1 respectively. In the same fashion u_2 and u_3 shall be assigned colours p_3 and p_2 respectively. Hence the Petersen graph is 3-colurable. These colouring can be summarised in following table:

Vertices Colour	v_1 p_1	v_2 p_2	v_3 p_1	v_4 p_2	v_5 p_3
Vertices Colour	u_1 p_2	u_2 p_3	u_3 p_3	u_4 p_1	u_5 p_2

Table 6.1

(c) Grotzsch Graph

Consider the Grotzsch graph shown in the Fig. 6.6. The outer edges of the graph consists of vertices v_1, v_2, v_3, v_4 and v_5. We can observe that this graph consists of 5 cycles (odd number of cycles) and they require 3 colours. Let the vertices v_1, v_2, v_3, v_4 and v_5 are assigned with P_1, P_2, P_1, P_2 and P_3 colours respectively. Now consider the inner edges. Since u_1 is adjacent to u_2 and u_5, and they are coloured with p_2 and p_3 respectively. Let u_1 is coloured with p_1. Continuing with this fashion we can colour u_2, u_3, u_4 and u_5 with colours p_2, p_3, p_1 and p_3 respectively. But vertex w is adjacent to each of the vertex u_1, u_2, u_3, u_4 and u_5, hence it cannot be coloured with any of the colours used for u_i (i = 1...5). Let w is coloured with p_4. Hence the Grotzsch graph is 4-chromatic.

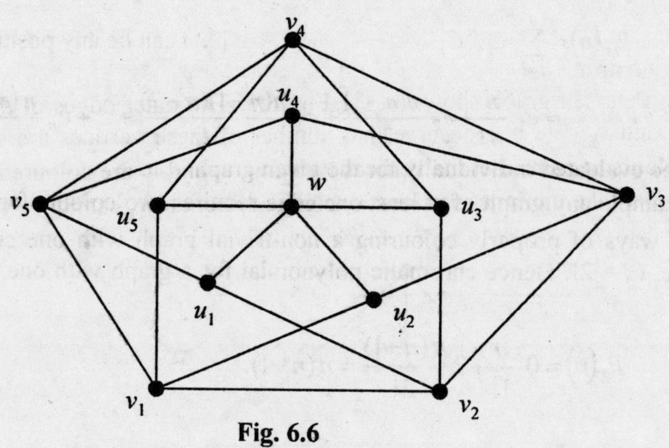

Fig. 6.6

Example 6.4

Show that following statements are equivalent:

(i) A graph G is 2-colourable.

(ii) G is bipartite.

(iii) Every cycle of G has even length.

Proof

(i) → (ii). Suppose that G is 2-colourable. Let M and N be set of vertices coloured by first and second colour respectively. Hence G is bipartite.

(ii) → (iii). Suppose that G is bipartite graph and let M and N form a bipartite partition of the vertices of G. Suppose cycle starts at a vertex $v \in M$, then it will go to a vertex in the partition N, and back to a vertex in M then to N and so on. Hence, when cycle returns to v it must be of even length.

Example 6.5

Show that a simple graph with chromatic number 2 is bipartite.

Solution

Since the chromatic number of the given graph is 2, hence its vertices set V can be partitioned into two disjoint subsets, say V_1 and V_2. These two vertices set can receive two different colours, say 1 and 2 and there will be no edge connecting any two vertices in V_1 or V_2. Hence the given graph is bipartite.

6.4 CHROMATIC POLYNOMIAL

Consider a graph of m vertices and let us assume that it can be properly coloured in many different ways using different colours. This property of graph can be expressed in simple and effective way by means of a polynomial. The **chromatic polynomial** gives the number of ways in which the given graph can be coloured by n or fewer number of colours and is denoted by $P_m(n)$. Birkhoff in 1912 gave this concept.

Let the graph G is properly coloured using i different colours in c_i number of ways. Since there are ${}^nC_i$ different ways of choosing i colours from n colours, hence there are ${}^nC_i$ different ways to properly colour G using i colours out of n colours. Hence, the chromatic polynomial is given by sum of these terms

$$P_m(n) = \sum_{i=1}^{n} c_i . {}^nC_i \qquad [\because i \text{ can be any positive integer from 1 to } n]$$

$$= c_1 . \frac{n}{n!} + c_2 \frac{n(n-1)}{2!} + c_3 \frac{n(n-1)(n-2)}{3!} + ... + c_n \frac{n(n-1)(n-2)....(n-m+1)}{n!}$$

Each c_i has to be evaluated individually for the given graph.

For example any graph of at least one edge requires two colours for proper colouring. So, c_l the number of ways of properly colouring a non-trivial graph with one colour is 0 and with two colours is 2, i.e. $c_2 = 2!$. Hence chromatic polynomial for a graph with one edge and two vertices is given by

$$P_2(n) = 0 \cdot \frac{n}{1!} + 2\frac{n(n-1)}{2!} = n(n-1).$$

NOTE

1. Other values of c_i depend on the nature of the graph.
2. For complete graph K_m with m vertices each vertex has (n -1) adjacent vertices, i.e. the graph can be coloured using m different colours in m! ways for proper colouring.
   Hence $c_m = m!$ and for $i < m$, $c_i = 0$.

THEOREM 6.1

The chromatic polynomial of a complete graph of m vertices is given by

$$P_m(n) = n(n-1)(n-2) \dots (n-m+1) \text{ and vice versa.}$$

PROOF

Consider a graph G with m vertices. There are n different ways of colouring of any vertex (first vertex) of graph. A second vertex can be coloured in (n -1) ways, the third in (n - 2) ways and so on. Similarly the mth vertex can be coloured properly in (n - m + 1) different ways iff every vertex is adjacent to each others. Therefore, chromatic polynomial is

$$P_m(n) = n(n-1)(n-2) \dots (n-m+1).$$

Hence the graph is a complete graph and vice versa.

Example 6.6

Find the chromatic polynomial of a connected graph on three vertices.

Solution

Since the given graph is a connected graph it contain at least one edge. Hence minimum two colours are required for any proper colouring of G. Thus $c_1 = 0$.

The number of ways a graph of m vertices with m different colours can be properly coloured is m! ways. Hence for graph on 3 vertices $c_3 = 3! = 6$.

Now following two cases possible:

(i) **If G is a triangle**

In this case G cannot be coloured with two colours. Hence $c_2 = 0$.

Thus $$P_m{}^{(n)} = \sum_{i=1}^{m} c_i \; {}^nC_i$$

$$= 0 + 0 + 6 \cdot {}^nC_3 \qquad [\because c_1 = 0,\ c_2 = 0]$$
$$= \frac{6\, n!}{3!\,(n-3)!} = \frac{n(n-1)(n-2)!}{(n-3)!}$$
$$= n(n-1)(n-2).$$

(ii) If G is a path.

In this case the end vertices can be coloured with only two ways with two colours and for each choice of end vertex only one choice of other colour is possible for the middle vertex. Thus c_2 = 2 and c_3 = 3! = 6. Hence

$$P_m^{(n)} = \sum_{i=1}^{m} c_i\, {}^nC_i$$
$$= 0 + 2\, {}^nC_2 + 6\, {}^nC_3$$
$$= 2 \cdot \frac{n(n-1)}{2!} + 6 \cdot \frac{n(n-1)(n-2)!}{3!}$$
$$= n(n-1) + n(n-1)(n-2)$$
$$= n(n-1)(1 + n - 2) = n(n-1)^2.$$

Example 6.7

Find the chromatic polynomial of K_4 complete graph of 4 vertices.

Solution

Consider a complete graph G with 4 vertices represented by K_4. Since minimum m colours required for proper coluring of complete graph K_m of m vertices. Therefore, c_i = 0, for i = 1, 2, 3, ..., m - 1. Hence for proper colouring of K_4, we require 4 colours. Thus $c_1 = c_2 = c_3 = 0$ and c_4 = 4! = 4·3·2·1 = 24

$$\therefore P_4(n) = c_1 \frac{n}{1} + c_2 \frac{n(n-1)}{2!} + c_3 \frac{n(n-1)(n-2)}{3!} + c_4 \frac{n(n-1)(n-2)(n-3)}{4!}$$
$$= n(n-1)(n-2)(n-3) \qquad (\because c_1 = c_2 = c_3 = 0, c_n = 4!)$$

Example 6.8

Find the chromatic polynomial of graph shown in Fig. 6.7.

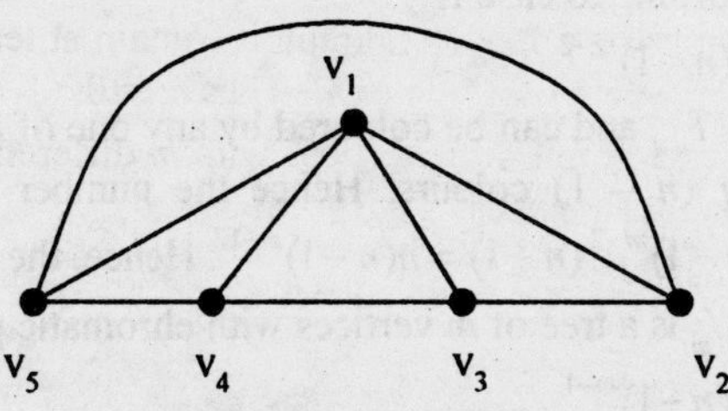

Fig. 6.7

Solution

The graph in Fig. 6.7 consists of triangles. Since a triangle requires minimum three colours for proper colouring. So $c_1 = c_2 = 0$. Now consider three colours p_1, p_2 and p_3 for colouring the vertices

v_1, v_2, v_3 in $3! = 6$ differents ways. So $c_3 = 6$. Similarly, when we have four colours p_1, p_2, p_3 and p_4 then the vertices v_1, v_2, v_3 and v_4 can be properly coloured in $4! = 24$ ways and v_5 vertex can be coloured with the colour of either v_2 or v_3. Thus $c_4 = 2 \times 24 = 48$. And $c_5 = 5! = 120$. Hence,

$$P_5(n) = \sum_{i=1}^{n} c_i {}^nC_i = c_1 {}^nC_1 + c_2 {}^nC_2 + c_3 {}^nC_3 + c_4 {}^nC_4 + c_5 {}^nC_5$$

$$P_5(n) = 6\frac{n(n-1)(n-2)}{3!} + 48\frac{n(n-1)(n-2)(n-3)}{4!} + 120\frac{n(n-1)(n-2)(n-3)(n-4)}{5!}$$

$$[\because c_1 = c_2 = 0]$$

$$= n(n-1)(n-2) + 2n(n-1)(n-2)(n-3) + n(n-1)(n-2)(n-3)(n-4)$$

$$= n(n-1)(n-2)[1 + 2(n-3) + (n-3)(n-4)]$$

$$= n(n-1)(n-2)(n^2 - 5n + 7).$$

THEOREM 6.2

An m-vertex graph is a tree iff its chromatic polynomial is giving by

$$P_m(n) = n(n-1)^{m-1}.$$

PROOF

We will prove the theorem by induction method.

The theorem holds for:

Case I: $m = 1,2$ is immediately evident.

Since for $m = 1$, graph G contains only one vertex which can be coloured in n-different ways. Thus the result is true for $n = 1$. When $m = 2$, then the graph G contains one edge. Hence it requires only two colours for proper coloring of the graph. Hence $c_1 = 0$ and two colours can be used in two different ways for the vertices of the graph, therefore $c_2 = 2$. Thus $P_n(m) = 0 + n(n-1) = n(n-1)$. Hence the result holds for $m = 2$.

Case II: Let us suppose that the chromatic polynomial of a tree of $(m-1)$ vertices is given by

$$P_{m-1}(n) = n(n-1)^{m-2}.$$

Case III: Now we wish to prove that a tree of m vertices have the chromatic polynomial given by

$$P_m(n) = n(n-1)^{m-1}.$$

Suppose T_m be the tree of m vertices with a pendant vertex v. Let $T_{n-1} = T_n - v$, then T_{n-1} is a tree of $(n-1)$ vertices and according to case II

$$P_{m-1}(n) = n(n-1)^{m-2}.$$

Let v' be the neighbour of v in T_{n-1} and can be coloured by any one of n colours, i.e. v can be coloured by any one of the remaining $(n-1)$ colours. Hence the number of different ways of properly colouring of T_n is $P_m(n) = n(n-1)^{m-2}(n-1) = n(n-1)^{m-1}$. Hence, the result follows by induction.

Conversely, suppose that the T'_m is a tree of m vertices with chromatic polynomial

$$P_m(n) = n(n-1)^{m-1}$$

then by induction we will prove that T'_m is a class of trees.

Case I: For $m = 2$, $P_2(n) = n(n-1)$.

Case II: Assume that T_{m-1} be a graph of $(m-1)$ vertices with a chromatic polynomial is a tree.

$$P_{m-1}(n) = n(n-1)^{m-2}$$

Case III: We wish to prove that the graph T_m of m vertices with a chromatic polynomial $P_m(n) = n(n-1)^{m-1}$ is also a tree. Let T_{m-1} consists of the vertices $v_1, v_2, v_3, \ldots v_{m-1}$ and $(m-2)$ edges and $T_m = T_{m-1} + v_m$. Now following two cases arise:

(i) v_m must have an edge with at least one of the vertices of T_{m-1}.

(ii) v_m is an isolated vertex and can be coloured in n ways. Hence the chromatic polynomial of T_m is

$$P_m(n) = n.\ P_{m-1}(n) = n^2(n-1)^{m-2}$$

But this contradicts the hypothesis. Hence, there exists at least one vertex in T_{m-1} which is connected to v_m with an edge (v_i, v_m) in T_{m-1}. Then T_m is connected with $(m-1)$ edges, i.e. $((m-2) +$ edge $(v_i\ v_m)$. Hence T_m is a tree. Since v_m can be coloured properly in $(n-1)$ ways. Therefore, the chromatic polynomial of T_m is given by

$$P_{m-1}(n).(n-1) = n(n-1)^{m-1}.$$

If any other vertex is joined to v_m, it gives a cycle and hence can not have the chromatic polynomial $P_m(n) = n(n-1)^{m-1}$. Therefore, T_m belongs to the class T'_m. Hence the theorem.

Example 6.9

How many ways a tree on 5-vertices can be properly coloured with at most 4 colours?

Solution

A tree with n vertices can be properly coloured with at most m colours in $n(n-1)^{m-1}$ ways. Here, $n = 5$, hence vertices can be properly coloured with at most $m = 4$ colours in $n(n-1)^{m-1} = 4(3)^4 =$ 324 ways.

THEOREM 6.3

For graph G and complement graph $\overline{G}$, the $\chi(G) = k$ and $\chi(\overline{G}) = k'$ satisfy the relations

(i) $2\sqrt{m} \le k + k' \le m+1$

(ii) $m \le kk' \le \left(\dfrac{m+1}{2}\right)^2$.

where m is number of vertices in G.

PROOF

Let G is k-chromatic. This is evident if we classify the vertices into k classes each with the same colour, and select one from each class. These k vertices are adjacent in G and non-adjacent in $\overline{G}$. Suppose these k vertices are coloured by same colour then in $\overline{G}$, we need $(k'-1)$ different colours which are to be applied on $(m-k)$ vertices. Because number of colours cannot exceed the number of vertices, hence $k' - 1 \le m - k$ or $k + k' \le m + 1$.

Suppose the vertices of G be classified into classes of the same colour with K_{max} being the largest number of vertices n i.e. the number of classes being k, $n\,\chi(G) \ge m$. Now all the vertices of the class K_{max} have the same colour in G which are non-adjacent. So, in $\overline{G}$ they need n different colours for adjacent vertices. Hence in $\overline{G}$, they are the vertices of a complete subgraph K_n.

$\Rightarrow\ \chi(\overline{G}) \ge n$. Hence $\chi(G).\ \chi(\overline{G}) \ge n.\chi(G) \ge m$.

It is evident that $(k - k')^2 \geq 0$ and $(k + k')^2 \geq 4kk'$.

$\Rightarrow \quad ((k + k')/2)^2 \geq kk'$.

But $\quad (m + 1)/2 \geq (k + k')/2$.

Hence $\quad ((m + 1)/2)^2 \geq ((k + k')/2)^2 \geq kk' \geq m$.

Hence, from $\quad ((k + k')/2)^2 \geq m$, we get

$$k + k' \geq 2\sqrt{m}.$$

6.5 WETCH-POWELL ALGORITHM FOR VERTEX COLOURING OF A GRAPH

Wetch-powell algorithm gives upper bound of chromatic number, it does not yield the minimum number of colours required to paint the graph. Wetch-powell algorithm is described below:

Step 1: Order the vertices of G in decreasing degree. (Such an ordering may not be unique since some vertices may have the same degree).

Step 2: Use first colour p_1 to the first vertex and then in the sequential order assign p_1 to each vertex which is not adjacent to a previous vertex which was assigned p_1.

Step 3: Repeat step 2 with a second colour p_2 and sequence of non-coloured vertices.

Step 4: Continue repeating with additional colours p_3, p_4, so on until all the vertices are coloured.

Example 6.10

Use Wetch-Powell algorithm to colour the graph shown in Fig.6.8. Find also the $\chi(G)$.

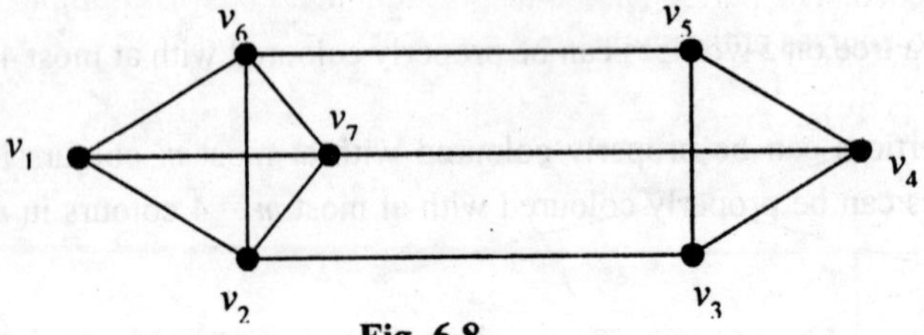

Fig. 6.8

Solution

First list the vertices of graphs G in decreasing degrees:

Vertex	v_6	v_2	v_7	v_1	v_3	v_4	v_5
Degree	4	4	4	3	3	3	3

Assign colour p_1 to the vertex v_6. Now find the vertices not adjacent to v_6 in the list. The next and only vertex on the list not adjacent to v_6 is v_3, assign same colour p_1 to v_3. Now, move to next uncoloured vertex v_2 in the list, assign second colour p_2 to v_2. At this stage we observe, v_5 is the only non-coloured vertex not adjacent to v_2 in the list. Colour this with p_2. Now we are left with no uncoloured vertices which are non-adjacent to either vertex v_2 or v_5. Move to colour p_3. Assign colour p_3 to v_7. There are no uncoloured vertices adjacent to v_7 in the list. Now assign fourth colour p_4 to remaining uncoloured vertices. Finally, we get the following table:

Vertex	v_6	v_3	v_2	v_5	v_7	v_1	v_4
Degree	p_1	p_1	p_2	p_2	p_3	p_4	p_4

Now, we see all the vertices are coloured and four colours were used. Hence, $\chi(G) \leq 4$. But the vertices v_1, v_2, v_6 and v_7 are adjacent to each other, hence at least four colours are required to colour these vertices. Thus $\chi(G) = 4$.

Example 6.11

An university offers seven courses coded as A, B, C, D, E, F, G. There are 12 students S1, S2, ..., S12, who intend to take these courses as given below:

S1: A, B, D; S2: A, B, G; S3: B, D; S4: A, C;
S5: C, D; S6: A, B; S7: D, F; S8: B, C;
S9: A, E; S10: A, E, G; S11: A, G; S12: A, E.

No student can appear for more are than one course in a given time period, but courses can run in parallel sessions. How many time periods are required?

Solution

Period (1) Course A: S1, S2, S4, S6, S9, S10 /Course D: S3, S5, S7
Period (2) Course B: S2, S3, S6, S8/ Course E: S1, S9, S10
Period (3) Course C: S1, S4, S5, S8/ Course G: S2, S10, S11
Period (4) Course F: S7, S11, S12.

Let the periods are shown as P_1, P_2, P_3 and P_4 in the Fig. 6.9. Here, to plan the teaching of the courses one must take care that when two or more courses are to be taught in the same period, different class rooms are needed. The period should be programme in non-overlapping time intervals. No student can appear in two courses at the same time.

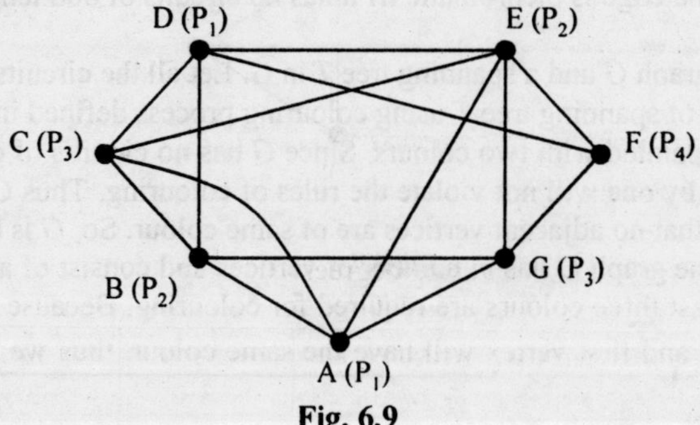

Fig. 6.9

6.6 BICHROMATIC GRAPH

If a graph G requires two colours for proper colouring then it is called **bichromatic graph**. In other words, if the graph can be partitioned into two non-empty subsets, such that each edge connects a vertex from each subset, then it can be painted with two different colours and called **bichromatic.**

Example 6.12

Consider famous three utilities puzzle problem. The objective of the puzzle is to supply three utilities gas, electricity and water to three houses x, y and z. Show that the problem is represented by bichromatic graph.

Solution

The vertices x, y and z show three houses and u_1, u_2 and u_3 represent the utilities gas, electricity and water respectively in graph G in Fig. 6.10. Let the vertices x, y and z are painted by

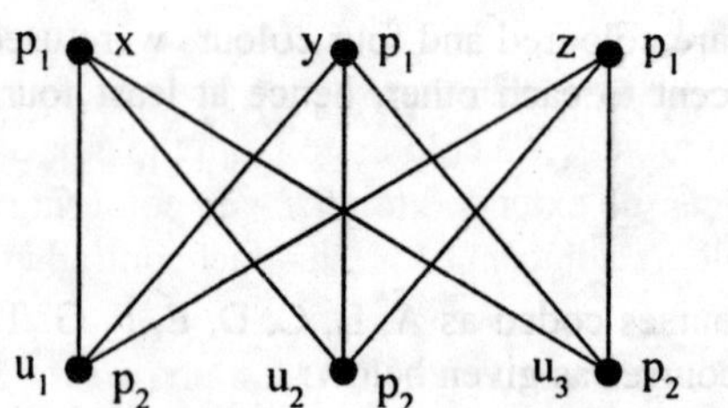

Fig. 6.10

colour p_1 and utilities by p_2. No two houses and utilities are adjacent. Hence, we need only two colours to represent the puzzle. So the chromatic number of graph G is 2. i.e. $\chi(G) = 2$. Thus the utility puzzle problem is bichromatic.

THEOREM 6.4

Every tree with two or more vertices is bichromatic.

PROOF

Let us consider a tree T. Choose any vertex v in the given tree T say root. Paint alternate vertices with two different colours in such a way that one colour is used for those vertices which are at odd distance from v and another colour is used for those which are at even distances from root. Thus tree T can be properly coloured with two colours. Hence it is bichromatic.

THEOREM 6.5

A graph with at least one edge is bichromatic iff it has no circuits of odd length.

PROOF

Consider a connected graph G and a spanning tree T in G. Let all the circuits in G are of even lengths. Now, paint the vertices of spanning tree T using colouring process defined in the theorem 6.1. We find that T can be properly painted with two colours. Since G has no circuits of odd length, hence on joining the vertices of T one by one will not violate the rules of colouring. Thus G can be coloured with two colours in such a way that no adjacent vertices are of same colour. So, G is bichromatic.

Conversely, consider the graph G has m number of vertices and consist of at least one circuit of odd length. In this case at least three colours are required for colouring. Because when m (number of vertices in G) is odd, the mth and first vertex will have the same colour, thus we need third colour for proper colouring.

6.7 CRITICAL GRAPH

A graph G is said to be ***n*-critical** if $\chi(G) = n$ and $\chi(G - v) < \chi(G)$ for each vertex v of G.

REMARK

An n-chromatic graph has n-critical subgraph.

THEOREM 6.6

If graph G is n-critical. Then

(a) G is connected.

(b) The degree of each vertex of G is at least $n - 1$.

(c) G can not be partitioned.

PROOF

(a) Let K be any component of disconnected graph G with chromatic number n. Suppose that v be any vertex of G not in K. Then $\chi(G - v) \geq n$. But this contradict the fact of n-critical graph. Hence G is connected.

(b) As given G is n-critical graph. For each vertex v of G, $\chi(G - v) \leq n - 1$. If degree of vertex $d(v) < n - 1$, we need at the most $n - 2$ colours to paint the adjacent node of v. It implies that $(n - 1)$ colouring of $G - v$ can be extended to a $(n - 1)$ colouring of G by colouring v by the colour distinct from the colours adjacent of v. This fact contradict the $\chi(G) = k$. So degree of vertex $d(v) \geq n - 1$.

(c) Let G_1 and G_2 are subgraphs of G then $G = G_1 \cup G_2$ and $G_1 \cap G_2 = K_r$ then $\chi(G_1) \leq k - 1$ and $\chi(G_2) \leq k - 1$. But, the vertices of $G_1 \cap G_2$ can be coloured with same colour in both sub-graphs. Hence, on combining they give $(k - 1)$ colouring of G. But this contradicts the fact $\chi(G) = k$. Hence, G can not be partitioned.

THEOREM 6.7

Suppose G is a graph with $\chi(G) > n$ and can be partitioned in $G(x)$ and $G(y)$ in such a way that they are n-colourable, then the edge cut (x, y) has at least n edges.

COROLLARY

The degree of vertex is greater than or equal to $m - 1$ if m chromatic graph has at least m vertices.

PROOF

Suppose K is an m-critical subgraph of m-chromatic graph G. According to theorem 6.6, degree of every vertex of K is at least $m - 1$ and so also in G. Clearly, K has at least m vertices because it is m-chromatic.

NOTE

1. For any graph G with maximum degree Δ of any vertex, chromatic number is defined as $\chi(G) \leq 1 + \Delta(G)$.
2. For any graph G, with minimum degree δ of any vertex, chromatic number is given by $\chi(G) \geq m/(m - \delta(G))$, where m is the number of vertices of G.

Example 6.13

Find the chromatic number of graph G shown in Fig.6.11.

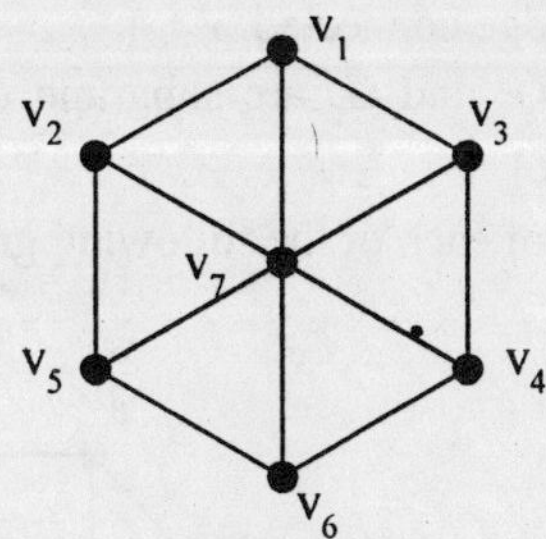

Fig. 6.11

Solution

For graph G shown in Fig. 6.11, $d(v_7) = 6$ and degree of all other vertices is 3. The maximum degree of G, $\Delta(G) = 6$ and minimum degree, $\delta(G) = 3$. Also by above note 1, we have $\chi(G) \leq 1 + \Delta(G)$, i.e. in this case $\chi(G) \leq 7$.

Since there are simple circuits of length 3, therefore χ (G) $\geq$ 3. Now, if χ(G) $\geq$ 6, then by above corollary there will be at least 6 vertices of degree greater than or equal to 5. But this is a contradiction. Hence $\chi(G) \leq 5$.

It implies $3 \leq \chi(G) \leq 5$. But still we do not know the exact chromatic number. Let us start, painting the triangle v_4, v_3, v_7 with three different paints p_1, p_2, p_3 respectively. Now, since v_1 is adjacent to v_3 and v_7, we can not paint it with p_2 and p_3. So we paint v_1 by colour p_1. Similarly v_2 by p_2, v_5 by p_1 and v_6 by p_2. Thus the graph can be coloured by three different colours. So $\chi(G) = 3$.

EXERCISE 6.1

1. Define the following terms:
   (a) chromatic number
   (b) critical graph
   (c) bichromatic graph
   (d) *n*-colouring
2. Find the chromatic number of the "wheel" graph in Fig. 6.12

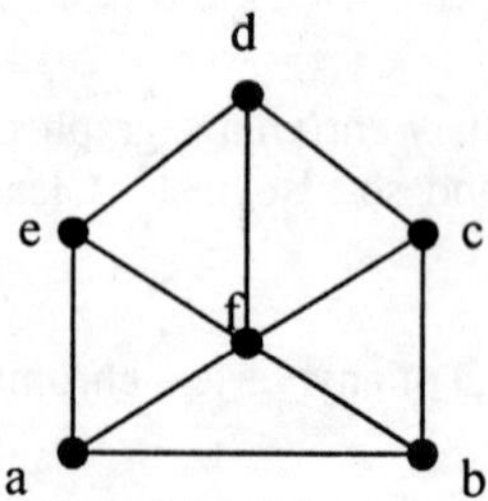

Fig. 6.12

3. Determine the minimum number of edges in a connected *m*-vertex graph with chromatic number *n*.
4. Every *n*-critical graph is $\overline{n-1}$ - edge connected.
   [Hint Let *G* is *n*-critical, and G_1 and G_2 are subgraph of *G* such that $\chi(G_1) \leq n - 1$ and $\chi(G_2) \leq n - 1$. Apply theorem 6.6.]
5. Determine the chromatic number of each of the following graphs:

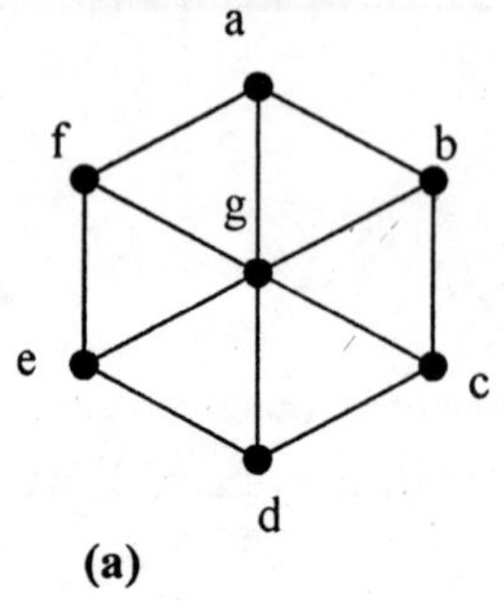

(a)

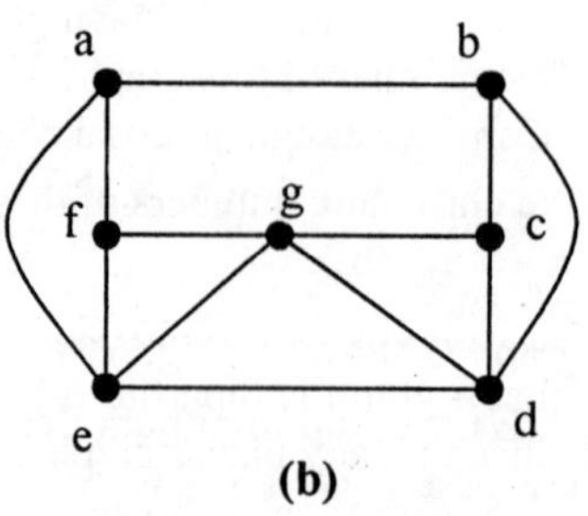

(b)

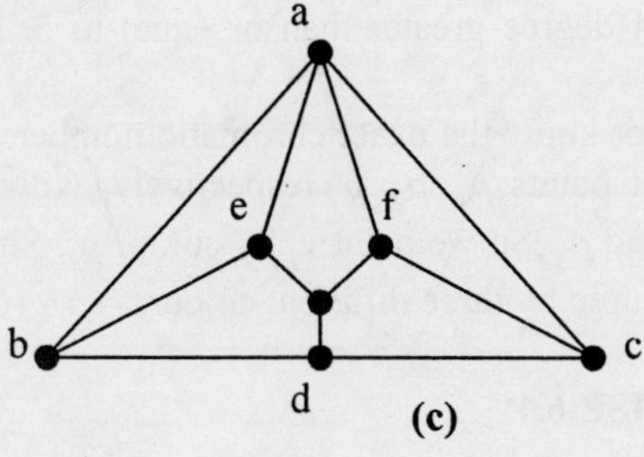

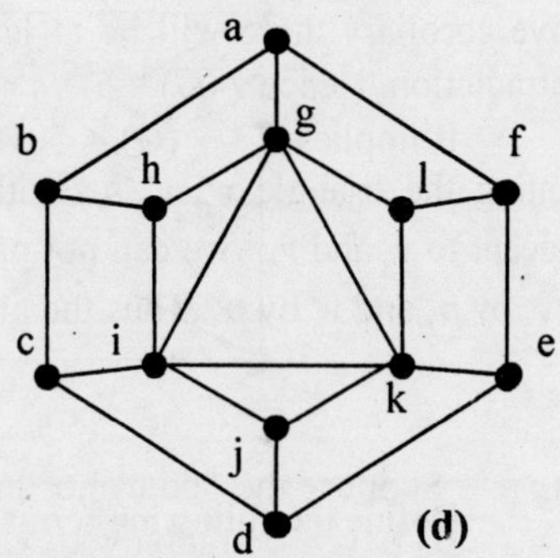

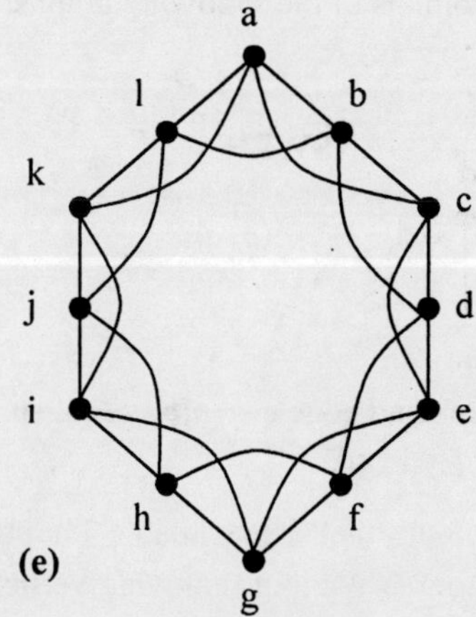

Fig. 6.13

6. What is the chromatic number of wheel W_n.
7. Prove that a wheel graph W_n, where n is odd, is a 4-critical graph.
8. Prove that any k-chromatic graph contains k-critical subgraph.
9. If a graph is a tree, what is its chromatic number? Explain your answer.
10. Show that a graph is bipartite if and only if it has no circuit of odd length.
11. Show that there is a graph having 30 edges in which every vertex has degree 5. What is the chromatic number of this graph?
    [**Hint** Consider a regular icosahedron]
12. Find the chromatic numbers of the following graphs:
    (a) K_n (b) $\bar{K}_n$ (c) $K_{m,n}$, $m,n \geq 1$
    (d) K_n with one edge removed.
13. Show that a graph is bipartite if and only if it can be coloured with two colours. Generalize and deduce that any planar graph is quadripartite.
14. For each of the graph G shown in Fig. 6.14 compute $\chi(G)$ and find $\chi(G)$-critical subgraph.

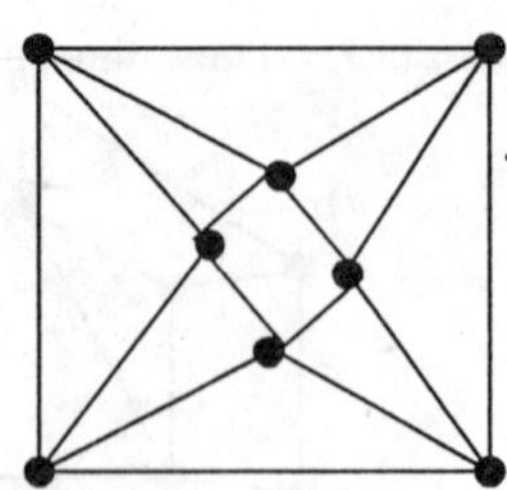
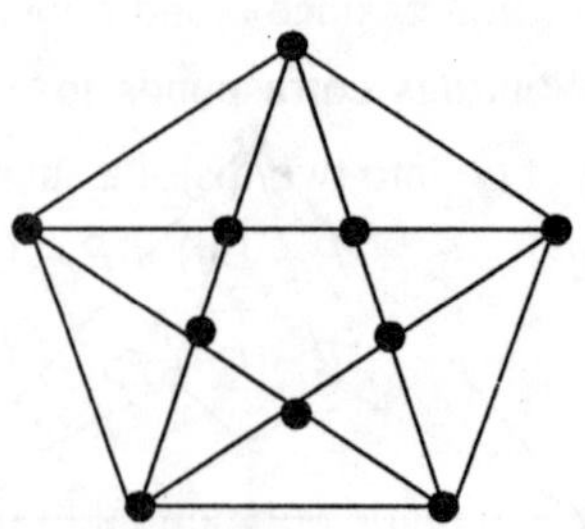

Fig. 6.14

14. Suppose the odd cycles in G are pair-wise intersecting meaning that every two odd cycles in G have a common vertex. Prove that $\chi(G) \leq 5$.
15. Sketch two different graphs that have the same chromatic polynomial.
16. Determine the chromatic polynomials of the following graphs:
(a) K_n (b) $\bar{K}_n$ (c) $K_{3,3}$

ANSWER

5. (a) 3 (b) 4 (c) 4 (d) 2 (e) 4.
9. 2.

THEOREM 6.8 (Decomposition Theorem)

If $G = (V, E)$ is a connected graph and edge $e = (a\ b) \in E$, then

$$P_{G_e}(m) = P_G(m) + P_{G'_e}(m)$$

Where m is the number of colours that are available for proper colouring, G_e denote the subgraph of G obtained by deleting e from G without removing vertices a and b, i.e. $G_e = G - e$ and G'_e is a second graph of G obtained from merging the vertices a and b, and the number of ways to properly colour the vertices in G_e with at most m colours is $P_G(m)$.

PROOF

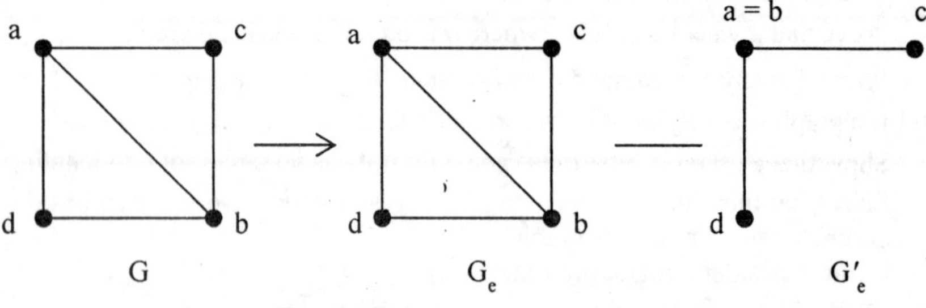

Fig. 6.15

The number of ways to properly colour the vertices in $G_e = G - e$ with at most m colours is $P_{G_e}(m)$. There are two possibilities of colouring. Those for which a and b have different colours and those for which a and b have same colour. We can observe that colouring of first type is also proper colouring of G. On the other hand the colourings of G_e that are not proper colourings of G occur when a and b

have same colour. Infact, since a and b merged are in G_e', they must have same colour there. Hence, each of these colourings corresponds to a proper colouring of G_e'. Thus, the partition of $P_{G_e}(m)$ proper colouring of G_e into two disjoint subsets results in the equation

$$P_{G_e}(m) = P_G(m) + P_{G_e'}(m)$$

or

$$P_G(m) = P_{G_e}(m) - P_{G_e'}(m)$$

Example 6.14

Find the chromatic polynomial of the graph shown in Fig.6.16 using decomposition theorem.

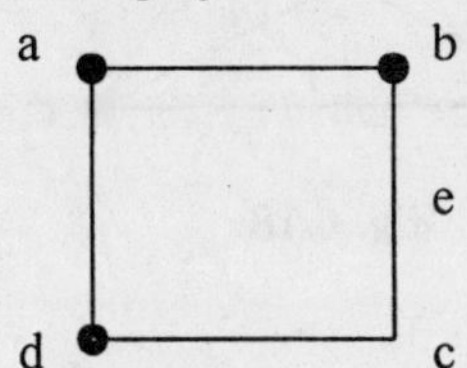

Fig. 6.16

Solution

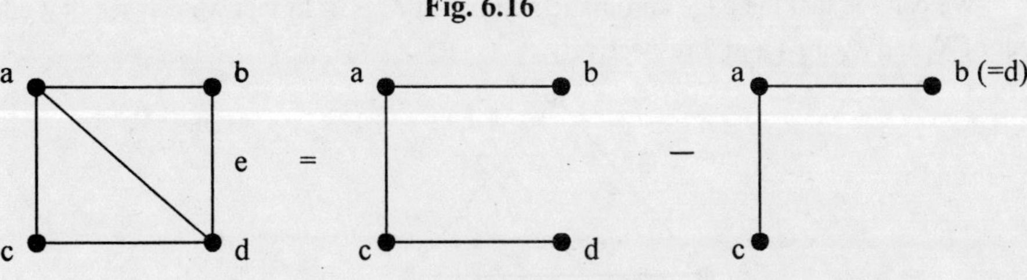

Fig. 6.17

Deletion of edge e give G_i. The chromatic polynomial of G_e is given by

$$P_{G_e}(m) = (m)(m-1)(m-2)$$

By merging the endpoints b and d of edge e, we get G'_e, then

$$P_{G_e'}(m) = (m)(m-1)^3$$

Hence, by decomposition theorem, the chromatic polynomial of G is given by

$$P_{G_e'}(m) = (m)(m-1)^3 - (m)(m-1)(m-2)$$

$$= (m)(m-1)\{(m-1)^2 - (m-2)\}$$

$$= (m)(m-1)\{m^2 - 2m + 1 - m + 2\}$$

$$= m^4 - 4m^3 + 6m^2 - 3m .$$

6.8 EDGE COLOURING

Similar to vertex colouring of a graph an edge colouring is assumed to be proper colouring of the edges which means no two adjacent edges of the graph are assigned same colour. A proper edge colouring with m-colours is called a (proper) ***m*-edge colouring**. A graph that can be assigned a (proper) k-edge colouring is k-edge colourable.

The minimum number of colours required for proper colouring of edges of a graph is called **edge chromatic** number and denoted by $\chi'(G)$. It is sometime also called as **chromatic index.** A graph is m-edge chromatic if its chromatic index is exactly k.

For example edge chromatic number $\chi'(G)$ of graph G shown in Fig. 6.18 is 4.

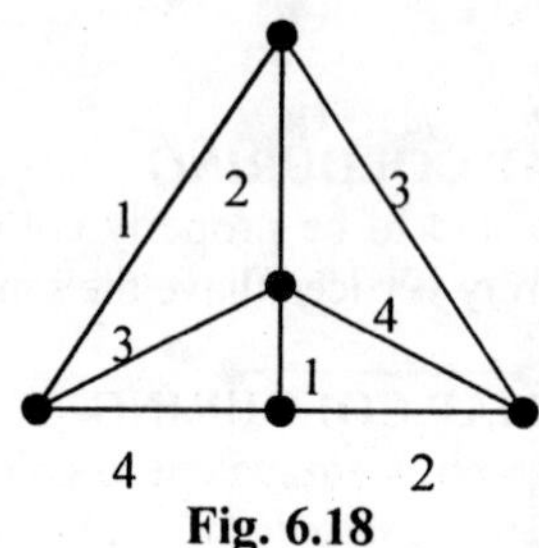

Fig. 6.18

Example 6.15

Edge chromatic number of K_3 is 3.

Solution

We can see that the edge chromatic number of K_3 is 3. In fact we can see that edge chromatic number of K_4 and K_5 are 4 and 5 respectively.

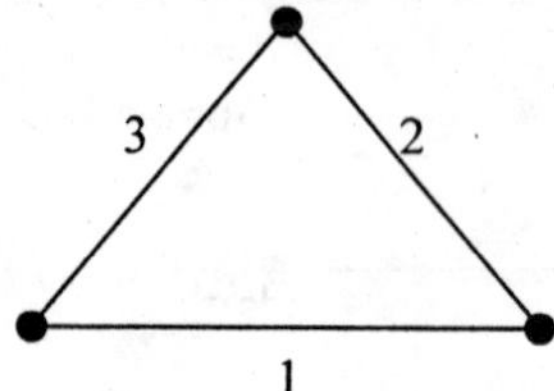

Example 6.16

Find the edge chromatic number of a cycle graph having n edges C_n.

Solution

When n is even: we can colour the edges alternately with two colours only. Therefore for even value of n, the edge chromatic number is 2.

When n is odd, say $2m+1$: In this case $2m$ edges can be properly coloured alternately with two colours, say colour 1 and 2. The first edge and the $2m^{\text{th}}$ edges having been coloured with colour 1 and colour 2, hence we require a third colour for the $(2m + 1)^{\text{th}}$ edge. Therefore, for odd value of n, the edge chromatic number is 3.

THEOREM 6.9 (Vizing 1964)

If G is a simple graph with largest vertex degree Δ, then $\Delta \leq \chi'(G) \leq \Delta + 1$.

THEOREM 6.10

$\chi'(K_n) = n$ if n is odd $(n \neq 1)$ and $\chi'(K_n) = n - 1$ if n is even.

PROOF

The result is clear for $n = 2$. We therefore assume $n \geq 3$.

When n is odd:

In this case we can n-colour the edges of K_n by keeping the vertices of K_n in the form of a regular n-gon*. After representing K_n as regular n-gon, with all diagonals drawn colour the edges around the boundary with different colour for each edge, and then colouring each remaining edge with the colour used for the boundary edge parallel to it. One can observe that the largest possible number of edges in K_n which can have same colour is $(n - 1) / 2$. Hence K_n is not $(n - 1)$ colourable. There are $(1/2)\, n(n - 1)$ edges in K_n.

6.9 COLOURING OF REGIONS OR MAP COLOURING

The regions of a planar graph are said to be properly coloured if no two adjacent regions (regions having a common edge and not merely vertices) have the same colour.

6. 10 VERTEX COLOURING VERSUS MAP COLOURING

Colouring of vertices in a planar graph is equivalent to colouring the map of the graph. We have seen in chapter 4 that a graph has a dual if and only if it is planar. Therefore, a planar graph can be converted into its dual graph. The vertices of the dual will correspond to the regions of original graph. Thus colouring the regions of the planar graph is equivalent to colouring the vertices of its dual and vice versa.

6.11 FOUR COLOUR PROBLEM

Till now, colouring of graph refers to proper colouring of its vertices and edges. Now, we consider a map in which the adjacent regions; not coloured with same paint will be termed as **proper colouring** of the map. If n colours are used to colour a map properly, it is called ***n*-colouring of map**. The proper colouring of regions is also called **map colouring**. Here we are not only interested in just proper colouring the regions of a given graph but our aim is also to use minimum number of colours while painting the regions. *Four colour conjecture** leads that every planar graph or map can be properly coloured with four colours.

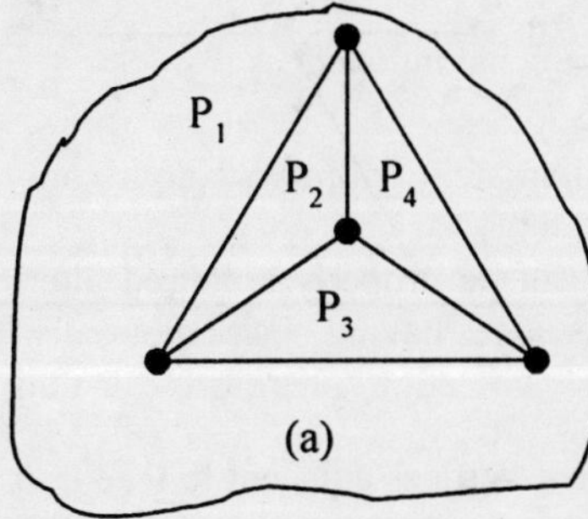

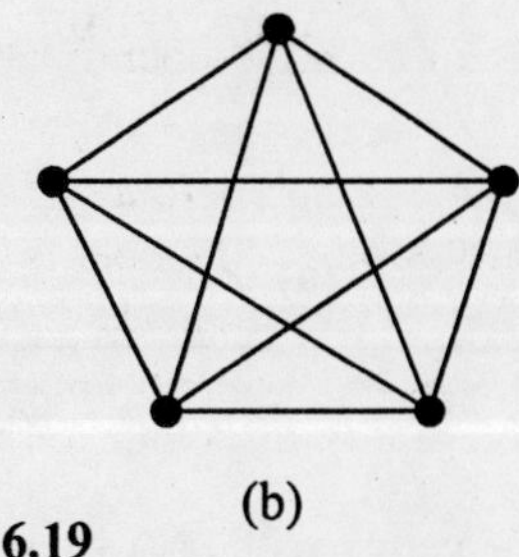

Fig. 6.19

REMARK

Non-planar surfaces can not be properly coloured. For instance, K_4 the complete graph on four vertices shown in Fig.6.19(a) can be coloured with four colours. But the graph shown in Fig. 6.19 (b) is not a planar graph, hence it requires more than four colours for proper colouring.

---

* A polygon with n sides is called n-gon. A polygon which has all sides mutually congruent and all angles mutually congruent is called a regular polygon.

* The **four colour conjecture** states that four colours are sufficient for colouring any map on a plane.

Example 6.17

Show that the vertices of a planar graph with less than 30 edges is 4-colourable.

Solution

To show this, let us assume that each vertex i has degree greater than or equal to 5, i.e.

$$2e = \sum_{i=1}^{n} d(i) \geq 5m$$

where m is number of vertices.

$$\text{or} \qquad m \leq \frac{2e}{5}$$

Also, each region is bounded by at least 3 edges. Hence,

$$2e = \sum_{i=1}^{r} R_i$$

where R_j is the degree of region j. Thus, $2e \geq 3r$ or $r \leq 2e/3$.
From Euler's formula

$$e = r + m - 2 \leq \frac{2e}{3} + \frac{2e}{5} - 2 = \frac{16e}{15} - 2$$

or $e \geq 30$ which is a contradiction to $e < 30$.

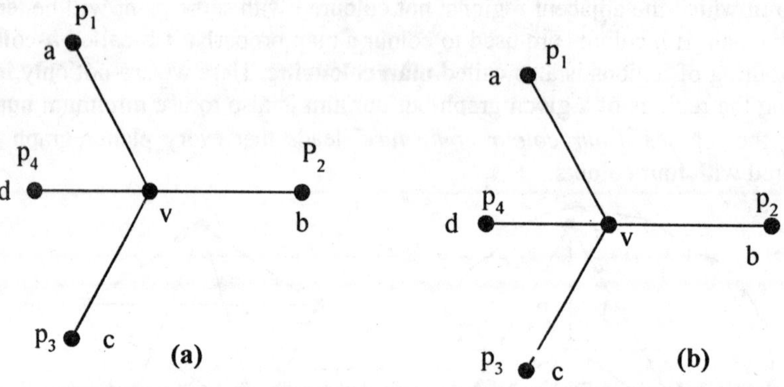

Fig. 6.20

Now let v be the vertex with $d(v)$ =4 and let the 4 vertices adjacent to v be a, b, c, d coloured are coloured are coloured by paints by paints p_1, p_2, p_3, and p_4 respectively. See Fig. 6.20. Now, assume that the result is true for $(m - 1)$. Then $G - v$ by induction hypothesis is 4-colourable, and hence a, b, c, d can be coloured with exactly 4 colours. If we assume there is no cycle between a and c then we can give the same colour p_1 to a and c and paint v with p_3 to v. See in Fig. 6.20(b). Now suppose that there exists a cycle between a and c, i.e. $v\ a\ v_1\ v_2\ c$. Since G is planar, vb cannot intersect the cycle. Now, start with b and interchange the colours p_1 to b p_2 to v and p_1 to a and so on while d retains colour p_4 (see in Fig 6.21). Hence we can colour v with colour p_2. Thus the theorem is true for m.

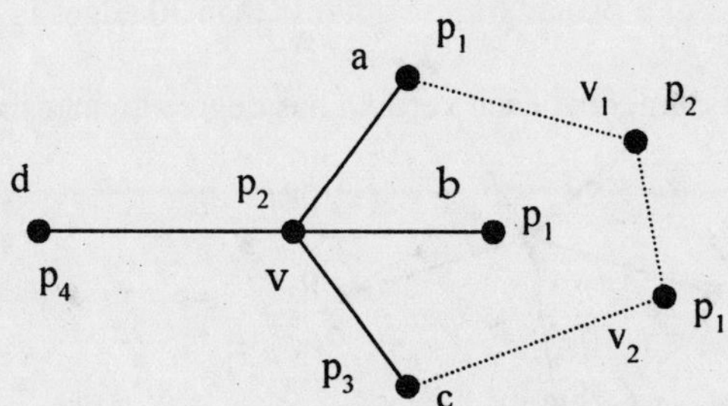

Fig. 6.21

Example 6.18

Show that a simple connected planar graph with 8 vertices and 13 edges cannot be 2-coloured.

Solution

If graph G is 2-colourable, then every cycle in G is of even length. By Èuler formula, $r = e - v + 2 = 13 - 8 + 2 = 7 = \bar{v}$, where v, e and r are the number of vertices, edges and regions of the graph respectively. Hence the dual $\bar{v}$ of G has 7 vertices and 13 edges, with each vertex of degree ≥ 3, but since the degree of each vertex is even deg $(\bar{v}) \geq 4$ for each vertex. Hence, $26 = 2\bar{e} \geq 4\ \bar{v} = 28$, which is a contradiction. Hence G is not 2- colourable.

6.12 THE FIVE COLOUR THEOREM

THEOREM 6.11

Any finite planar graph G can be coloured using five colours (or fewer).

PROOF

To prove the theorem, we use induction on the number of vertices of the graph. Let G be a planar graph with m vertices. Since G is a planar graph and connected, hence there exist at least one vertex of degree less then or equal to five. Let v be the vertex of degree ≤ 5, and consider the induced subgraph G' of G on $G - v$. According to induction hypothesis, induced subgraph G′ requires no more than five colours. If the degree of vertex v is less than 4, then there is no difficulty for proper colouring. But the vertex v has exactly five neighbours and they are all differently coloured as shown in Fig. 6.22(a).

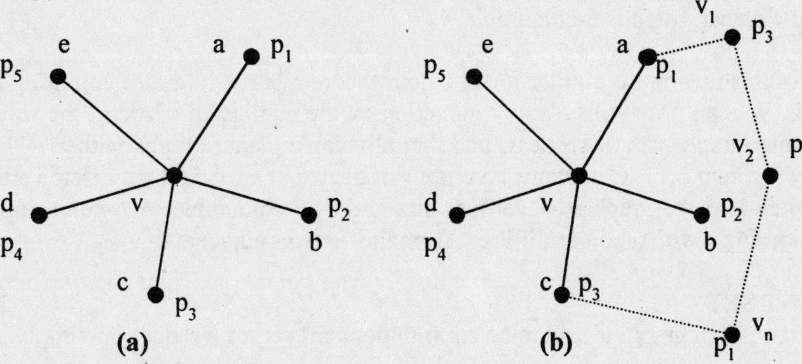

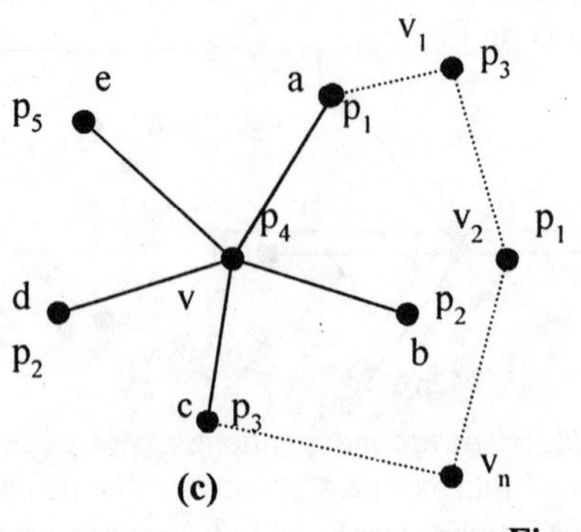

Fig. 6.22

Suppose there is a path between vertices a and c, say $a\ v_1\ v_2 \ldots\ v_n\ c$ [see fig.6.22(b)]. Colour alternatively a by p_1, v_1 by p_2, v_2 by p_1 and so on. Also, according to the property of planar graph there is no similar path between b and d (because it will intersect the path a and c). Hence there is no path of alternating colours p_2 and p_4 through vertices b to d. So, vertex d can be painted with colour p_2 and vertex b is still with colour p_2. Now we have colour p_4 left over with which we can paint vertex v.

Similarly, if we assume that there is no path between a and c and a path between b and d, then we can paint v with colour p_1 or p_3 instead of p_4. Hence the theorem.

THEOREM 6.12

It G is a simple graph with largest vertex degree Δ, then G is $(\Delta + 1)$ colourable.

PROOF

We prove the theorem by induction on the number of vertices. Suppose G be a simple graph with m vertices. If we delete the vertex v and its incident edges, then the graph will remain a simple graph with $m - 1$ vertices and largest-degree at most Δ. By our induction hypothesis, this graph is $(\Delta + 1)$ colourable. Now, colour the vertex v with different from the vertices adjacent to v for $(\Delta + 1)$-colouring of graphs G.

6.13 THE SIX COLOUR THEOREM

THEOREM 6.13

Every simple planar graph is 6-colourable.

PROOF

The proof of this theorem is similar to six colour theorem based on proof colouring of edges that of theorem 6.12. We can prove this also by induction on the number of vertices, the result being trivial for simple planar graph with m vertices, and that all simple planar graphs with $m - 1$ vertices are 6-colourable. By theorem 6.11, G contains a vertex v of degree at most 5. It we delete vertex v and its incident edges, then also the graph n -1 vertices and is thus 6-colourable. A 6-colouring of G is then obtained by colouring v with a colour different from the vertices adjacent to v.

6.14 INDEPENDENT SET

A set of vertices in a graph is said to be an **independent set** of vertices or simply an independent set if no two vertices in the set are adjacent. A single vertex in any graph constitutes an independent set. Consider graph *G* shown in Fig. 6.23.

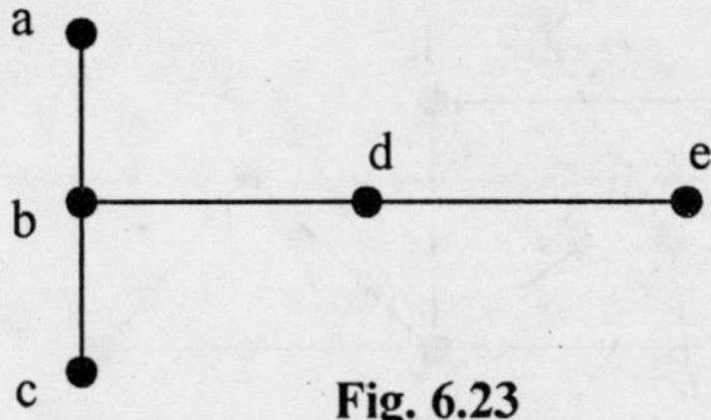

Fig. 6.23

The set $\{a\}$, $\{a, c, d\}$, $\{a, c, e\}$,$\{b, e\}$ are independent sets of *G*.

An independent set of graph *G* that is not a proper subset of any independent set of *G* is called **maximal independent set** of *G*. The set $\{a, c, d\}$ and $\{b, e\}$ are maximal independent sets of the graph. The set $\{a, c\}$ is not a maximal independent set of the graph *G* since it is a proper subset of the independent set $\{a, c, e\}$. The number of vertices in the largest independent set of *G* is called the **independence number** of *G*. It is denoted by α (*G*).

The minimum number of disjoint maximal independent sets of graph *G* whose union is the vertex set of graph *G* is equal to the chromatic number of the graph *G*.

6.15 CHROMATIC PARTIONING

Let *S* is the set of all disjoint independent sets of *G* and *V* is the union of *S*. Then *S* is the **partition** of the vertex set *V* of *G*. Out of the all possible subsets the one which has the minimum cordinately is called partitioning of graph *G*. In other words each *n*-colouring of graph *G* partitions *V*(*G*) into *n* -independent sets called **colour classes**. Such a partitioning induced by χ (*G*) colouring of graph *G* is called **chromatic partitioning**.

Consider the graph shown in the Fig. 6.24.

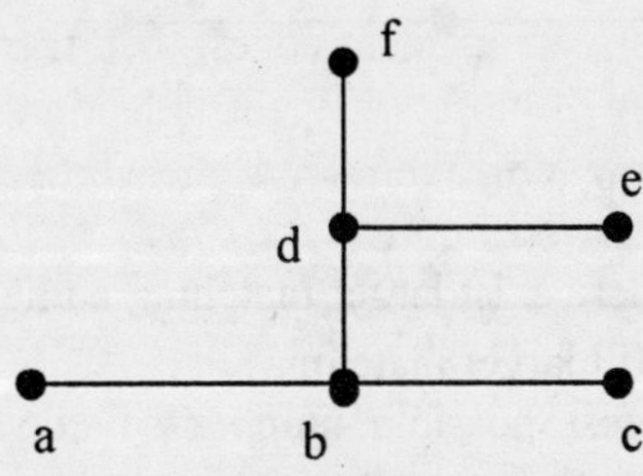

Fig. 6.24

Independent sets that are partitions of graph *G* are

$\{a, f\}$, $\{b, e\}$ and $\{d, c\}$
$\{a, c, e, f\}$, $\{b\}$and $\{d\}$
$\{a, c, d\}$ and $\{b, e, f\}$

Out of these independent sets $\{\{a, c, d\}, \{b, e, f\}\}$ contains only two sets and this is the set having minimum coordinately. Therefore, it is chromatic partition of *G*. Since this set contains two elements hence the chromatic number of this graph is two. The graph that has only one chromatic partition is called **uniquely colourable**.

Example 6.19

Find the chromatic partitioning of graph G shown in Fig. 6.25.

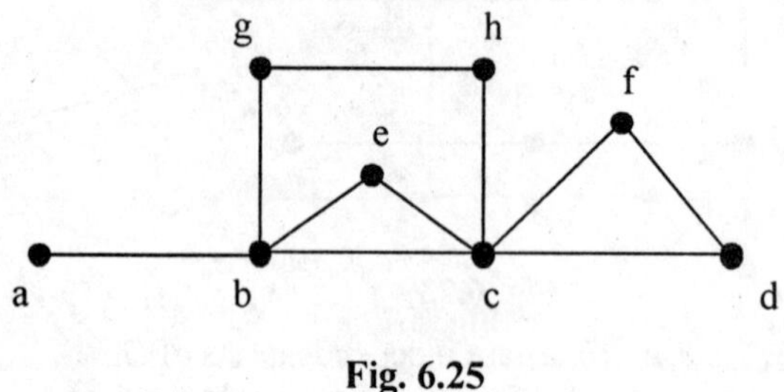

Fig. 6.25

Solution

$\{a, c, g\}$, $\{b, f, h\}$and $\{e, d\}$ is a chromatic partition of graph G which has chromatic number 3.

6.16 MATCHING

Matching problems arise in many applications. For example assignment problems involving resource allocation, balancing the traffic load among servers on the Internet, dating services want to pair up compatible couples etc. Matching M in a graph G is a set of pairwise non-adjacent edges. A single edge in a graph G is obviously a matching. A vertex is said to be **matched** if it is incident to an edge in the matching, otherwise vertex is unmatched. For example in the graph shown in Fig. 6.26,

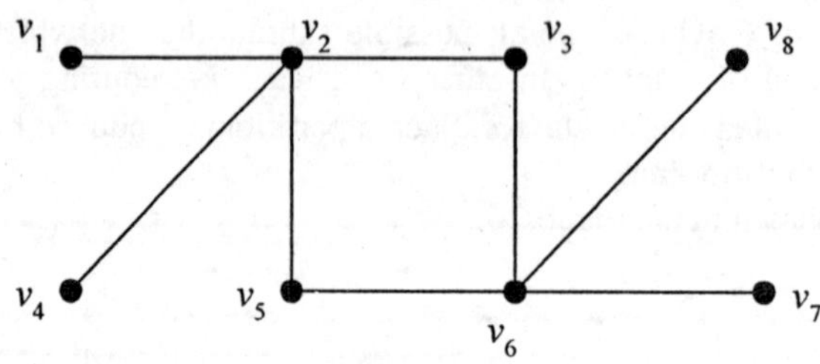

Fig. 6.26

$\{v_2v_5, v_3v_6\}$ is a matching of size two. But there exist larger matching namely: $\{v_1v_2, v_3v_5, v_6v_7\}$ which is a matching of size three. Can there exist still larger matching?

A matching M that contains largest possible number of edges is called a **maximum matching**. In other words a matching of a graph G which is not proper subset of any matching of the graph G is called a maximum matching. We can have more than one maximum matchings with different size.

A matching M of a graph G with the property that if any edge not in M is added to M, it is no longer a matching is called a **maximal matching**. For example in a complete graph of three vertices (triangle) any single edge is a maximal matching. It is to be noted that every maximum matching is maximal, but every maximal matching is not a maximum. A graph may have many different maximal matchings.

In other words any set M of independent lines of graph G is called a matching of G. If $uv \in M$, we say that u and v are M-saturated. A matching M is called **perfect matching** if every vertex of G is M-saturated, Every perfect matching is both maximum and maximal.

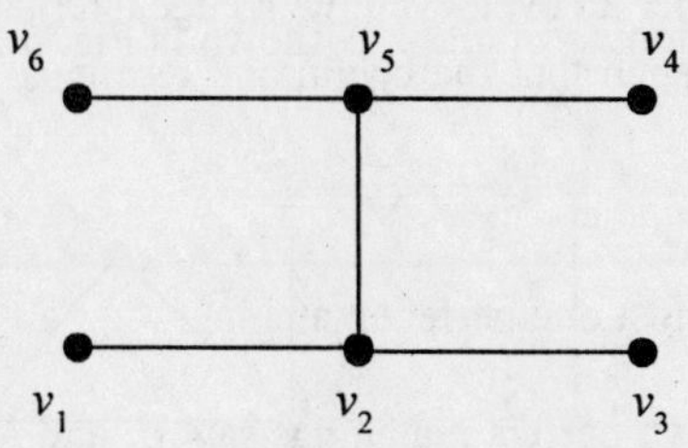

Fig. 6.27

Consider the graph shown in Fig. 6.27

Consider the sets $\{v_1\ v_2\}$, $\{v_2\ v_3\}$, $\{v_2\ v_5\}$, $\{v_4\ v_5\}$, $\{v_5\ v_6\}$, $\{v_1v_2,\ v_4v_5\}$, $\{v_2v_3,\ v_5v_6\}$, $\{v_1v_2,\ v_5v_6\}$ and $\{v_2v_3,\ v_4v_5\}$. We can observe here the matching $\{v_1v_2\}$ is a proper subset of the matching $\{v_1\ v_2,\ v_4\ v_5$. Hence $\{v_1\ v_2\}$ is not a maximal matching. But $\{v_2\ v_5\}$, $\{v_1\ v_2,\ v_4v_5\}$, $\{v_2\ v_3,\ v_5\ v_6\}$ and $\{v_2\ v_3,\ v_4v_5\}$ are not a proper subset of any other matching of the graph G. Hence these are maximal matchings.

If M be a matching of a graph G then a path in which the edges belong alternatively to the matching is called an **alternating path**. An alternating path whose origin and terminus both unsaturated is called an **augmenting path**. For the graph shown is Fig.6.28 (G_1) $P_1 = \{v_3, v_4, v_5, v_6\}$ an alternating path in G_1.

Consider the graph G_1 shown in Fig. 6.28. The matchings in G_1 are

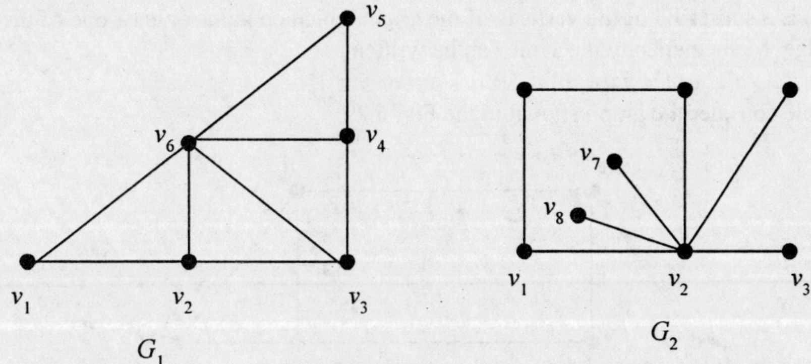

Fig. 6.28

$M_1 = \{v_1v_2,\ v_3v_6,\ v_4v_5\}$ and $M_2 = \{v_2v_6,\ v_4v_5\}$ are matchings in G_1. But M_2 is not a perfect matching. The vertices v_2 and v_6 are not M_2-saturated. For graph G_2, $M = \{v_2v_3,\ v_1v_6\}$ is a maximum matching but not a perfect matching.

THEOREM 6.14

A matching M in a graph G in maximum matching if and only if G contains no M-augmenting path.

THEOREM 6.15

Let M_1 and M_2 be two matchings in graph G. Let $M_1\ \Delta\ M_2 = (M_1 - M_2) \cup (M_2 - M_1)$ be symmetric difference of M_1 and M_2. Let $H = G\ (M_1 \Delta M_2)$ be the subgraph of G induced by $M_1 \Delta M_2$.

Then each component of H is either an even cycle with edge alternately in M_1 and M_2 or a path P with edges alternately in M_1 and M_2 such that the origin and terminus of P are unsaturated in M_1 or M_2 augmenting path.

Example 6.20

Show that a tree has at most one perfect matching.

Solution

Consider a tree T with two perfect matchings say M_1 and M_2. Then degree of every vertex in $H = T(M_1 \Delta M_2)$ is 2. Thus every component in H is an even cycle with edges alternately in M_1 and M_2. But this contradicts the fact that tree T has no cycle. Hence T has at most one perfect matching.

THEOREM 6.16

Let G be a k-regular bipartite graph with $k > 0$. Then G has a perfect matching.

THEOREM 6.17

A complete matching of the vertex set V_1 into V_2 in a bipartite graph exists if and only if every subset of v vertices in V_1 is collectively adjacent to V or more vertices in V_2 for all values of V.

6.17 COVERING

Consider a graph G. A covering of graph G is a set of vertices (or edges) so that elements of the set are adjacent to all edges (or vertices) of the graph. Specifically a vertex cover of a graph $G = (V, E)$ is a subset V' of the vertices of the graph which contains at least one of the two end points of each edge. Mathematically the same can be written as

$$V' \subseteq V : \forall \{a, b\} \in E : a \in V' \text{ or } b \in V'.$$

For example consider the graph shown in the Fig. 6.29.

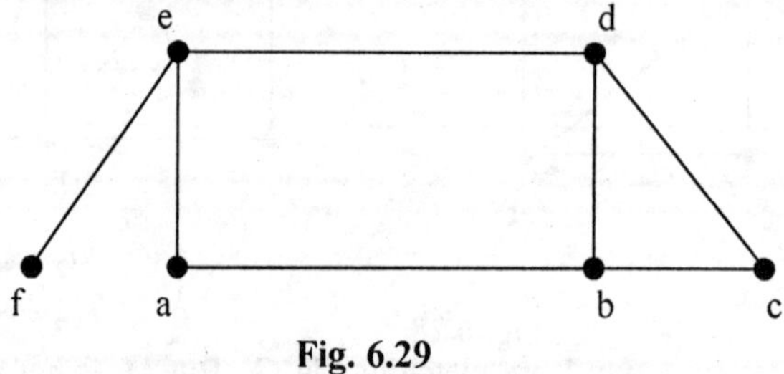

Fig. 6.29

The vertex sets $\{c, a, d, f\}$ and $\{b, e, d\}$ are examples of vertex cover. For any graph G the set of its vertices is trivially a vertex covering.

The minimum vertex covering is the smallest vertex cover. The vertex covering number $w_v(G)$ for a graph G is the size of the minimum edge covering. It is to be noted that when we speak covering we usually refer to vertex covering.

An edge covering of graph G is the set of edges E so that every vertex of G is adjacent to at least one edge E. The minimum edge covering is the smallest edge cover. This is also said to be E covers the vertices of the graph. The edge covering number $w_E(G)$ for a graph G is the size of the minimum edge covering.

A covering of a graph such that no proper subset of it is a cover of G is called **minimal covering** of G.

For any graph G, its set of edges is trivially an edge covering.

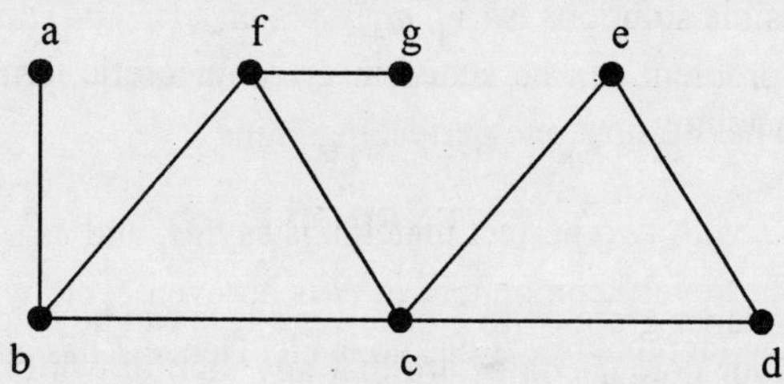

Fig. 6.30

In the graph G shown in Fig.6.30 the edge sets $\{gf, ab, ce, de\}$, $\{gf, ab, cf, de\}$, $\{gf, ab, ce, cd\}$ are edge covering of G. These coverings are minimal coverings of G. The smallest number of edges in a covering of G is four. Hence edge covering number $w_E(G)$ of this graph is four.

A complete bipartite graph $k_{m,n}$ has $w_V(k_{m,n}) = \min\{m, n\}$ and $w_E(k_{m,n}) = \max\{m, n\}$.

OBSERVATIONS

1. Every pendent edge in a graph is included in every covering of the graph.
2. A covering exists for a graph if and only if the graph has no isolated vertex. Thus covering exists for every connected graph.
3. A covering of a n-vertex graph will have at least $(n/2)$ edges.
4. Every covering contains minimal covering.
5. No minimal covering can contain a circuit. Hence, a minimal covering of an n-vertex graph can contain not more than $n - 1$ edges.
6. The edge covering number of a complete graph on n-vertex is $\frac{n+1}{2}$.

6.18 MARRIAGE PROBLEM (Hall's marriage problem 1935)

Marriage problem basically requires to match girls with the set of n boys. Each woman would happily marry some subset of the men, and any man would be happy to marry a woman who wants to marry him. Since polygamy and polyandry is not allowed, every woman can be married to at most one man, and every man to at most one woman. The problem whether it is possible to pair up the man and women so that every person is happy is called **marriage problem**.

For example, consider there are four unmarried women $\{w_1, w_2, w_3, w_4\}$ and five unmarried men $\{m_1, m_2, m_3, m_4, m_5\}$. Their friendship is shown in the following table.

Women	Men		
w_1	m_1	m_4	m_5
w_2	m_1		
w_3	m_2	m_3	m_4
w_4	m_2	m_4	

This situation can be represented by bipartite graph in which the vertex classes are the set of four women and five men, and a women $w_{i;}$ $i = 1, 2, 3, 4$ is joined by an edge to a men m_j, ($j = 1, 2, 3, 4, 5$) if they like each other.

One of the possible solutions is (w_1, m_4), (w_2, m_1), (w_3, m_3) and (w_4, m_2).

The marriage problem can be stated in graph-theoretic terms as asking if a given bipartite graph G has a perfect matching.

EXERCISE 6.2

1. Show that any planar graph with 2 or fewer edges can be coloured by four colours.
2. Deduce from four colour conjecture that any map drawn on the surface of a sphere can be coloured using four colours (or fewer).
3. If G has a Hamiltonian cycle then show that regions of plane graph can be 4-coloured, [**Hint** Hamiltonian cycle divides the regions into interior and exterior one].
4. Find the chromatic polynomial of graph in Fig. 6.31.
5.

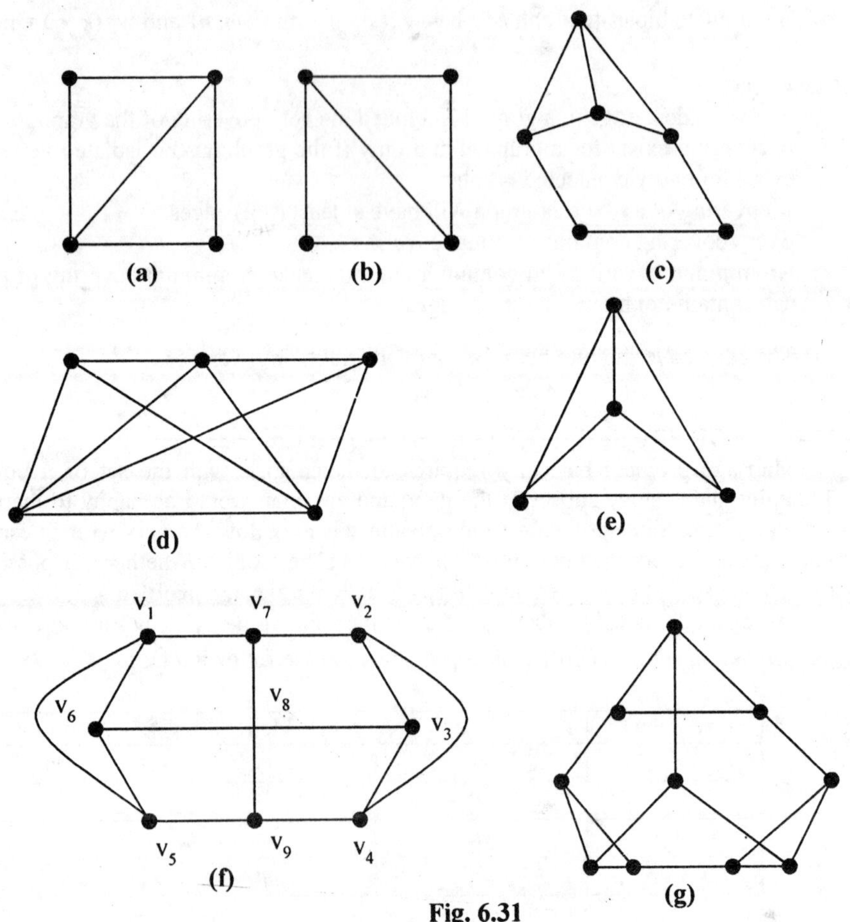

Fig. 6.31

6. If δ is the minimum number of degrees of any vertex in G with m vertices then,

$$\chi(G) \geq m/(m - \delta(G)).$$

[**Hint** Since $\delta(G)$ is the minimum number of degree of v, then at most $(m - \delta(G))$ vertices will be having the same colour as v. If we place vertices of same colour in one class then there are exactly $\chi(G)$ classes. Each class contains at most $m - \delta(G)$ or $\chi(G).(m - \delta(G)) \geq m$ or $\chi(G) \geq m/(m - \delta(G))$].

7. How many colours are required to colour the map shown in Fig. 6.32 and what shape it will have in 3D?

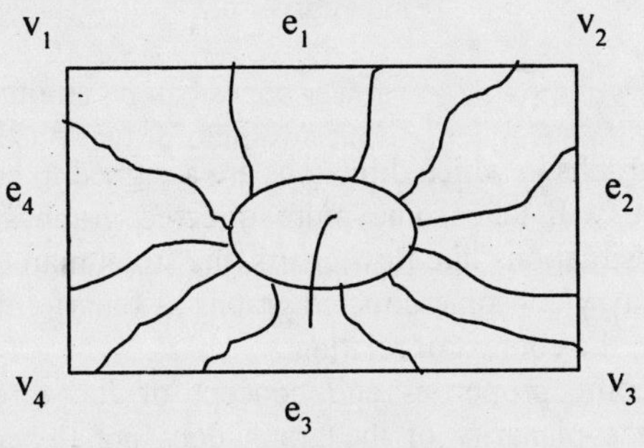

Fig. 6.32

[**Hint** Torus (ring or quirt) may be formed from a rectangular piece of sheet shown in Fig. 6.32 by first joining the edges e_1 and e_3 and then e_4 and e_2].

8. Prove that a 2-regular graph G has a perfect matching if and only if every component of G is an even cycle.

9. Give an example of 2-regular graph, which has no perfect matching.

7 DIRECTED GRAPHS

7.0 INTRODUCTION

Directed graphs are graphs in which directions are assigned to edges as it was appeared in the chapter 2. In this chapter we will take some more directed graphs. Such graphs are frequently encountered in many physical situations like flow charts, the street map of a city with one way streets, electrical network, etc. One may view some directed graphs as binary relations. For this reason, some texts discuss directed graphs in the context of relation.

Most of the definitions, properties and concept of the directed graphs are similar to undirected graphs. For instance planarity of the graph does not depend on whether the *graph is directed or undirected.* In this chapter we shall study mainly those properties of directed graphs, that are not shared by undirected graphs.

7.1 DIRECTED GRAPH OR DIGRAPH

A directed graph (or digraph) G consists of a vertex set $V(G)$, an edge set $E(G)$ and a mapping f that maps every edge onto ordered pair of vertices (v_i, v_j). We write $v_i v_j$ for the edge (v_i, v_j) with v_i being the tail and v_j being the head. Directed edges are shown by means of arrows diagrammatically. In the case of directed edges, the two possible edges between a pair of vertices which are opposite in direction are considered distinct. If pair of vertices joined by more than one edges are in the same direction, such edges are called **parallel edges**. Two vertices v_i and v_j of a digraph are adjacent if there is an edge between v_i and v_j of the form $v_i v_j$ or $v_j v_i$. The vertices v_i and v_j **are incident** to such an edge. A directed graph is said to be **finite** if it has finite number of edges and vertices. For example the graph shown in Fig.7.1is finite. It has four vertices and eight edges. If the vertices and/or edges of a directed graph G are labeled (with some type of data), then G is called a **labeled directed graph**.

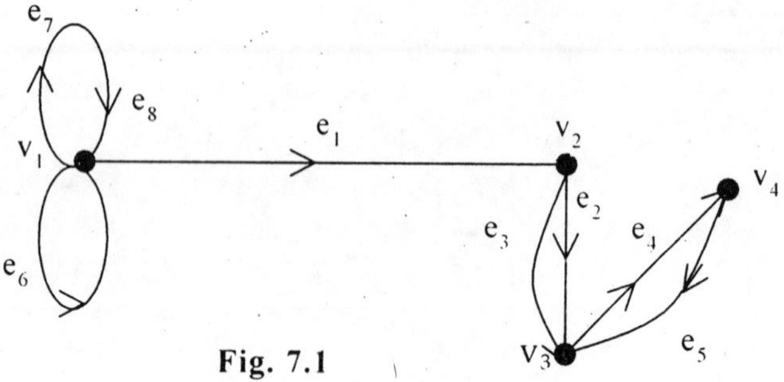

Fig. 7.1

7.2 DEGREE OF A DIRECTED GRAPH

In a directed graph for any vertex v_i the number of edges which have v as the initial vertex is called **outdegree** of the vertex v denoted by $d^+(v)$. The number of edges which have v as **their** terminal vertex is called the **indegree** of v denoted by $d^-(v)$. A loop at a vertex contributes 1 to the indegree and 1 to the outdegree of the vertex. For example, consider the graph shown in the Fig.7.2, here

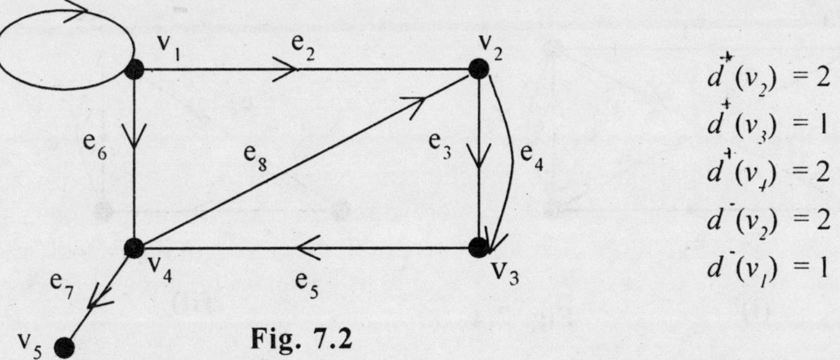

Fig. 7.2

A vertex with zero indegree is called a **source** and a vertex with zero outdegree is called a **sink**.

THEOREM 7.1

Show that the sum of the indegrees of all the vertices of a simple directed graph G is equal to the sum of the out degrees of all its vertices and this sum is equal to the number of edges of the graph.

PROOF

Consider a directed graph $G = (V, E)$. Let for each indegree of a vertex v_i, there corresponds an outdegree of the vertex, say v_j i.e. $v_i\, v_j$ is an edge. Hence applying the same logic to each vertex, we observe that the sum of all indegrees of all the vertices of a graph is equal to sum of outdegrees of all the vertices of the graph. It is clear now that each edge contributes to one indegree and one outdegree to the sum of degrees, and the number of edges is obviously half of the sum of all the indegrees and outdegrees.

$\therefore$ Sum of the indegrees = Sum of the outdegree = Number of edges of any directed graph.

7.3 SIMPLE DIRECTED GRAPH

A graph G is said to be **simple digraph** if the edges of G are all distinct and there is not loop. In other words in a simple directed graph each ordered pair of vertices occurs one and only one time as an edge. For instance graph shown in Fig.7.3(a) is a simple directed graph.. It is to be noted that underlying graph of a simple digraph need not be simple graph. For example consider the graph shown in Fig.7. 3 (b) and (c). Here 7.3 (b) is simple diagraph whereas 7.3(c) is not a simple graph.

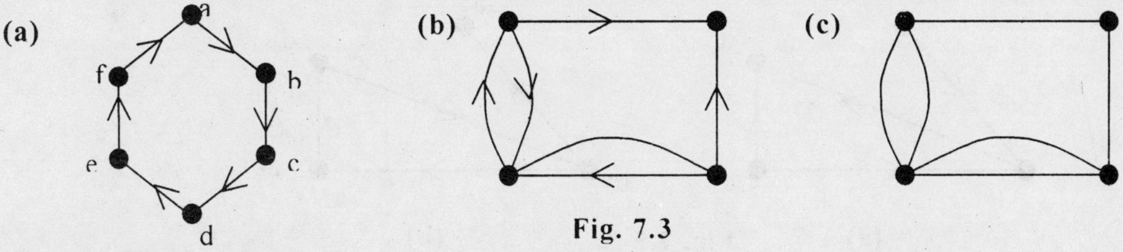

Fig. 7.3

7.4 SUBGRAPHS

Consider a digraph $G = (V, E)$. A directed graph $g = (v, e)$ is said to be a subgraph of G if e is a subset of E and v is a subset of V. For example consider the graph shown in Fig. 7.4. Here, graph (ii) is the subgraph of (i) where $v = \{b, c, d\}$ and $e = \{e_2, e_3, e_5, e_7\}$.

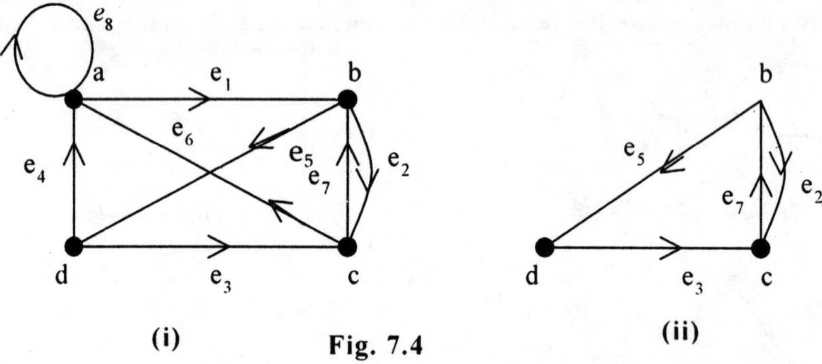

Fig. 7.4

7.5 SIMPLE COMPLETE DIGRAPH

A simple digraph $G(V, E)$ is said to be **complete** if every vertex is adjacent to all other vertices of the graph.

Example 7.1

Show that a simple complete digraph with n vertices has the maximum number of edges viz., $n(n - 1)$ edges, assuming that there is no loop.

Solution

Since any vertex out of n vertices can be chosen in $C(n, 1) = n$ ways. Let us take a vertex v_1. Since each vertex is connected to every other vertices, here vertex v_1 is connected to vertices, say v_2, v_3, v_4, . . . , v_n by an edge in $(n - 1)$ ways. As there are n vertices, there will be $C(n - 1)$ edges. As the graph taken is simple without any loop, hence $n(n - 1)$ is the maximum number of possible edges.

REMARK

If complete graph contains n loops then there will be $n + (n - 1)n = n^2$ edges considering each loop as a single edge.

7.6 CONVERSE OF A DIRECTED GRAPH

The converse of a directed graph $G = (V, E)$ is denoted by G^c, is again a directed graph (V, e) where an edge $(u, v) \in e$ if and only if $(v, u) \in E$. For example graph shown in Fig. 7.5(b) is converse of graph shown in Fig.7.5 (a).

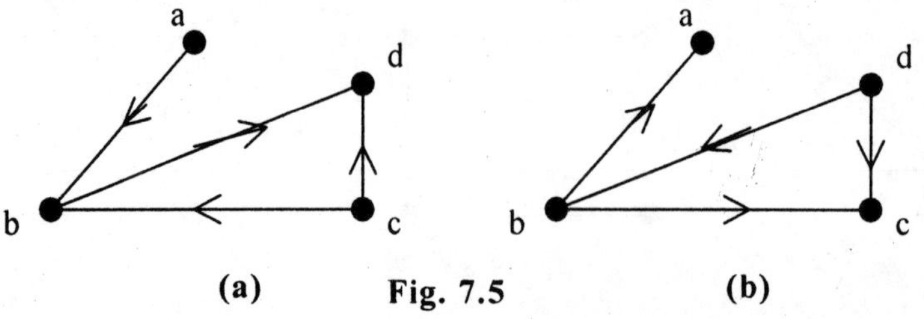

Fig. 7.5

Example 7.2

Find the converse graph of the following graphs.

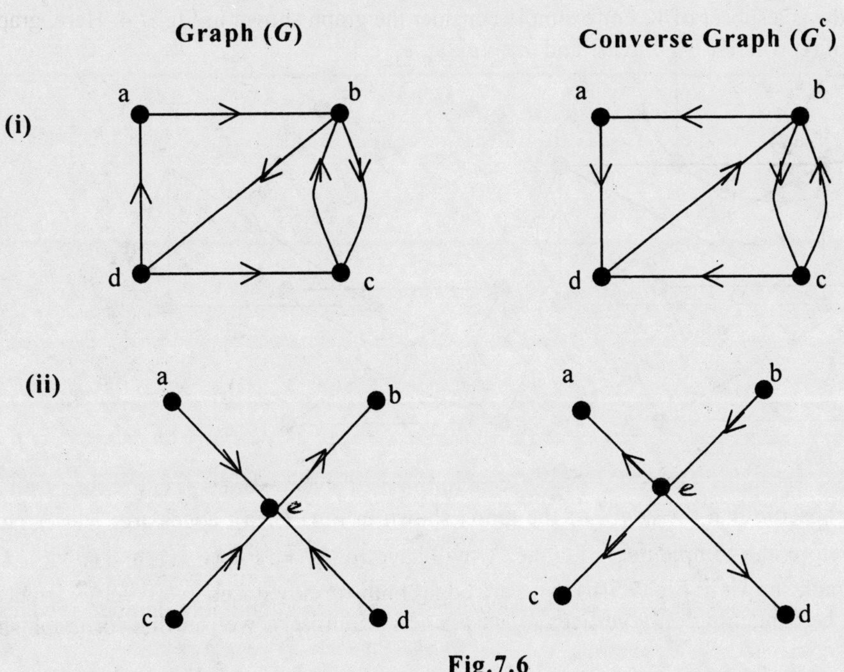

Fig.7.6

7.7 COMPLEMENT OF A SIMPLE DIRECTED GRAPH

The complement of simple digraph $G = (V, E)$ is the directed graph $\bar{G} = (V, \bar{E})$, where $\bar{E} \subset V \times V - E$, i.e. an edge exist in $\bar{G}$ if and only iff it does not exist in G.

Example 7.3

Find the complement of the following graphs.

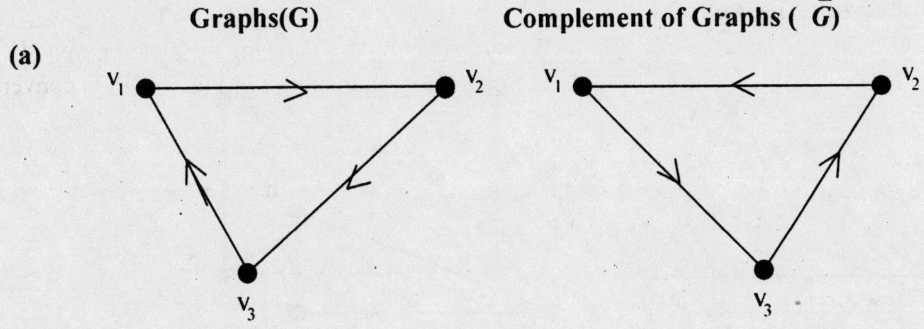

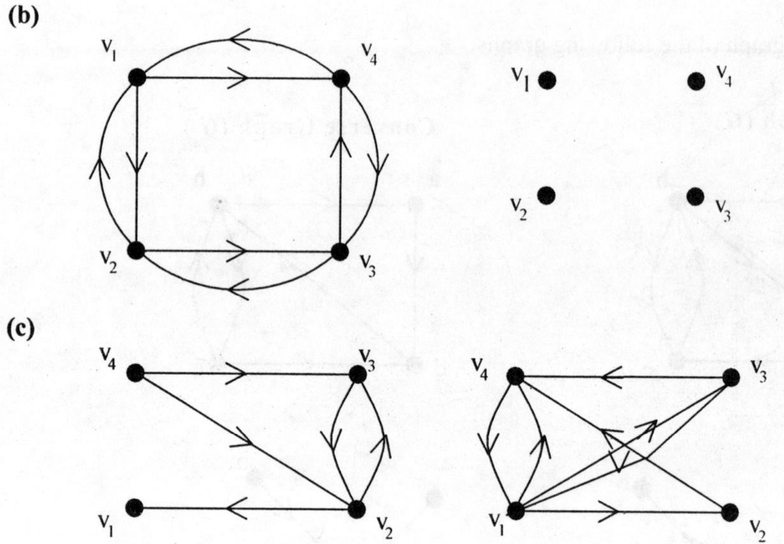

Fig. 7.7

We can observe that the edges of the graph shown in Fig. 7.7(a), edges are (v_1, v_2), (v_2, v_3) and (v_3, v_1). Therefore, the complement of G, i.e. $\bar{G}$ will have edges (v_2, v_1), (v_3, v_2) and (v_1, v_3).

In the graph shown in Fig. 7.7(b) there are edges both ways between every vertex. Hence the complement will be containing only vertices v_1, v_2, v_3 and v_4. Similarly, we can find for graph shown in Fig. 7.7 (c).

7.8 TOURNAMENT GRAPHS

(i) A **tournament graph** (or tournament) is a directed graph with the property that no edge connects a vertex to itself and between any two vertices there exist one and only one edge. Graphs shown in the Fig.7.8 (a), (b) and (c) are tournament graphs whereas Graph shown in Fig.7.8 (d) is not a tournament graph.

(ii) A tournament where every team play with each other teams exactly once is called **round robin tournament**. This situation can be expressed in terms of directed graphs. A tournament graph with the property that between any two distinct vertices there is exactly one edge is called a **complete tournament (or round robin) graph**.

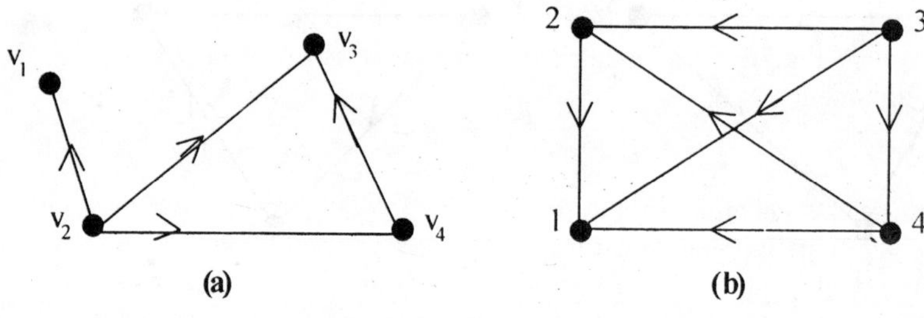

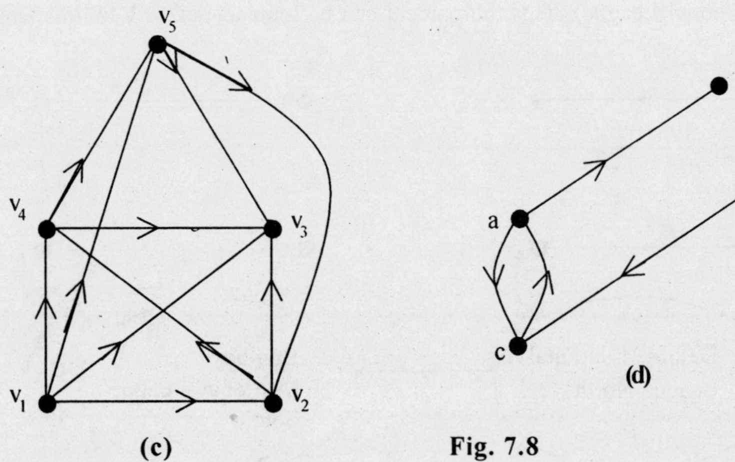

Fig. 7.8

Graphs shown in Fig.7.8(b) and (c) are complete tournament graphs. It is to be noted here that (a, b) is an edge if team a beats team b.

(iii) A tournament where a team which is defeated by another team is knocked out of the tournament and the winning team plays with another team and again out of these two teams, the winner precedes in the tournament whereas defeated team getting out of the tournament is called a *single elimination tournament.* **A single-elimination tournament graph** is again a tournament graph with the following properties:

(a) One vertex has no edge terminating at it (this vertex is called **champion vertex**) and at least one edge initiating from it.

(b) Every other vertex is the terminal vertex of exactly one edge.

(c) There is a path from the champion vertex to every other vertex.

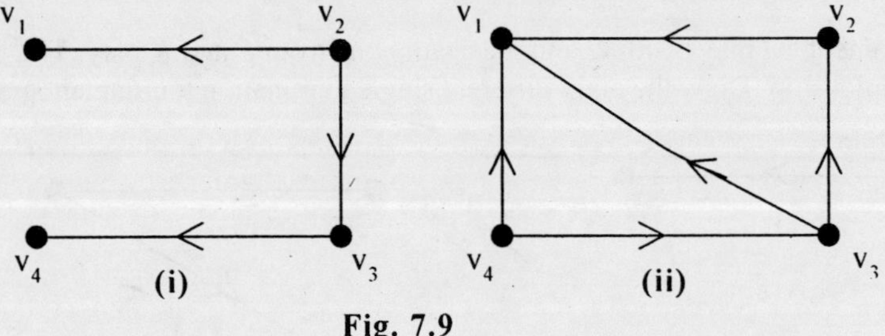

Fig. 7.9

Graphs shown in Fig.7.9 (i) is a single elimination tournament graph whereas (ii) is not a single elimination tournament graph.

Example 7.4

In a National cricket championship it is decided to go with the single elimination tournament. The two divisional champions in the national league (North and south) compete in a series of games. The loser is eliminated while winner competes against the winner of the national league series (which will be decided in the National League). This tournament can be represented with tournament graph as shown in the Fig. 7.10.

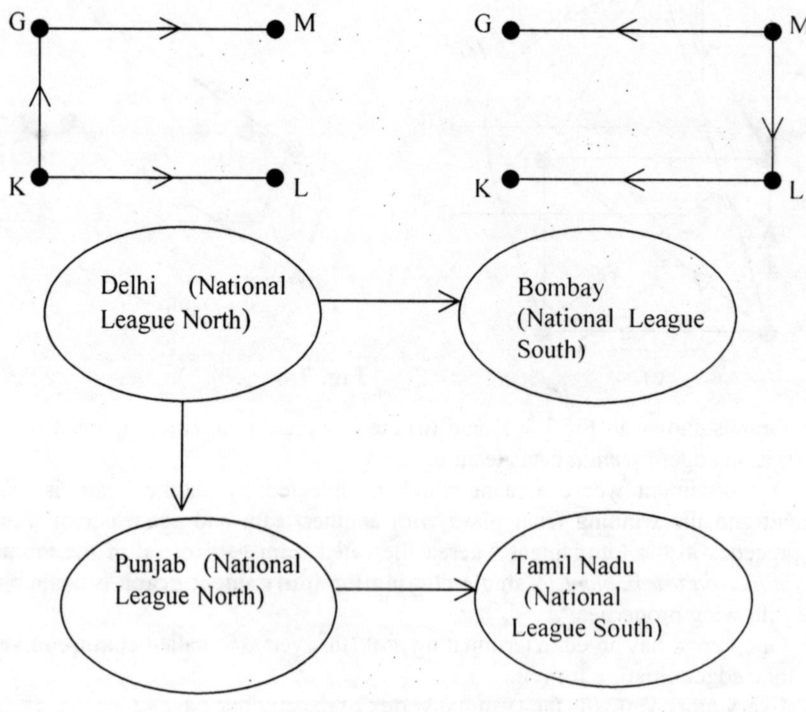

Fig. 7.10

Example 7.5

In a final tournament, M plays with G and the champions from L and K play. The winners of the games play for championship. Draw the eight different single elimination tournament graphs.

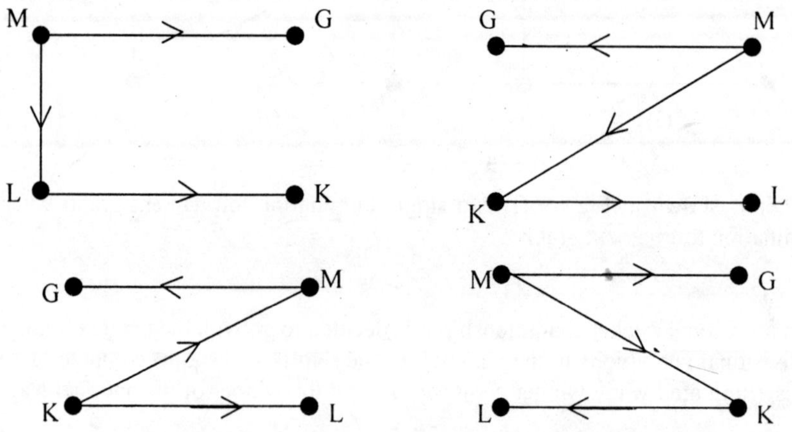

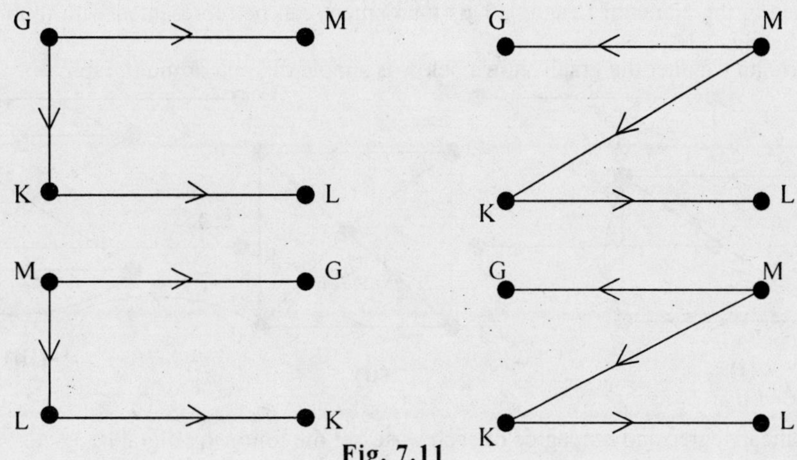

Fig. 7.11

Example 7.6

How many edges does a single tournament graph with n vertices have?

Solution

Since the only one champion vertex has no indegree and every other vertex has exactly one outdegree therefore, it is $(n - 1)$.

Example 7.7

How many edges, does a complete tournament graph with n vertices have?

Solution

As two vertices can be selected from n vertices in $C(n, 2)$ ways and since there is exactly one edge between any two distinct vertices, hence it can be done in $n.(n - 1)/2$ ways.

7.9 LINE DIGRAPH AND LINE GRAPH

A **line digraph** or an **edge digraph** written $L(G)$ of a digraph is defined as follows:
There is exactly one vertex v_i in $L(G)$ for every edge e_i in G.

(i) Whenever edges e_i and e_j (for a self-loop $e_i = e_j$) are such that e_i is incident into a vertex v and e_j is incident out of the same vertex v, an edge is drawn from the corresponding v_i to v_j in $L(G)$.

Example 7.8

Consider the graphs shown in the Fig. 7.12. Here in the graph the edges e and f share a vertex; for digraphs the head of e must be the tail of f.

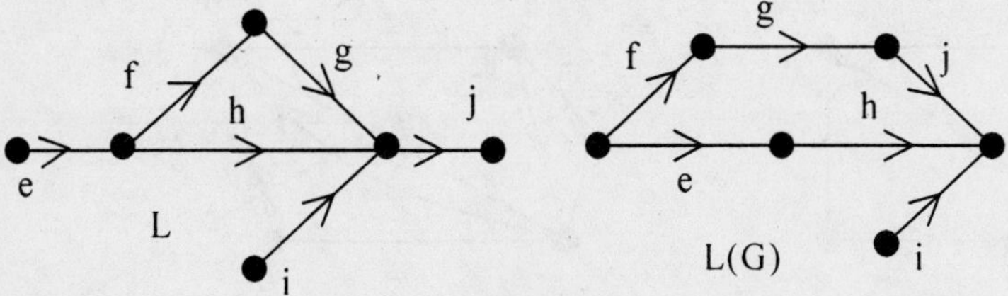

Fig. 7.12

EXERCISE 7.1

1. Determine whether the graph shown below is simple digraph or multigraph.

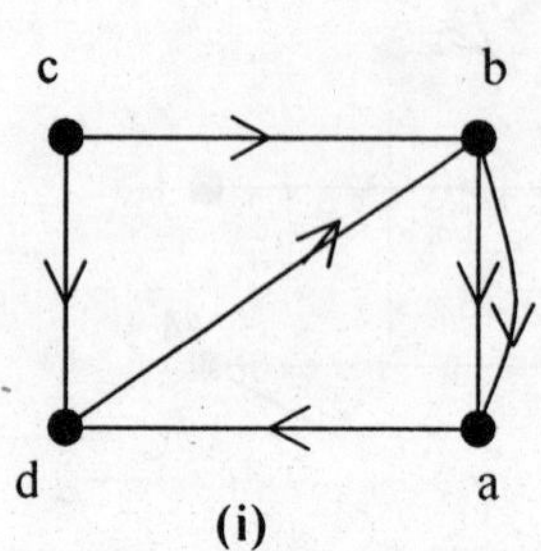

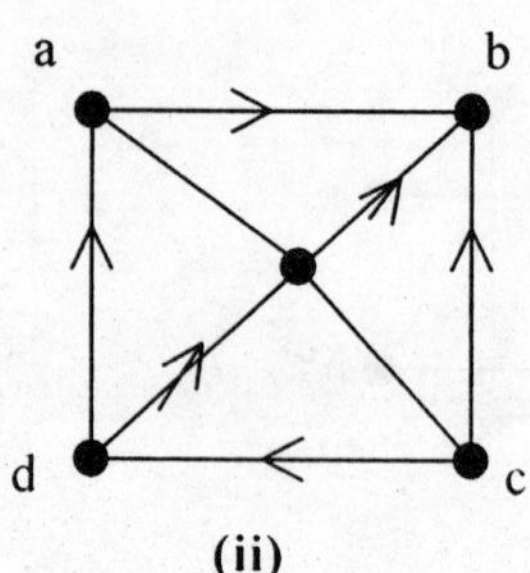

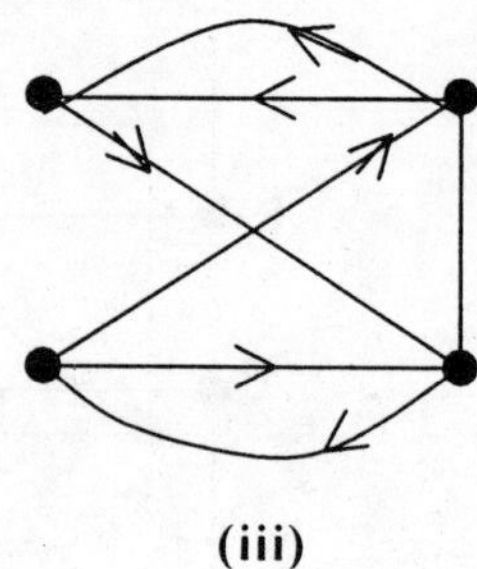

2. Find the indegree and outdegree of each vertex of the following digraphs.

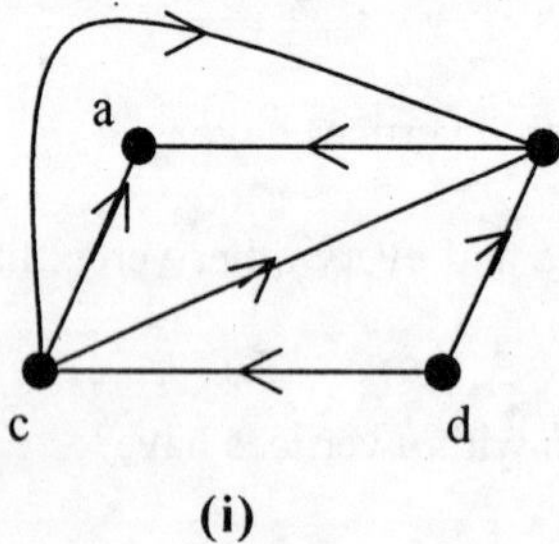

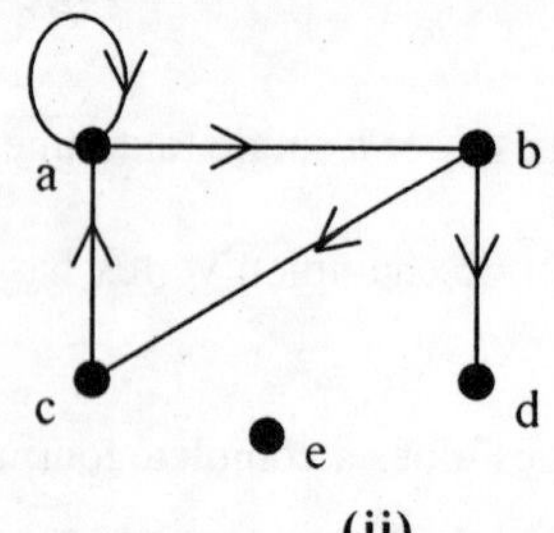

Is there any source or sink?

3. Find the subgraph of the following graph determined by vertex set $v = \{a, b, c\}$.

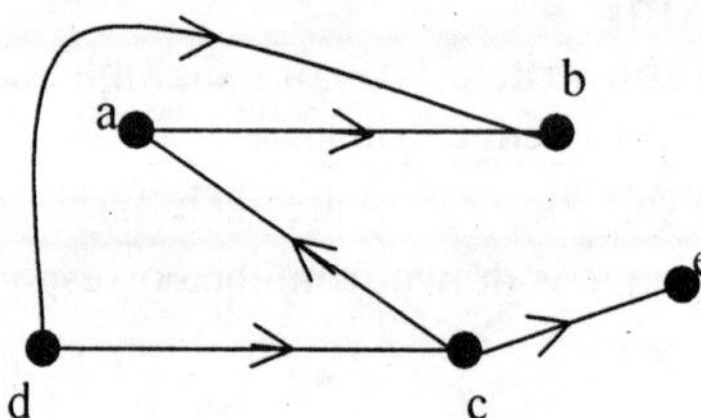

4. Find the converse graph of the following graphs:

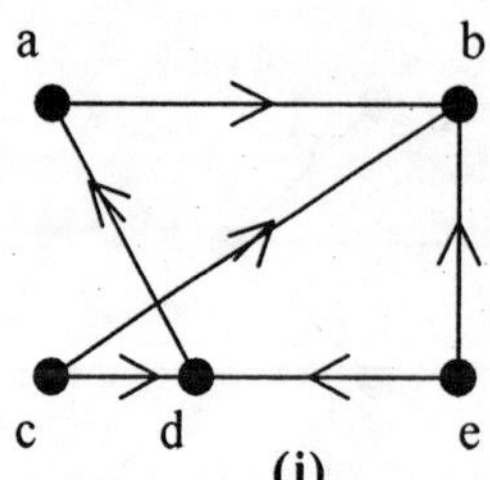

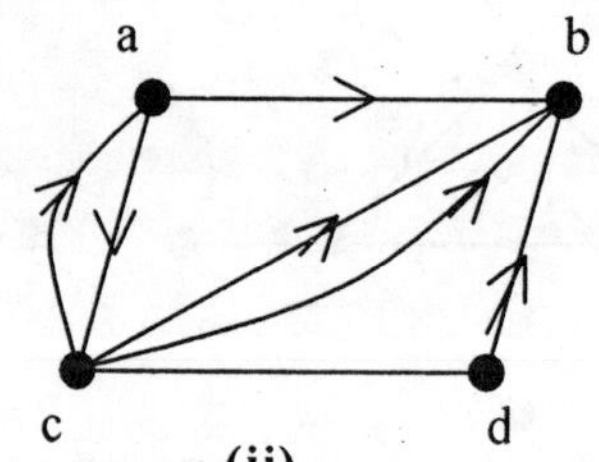

5. Find the complement digraph of the following simple directed graphs :

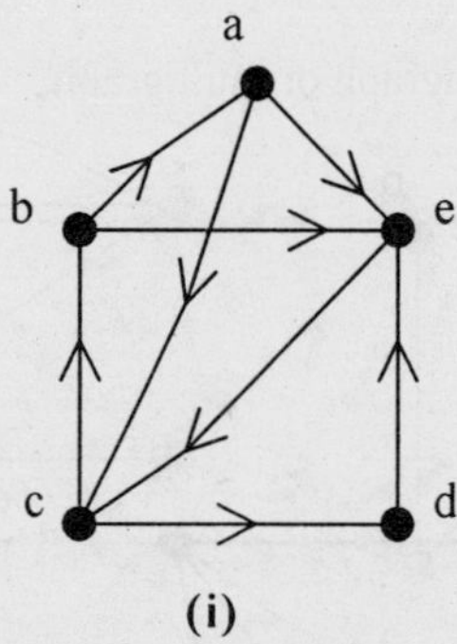

(i)

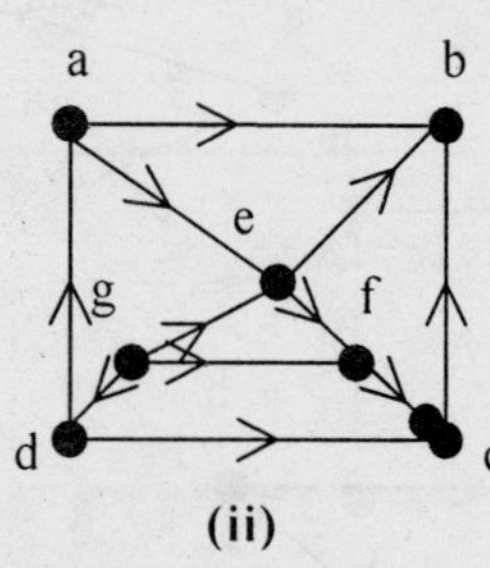

(ii)

6. How many different tournaments are there with *n* vertices?
7. What is the sum of the in-degree and out-degree of a vertex in a tournament graph?
8. How many edges does a single elimination tournament graph with *n* vertices have?
9. How many edges does a complete tournament graph with n vertices have?
10. Find the indegrees and outdegrees of the vertices v_4 and v_5 of the following graph.

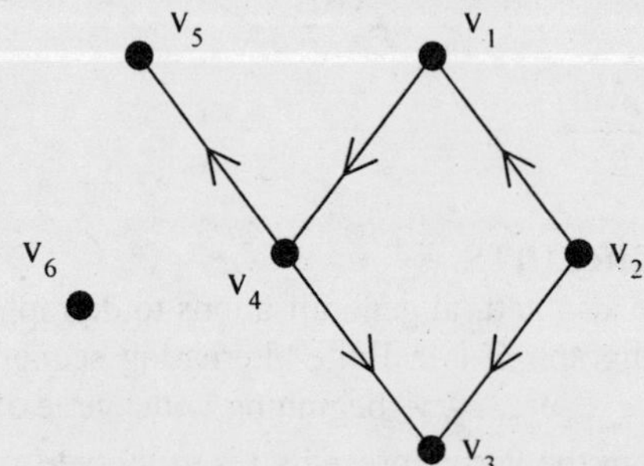

11. (a) How many edges does a single elimination tournament graph with *n* vertices have?
(b) How many edges does a complete tournament graph with *n* vertices have?

ANSWERS

1. (i) Multigraph
(ii) Simple digraph
(iii) Simple
2. (a) (i) $d^{+}(a) = 0,\ d^{-}(a) = 2,\ d^{+}(b) = 1,\ d^{-}(b) = 3$
$d^{-}(d) = 0,\ d^{+}(d) = 0,\ d^{+}(d) = 2,\ d^{-}(c) = 1,$
$d^{+}(c) = 3$, Source is *d* and sink is *a*.
(ii) $d^{+}(a) = 2,\ d^{-}(a) = 2,\ d^{+}(b) = 2,\ d^{-}(b) = 1,$
$d^{+}(c) = 1,\ d^{-}(c) = 1,\ d^{+}(d) = 0,\ d^{-}(d) = 0,$
$d^{+}(e) = 0,\ d^{-}(e) = 0$

3.

4.

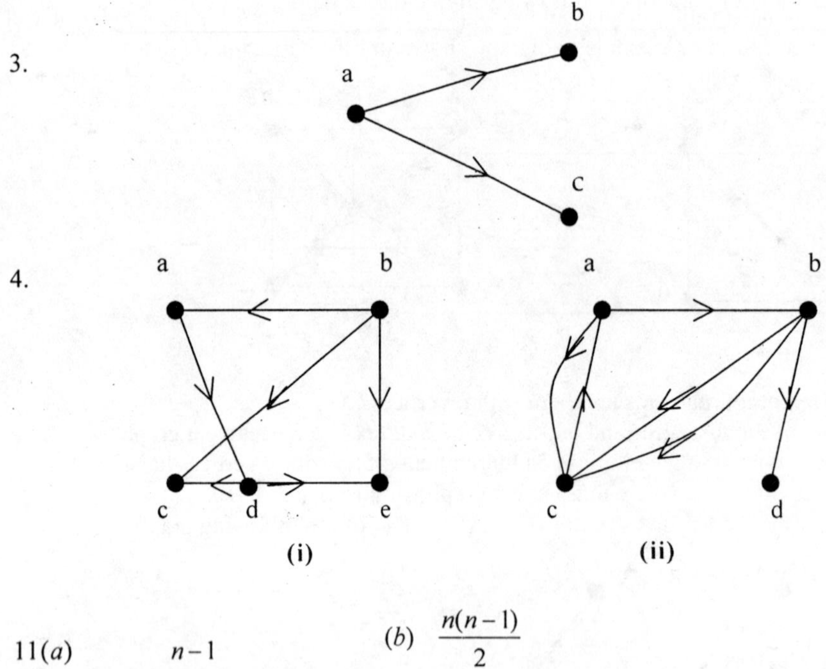

11(*a*) $n-1$ (*b*) $\frac{n(n-1)}{2}$

7.10 DIRECTED WALK, PATHS AND CIRCUITS

Directed walk, paths and circuit are also natural generalisations to digraphs of the definitions given in chapter 2. A **directed walk** in a digraph G is a finite alternating sequence of vertices and directed edges of the form $v_0, e_1, v_1, e_2, \ldots, e_{n-1}, v_{n-2}, e_n, v_n$ beginning with vertex v_0 and ending with vertex v_n such that each edge is oriented from the vertex preceding it to the vertex following it. This alternating sequence is called a **directed walk** from v_0 to v_n. For example in digraph shown in Fig. 7.13, $a, e_2, b, e_6, d, e_7, a, e_4, b$ is a walk.

A **semi walk** in a digraph is a walk in the corresponding undirected graph, but it is not a directed walk. For example in digraph shown in Fig. 7.13, $d, e_6, b, e_3, a, e_2, b$ is a semi walk and not a directed walk as the edge e_6 is not oriented from d to b but from b to d.

An **open directed walk** is a directed walk whose initial and final vertices are distinct. A **closed directed walk** is a directed walk whose initial and final vertices are same.

In the analogous way as defined for undirected graphs, we can define **circuit, directed trails, directed path** and **directed cycle**.

A **circuit** in a directed graph is a closed directed walk in which no directed edge is repeated, it is a close directed trail.

A **directed path** or path in a directed graph is a trail with no repeated vertex. It is an open directed walk in which neither any directed edge nor any vertex is repeated. A **cycle** in a digraph is a directed circuit in which no vertex is repeated except the initial vertex that is also the end vertex of the sequence. For example in digraph shown in Fig. 7.13, cycle: $d, e_7, a, e_2, b, e_6, d$ is a directed cycle and $d, e_6, b, e_3, a, e_7, d$ is a semi cycle.

It is to be noted that a trail cannot contain a given edge ($a\ b$) more than once, but it can contain both edges (ab) and (ba). For example in digraph shown in Fig. 7.13, *cbabd* is a trail.

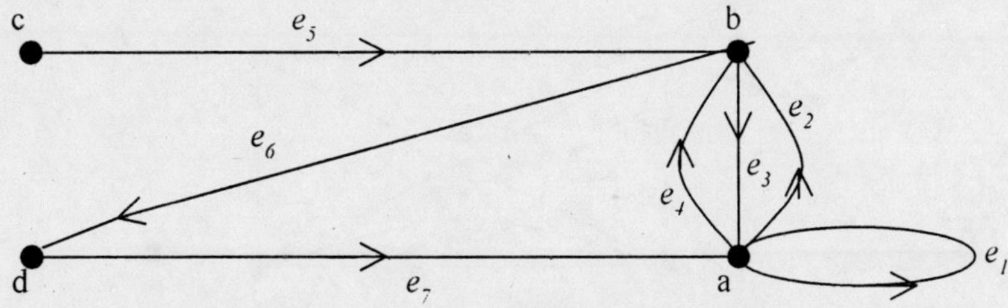

Fig. 7.13

A path in a digraph is said to be **simple** if all the edges are distinct. If in a path all the nodes through which it travels are distinct, then it is said to be **elementary**. A cycle is said to be **simple cycle** if its path is simple.

An elementary cycle is a cycle if its path is elementary (but the initial node is the only which appears twice).

7.11 LENGTH OF A PATH

The number of edges appearing in the sequence of a path is called the **length of the path** viz. an edge if its not a self loop is a path of length one. The length of a path is one less than the number of vertices in the path.

Example 7.9

In the graph shown in the Fig.7.14 some of the paths originating in vertex a and ending in vertex c are

$P_1 = (ab, bc)$
$P_2 = (ad, dc)$
$P_3 = (ab, bd, dc)$
$P_4 = (ab, bd, da, ab, bc)$
$P_5 = (ab, bd, da, ad, dc)$
$P_6 = (aa, aa, \ldots\ldots, ab, bc)$

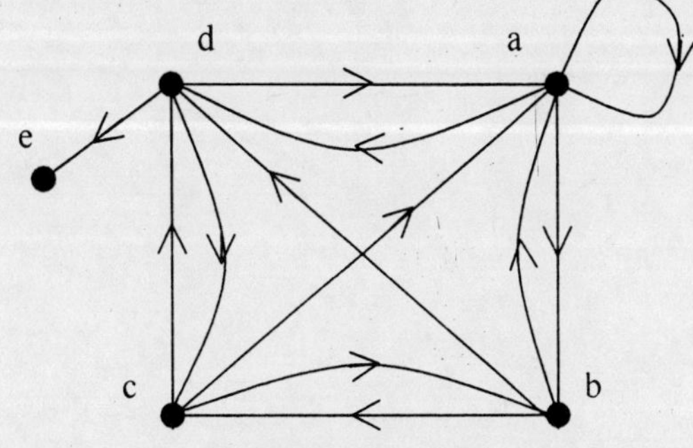

Fig. 7.14

It is to be noted here that the paths p_1, p_2 and p_3 are elementary while p_5 is simple but not elementary. It can be shown that if there exists a path from a there must also be an elementary path from v_i vertex say v_i to another vertex v_j, then to v_j.

Example 7.10

Determine simple paths, elementary paths and the path that is neither simple nor elementary originating from vertex *a* to vertex *d* of the following graph.

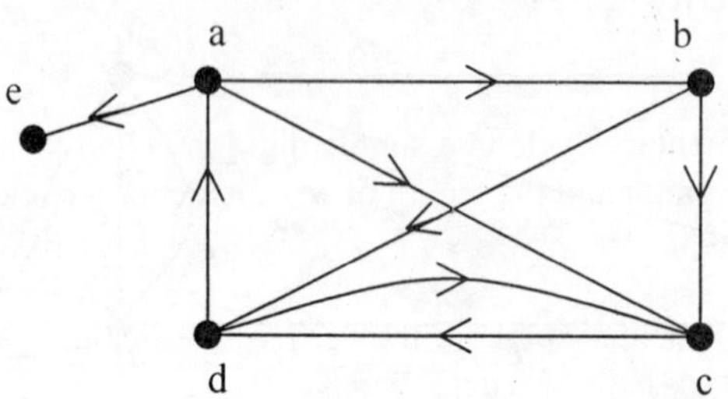

PATH	TYPE	LENGTH
$P_1 = (ab)$	elementary	1
$P_2 = (ab, bd)$	elementary	2
$P_3 = (ab, bc, cd)$	elementary	3
$P_4 = (ab, bd, dc, cd)$	simple	4
$P_5 = (ab, bc, cd, dc, cd)$	Neither simple nor elementary	5

THEOREM 7.2

There is a path from vertex v_1 to a vertex v_2 iff there exists a simple path from vertex v_1 to v_2.

Example 7.11

Find some elementary and simple cycles in the following graph.

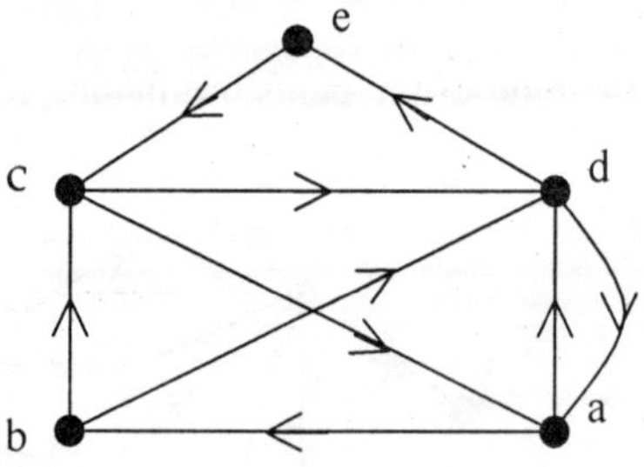

Fig. 7.15

Cycle	Type
C_1: *acd*	elementary
C_2: *abca*	elementary
C_3: *abcda*	elementary
C_4: *dabcad*	simple
C_5: *cdec*	elementary
C_6: *decade*	elementary
C_7: *adecda*	simple

Here, one can observe that any path which is not elementary, contains cycles traversing through those vertices, which appears more than once in this path. If we delete such cycles we obtain elementary paths. Because of this property some authors use the term path to mean only elementary paths. Readers are advised to verify it.

THEOREM 7.3

The length of any elementary cycle in a simple digraph is less than or equal to $n - 1$, where n is the number of vertices in the graph and the length of any elementary cycle does not exceed n, where n is the number of vertices.

PROOF

We know that in an elementary path all the vertices are distinct. This implies the number of distinct vertices in any elementary path of length m is $m + 1$. Therefore, the length of any elementary path having all n vertices of the graph cannot be greater than $n - 1$.

Similarly, in an elementary cycle all the vertices are distinct (except for initial and final). Hence number of distinct vertices in the cycle of length m is m. Therefore, the length of an elementary cycle having n vertices cannot exceed n.

7.11-1 CONNECTIVITY, CONNECTED COMPONENTS

A directed graph is said to be **connected** if the graph when treated as an undirected graph is connected. A digraph G is said to be **weekly connected** if there exist a semi path between any pair of vertices v_i and v_j in G. Graph shown in Fig.7.16 (a) is weekly connected.

A digraph is said to be **unilaterally connected** if for any pair of vertices v_i and v_j of the graph at least one of the vertices of the pair is reachable from the other vertex, i.e. there exist a directed path either from v_i to v_j or from v_j to v_i. Graph shown in Fig.7.16 (b) is unilaterally connected. One can observe in this graph that the vertex v_1 is reachable from v_2 but the vertex v_2 is not reachable from v_2.

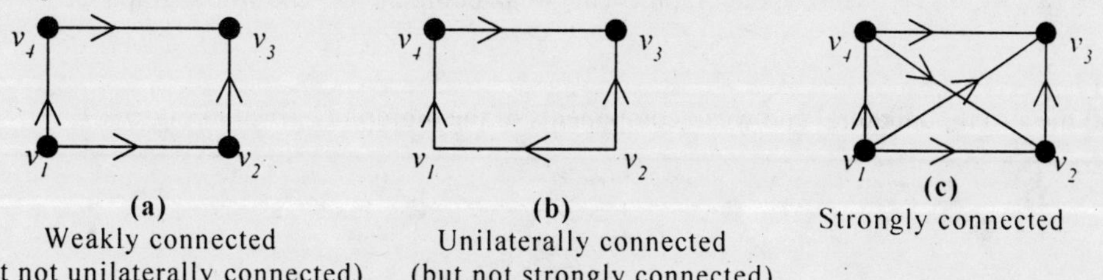

Weakly connected (but not unilaterally connected)

Unilaterally connected (but not strongly connected)

Strongly connected

Fig. 7.16

If for any pair of vertices v_i and v_j of graph both the vertices of the pair are reachable from one another, then the graph is called **strongly connected**. In other words a digraph G is said to be strongly connected if for any pair of vertices v_i and v_j in G there exist at least one path directed from v_i to v_j and also a path directed from v_j to v_i. Graph shown in Fig.7.16(c) is strongly connected.

A graph, which is not connected, is said to be **disconnected**. Here, one can observe that strongly connected graph $\Rightarrow$ unilaterally connected $\Rightarrow$ weakly connected.

A null graph of more than one vertex is disconnected.

7.12 MAXIMAL

A subgraph S is said to be **maximal** with respect to some property if no other subgraph has the property and also includes S. For a simple digraph G a maximally strongly connected subgraph is called a **strong component**. A maximal unilateral (weak) connected subgraph is called **unilateral (weak) component**. In Fig.7.17 a disconnected graph with two components are shown.

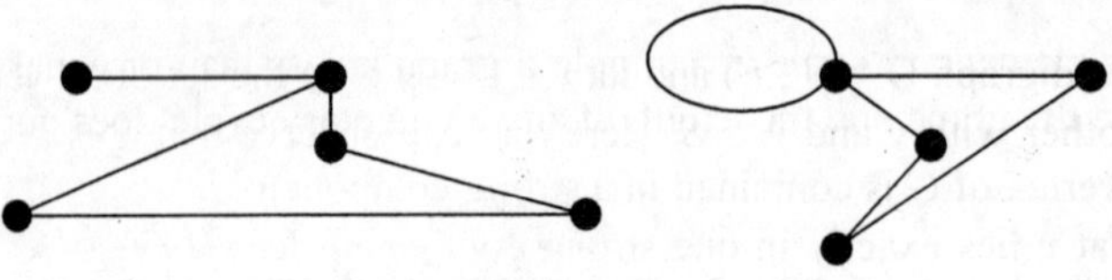

Fig. 7.17

Example 7.12

Find the strong, unilateral and weak components of the digraph shown in Fig. 7.18.

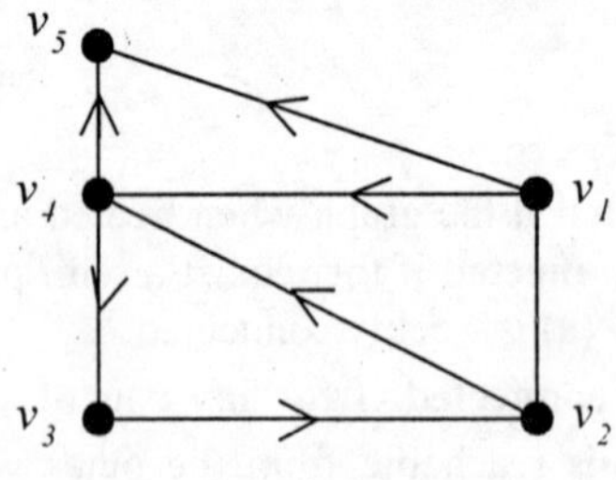

Fig. 7.18

Solution

From the figure one can observe (i) strong components: $\{v_1, v_2, v_3, v_4\}$, $\{v_5\}$ (ii) unilateral component $\{v_1, v_2, v_3, v_4, v_5\}$ = the whole graph G (iii) weak component = the whole graph G.

Example 7.13

Find the strong, unilateral and weak components of the digraph G given in Fig. 7.19.

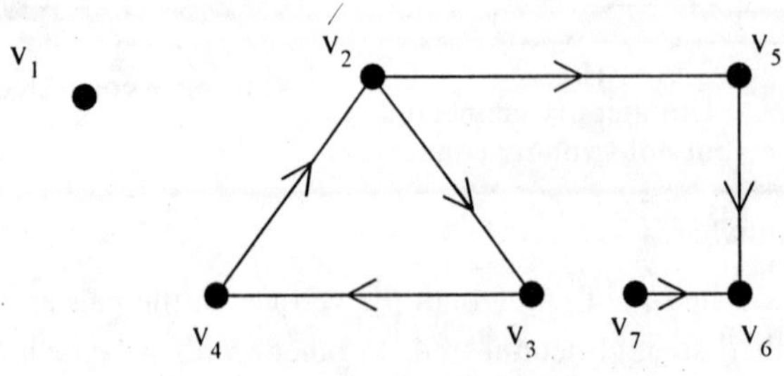

Fig.7.19

From the Fig. 7.19 one can observe easily that:

(i) strong components: $\{v_2\, v_3\, v_4\}$, $\{v_5\}$, $\{v_6\}$, $\{v_7\}$, $\{v_1\}$

(ii) unilateral components: $\{v_3\, v_4\, v_2\, v_5\, v_6\}$, $\{v_7\}$, $\{v_1\}$

(iii) Weak components: $\{v_3\, v_4\, v_2\, v_5\, v_7\}$, $\{v_1\}$.

THEOREM 7.4

In a simple digraph G, every vertex of the digraph lies in exactly one strong component. Also every node and edge of simple digraph is contained in:

(i) exactly one weak component

(ii) at least one unilateral component.

PROOF

Consider a simple digraph $G = (V, E)$ and let $v \in V$. Let S be the set of all vertices of G which are reachable from each other with v and $v \in S$. Here one can observe that S is a strong component which implies that every vertex of G is contained in a strong component.

Now to prove that v lies exactly in one strong component let $v_i\ (\neq v) \in S_1$ and $v_j\ (\neq v) \in S_2$. Further, consider v_i is mutually reachable from v and v_j is also mutually reachable from v. This implies that v_j is mutually reachable from v and v is mutually reachable from v_i. Hence v_i is mutually reachable from v_j.

$\therefore v_i \in S_1$ and S_2

But this contradicts the maximal property of the component. Therefore, every node is contained in exactly one strong component.

This theorem may also be stated as: strong components of a digraph G partitions the vertex set V of G.

Now, we prove the following results:

1. Any edge $e \in E$ of a simple digraph G may or may not be contained in a strong component.

PROOF

Consider the graph shown in the following Fig. 7.20. Strong components of graph G are $\{v_3, v_4, v_2\}, \{v_5\}, \{v_6\}, \{v_7\}$.

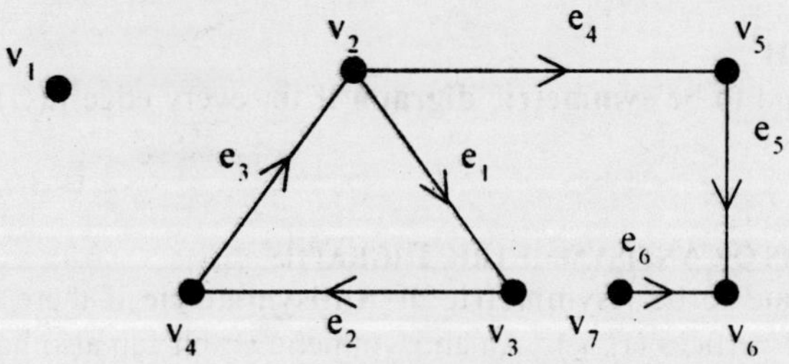

Fig. 7.20

e_2, e_3, e_1 are in strong component whereas the vertices v_2 and v_5 of the edge e_4 lie in different strong components.

2. In a graph $G = (V, E)$ if an edge $e \in E$ is in strong component then 'e' must be a part of a cycle of G.

PROOF

Consider graph $G = (V, E)$. Let v_i and v_j belong to a strong component of G and $e_k = v_i v_j$.

As considered component is strong, one vertices v_i and v_j are reachable to each other. Therefore, $v_i v_j$ is a part of cycle of G.

It is to be noted that e_1, e_2, e_3 belongs to a cycle in graph G given in the result 1.

It can be proved that every node of a simple digraph is contained in exactly one weak component as we have proved for strong component case.

Now to prove that every edge of a simple digraph is contained in exactly one weak component; one can proceed as follows:

Let $e_k \in G$ lie in two distinct weak components R and S. Let v_i and v_j be two terminal vertices of e_k. Since e_k lies in two weak components, this implies that v_i and v_j lie in two weak components.

But this is impossible as it contradicts the maximality property of the weak components.

7.13 ORIENTABLE GRAPH

An undirected graph is said to be **orientable** if direction can be assigned to each of the edges of the graph so that resultant graph is strongly connected. For example, if G is the graph shown in Fig. 7.21(i), then G is orientable, since it can be directed to give the strongly connected digraph as shown in Fig.7.21 (ii).

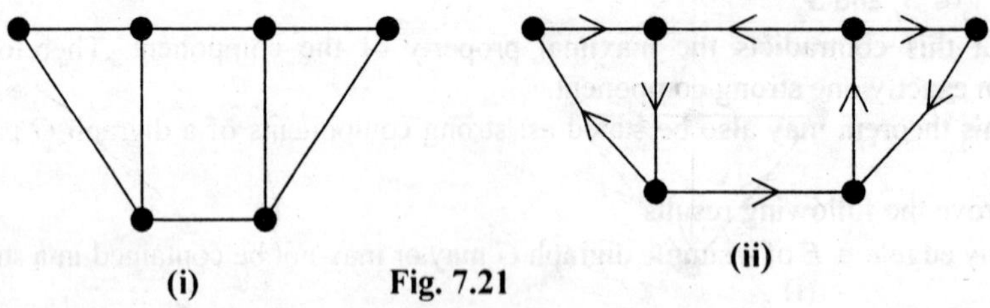

Fig. 7.21

It is to be noted here that any Eulerian graph is orientable, since we simply follow any Eulerian trail, directing the edges in the direction of the trail as we go.

7.14 SYMMETRIC DIGRAPH

A directed graph is said to be **symmetric digraph** if for every edge $E(G)$, say (v_i, v_j) there exist also an edge (v_j, v_i).

7.14-1 ASYMMETRIC DIGRAPHS OR ANTISYMMETRIC DIGRAPHS

A directed graph is said to be **Asymmetric** or **Antisymmetric** if there exist at most one directed edge between a pair of vertices (v_i, v_j). An antisymmetic graph can also have self-loops. For instance graph shown in the Fig. 7.22 is an antisymmetic digraph.

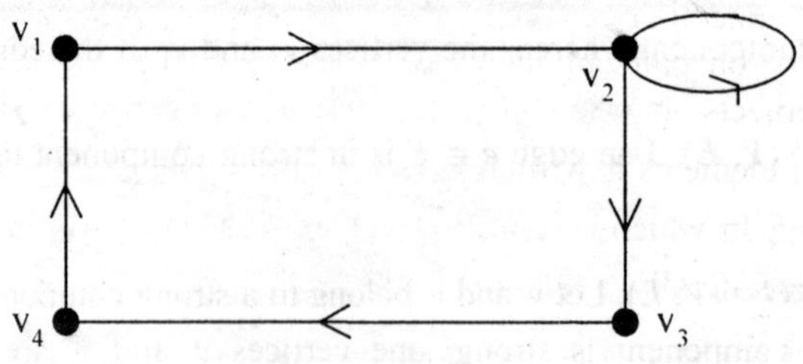

Fig. 7.22

7.14-2 Simple Symmetric Digraph

A digraph that is both simple and symmetric is called a **simple symmetric digraph**. A graph that is both simple and asymmetric is called **simple asymmetric**.

7.15 COMPLETE DIGRAPH

We have already defined a complete undirected graph as a graph in which every vertex is joined to every other vertex by one edge. But here in digraph we have following two kinds of complete graphs.

A simple digraph in which there exist exactly one edge directed from every vertex to every other vertex is said to be **complete symmetric digraph**.

If a digraph has exactly one edge directed between every pair of vertices then it is called **complete asymmetric digraph**. Graphs shown in Fig. 7.23 (i) and (ii) are complete symmetric and complete asymmetric respectively.

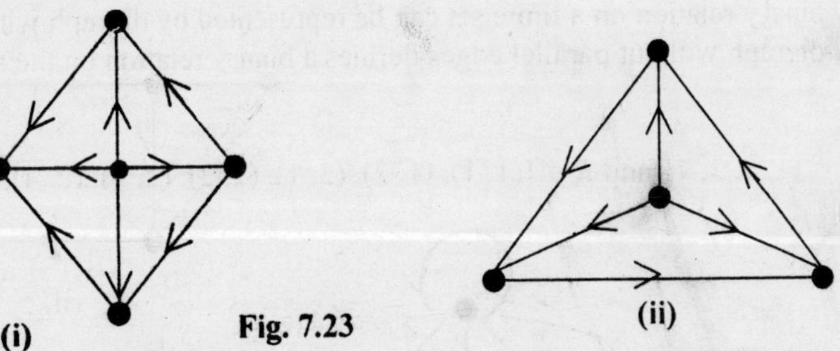

Fig. 7.23

REMARK

A complete asymmetric graphs of m vertices contains $m(m-1)/2$ edges.

7.16 PSEUDOSYMMETRIC DIGRAPHS

If indegree of a vertex in a digraph is equal to outdegree, i.e. $d^{+}(v_i) = d^{-}(v_i)$, then the digraph is said to be **pseudosymmetric** or **isograph** or **balanced**. A pseudo-symmetric digraph is said to be **regular** if every vertex has the same in-degree and out-degree as every other vertex.

7.17 BINARY RELATIONS AND DIRECTED GRAPHS

The calculus of binary relations and theory of graphs are closely related. We are familiar with many relations which are used in computer science and mathematics, e.g. "less than", "is subset of", "is parallel to", "is orthogonal to", and so on. These relations consider the existence or non-existence of a certain connection between pairs of objects taken in a definite order. Each of these relations is defined only on pairs of numbers of the set and this is why they are called **binary relation**. In a set of objects A, where $A = \{a_1, a_2, a_3,...\}$a **binary relation** R or simply relation between ordered pair of elements is written as (a_i, a_j), where a_i is designated as the first element and a_j as the second element. In which case $a_i \; R \; a_j$ and we say that R is a binary relation on A or a_i has relation R on a_j. More specifically, a relation can be defined as follows:

Let A and B be nonempty sets. A **relation** R from A to B is a subset of $A \times B$. A digraph is most natural way of representing a binary relation on a set A. We can represent each $a_i \in A$ by a vertex

a_i. If relation R between a_i and a_j is specified, a directed edge is drawn from vertex a_i to a_j for every pair (a_i, a_j). Graph shown in Fig. 7.24 represents the relation "is less than" on a set consisting of {1, 2, 5, 8, 9}.

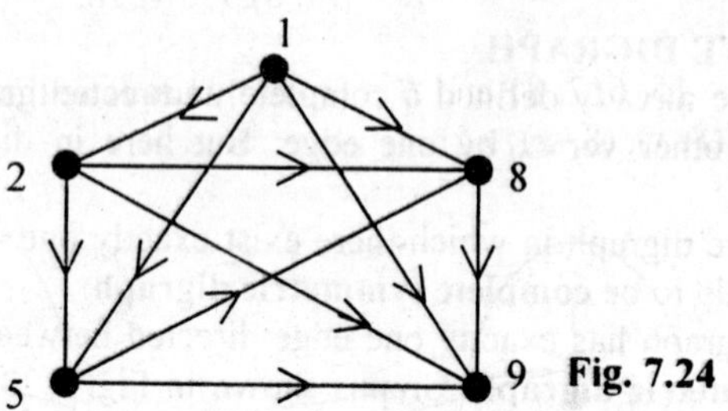

Fig. 7.24

REMARK

1. Every binary relation on a finite set can be represented by digraph without parallel edges.
2. Every digraph without parallel edges defines a binary relation on the set of its vertices.

Example 7.14

Let $A = \{1, 2, 3, 4\}$ and $R = \{(1, 1), (1, 2), (2, 1), (2, 2), (2, 3), (2, 4), (3, 4), (4, 1)\}$. Then the digraph of R is

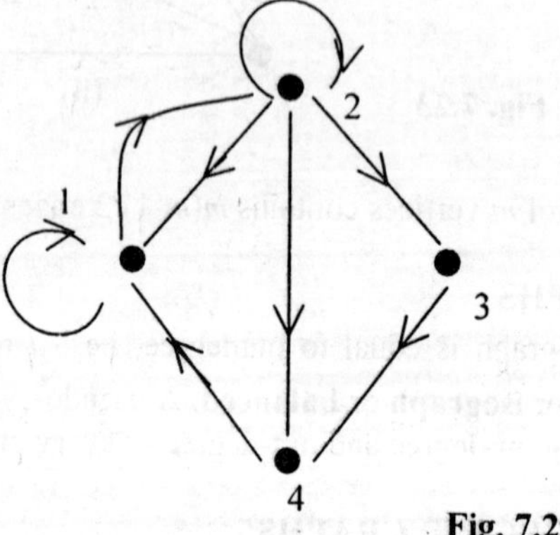

Fig. 7.25

Example 7.15

Find the relation determined by the graph shown in Fig. 7.26.

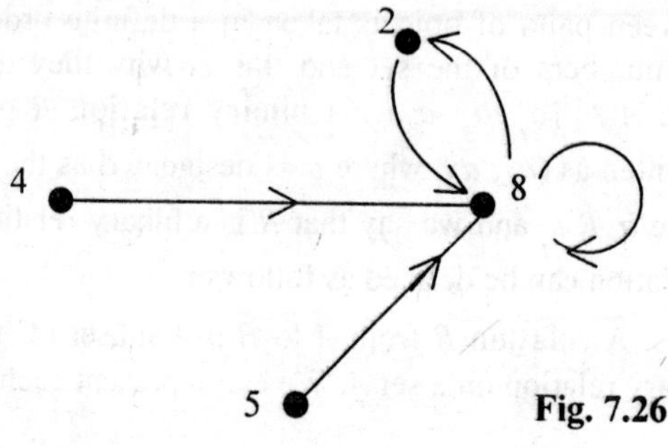

Fig. 7.26

Solution

Since $a_i R a_j$ iff there is an edge from a_i to a_j, we have

$$R = \{(4, 8), (2, 8), (8, 2), (8, 8), (5, 8)\}.$$

Example 7.16

Given $A = \{5, 6, 7, 8\}$ and $B = \{x, y, z\}$. Let R be the relation from A to B defined as $R = \{(5, y), (5, z), (7, y), (8, x), (8, z)\}$.

Solution

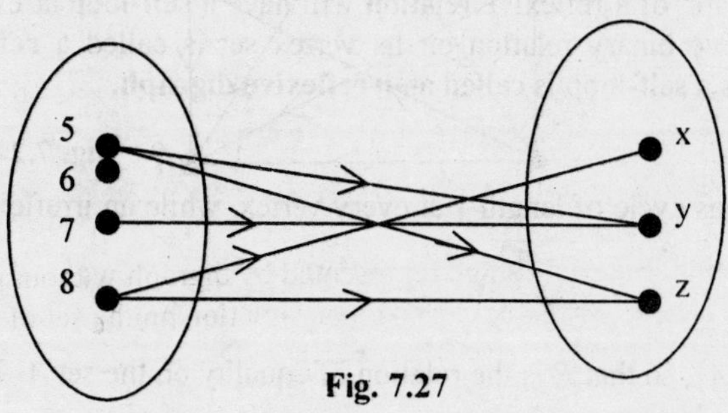

Fig. 7.27

Example 7.17

Let $A = \{1, 2, 3, 4, 6\}$, and let R be the relation on A defined by "a divides b", written by b/a. (Note b/a iff there exists an integer c such that $ac = b$). Write R as a set of ordered pairs and draw its directed graph.

Solution

Find those numbers in A divisible by 1, 2, 3, 4, and then 6. These are

1/1, 2/1, 3/1, 4/1, 6/1, 2/2, 4/2, 6/2, 3/3, 6/3, 4/4, 6/6.

Hence $R = \{(1, 1), (1, 2), (1, 3), (1, 4), (1, 6), (2, 2), (2, 4), (2, 6), (3, 3), (3, 6), (4, 4), (6, 6)\}$.

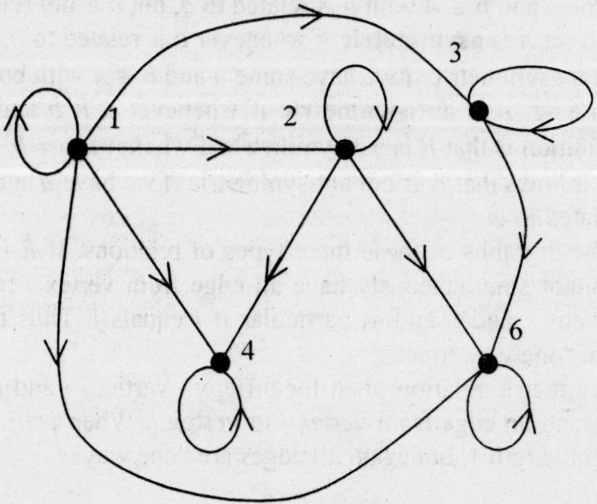

Fig. 7.28

7.18 PROPERTIES OF RELATION

7.18-1 REFLEXIVE AND IRREFLEXIVE RELATION

A relation R on a set A is **reflexive** if $(a, a) \in R$ for all $a \in A$, i.e. if $a\ R\ a$ for all $a \in R$. A relation R on a set is **irreflexive** if $a \not R a$ for every $a \in A$.

Thus R is reflexive if every element is related to itself and it is irreflexive if no element is related to itself. Thus the digraph of a reflexive relation will have a self-loop at every vertex. Such a digraph representing a reflexive binary relation on its vertex set is called a **reflexive digraph.** A digraph in which no vertex has a self-loop is called an **irreflexive digraph.**

NOTE

A reflexive relation has cycle of length 1 at every vertex, while an irreflexive relation has no cycle of length 1.

Example 7.18

1. Let $R = \{(a, a) | \ a \in A\}$, so that R is the relation of equality on the set A. Then R is reflexive, since $(a, a) \in R$ for all $a \in A$.

2. Let $X = \{2, 3, 4\}$ and let $R = \{(2, 2), (2, 3)\}$. Then R is not reflexive since $(3, 3) \notin R$ and $(4, 4) \notin R$. Also R is not irreflexive, since $(2, 2) \in R$.

3. Let R $= \{(x, y) \in x \times x \mid x \neq y\}$, so that R is the relation of inequality on the set X. Then R is irreflexive, since $(x, x) \notin R$ for all $x \in A$.

7.18-2 SYMMETRIC, ASYMMETRIC AND ANTI-SYMMENTRIC RELATIONS

A relation R on a set A is **symmetric** if whenever $a\ R\ b$, then $b\ R\ a$. It follows that R is not symmetric if we have some a and $b \in A$ with a is related to b, but b is not related to a.

A relation R on a set A is **asymmetric** if whenever a is related to b, then b is not related to a. It then follows that R is not asymmetric if we have some a and $b \in A$ with both $a\ R\ b$ and $b\ R\ a$.

A relation R on a set A is **antisymmetric** if whenever $a\ R\ b$ and $b\ R\ a$, then $a = b$. The contrapositive of this definition is that R is antisymmetric if whenever $a \neq b$, then a is not related to b or b is not related to a. It follows that R is not antisymmetric if we have a and b in A, $a \neq b$, and both a is related to b and b is related to a.

Now consider the digraphs of these three types of relations. If R is an asymmetric relation, then the digraph of R cannot simultaneously have an edge from vertex i to vertex j and vertex j to vertex i. This is true for any i and j, and in particular if i equals j. Thus there can be no cycles of length 1, and all edges are "one way street".

If R is an antisymmetric relation, then for different vertices i and j there cannot be an edge from vertex i to vertex j and an edge from vertex j to vertex i. When $i = j$, no condition is imposed. Thus there can be cycles of length 1, but again all edges are "one way".

Example 7.19

Graphs shown in Fig.7.29 and 7.30 are graphs of symmetric binary relation R.

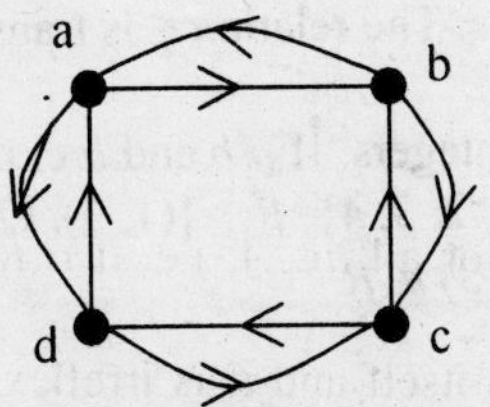

Fig. 7.29 Digraph of R

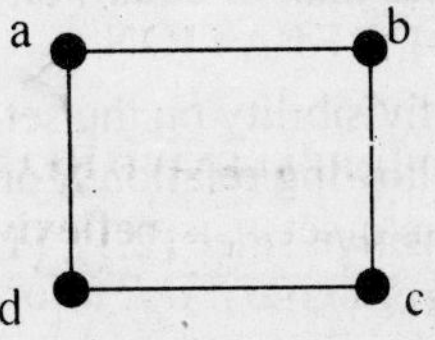

Fig. 7.30 Graph of R

One can note here that the digraph of symmetric relation R has the property that if there is an edge from vertex i to vertex j, then there is an edge from vertex j to vertex i. Thus, if two vertices are connected by an edge, they must always be connected in both directions. Because of this reason, it is possible to give a different representation of a symmetric relation. The vertices are kept as they appear in the digraph, but if two vertices a and b are connected by edges in each direction, we replace these two edges with one undirected edge or a "two-way street." This undirected edge is just a single line without arrows and connects a and b.

NOTE

1. Every undirected graph is a representation of some symmetric binary relation on the set of its vertices.
2. Every undirected graph with n edges can be thought of as a symmetric digraph with $2n$ directed edges.

Example 7.20

Let $B = Z$, the set of integers and let $R = \{(x, y) \in B \times B \mid a < b\}$ so that R is a relation "is less than". Is R symmetric, asymmetric, or antisymmetric?

Solution

If $a < b$, then it is not true that $b < a$, so R is not symmetric.

If $a < b$, then $b \not< a$, so R is asymmetric.

If $a \neq b$, then either a is not less than b or b is not less than a, so that R is antisymmetric.

Example 7.21

Let $A = \{5, 6, 7, 8\}$, and let $R = \{(5, 6), (6, 6), (7, 8), (8, 5)$.

Solution

One can observe here that R is not symmetric, since $(5, 6) \in R$, but $(6, 5) \notin R$. Also R is not asymmetric, since $(6, 6) \in R$.

7.18-3 Transitive Relations

A relation R on a set A is **transitive** if whenever $a\ R\ b$ and $b\ R\ c$, then $a\ R\ c$. A relation R on A is **not transitive** if there exist a, b and c in A so that $a\ R\ b$ and $b\ R\ c$, but a is not related to c.

The digraph of a transitive but irreflexive and asymmetric binary relation is shown in Fig. 7.24. A digraph representing a transitive relation on the set of vertices is called **transitive directed graph**.

Example 7.22

Consider the following relation R.

(a) Relation $R \leq$ (less than or equal) on set Z of integers. The relation $\leq$ is transitive since if $a \leq b$ and $b \leq c$, then $a \leq c$.

(b) Relation '/' of divisibility on the set N of positive integers. If a/b and b/c, then a/c.

(c) Consider the following relation R on the set $A = \{1, 2, 3, 4\}$. $R = \{(1, 3), (2, 1)\}$

The relation R is not transitive since $(2,1), (1, 3) \in R$ but $(2, 3) \notin R$.

THEOREM 7.5

A relation R is transitive iff there is a path of length greater than 1 from vertex a to vertex b.

7.18-4 EQUIVALENCE RELATION

A relation R on a nonempty set A is called an equivalence relation if it is reflexive, symmetric and transitive. That is, R is an equivalence relation on A if it has the following three properties:

(i) For every $a \in A$, $a R a$

(ii) If $a R b$, then $b R a$.

(iii) If $a R b$ and $b R c$, then $a R c$.

The graph representing an equivalence relation is called an **equivalence graph.**

Example 7.23

(i) Let $A = \{1, 2, 3, 4\}$ and Relation $R = \{(1, 1), (1, 2), (2, 1), (2, 2), (3, 4), (4, 3), (3, 3), (4, 4)\}$ is an equivalence relation.

(ii) The classification of animals by species, i.e. the relation "is of the same species as" is an equivalence relation as the set of animals.

(iii) The relation $\subseteq$ of set inclusion is not an equivalence relation. It is reflexive and transitive but it is not symmetric since $A \subseteq B$ does not imply $B \subseteq A$.

Example 7.24

Let $A = Z$ and consider the relation

$R = \{(a, b) \leq A \times A\}$ $a \equiv r \pmod 2$ and $b \equiv r \pmod 2)\}$

i.e. $a R b$ iff a and b yield same remainder, r, when divided by 2. In this case we write $a \equiv b \pmod 2$, read a is congruence to b mod 2". Show that congruence mod 2 is an equivalence relation.

Solution

Since $a \equiv a \pmod 2$. Thus R is reflexive. If $a \equiv b \pmod 2$, then $a \equiv r \pmod 2$ and $b \equiv r$ (mod 2). Then, $a \equiv r \pmod 2$, $b \equiv r \pmod 2$ and $c \equiv r \pmod 2$, i.e. all three yield the same remainder when divided by 2. Thus, $a \equiv c \pmod 2$. Hence congruence mod 2 is an equivalence relation.

Example 7.25

Consider the relation R "is congruent to modulo 3" defined on the set of 11 integers, 10 through 20. The graph is shown in Fig.7.31 is equivalence class. It is to be noted in the graph that each undirected edge represents two parallel but oppositely directed edges. It can be observed that the vertex set of the graph is divided into three disjoint classes each in separate component. Each component is an undirected subgraph due to symmetry with a self-loop at each vertex due to reflexivity. And in each component every vertex is related to every other vertex.

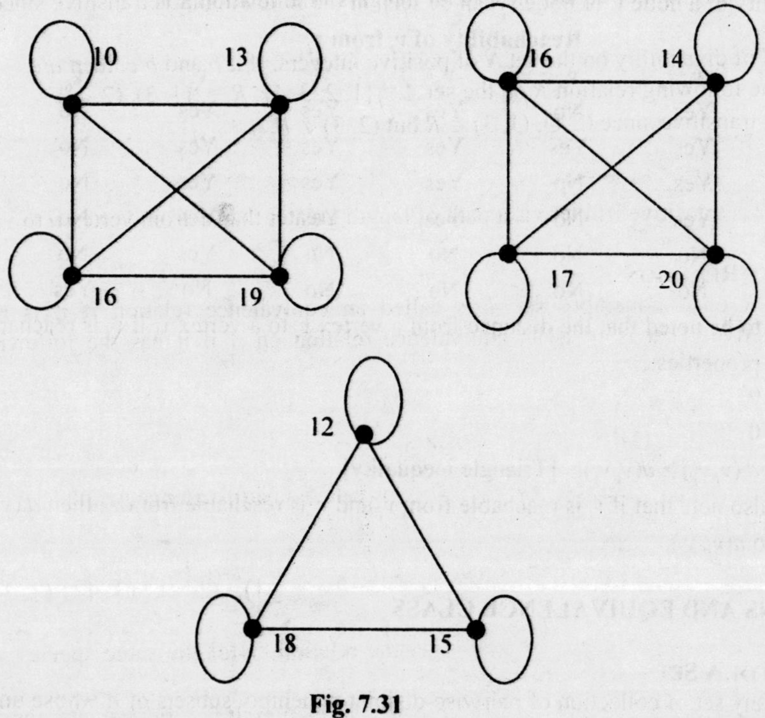

Fig. 7.31

7.19 REACHABLE SET

The set of vertices which are reachable from a given vertex v is called reachable set of v.

In a digraph $G = (V, E)$ a subset A of V is called a **node base** if its reachability set is V and if no proper subset of A has this property.

Example 7.26

Determine the reachability of each node corresponding to every other node for the graph shown in Fig. 7.32.

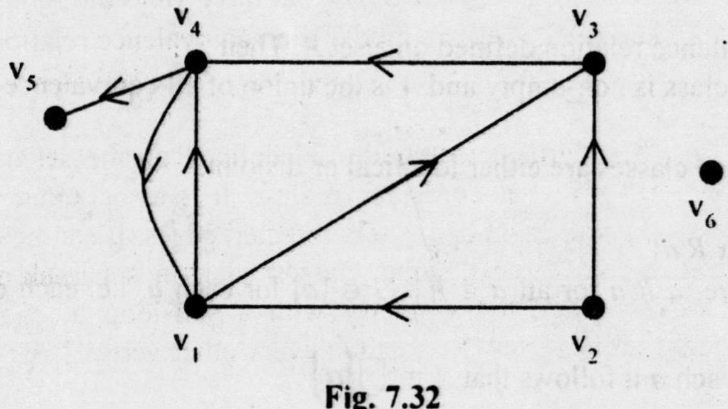

Fig. 7.32

Reachability of a node v_i to node v_j can be seen in the following table.

Reachability of v_j from v_i

v_i / v_j	v_1	v_2	v_3	v_4	v_5	v_6
v_1	Yes	No	Yes	Yes	Yes	No
v_2	Yes	Yes	Yes	Yes	Yes	No
v_3	Yes	No	Yes	Yes	Yes	No
v_4	Yes	No	Yes	Yes	Yes	No
v_5	No	No	No	No	Yes	No
v_6	No	No	No	No	No	Yes

Here it is to be noted that the distance from a vertex v_i to a vertex v_j if v_j is reachable from v_i satisfies following properties:

(i) $d(v_i, v_j) \geq 0$

(ii) $d(v_i, v_i) = 0$

(iii) $d(v_i, v_j) + d(v_j, v_k) \geq d(v_i, v_k)$ [Triangle inequality]

One can also note that if v_i is reachable from v_j and v_j is reachable from v_i, then $d(v_i, v_j)$ is not necessarily equal to $d(v_j, v_i)$.

7.20 PARTITIONS AND EQUIVALENCE CLASS

7.20-1 Partition of a Set

Let A be any set. A collection of pairwise-disjoint nonempty subsets of A whose union is A is called **partition** of A.

Example 7.27

For any non-empty set A the collection of all singleton subsets of A is partition of A. If $S = \{1, 3, 5, 7 \ldots\}$ and $T = \{2, 4, 6, 8 \ldots\}$, then (S, T) is a partition of N.

7.20-2 Equivalence Class

Let R be an equivalence relation defined on a set A. For any $a \in A$, the set $[a] = \{x \in A \mid x R a\}$ is called **equivalence class** determined by A. It is also denoted by $R(a)$.

Example 7.28

Let R be an equivalence relation defined on a set A. Then

(i) Each equivalence class is non-empty and A is the union of all equivalence classes.

(ii) $a R b \Rightarrow [a] = [b]$

(iii) any two equivalence classes are either identical or disjoint.

Solution

Let $[a] = \{x \in A \mid x R a\}$.

(i) Since R is reflexive, $a R a$ for all $a \in A \mid G \in [a]$ for each a, i.e. each equivalence class is nonempty.

Since $a \in [a]$ for each a it follows that $A = \bigcup_{a \in A} [a]$.

(ii) Suppose $a\ R\ b$. To prove $[a] = [b]$ let $x \in [a]$ then $x\ R\ a$. Since $a\ R\ b$, $x\ R\ a$, $a\ R\ b \Rightarrow x\ R\ b$, hence R is transitive and $x \in [b]$ $\therefore$ $[a] \subseteq [b]$. Similarly $[b] \subseteq [c]$ so that $[a] = [b]$. Suppose $[a] = [b]$. Since $a \in [a]$, $a \in [b]$ so that $a\ R\ b$ from the definition $[b]$.

(iii) Suppose $[a]$, $[b]$ are two equivalence classes. Now, to prove either $[a] = [b]$ or $[a] \cap [b] = \phi$, suppose $[a] \cap [b] \neq \phi$. Then there exist $x \in [a] \cap [b]$.

$\therefore$ $x \in [a]$ and $x \in [b]$ so that $x\ R\ a$ and $x\ R\ b$, i.e. $a\ R\ x$ and $x\ R\ b$ [$\because$ R is symmetric] i.e. $a\ R\ b$ [$\because$ R is transitive] and hence by (ii) $[a] = [b]$.

THEOREM 7.6

Let R be an equivalence relation on a set A and let $a \in A$ and $b \in A$, then $A\ R\ b$ iff $R(a) = R(b)$.

PROOF

Suppose that $R(a) = R(b)$. Since R is reflexive, $b \in R(b)$ therefore, $b \in R(a)$, so $a\ R\ b$. Conversely, consider that $a\ R\ b$. Then note that

(i) $b \in R(a)$ by definition, therefore, R is symmetric and

(ii) $a \in R(b)$ by the fact that symmetry of R means $a \in R(b)$ iff $b \in R(a)$.

Now to show that $R(a) = R(b)$ first choose an element $x \in R(b)$. Since R is transitive, the fact that $x \in R(b)$ together with (i) implies by the fact that transitivity of R means if $b \in R(a)$ and $c \in R(b)$, then $c \in R(a)$.

$\Rightarrow$ that $y \in R[b]$. Thus $R(a) \subseteq R(b)$, so we must have $R(a) = R(b)$.

NOTE

An equivalence relation on a set, partitions the elements of the set into class such that two elements are in same class iff they are related.

EXERCISE 7.2

1. Suppose R is a relation on a set A. A **path of length n** is R from a to b is a finite sequence $\Pi : a, x_1, x_2, \ldots, x_{n-1}, b$ beginning with a and ending with b, s.t. $a\ R\ x_1$, $x_1\ R\ x_2$, ..., $x_{n-1}\ R\ b$. A path of length n involves $n+1$ elements of A, although they are not necessarily distinct. Consider the graph shown in the Fig. 7.33.

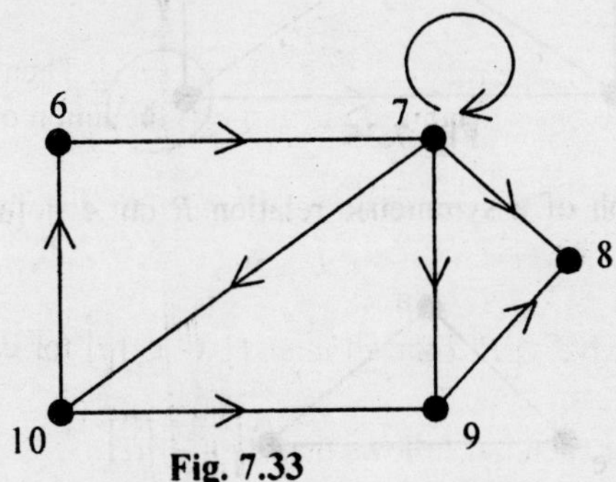

Fig. 7.33

P_l: 6, 7, 10, 9, 8 is a path of length 4.

Now consider the graph shown in the Fig.7.34

(i) List all paths of length 1.

(ii) Find a cycle starting at vertex 6.

(iii) List all paths of length 2 starting from vertex 2.

(iv) Draw the digraph of R^2. ($x\ R^n\ y$ means that there is a path of length n from x to y in R where n in a positive integer.

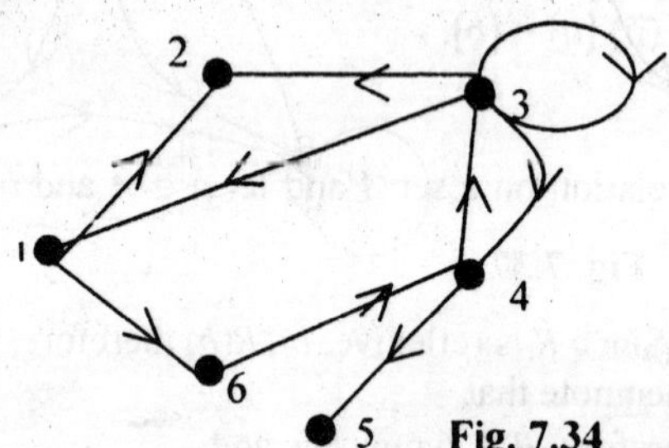

Fig. 7.34

2. Let $A = \{1, 2, 3, 4, 5\}$ and R be the relation defined by $a\ R\ b$ iff $a < b$. Compute R^2 and R^3.
3. Let $A = \{5, 6, 7, 8\}$. Determine whether the relation is reflexive, irreflexive, symmetric, asymmetric, antisymmetric or transitive.

(i) $R = \{(5, 5), (5, 6), (6, 5), (6, 6), (7, 7), (7, 8), (8, 7), (8, 8)\}$

(ii) $R = \{(5, 6), (5, 7), (5, 8), (6, 7), (6, 8), (7, 8)\}$

(iii) $R = \{(5, 7), (5, 5), (7, 5), (5, 6), (7, 7), (8, 8)\}$

(iv) $R = \phi$

(v) $R = \{(5, 6), (5, 7), (7, 5), (5, 5), (7, 7), (7, 6), (5, 8), (8, 6), (7, 8)\}$

4. Determine whether the relation R whose digraph is given is reflexive, irreflexive, symmetric, asymmetric, antisymmetric or transitive.

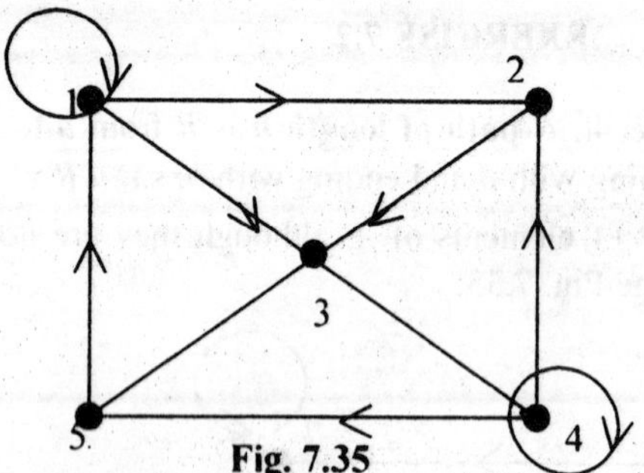

Fig. 7.35

5. Consider the graph of a symmetric relation R on $A = \{a, e, i, o, u\}$ shown in Fig. 7.13. Determine R.

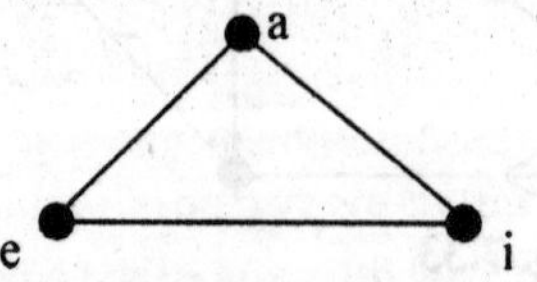

Fig.7.36

6. Determine whether the relation R whose digraph is shown in Fig.7.37 is an equivalence relation.

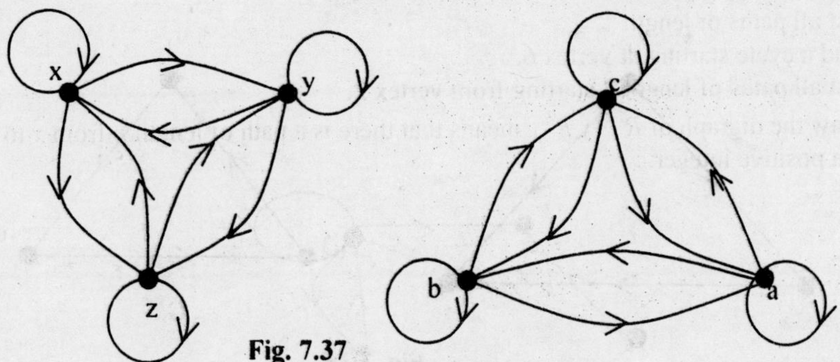

Fig. 7.37

7. Determine whether the relation R on the set A is an equivalence relation $A = \{1, 2, 3, 4\}$, $R = \{(1, 1), (2, 1), (2, 2), (3, 3), (4, 4), (4, 3)\}$
8. If $\{\{a, c, e\}, \{b, d, f\}\}$ is a partition of the set $A = \{a, b, c, d, e, f\}$, determine the corresponding equivalence relation R.
9. Sketch all different nonisomorphic simple digraphs with 1, 2 and 3 vertices.
10. An irreflexive, asymmetric, transitive relation on a set is called a **strict partial order**. Give two examples of strict partial orders. Show that the digraph of **strict partial order** is **acyclic.**
11. Determine whether the relation R on the set A is an equivalence relation where A = the set of all triangles in the plane; $a\ R\ b$ if and only if a is similar to b.
12. The combinations of reflexivity, symmetry and transitivity define ($2^3 = 8$) types of binary relations. Two such relations are equivalence and partial order. List the other six and sketch a digraph for each.

ANSWERS

1. (i) 1, 2; 1, 6; 2, 3; 3, 3; 3, 4; 4, 3; 4, 5; 4, 1; 6, 1.
   (iii) One is 6, 4, 1, 6.
3. (i) Reflexive, symmetric, transitive
   (i) Irreflexive, symmetric, asymmetric, anti-symmetric, transitive.
   (ii) Transitive
4. Anti-symmetric, transitive
6. No
7. No
8. $\{(a, a), (a, c), (a, e), (c, a), (c, c), (c, e), (e, a), (e, c), (e, e), (b, b), (b, f), (b, d), (d, b), (d, d), (d, f), (f, b), (f, d), (f, f)\}$.

7.21 ISOMORPHISM IN DIGRAPH

For this one can imitate the definition of isomorphism given in chapter 2. However, by preserving adjacency in digraph $G = (V, E)$, we mean for every pair of vertices v_i and v_j in V (v_i, v_j) it is in E if and only if $\{f(v_i), f(v_j)\}$ is in E and preserves the direction of the edges.

Example 7.29

Test for isomorphism the following digraphs G and G'.

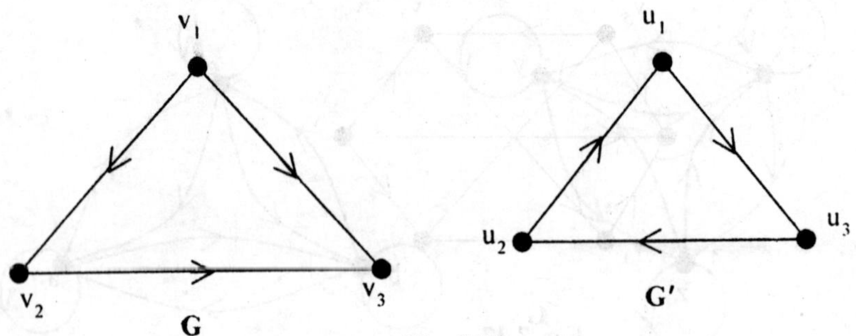

Fig. 7.38

From the graphs G and G' one can observe that G has a vertex with two edges terminating (incoming) at vertex v_3, whereas G' has no vertex with two edges terminating (incoming) at any vertex, hence G and G' are not isomorphic.

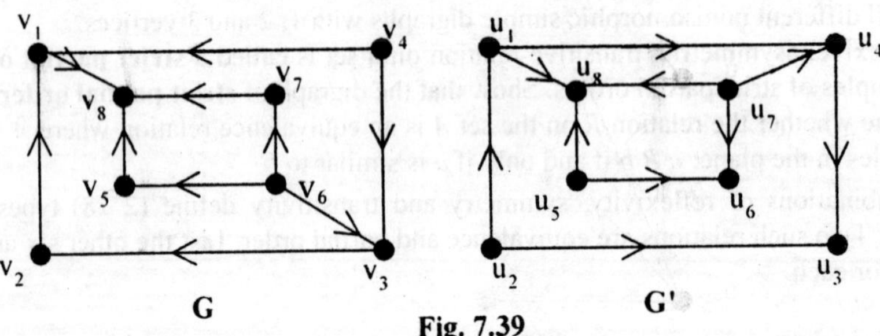

Fig. 7.39

Consider graphs shown in Fig.7.39. We can see that G is not isomorphic to G' since the vertex v_6 with outdegree 3 is not adjacent to the vertex v_8 with indegree 3 in G whereas the vertex u_7 with outdegree 3 is adjacent to the vertex u_8 with indegree 3 in G'.

Example 7.30

Show that the following digraphs G_1 and G_2 are isomorphic.

(i)

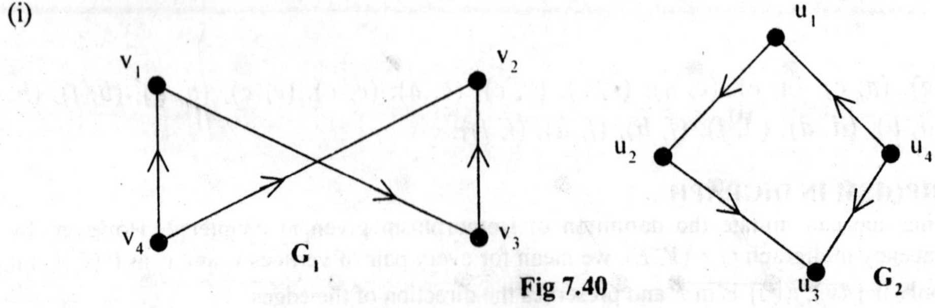

Fig 7.40

Graph shown in G_1 and G_2 are isomorphic since $v_1 \to u_1$; $v_2 \to u_3$; $v_3 \to u_2$; $v_4 \to u_4$.

(ii)

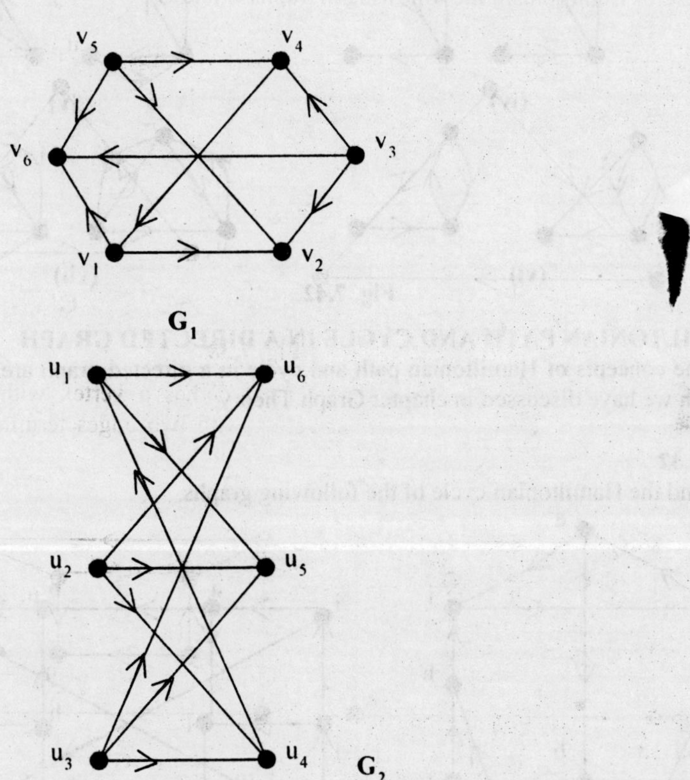

Fig 7.41

Graph G_1 and G_2 are isomorphic since $v_1 \to u_1$; $v_2 \to u_6$; $v_3 \to u_3$; $v_4 \to u_4$; $v_5 \to u_2$; $v_6 \to u_5$.

Example 7.31

Draw all possible simple non-isomorphic digraphs having three vertices assuming that there are no loops.

Solution

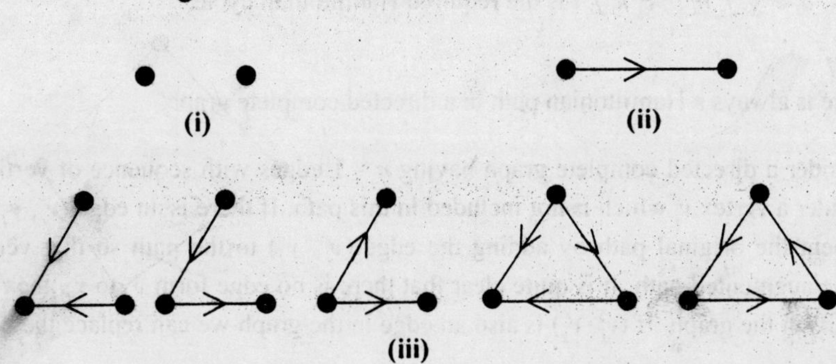

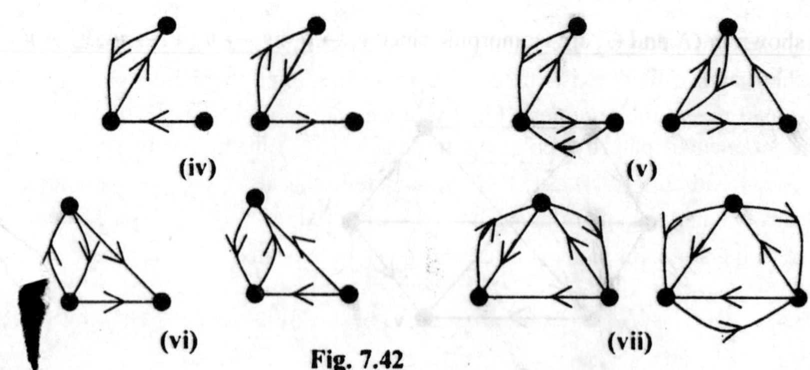

Fig. 7.42

7.22 HAMILTONIAN PATH AND CYCLE IN A DIRECTED GRAPH

The concepts of Hamiltonian path and cycle in a directed graph are analogues to undirected graph which we have discussed in chapter Graph Theory.

Example 7.32

Find the Hamiltonian cycle of the following graphs.

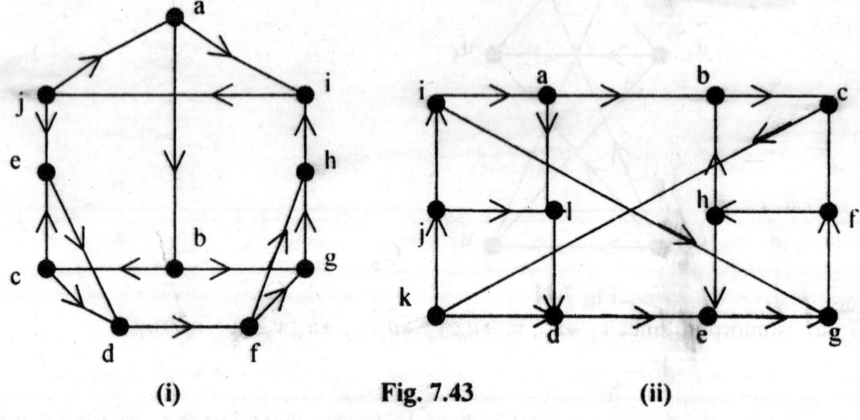

Fig. 7.43

Solution

(i) *a, b, c, e, d, f, g, h, i, j, a* is the required Hamiltonian cycle.

(ii) *i, a, l, d, e, g, f, h, b, c, k, j, i* is the required Hamiltonian cycle.

THEOREM 7.7

There is always a Hamiltonian path in a directed complete graph.

PROOF

Consider a directed complete graph having $n - 1$ edges with sequence of vertices (v_1, v_2, v_3 ,...., v_n). Consider a vertex v_i which is not included in this path. If there is an edge (v_i, v_1) in the graph, we can augment the original path by adding the edge (v_i, v_1) to the path so that vertex v_i will be included in the augmented path. It is quite clear that there is no edge form v_i to v_1, then there must be an edge (v_1, v_i) in the graph. If (v_i, v_2) is also an edge in the graph we can replace the edge (v_1, v_2) in the original path with two edges (v_1, v_i) and (v_i, v_2) so that the vertex v_i is included in the augmented path. On the other hand, if there exists no edge from v_i to v_2, then there must exist an edge (v_2, v_i) in the path and we can repeat the argument. Finally, if we can find that it is not possible to include the vertex v_i in the augmented path by replacing an edge (v_j, v_{j+1}) in the original path with two edges (v_j, v_x) and (v_x, v_{j+1}) with $1 \leq j \leq n-1$, then we conclude that there must be an edge (v_n, v_i) in the graph. This means, we can augment the original path by adding to its edge (v_n, v_i) so that the vertex v_i will be included in the augment path. This argument can be repeated until all vertices in the graph are included in a path.

THEOREM 7.8

Let G be a strongly connected digraph with n number of vertices. If outdegree of $v \geq n/2$ and indegree of $v \geq n/2$ for each vertex v, then G is Hamiltonian.

PROOF

One can observe that such results cannot be obtained easily, therefore let us consider the types of digraphs that are Hamiltonian. Let us take the tournament graph to prove this. As we know a tournament graph is a digraph in which any two vertices are joined by exactly one edge. Because tournament graphs may be having sources or sinks, they are in general not Hamiltonian. But the following theorem shows that every tournament graph is nearly Hamiltonian.

THEOREM 7.9

(i) Every non-Hamiltonian tournament graph is semi-Hamiltonian.

(ii) Every strongly connected tournament graph is Hamiltonian.

PROOF

(i) As the statement is true when the tournament has less than four vertices. Let us prove this by induction on the number of vertices, and assume that every non-Hamiltonian tournament on n vertices is Semi-Hamiltonian (A non Hamiltonian graph G is said to be semi-Hamiltonian if there exists a path passing through every vertex. Graph in Fig. 7.44 is semi-Hamiltonian).

Consider P to be a non-Hamiltonian tournament graph on $n + 1$ vertices, and let P' be the tournament on n-vertices obtained by deleting from P a vertex v and its incident edges. By the hypothesis of induction, P' has a semi Hamiltonian path $v_1 \rightarrow v_2 \rightarrow v_3 \rightarrow \ldots \rightarrow v_n$. Now,

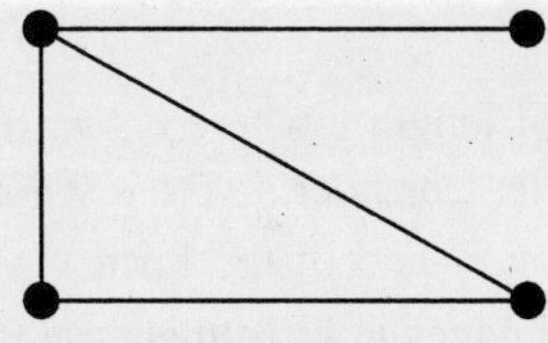

Fig. 7.44

there arises following three cases:

(a) If $(v\, v_1)$ is an edge in P, then the required path is $v \rightarrow v_1 \rightarrow v_2 \rightarrow \ldots v_n$.

(b) If $v\, v_1$ is not an edge in P, which implies that $v_1\, v$ is, and if there exists a j such that $v\, v_j$ is an edge in P, then if we choose j to be the first such, the required path is $v_1 \rightarrow v_2 \rightarrow \ldots \rightarrow v_{j-1} \rightarrow v \rightarrow v_j \rightarrow \ldots \rightarrow v_n$, which is shown in Fig. 7.45.

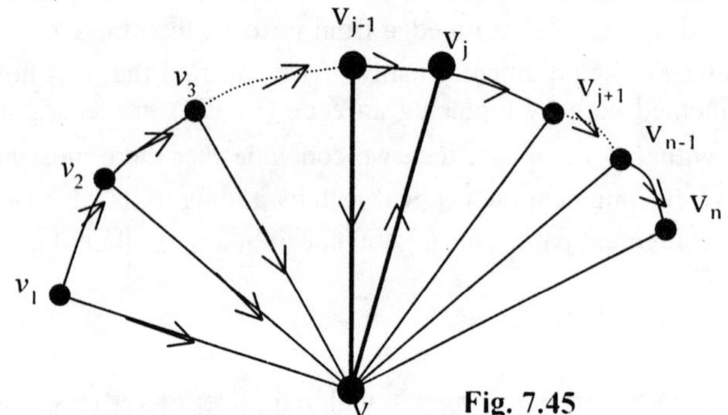

Fig. 7.45

(c) The third possibility is that there is no edge in P of the form $v\,v_j$, then the required path will be $v_1 \to v_2 \to \ldots \to v_n \to v$.

(iii) Let us prove that a strongly connected tournament graph P on n vertices contains cycle of length 3, 4, ... n. In order to show that P contains a cycle of length 3, let v be any vertex of P, and let T be the set of all vertices t such that t', w' is an edge in P and W be the set of all vertices such that wv is an edge. As we have assumed P to be strongly connected, T and W must both be non-empty, there must be an edge in P of the form $t'w'$, where t' is in T and w' is in W (See Fig. 7.46). The required cycle of length 3 is then $v \to t' \to w' \to v$.

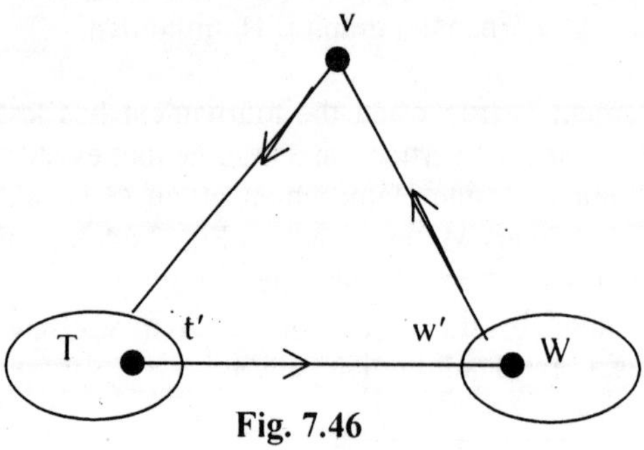

Fig. 7.46

Let us now show that there exist a cycle of length r where $r \leq n$, then there is one cycle of length $r + 1$. Let $v_1 \to \ldots \to v_r \to v_1$ be such a cycle. Consider a vertex v not contained in this cycle, and is such that there exist edges in P of the form $v\,v_j$ and of the form $v_j\,v$. This implies that there must be a vertex v_i such that both $v_{i-1}\,v$ and $v\,v_i$ are edges in T. In this case the required cycle will be $v_1 \to v_2 \to \ldots v_{i-1} \to v \to v_i \to \ldots v_r \to v_1$ (see Fig. 7.47).

If there exist no vertex with the above mentioned property, then the set of vertices not contained in the cycle may be divided into two disjoint sets T and Z, where T is the set of vertices t such that $t\,v_i$ is an edge for each i. since P is strongly connected T and W must both be nonempty, and there must be an edge in P of the form $t'w'$, where $t'$is in T and w' is in W. The required cycle is then (See Fig. 7.48).

$v_l \to t' \to w' \to v_3 \ldots \to v_r \to v_l$

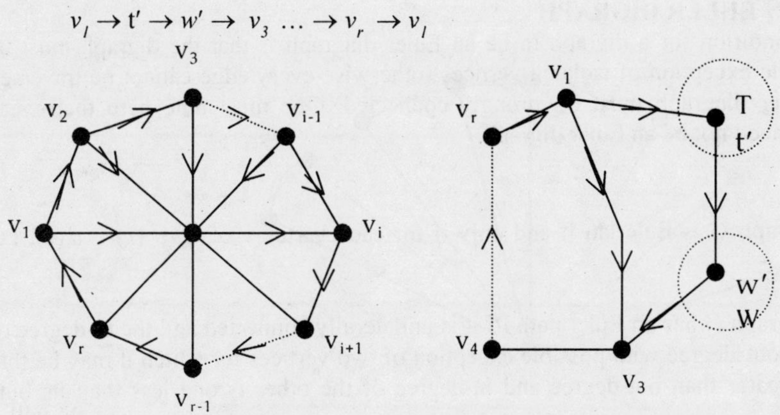

Fig. 7.47 Fig.7.48

7.23 EULERIAN DIGRAPHS

In this section we extend our study of Euler graph to digraphs.

If some closed walk in a digraph contains all the vertices and edges which traverses every edge of *G* exactly once, then the closed walk is called **directed Euler line**. A digraph containing a **directed Euler Line** is called an **Euler digraph**. For example, the digraph in Fig. 7.49 is not Eulerian, although its underlying graph is a Eulerian graph. The digraph in Fig. 7.50 is an Eulerian digraph, and the walk *u v w x y z* is an Euler line.

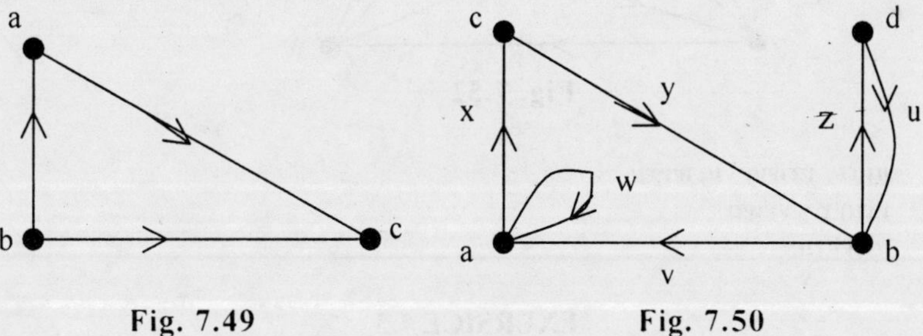

Fig. 7.49 Fig. 7.50

Example 7.33

The graph shown in Fig. 7.50 is an Euler digraph and $e_1\ e_2\ e_3\ e_4\ e_5\ e_6\ e_7$ is an Euler line.

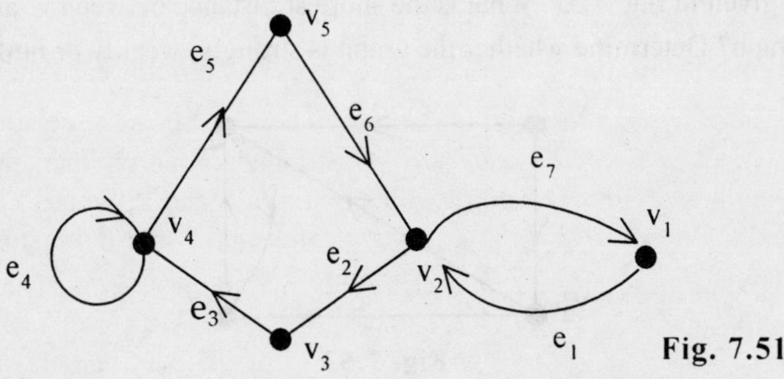

Fig. 7.51

7.24 EXISTENCE OF AN EULER DIGRAPH

The necessary condition for a digraph to be an Euler diagraph is that the digraph must be connected with the possible exception of isolated vertices (otherwise every edge cannot be traversed in one walk) and an Euler diagraph must be strongly connected. One must note here that every strongly connected graph need not be an Euler digraph.

THEOREM 7.10

A connected digraph G is Eulerian if and only if for each vertex v of G, $d^{+}(v) = d^{-}(v)$, i.e. balanced.

THEOREM 7.11

A directed multigraph G has an Euler path iff it is unilaterally connected and the in-degree of each vertex is equal to its out-degree with possible exception of two vertices for which it may be that indegree of one is one greater than out-degree and in-degree of the other is one less than its out-degree.

Example 7.34

In an tournament graph shown in the Fig. 7.52, find (i) cycles of length 3, 4, 5 (ii) an Eulerian trail (iii) a Hamiltonian cycle.

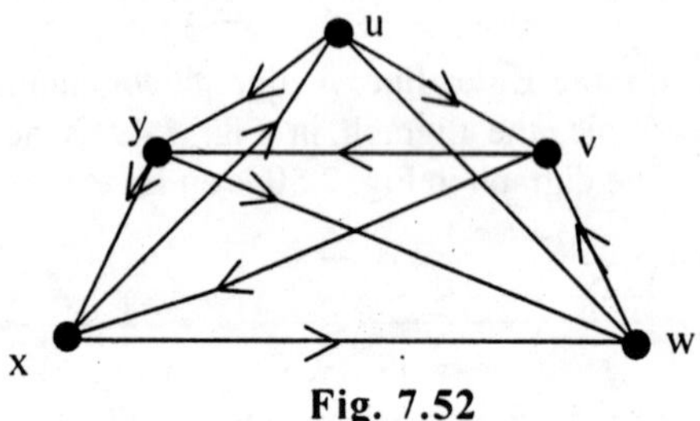

Fig. 7.52

Solution

(i) *uyxu, yxwvy, uywvxu.*
(ii) *yxuvx wvywu*
(iii) *uywvxu.*

EXERSICE 7.3

Define the path length in a graph G. Give three different elementary paths from v_1 to v_2 for digraph given in Fig. 7.53. What is the shortest distance between v_1 and v_2? Is there any cycle in the graph? Determine whether the graph is strongly, weakly or unilaterally connected.

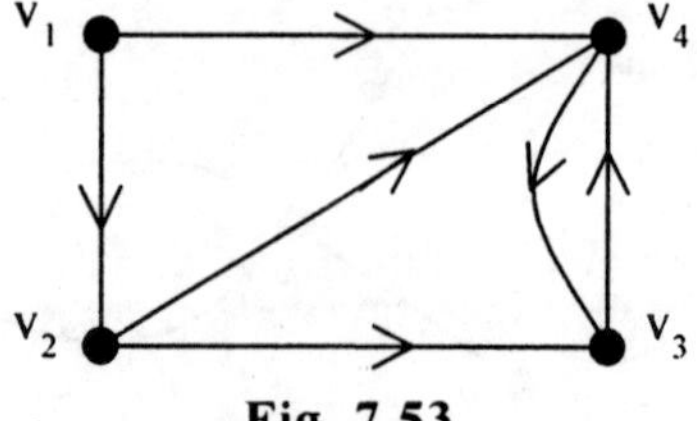

Fig. 7.53

2. Show that a simple digraph is strongly connected iff there is a cycle in G, which includes each node at least once and no isolated node.
3. (a) Show that the sum of the in-degrees over all vertices is equal to the sum of the out-degrees over all vertices in any directed graph.
   (b) Show that the sum of the squares of the in-degree over all vertices is equal to the sum of the squares of the outdegrees over all vertices in any directed complete graph.
4. Give three different elementary paths from v_1 to v_2 for the digraph given below:

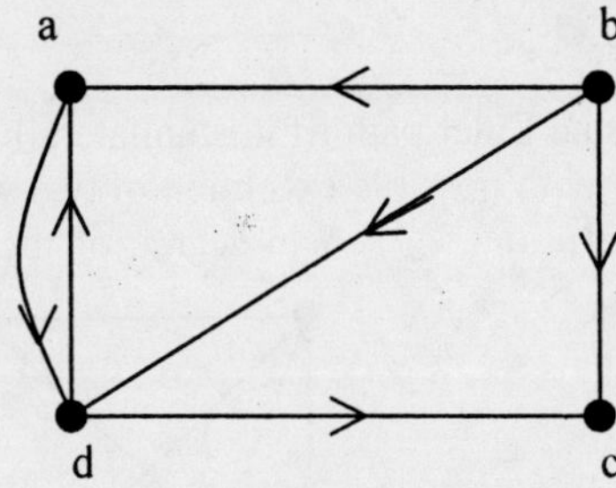

Fig. 7.54

5. Prove that in a simple digraph G every vertex of the digraph lines in exactly one strong component.
6. Give an example of a digraph which is unilaterally connected but not weekly connected.
7. Give an example of a digraph which is unilaterally connected but not strongly connected.
8. Sketch all distinct (nonisomorphic) orientations of a complete graph of four vertices.
9. (i) Show that the graph shown in Fig.7.55 is orientable.
   (ii) Show that any graph with Eulerian circuit is orientable.
   (iii) Show that any graph with Hamiltonian circuit is orientable.
   (iv) Show that a connected undirected graph is orientable iff each edge of the graph is contained in at least one circuit.

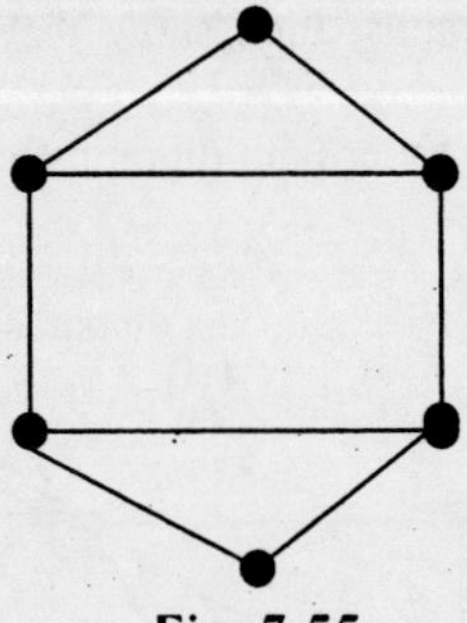

Fig. 7.55

10. Define isomorphism of directed graphs. Sketch all different (nonisomorphic) simple digraphs with 1, 2 and 3 vertices. Determine whether the given pair of directed graphs is isomorphic.

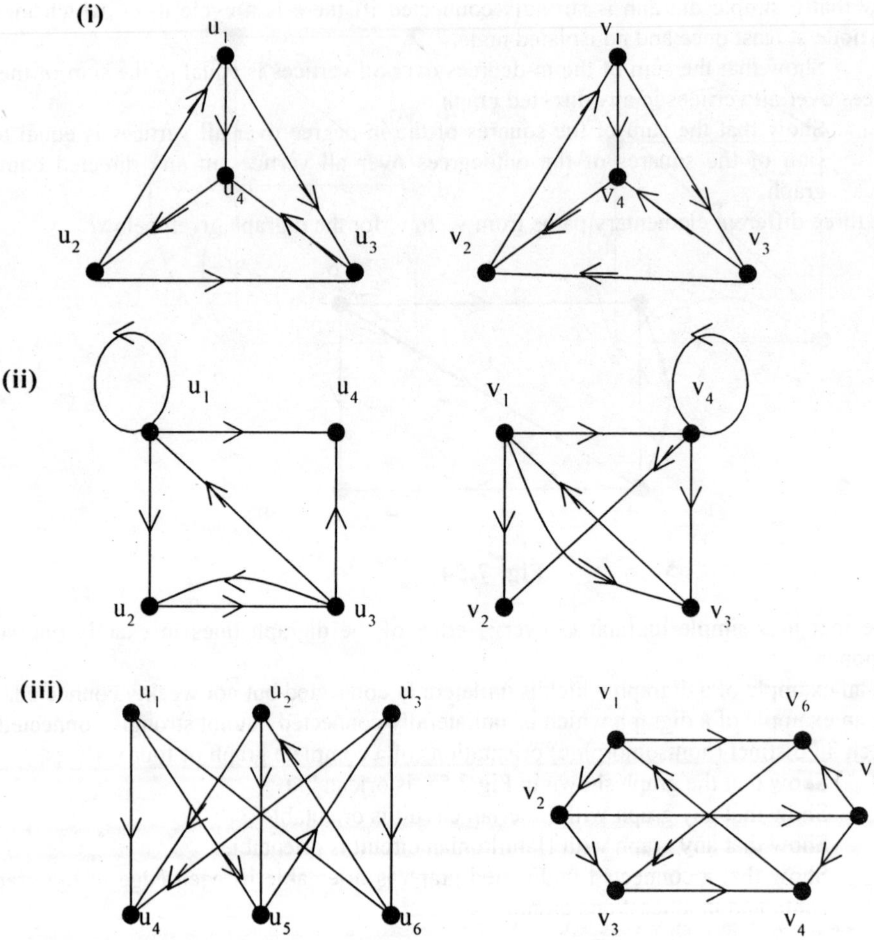

(i)

(ii)

(iii)

11. If G and H are isomorphic digraphs, then show that the converse of G and H are also isomorphic.

12. Determine whether the following Euler path digraphs has an (a) Euler Circuit (b) If exist, Construct.

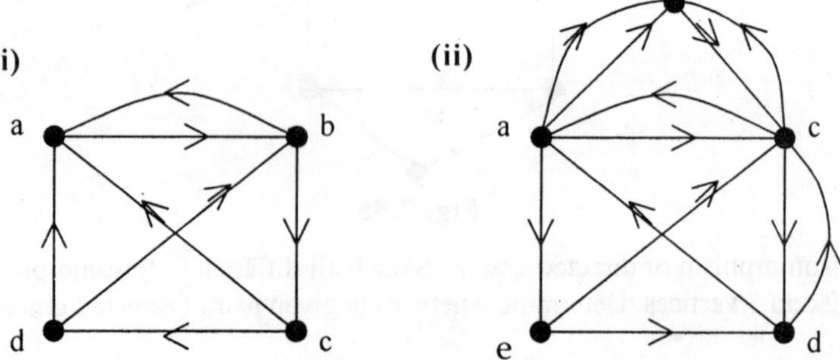

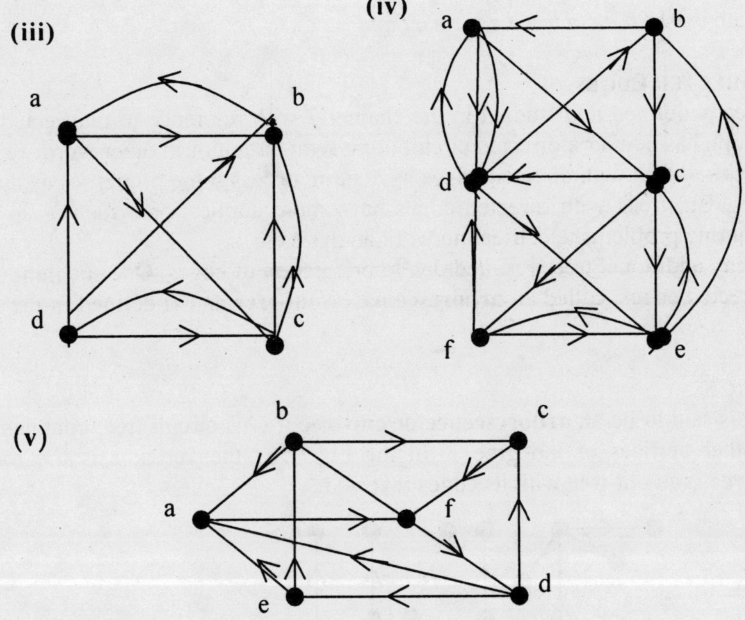

13. Prove that every Euler digraph without isolated vertices is strongly connected. With the help of a counter example show that the converse is not true.

ANSWERS

7. Ans.

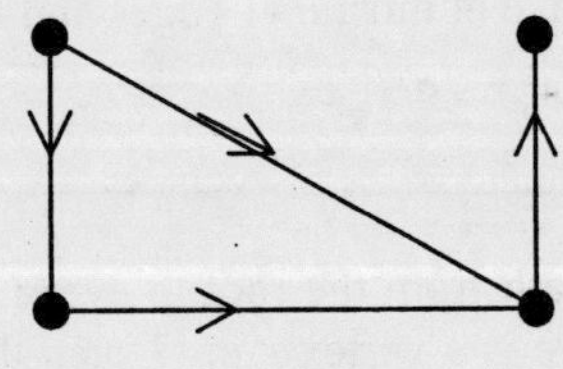

8. **Ans.**

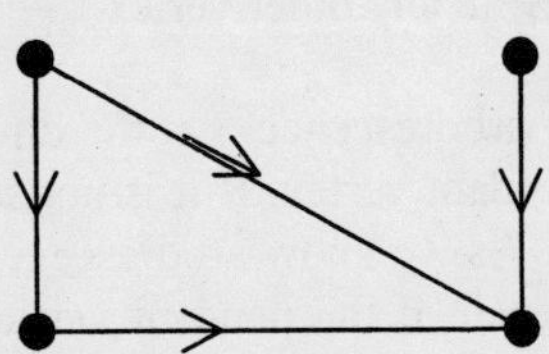

10. (i) isomorphic
    (ii) isomorphic
12. (i) No Euler circuit
    (ii) No Euler circuit

(iii) Euler path exist $d\ a\ b\ ac\ b\ c\ d\ b$

(iv) Euler path exist $f\ d\ a\ c\ d\ b\ a\ d\ e\ b\ c\ e\ c\ f\ e\ f$.

7.25 TREES WITH DIRECTED EDGES

A tree is already defined and studied in the chapter 3 with reference to undirected graph. A tree in a connected graph has neither a directed circuit nor a semi circuit or in other word we can say it has no circuit. Properties of tree with directed edges have more or less same properties as that of trees with undirected edges. But trees with directed edges have more applications, such as in computer programming and counting problems, electrical network analysis etc.

There are many additional trees resulted due to orientation of edges. One such important tree is rooted tree with directed edges, called an **arborescence** or **out-tree** and is defined in the following section.

7.26 ARBORESCENCE

A digraph G is said to be an **arborescence** or **out-tree** if G is circuit free, contains a root of indegree 0 and all other vertices of indegree 1. In the Fig.7.56 the vertex r is the root of the arborescence. An **in-tree** is an out-tree with its edges reversed.

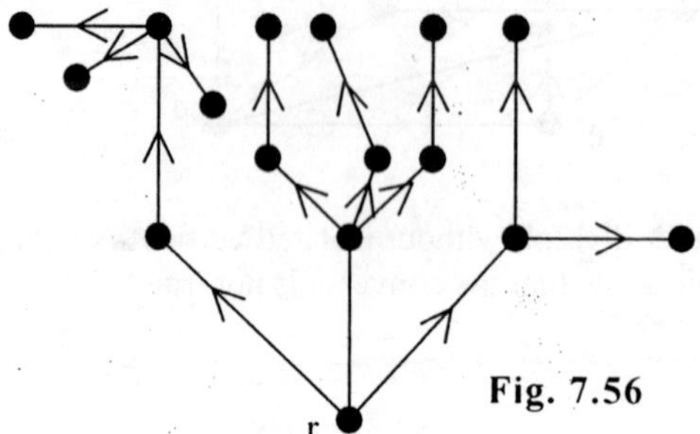

Fig. 7.56

NOTE

An arborescence with n vertices can have at most n - 1 edges and therefore some of indegrees of all vertices of G, i.e. $d^-(v_1) + d^-(v_2) + \ldots + d^-(v_n) \le n-1$.

THEOREM 7.12

An arborescence contains a directed path from the root to every other vertex. Conversely, a circuit free digraph G is an arborescence if there is a vertex r in G such that every other vertices can be accessed from r while r can not be accessed from any other vertex.

PROOF

Let us consider a directed path d in an arborescence. Let the directed path starts from root r and continue as far as possible and ends at a pendant vertex. If it is not so we get a vertex whose in-degree is two or more, which is a contradiction. As we know arborescence is connected, every vertex lies on some directed path from the root r to each of the pendant vertices. Conversely, G is a tree because every vertex in G can be accessed from r, and G has no circuit. Further, as r is not accessible from any other vertex, indegree of vertex r = 0. Since every other vertices are accessible, therefore indegree of each other vertices will be at least one. But indegree cannot be greater than one because if G contains n vertices than there will be e - 1 edges in G.

7.26-1 Application of Arborescences

Here we discuss an important application of arborescences to the theory of computer algorithm.

7.26-2 Polish Notation

Consider an algebraic expression involving binary operations, i.e. addition, subtraction, multiplication and division.

$$x + y - \frac{pq}{r^a - s} \qquad \dots (1)$$

If we want this expression be evaluated by the computer, it must perform the operation in certain order or in other words we can say to evaluate such an expression, the machine will have to scan the expression to find the sequence of operations to be performed.

The process of writing the operators of an expression before their operands (or after their operands) is called the **Polish notation** in honour of its discoverer, the polish mathematician Luksiewicz. The fundamental property of polish notation is that the order in which the operations are to be performed is completely determined by the positions of the operators and operands in the expression. Accordingly one never needs parenthesis while writing expression in polish notation. The expression (i) in procedural language may be written as

$$x + y - p\,q \div (r \uparrow a - s) \qquad \dots (2)$$

The translation of expression from procedural language into polish notation is extremely important in compiling and can be done by first representing the given expression by means of an arborescence as follows:

(i) Represent each variable (or constant) of the expression by a pendant vertex.

(ii) Each internal vertex represents a binary operator having the two sub arborescences as its operands.

7.26-3 How To Obtain The Expression In Polish Notation

In order to obtain the expression in polish notation, we traverse the arborescence starting from the root from left to right and from top to bottom as shown in the Fig.7.57 by the dotted line. Whenever we encounter a vertex that has not been traversed before, we append its label to the existing string.

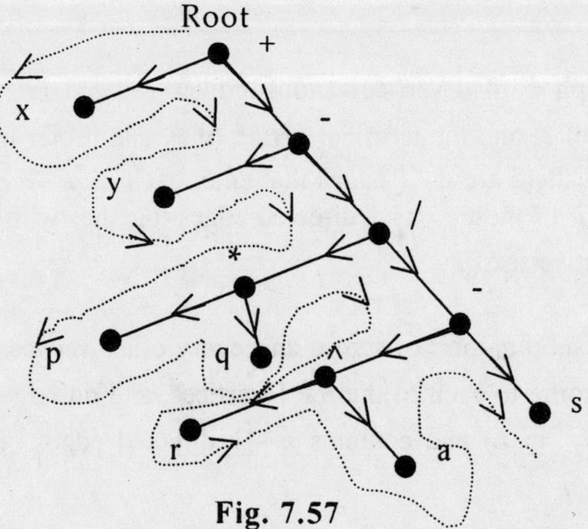

Fig. 7.57

Adopting this process, the Fig.7.57 gives

$$+ x - y \div pq - \uparrow r\, a\, s \qquad \ldots (3)$$

7.26-4 Evaluation of an Expression In Polish Notation

To evaluate an expression in polish notation follow following procedure:

Begin at the right extreme and move to the left. When we come across an operator, the operation is performed between the two operands immediately to the right of it. Now treat the resultant as one operand for the next operation. One can verify when we adopt this procedure, we find expression (3) equal to (2).

7.26-5 Spanning Arborescence

A spanning tree in an undirected graph consisting of n -1 edges is analogous to a spanning tree in an n-vertex connected digraph. A **spanning arborescence** in a connected graph is a spanning tree that is an arborescence. For instance, spanning arborescence in Fig.7.58 is $\{e_6, e_7, e_8\}$.

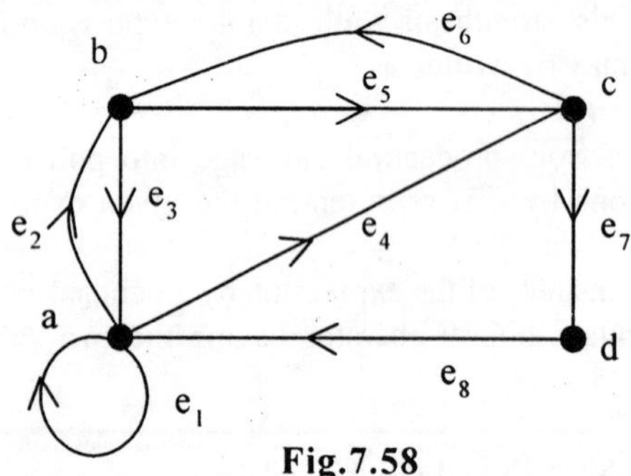

Fig.7.58

7.26-6 Relationship between a Spanning Arborescence and Euler Line

The relationship between a spanning arborescence and Euler line can be demonstrated with the help of following theorems:

Theorem 7.13

Consider a connected digraph G of n vertices and m edges. Let $L = (e_1, e_2, \ldots\ldots, e_m)$ be an Euler line. Let a is the initial vertex of e_1 and the terminal vertex of e_m (i.e. Euler line starts and ends at vertex a). Among the m edges in L there are n - 1 edges that enter each of n - 1 vertices, other than a for the first time. The subdigraph G_1 of these n - 1 directed edges together with the n vertices is a spanning arborescence of G_1 rooted at vertex a.

Proof

Indegree of vertex a in the subdigraph G_1 is zero and every other vertices is of indegree 1; since G_1 includes exactly one edge going to each of the n - 1 vertices, and no edge going to a. G_1 in connected (from the definition of G_1 in L) and contains n - 1 directed edges. Therefore, G_1 is a spanning arborescence in G rooted at a.

Example 7.35

In Fig.7.58 $L = \{e_7, e_8, e_1, e_4, e_6, e_3, e_2, e_5\}$ is an Euler Line, which is started and ended at c. The subdigraph $\{e_6, e_7, e_8\}$ is a spanning arborescence having root at c.

NOTE

If the digraph is Eulerian, theorem 7.12 gives a method of obtaining a spanning arborescence rooted at any specified vertex.

THEOREM 7.14

Consider a digraph G and a spanning in-tree T in G having root at vertex r. Then a directed walk D - $\{e_1, e_2, ..., e_n\}$ will be a directed Euler line, if it is constructed in the following way:

(i) No edge is used in D more than once, i.e. $e_k \neq e_j$ when $j \neq k$.

(ii) In existing a vertex the one edge that belongs to T is not used until all other outgoing edges have been traversed.

(iii) D determinates only when it cannot be continued by rule (i), i.e. the walk is ended only when a vertex is reached from where there is no edge left on which to exit.

PROOF

Since graph G is balanced ($\bar{d}(v) = \overset{+}{d}(v)$) hence all the vertices must have been entered as often as they have been left; therefore walk w must terminate at root r. Let us assume that there is an edge, say e in G and is not included in D. Let the terminal vertex of e be v. As G is balanced, there should be an edge say f, not included in D whose initial vertex is v. But by rule (i) edge f going out of v must be in T. It implies, this discarded edge leads to another discarded edge g in T, and so on. Finally we reach at r and find an outgoing edge there, which is not included in D. This is a contradiction of rule (i).

Example 7.36

The subdigraph shown in Fig.7.59 $\{e_2, e_3, e_5, e_{10}, e_{11}\}$ is a spanning in-tree.

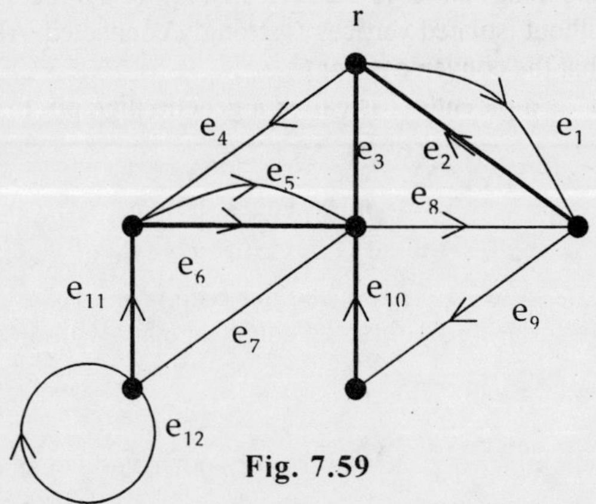

Fig. 7.59

THEOREM 7.15

If all the above three conditions of theorem 7.13 are satisfied then the number of different

Euler lines formed from a given in-tree T is $\prod_{i=1}^{n} [\overset{+}{d}(v_i) - 1]!$.

PROOF

If we start with edge e_l at r, the number of different Euler lines formed from a given in-tree T, can be computed by taking all the alternatives available at each vertex, after starting with e_l. Since at each vertex in T there is exactly one outgoing edge and this edge is to be selected last by rule (ii), the remaining $[\overset{+}{d}(v_i) - 1]!$ edges at vertex v_i can be selected in $[\overset{+}{d}(v_i)-1]!$ ways. But these are our independent choices, we have altogether $\prod_{i=1}^{n} [\overset{+}{d}(v_i)-1]!$ different Euler lines that satisfy the three rules in theorem 7.13.

Example 7.37

Let us find different directed Euler lines in Fig.7.21. By the application of above three rules in the in-tree starting with edge e_4, we will get two directed Euler lines, i.e. we see that the value of $\prod_{i=1}^{n} [\overset{+}{d}(v_i)-1]!$ for Fig.7.59 is 2.

NOTE

It is to be noted that the value 2 obtained in the above example are not all the directed Euler lines in the digraph, but these are generated by the specific in-tree in accordance with the rules in theorem 7.13.

EXERCISE 7.4

1. Prove that every edge in a digraph belongs either to a directed circuit or a directed cut-set.
2. Prove that every Euler digraph without isolated vertices is strongly connected. Also show, by constructing a counter example, that the converse is not true.
3. A digraph defined on the relation "is a parent of" is called a genetic digraph. Investigate the properties of **genetic digraph**.

8 ENUMERATION OF GRAPHS

8.0 INTRODUCTION

Enumeration of graphs is one of the most involved areas in graph theory and deserves an entire volume itself. In this chapter we will briefly discuss some enumerative techniques. The most important of them is Polya's counting technique.

8.1 TYPES OF ENUMERATION

Basically Graph enumeration can be divided into two categories:

(a) Counting particular kind of graph or digraph. For instance all connected graphs with 6 vertices and 2 cycles.

(b) Counting subgraphs of a graph of particular type in a given graph. For instance the number of edge disjoint paths of length k between vertices v_1 and v_2 in G.

Example 8.1

If we have n unordered pair of vertices and e edges then in the construction of all simple graphs there are $n(n - 1)/2$ unordered pair of vertices. In case the vertices are labelled, we will treat the vertices as separate from one another, then there are

$$\binom{n(n-1)/2}{e} \qquad \text{... (1)}$$

ways of selecting e edges from the graph. In other words we can say that the expression $n(n - 1)/2$ gives the number of simple labelled graphs with n vertices and e edges. The number of different unlabeled graphs of certain kind equals the number of equivalence classes under isomorphism of the labeled graphs.

THEOREM 8.1

The number of simple, labelled graphs of n vertices is $2^{n(n-1)/2}$.

PROOF

In the expression (1) if we put 0, 1, 2, 3, ...,$n(n - 1)/2$ for e, then the number of simple graphs of n vertices and 0, 1, 2, 3, ... $n(n - 1)/2$ edges can be obtained. The sum of all such numbers is the number of all simple graphs with n vertices. Using following identity one can prove the theorem

$$\binom{m}{0} + \binom{m}{1} + \binom{m}{2} + \ldots.. + \binom{m}{m-1} + \binom{m}{m} = 2^m.$$

THEOREM 8.2 (CAYLEY'S FORMULA)

There are m^{m-2} labelled trees with m vertices ($m \geq 2$).

PROOF

Let us label m vertices of tree T as 1, 2, 3, ..., m. If we go on removing the pendant vertex having smallest label and the edge joining it, we perform these operation m - 2 times, and finally we will be left with only two vertices when the process is terminated. Let $p_1, p_2, p_3, \ldots p_{m-2}$ are the pendant vertices with the lowest label when we are left with m - 1, m - 2, m - 3, ..., 2 vertices respectively after $1^{st}, 2^{nd}, 3^{rd}, \ldots .(m-2)^{th}$ steps and $q_1, q_2, \ldots . q_{m-2}$ are the vertices adjacent to $p_1, p_2, \ldots p_{m-2}$ respectively after $1^{st}, 2^{nd}, 3^{rd}, \ldots .(m-2)^{th}$ steps. Then the tree T defines the sequence

$(q_1, q_2, \ldots . q_{m-2})$ uniquely ... (1)

Conversely, consider the sequence $q_1, q_2, \ldots . q_{m-2}$ of m-2 labels, then a m vertex tree can be formed uniquely as follows:

Find the first number in the sequence

$1, 2, 3, \ldots, m$ ... (2)

which is not there in the sequence (1). Obviously, this number is p_1. At this stage edges (p_1, q_1) is defined. Delete q_1 from sequence (1) and p_1 from (2). Now, in the sequence (2) we find the first number that does not appear in the remainder of (1). This would be p_2, and thus edge (p_2, q_2) is defined. Continue this process till we are left with no element in the sequence (1).

Now for each of the m - 2 elements in sequence (1), we can select any one of m numbers, thus forming

m^{m-2} ... (3)

(m - 2) tuples, each defining a distinct labelled tree of m-vertices. As every tree defines one of these sequences uniquely,

$\Rightarrow$ there is a one-to-one correspondence between the trees and the m^{m-2} sequence. Hence the theorem.

Example 8.2

Consider a vertex labelled tree, which yields sequence (1, 1, 3, 5, 5, 5, 9). It is to be noted here that the vertex j appears in the sequence (1) iff it is not pendant.

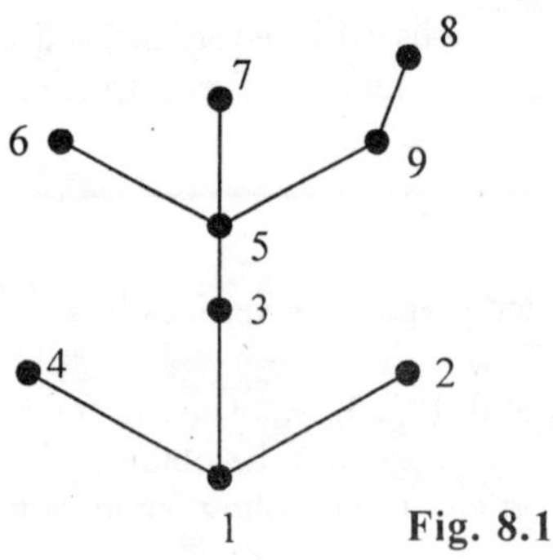

Fig. 8.1

Again consider a sequence (4, 4, 3, 1, 1). We can form a seven vertex tree as follows: (2, 4) is the first edge. Second edge is (5, 4), then (4, 3), (3, 1), (6, 1) and finally (7, 1). See Fig. 8.2.

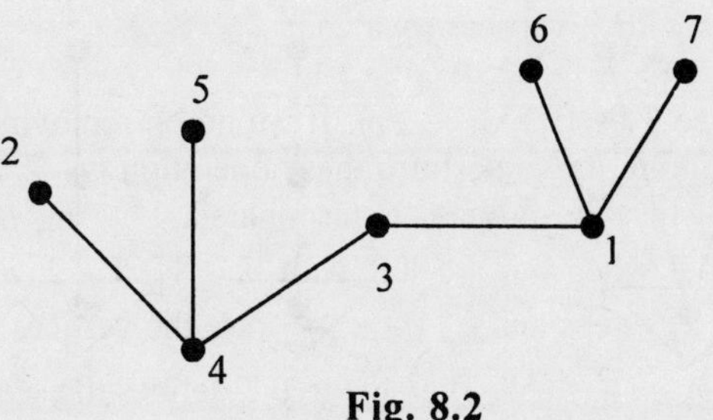

Fig. 8.2

REMARK

As we know in a rooted tree one node is marked as root. Hence, in this case, for each of the m^{m-2} labelled trees, we have m rooted labelled trees, because any of the m vertices can be made a root. Hence the number of different rooted labelled trees with m vertices is m^{m-1}. Rooted trees for $m = 1$, 2, and 3 are given in the Fig. 8.3.

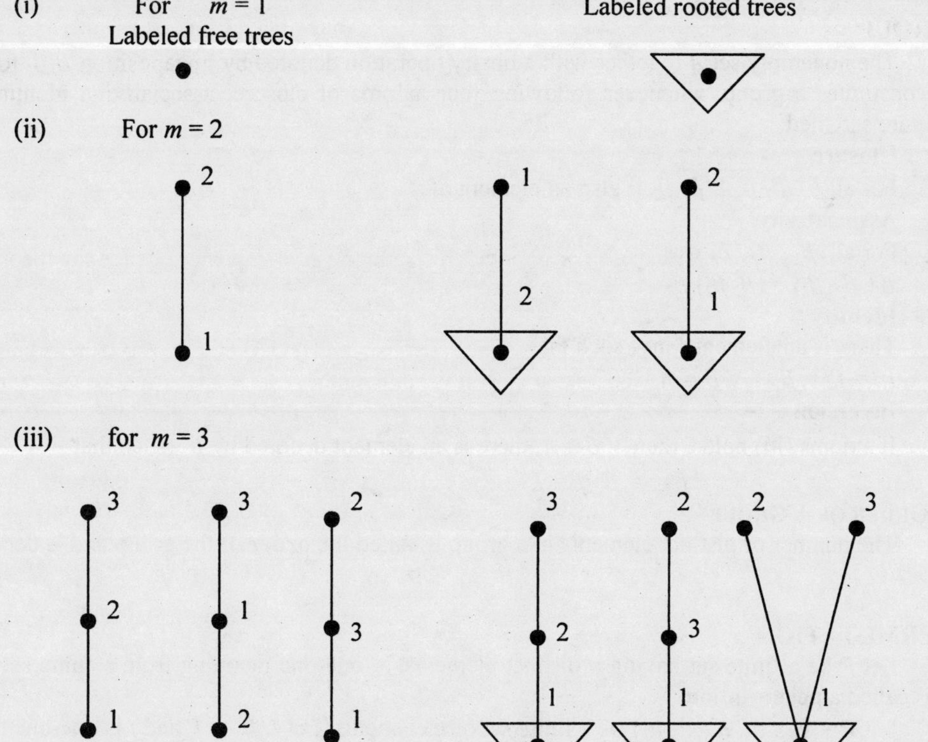

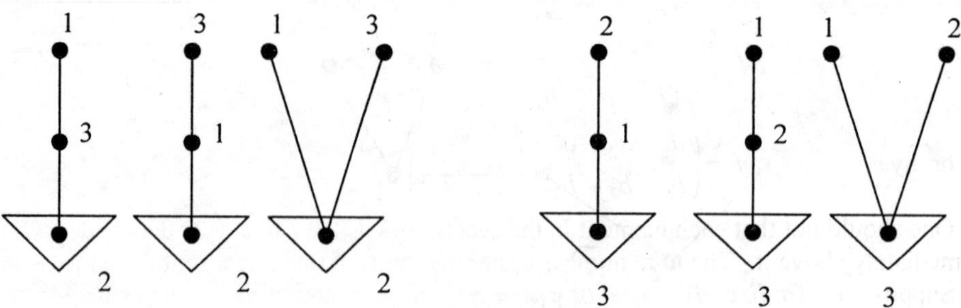

Fig. 8.3 (Rooted labeled trees of one, two and three vertices)

THEOREM 8.3

The number of trees with vertex set $[n]$ in which vertices have degree $d_1, \ldots, d_n$ respectively is $\dfrac{(n-2)!}{\prod(d_i - 1)!}$.

8.2 GROUP

The nonempty set A together with a binary operation denoted by juxtaposition $\beta_1\beta_2$ for β_1, β_2 in A, constitutes a group, whenever following four axioms of closure, associativity, identities and inverse are satisfied:

(i) **Closure**

For all $\beta_1, \beta_2 \in A$, $\beta_1 \beta_2$ is also an element of A.

(ii) **Associativity**

For all $\beta_1, \beta_2, \beta_3 \in A$

$\beta_1(\beta_2 \beta_3) = (\beta_1 \beta_2) \beta_3$.

(iii) **Identity**

There is an element I in A such that

$I\beta = \beta I = \beta \quad \forall \beta \in A$

(iv) **Inversion**

If axioms (iii) holds, then $\forall \beta \in A$, there is an element denoted by β^{-1} such that

$$\beta\beta^{-1} = \beta^{-1}\beta = I.$$

8.2-1 ORDER OF A GROUP

The number of distinct elements in a group is called the **order** of the group and is denoted by $O(G)$.

8.3 PERMUTATION

Let P be a finite set having n distinct elements. A one-one mapping from a finite set P onto itself is called a **permutation**.

Let $P = \{a_1, a_2, a_3, \ldots, a_n\}$ be a finite set of n elements. Let $f: P \to P$ and f is one-one then f is permutation of degree n. Let $f(a_1) = b_1$, $f(a_2) = b_2$, $\ldots, f(a_n) = b_n$, where $\{b_1, b_2, \ldots, b_n\} = \{a_1, a_2, \ldots, a_n\}$. Here $\{b_1, \ldots, b_n\}$ is nothing but arrangement of n elements of P.

The permutation can be symbolised by

$$f = \begin{matrix} (a_1 & a_2 & a_3 & \ldots\ldots & a_n) \\ \downarrow & \downarrow & \downarrow & \ldots\ldots & \downarrow \\ (b_1 & b_2 & b_3 & \ldots\ldots & b_n) \end{matrix}$$

or by $$f = \begin{pmatrix} a_1 & a_2 & a_3 & \ldots\ldots & a_n \\ b_1 & b_2 & b_3 & \ldots\ldots & b_n \end{pmatrix}$$

One should not that each element in the second row is the f-image of the element of first row lying immediately above it. The total number of permutations defined on a set of n symbols is $n!$.

Suppose $A = (a, b, c, d)$ is a set of 4 elements. A permutation on the elements denoted by f is represented as

$$f = \begin{matrix} (a & b & c & d) \\ \downarrow & \downarrow & \downarrow & \downarrow \\ (b & d & c & a) \end{matrix}$$

This permutation can also be represented with the help of a digraph as shown in Fig.8.4

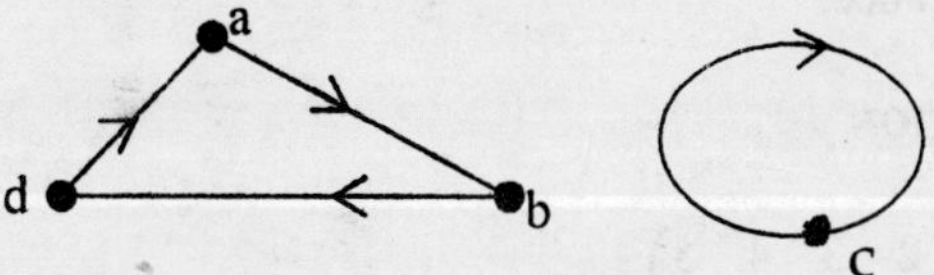

Fig. 8.4

8.3-1 CYCLIC PERMUTATION

A permutation on n symbol is said to be a cyclic permutation if out of total n elements, m elements can be arranged as the image of each element in this arrangement is its preceding element and the image of the last element is first element. The image of all the remaining $n - m$ element is self. Here m is known as **length of the cycle** in permutation.

Now consider the permutation $f = \begin{pmatrix} a & b & c & d \\ b & d & c & a \end{pmatrix}$.

This can also be represented as $f = \begin{pmatrix} a & b & d \\ b & d & a \end{pmatrix}\begin{pmatrix} c \\ c \end{pmatrix}$.

This permutation can further be represented as ((*abd*) (*c*)), where (*abd*) and (*c*) are called **cycles of permutation**. We can observe here that every permutation can be represented as a cyclic permutation. The number of edges in a cycle is called **length** of the cycle. In (*abd*), the length is 3 and in (*c*) it is one. Any cycle of length one is an **identity** permutation.

The number of elements in the object set A is called the **degree** of the permutation. The degree of the permutation (*abd*) (*c*) is 4. A permutation of degree k is said to be of type $(c_1, c_2, \ldots, c_k)$ if it has c_i cycles of length i, $i = 1, 2, \ldots, k$. The permutation ((*abd*)(*c*)) is of the type (1, 0, 1, 0), since we have one cycle (*c*) of length 1, 0 cycle of length 2, 1 cycle (*abd*)of length 3 and 0 cycle of length 4.

Another method to represent this type of permutation is by introduction of k dummy variable (also called the cycle structure) of the type y_i (i = 1, 2, 3,... k). In this method permutation is represented as

$$y_1^{c_1}, y_2^{c_2}, y_3^{c_3}, y_4^{c_4}, y_5^{c_5}, y_6^{c_6}, y_7^{c_7}, y_8^{c_8}, \ldots, y_k^{c_k}$$

Here the dummy variable y_i, has no significance except in this subscripts represents length of the cycles and the exponent represents the number of cycles attached.

We note that

$1\, c_1 + 2\, c_2 + 3\, c_3 + \ldots + k\, c_k = k$, the degree of set A.

For illustration see following example.

Example 8.3

Suppose $A = (a, b, c, d, e, f, g, h)$. If permutation f on A is $(abf)(c)\,(deh)\,(g)$ then it is of type (2, 0, 2, 0, 0, 0, 0, 0) as we have 2 cycles namely, (c) and (g) of length 1; 2 cycles, (abf) and (deh) of length three, and there are no cycles of length 2, 4, 5, 6, 7, 8. Introducing a dummy variable (also called the **cycle structure**) y_i, this can be represented as

$$y_1^2\, y_2^0\, y_3^2\, y_4^0\, y_5^0\, y_6^0\, y_7^0\, y_8^0 = y_1^2\, y_3^2$$

where index i of y denotes its length and the exponent the number of cycles of length i. The y's are dummies and have no other significance.

8.3-2 Composition of Permutation

Let

$$f = \begin{pmatrix} 1 & 2 & 3 & 4 & 5 \\ 2 & 1 & 4 & 5 & 3 \end{pmatrix}$$

and

$$g = \begin{pmatrix} 1 & 2 & 3 & 4 & 5 \\ 3 & 4 & 1 & 2 & 5 \end{pmatrix}$$

A composition of permutation fg is another permutation obtained by first applying f and then applying g on the resultant, thus

$$fg = \begin{pmatrix} 1 & 2 & 3 & 4 & 5 \\ 4 & 3 & 2 & 5 & 1 \end{pmatrix}$$

For example, $f(1) = 2$ and $gf(1) = g(2) = 4$.

8.3-3 Permutation Group

A collection of $n!$ permutations acting on a set A and given by $(f_1, f_2, \ldots, f_n)$ forms a **permutation group** if the axioms of closure, associativity, identity and inverse are satisfied. For example it can be shown that for a set $A = (a, b, c, d)$, the set of permutation $((a), (b), (c), (ac), (bd), (abcd)\,(adcb))$ forms a permutation group.

The number of elements in the permutation group is called the **order** while the number of elements of the object set A on which the permutation group is defined is called the **degree** of the permutation group. If a permutation group A acts on object set X, then $|A|$ is the order of the group and $|X|$ is the degree. It can be shown that the set of all $k!$ permutations on set A of k elements form a permutation group. Such a group of order $k!$ and degree k is called the **full symmetric** group, denoted by S_k.

8.3-4 Cycle Index of a Permutation Group

For a permutation group of order *n*, if we add cycle structure of all *n* permutations and divide the sum by its order *n*, then we get an expression called the cycle index *Z(P)*.

Example 8.4

Suppose we are given the cycle structures $y_1^3, y_1y_2, y_1y_2, y_1y_2, y_3y_3$ where $A = (a, b, c)$, then the cycle index is given by

$$Z(A) = (y_1^3 + 3y_1y_2 + 2y_3)/6.$$

Exactly in similar fashion, we get cycle index of permutation group of degree 4 and order 4, which is *(a) (b) (c) (d)*, *(ac) (bd)*, *(abcd) (abcb)*,

$$\frac{1}{4}\left(y_1^4 + y_2^2 + 2y_4\right)$$

Let us form a table for different permutation possible for full symmetric group of degree 4.

Permutation	Number of such Permutation	Cycle Structure
(4, 0, 0, 0)	1	y_1^4
(2, 1, 0, 0)	6	$y_1^2 y_2$
(1, 0, 1, 0)	8	$y_1 y_3$
(0, 2, 0, 0)	3	y_2^2
(0, 0, 0, 1)	6	y_4

If we want to get cycle index of S_4 from this table, we multiply the corresponding entries in the second and third columns, add the products, and then divide the sum by 4!, the order of the group. Thus

$$Z(S_4) = \frac{1}{24}(y_1^4 + 6y_1^2 y_2 + 8y_1 y_2 + 3y_2^2 + 6y_4)$$

In order to display the variables involved, the cycle index of a permutation group P is often written as

$$Z(P) = Z(P; y_1, y_2 \ldots, y_k).$$

8.3-5 Cycle Index of the pair Group

If we are interested in permutation of *n* vertices of graph *G*, the *n*(*n*-1)/2 unordered vertex pairs also get permuted. For instance, let $S = \{a, b, c, d\}$ be the set of vertices of a four-vertex graph.

The permutation

$$\beta = \begin{pmatrix} a & b & c & d \\ d & b & a & c \end{pmatrix}$$

on the vertices induce the following permutation on the six unordered vertex

$$\beta' = \begin{pmatrix} ad & ac & ad & bc & bd & cd \\ db & da & dc & ba & bc & ac \end{pmatrix}$$

The permutation β and the induced permutation are shown in the Fig. 8.5

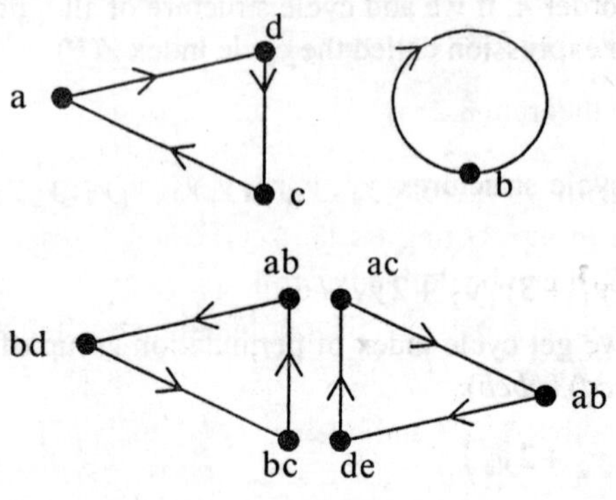

Fig. 8.5

NOTE

(i) Each of the $n!$ possible permutations on the n-vertices of the graph results in some permutation of the $n(n-1)/2$ unordered vertex pairs.

(ii) In case of digraph it will result in $n(n-1)$ ordered vertex pairs.

(iii) If a set of permutations on the vertices forms a group, the induced set of permutations on the pair of vertices also form a group.

8.3-6 PAIR GROUP

The full symmetric group S_n on n vertices of a graph induces a group R_n of $n!$ permutation on the pair vertices (except for $n = 2$, in which case the number of possible permutations on the pair is 1 rather than 2). Such an induced group is called the **pair group R_n**.

Example 8.5

Consider pair group R_4 induced by S_4 the full symmetric group on the vertices of a four-vertex graph. It is to be observed that identity permutation* on the four vertices of the graph yields an identity permutation on the six pairs of vertices. A permutation with two cycles of length one and one cycle of length two yields two cycle of length one and two cycles of length two and so on. The following table gives the cycle structures of permutation in S_4 and corresponding cycle structures of the induced permutations on the pair of vertices

Term in Z(S₄)	Induced term in Z(R₄)	No. of permutations
y_1^4	y_1^6	1
$y_1^2 y_2$	$y_1^2 y_2^2$	6
$y_1 y_3$	y_3^2	8
y_2^2	$y_1^2 y_2^2$	3
y_4	$y_2 y_4$	6

* Identity permutation is a permutation which takes every element to itself.

$\therefore$ the cycle index of the pair group R_4 (induced on the pairs of vertices by S_4) is

$$Z(R_4) = \frac{1}{24}\left(y_1^6 + 9y_1^2 y_2^2 + 8y_3^2 + 6y_2 y_4\right) \qquad ...(1)$$

For a general expression for $Z(R)$ see theorem 8.2.

8.3-7 Identical Permutation Group

If A and B are permutation groups acting on the sets X and Y respectively and if they are isomorphic, will be denoted by $A \cong B$. It can be defined as follows:

$A \cong B$ if there is a one-to-one map $f\colon A \leftrightarrow B$ between the permutations such that

$$\forall \beta_1, \beta_2 \in \mathrm{A}, \quad f(\beta_1\beta_2) = f(\beta_1)f(\beta_2) \qquad ...(1)$$

Identical permutation group is denoted by $A \equiv B$ and other than (1) it should have an one-to-one map $g : X \leftrightarrow Y$ between the objects such that $x \in X$ and $\alpha \in A$

$$g(\alpha x) = g(\alpha)\, g(x).$$

8.4 AUTOMORPHISM

An automorphism of a graph G is an isomorphism of G with itself. Thus each automorphism β of G is a permutation of the point set V which preserves adjacency.

Note

- Any automorphism followed by automorphism is also an automorphism hence automorphism of G form a permutation group $\Gamma(G)$, which acts on the points of G. It is known as the **group** of G or sometimes as the **point group** of G. The group $\Gamma(D)$ of a digraph is defined similarly.

8.5 POWER GROUP

Let A and B are permutation groups acting on the sets X and Y respectively. The **power group** denoted by B^A acts on Y^X, the set of all functions from X into Y. We will always assume that power group acts on more than one function.

For each pair of permutations α in A and β in B, there is a unique permutation, written β^α in B^A. Let us specify the action of β^α on any function φ in Y^Z by the following equation which gives the image of each $x \in X$ under the function $\beta^\alpha \varphi$:

$$(\beta^\alpha \varphi)(x) = \beta\, \varphi(\alpha x).$$

8.6 ORBIT

Consider a permutation group acting on objects of the set X. The **orbit** of x, for any element $x \in X$ denoted by $\sigma(x)$, is the subset of X which consists of all elements z in X such that for some permutation β in A, $\beta x = z$.

8.7 STABILIZER

The stabilizer of x denoted by $A(x)$, is the subgroup of A which consists of all the permutations in A that leave x fixed.

The result follows from an application of the formula $|\sigma(x)| . |A(x)| = |A|$ and its interpretation in the present context.

8.8 GENERATING FUNCTIONS

Generating function is one o the most important tools for enumerations. Basically the generating function is a method to solve the recurrence relations. A generating function $G(x)$ is define as a power series

$$G(x) = a_0 + a_1x + a_2x^2 + \dots \text{ in some dummy variable } x.$$

The coefficient a_k of xk is the number, which depends on a collection of k objects to be enumerated.

e.g. $$(1+x)^n = \binom{n}{0} + \binom{n}{1} + \binom{n}{2}x^2 + \binom{n}{n}x^n$$

$$= {}^nC_0x^o + {}^nC_1x^1 + {}^nC_2x^2 + {}^nC_nx^n$$

The coefficient of xk gives the numbers of distinct combination of n different objects taken k at a time. The variables x has no significance.

The generating function is also known as a **counting series** or an **enumerator**. Many important results regarding generating function is already given in the chapter Combinatorics.

8.9 PARTITIONS

Partition of a positive integer technique can also be applied to obtain enumeration. In this case a positive integer N is expressed as a sum of positive integers

$$N = p_1 + p_2 + \dots + p_n$$

such that $$p_1 \geq p_2 \geq p_3 \geq \dots \geq p_n \geq 1$$

The n-tuple is called a partition of integer N e.g., (6), (5, 1), (4, 2), (3, 3), (3, 2, 1), (2, 2, 2), (3, 1, 1, 1), (2, 2, 1, 1), (2, 1, 1, 1, 1) and (1, 1, 1, 1, 1, 1) are the different partitions of the integer 6.

The integer p_i's are called parts of the partitioned number N.

8.10 ROOTED UNLABELED TREES

Let a_n be the number of unlabeled rooted trees of n vertices, and let a_n (p) be the number of such rooted trees of n vertices in which the degree of the root is p. Then

$$a_n = \sum_{p=1}^{n-1} a_n(p) \qquad \dots(1)$$

As we know a rooted tree T of n vertices and with root q of degree p can be composed of p rooted subtrees each attached to q by means of an edge between it roots and q. e.g., 11-vertex, rooted tree is composed of four rooted subtrees.

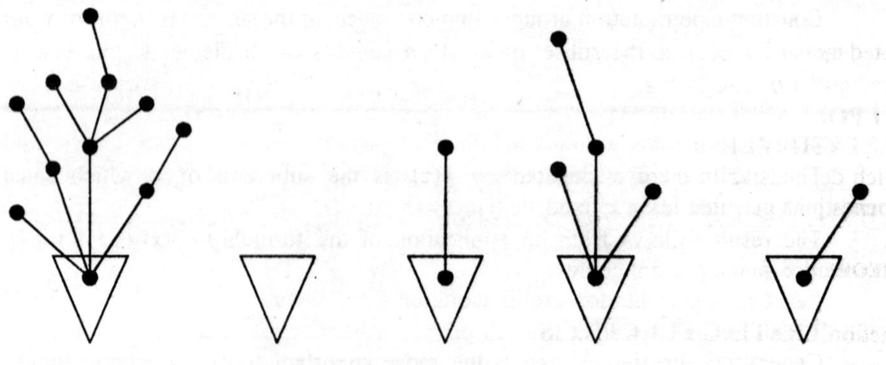

A tree T having n vertices, the n -1 vertices are distributed in the p subtrees and T defines an p-part partition of the number n -1. Let us consider that k_j is the number of such subtrees with j vertices. Then

$$k_1 + 2k_2 + 3k_3 + ... + (n-1)k_{n-1} = n-1 \qquad ... (2)$$

and
$$k_1 + k_2 + k_3 + ... + k_{n-1} = p \qquad ... (3)$$

The equations (2) and (3) represent an p-part partition of integer n - 1, in which integer i appears, k_i times ($0 \le k_i \le n - 1$).

$n = 11, p = 4.$

$k_1 = 1, k_2 = 2, k_5 = 1, k_3 = k_4 = k_6 = k_7 = ... = k_{10} = 0$

$\sum k_j = 4, \ \sum jk_j = 10.$

We form a_j different rooted trees with j unlabeled vertices. From these trees we select k_j trees to construct subtree of T. Because the same tree may appear more than once as a subtree of T. Here, we have a problem of finding the number of ways of selecting k_j objects out of a_j object with unlimited repetition. This number is

$$\binom{a_j + k_j - 1}{k_1} \qquad ...(4)$$

Since each selection is independent, hence the possible number of different trees for this particular partition is given by

$$a_n(k_1, k_2, ..., k_{n-1}) = \binom{a_1 + k_1 - 1}{k_1}\binom{a_2 + k_2 - 1}{k_2} ... \binom{a_{n-1} + k_{n-1} - 1}{k_{n-1}} \qquad ...(5)$$

Where a_n (k_1, k_2, ..., k_{n-1}) denotes the number of n vertex, rooted tree corresponding to the partition.

$$1^{k_1} 2^{k_2} 3^{k_3} ... (n-1)^{k_{n-1}}$$

In addition of a_n (k_1, k_2, ..., k_{n-1}) over all possible partitions of n -1 gives the total number of spanning trees. That is

$$a_n = \sum_{\text{partitions of } n-1} \binom{a_j + k_j - 1}{k_j} \qquad ...(7)$$

Equation (7) is a recurrence relation and the solution of this equation gives a_n the number of rooted trees of n vertices, in terms of a_1, a_2, ..., a_{n-1}.

8.11 POLYA'S THEOREM

Polya's showed how to get a formula which enumerates the orbits according to weight and which depends upon the cycle structure of permutations in the given group. Polya's theorem in turn depends on a generalization of a formula due to Burnside.

THEOREM 8.4

Let A be a permutation group acting on a set X with orbits σ_1, σ_2, σ_1..., σ_3, and let w be a function which assigns a weight to each orbit. Furthermore, w is defined on X so that $w(x) = w(\sigma_i)$ whenever $x \in \sigma$. Then the sum of the weights of the orbit is given by

$$|A|\sum_{i=1}^{n} w(\sigma_i) = \sum_{\beta \in A} \sum_{x=\beta} w(x) \qquad \text{... (i)}$$

PROOF

As we know the order | A| of the group A is the product | *A* (*x*) | . | *σ* (*x*) | for any *x* in *X*, where *A*(*x*) is the stabilizer of *x*. Further, since the weight function is constant on the elements in a given orbit,

$$\therefore |\sigma_i| w(\sigma_i) = \sum_{x \in \sigma_i} w(x)$$

for each orbit σ_i.

Summing up these facts, we get

$$|A| \; w(\sigma_i) = \sum_{x \in \sigma_i} |A(x)| w(x).$$

Summing over all orbits, we get

$$|A|\sum_{i=1}^{n} w(\sigma_i) = \sum_{i=1}^{n} \sum_{x \in \sigma_i} |A(x)| w(x),$$

From which (*i*) follows readily,

Let us now state the conventional form of Burnside's Lemma. For a permutation *β*, expressed as a product of disjoint cycles, let $t_k(\beta)$ denote the number of cycles of length *k*.

COROLLARY

Burnside's Lemma. The number of orbits of the permutation group *A* is

$$\frac{1}{|A|} \sum_{\beta \in A} t_1(\beta).$$

PROOF

Let *A* be a permutation group of order *n* and degree *d*. The cycle index *Z*(*A*) is the polynomial in *d* variables $a_1, a_2, \ldots a_d$, given by the formula

$$Z(A) = \frac{1}{|A|} \sum_{\beta \in A} \prod_{k=1}^{d} a_k^{t_k(\beta)} \qquad \text{... (2)}$$

Since, for any permutation *β*, the numbers $t_k = t_k(\beta)$ satisfy

$$1.t_1 + 2.t_2 + \ldots + d.t_d = d,$$

they constitute a partition of the integer *d*. Let us use the vector notation $(t) = (t_1, t_2, \ldots, t_d)$ in describing *β*. For instance, the partition 5 = 3 + 1 + 1 corresponds to the vector (*t*) = (2, 0, 1, 0, 0).

The general form of classical counting problems to which the Polya's theorem applies are same. Consider a given domain *D*, a range *R*, and a weight function *w* defined on *R*. For instance, consider a particular weight function say *w* assign to each ordered pair $w(r) = (w_1 r, w_2 r)$ of nonnegative integers. The objects to be counted will appear then as a function from *D* to *R*. Then to complete the statement of the problem we have to enquire when two functions in R^D are regarded the same. This can be done by taking a group A which acts on *D*, in such a way that the two functions are equivalent when they are in the same orbit of E^A, where E is the identity group of degree | *R* |.

ILLUSTRATION OF THE ABOVE IDEA

Consider a "necklace problem". Let the necklaces have four beads, some yellow and some black. Two such necklaces are treated as equivalent if they can be made congruent with respect to the preservation of colours of their beads. Here observe that

(i) domain D = {positions where beads are to be placed}

(ii) Range R = Set {yellow bead, black bead}

(iii) $\varphi \varepsilon R^D$ is an assignment of one bead to each place, giving a necklace.

It is to be noted in this example that A is the dihedral group D_4, and the weight function w can be taken as w(yellow bead) = (1, 0) and w(blue bead) = (0, 1).

Let us get back to the theorem. Following the intuitive terminology of Polya. We have

(i) domain elements are *places*,

(ii) range elements are *figures*,

(iii) functions are *configurations* and

(iv) Permutation group A is the *configuration group*.

Let us assign a weight $w(\varphi)$ to each $\varphi \in R^D$ by the equation

$$w(\varphi) = \prod_{d \in D} x^{w_1\varphi(d)} y^{w_2\varphi(d)}, \qquad \ldots(3)$$

One can observe it easily that every function in a given orbit of R^D under E^A has same weight, in other words we can say that the weight of the orbit is equal to the weight of any function in it.

Let there are c_{mn} figure of weight (m, n) in R and c_{mn} orbits (equivalence classes of configurations) of weight $x^m y^n$ in R^D. The figure counting series

$$c(x, y) = \sum C_{mn} x^m y^n \qquad \ldots(4)$$

enumerates the elements of R by weight, and the configuration series

$$C(x, y) = \sum C_{mn} x^m y^n \qquad \ldots(5)$$

is the generating function for equivalent classes of functions. Polya's theorem expresses $c(x, y)$ in terms of $C(x, y)$.

In case we write $Z(A) = Z(A;\, a_1, a_2 \ldots, a_d)$ in (2), then for any function $f(x, y)$, we define

$$Z(A, f(x, y) = Z(A; f(x, y), f(x^2, y^2), \ldots, f(x^d, y^d)) \qquad \ldots(6)$$

THEOREM 8.5 (POLYA'S ENUMERATION THEOREM)

The configuration counting series is obtained by putting the figure counting series into the cycle index of the configuration group,

$$C(x, y) = Z(A, c(x, y)).$$

PROOF

Consider β to be a permutation in A, and let $\overline{\beta}$ be the corresponding permutation in the power group E^A. Let us assume first that φ is a configuration fixed by $\overline{\beta}$ and C is a cycle of length k in the disjoint-cycle decomposition β. Then $\varphi(b) = \varphi(cb)$ for each element b in the representation of c so that all elements permuted by c should have the same image under φ. Conversely, if the elements of each cycle of the permutation $\overline{\beta}$ have same image under a configuration φ, then $\overline{\beta}$ fixes φ. This implies that all configuration fixed by β are obtained independently by selecting an element e in R for each cycle C of β and setting $\varphi(b) = e$, $\forall$ b permuted by C. Then if the weight $w(r)$ is (m, n), where $m = w_1 r$ and $n = w_2 r$ and C has length k, the cycle C contributes a factor of

$$\sum_{e\in R}\left(x^m y^n\right)^k \text{ to the sum } \sum_{\varphi=\beta\varphi} W(\varphi)$$

$$\therefore \quad \sum_{e\in R}\left(x^m y^n\right)^k = c\left(x^k, y^k\right), \qquad \dots (8)$$

And for each β in A, we have

$$\sum_{\varphi=\beta\varphi} W(\varphi) = \prod_{k=1}^{s} c\left(x^k, y^k\right)^{t_k(\alpha)} \qquad \dots (9)$$

Now, summing (9) over all permutations β in A or equivalently over all $\bar{\beta}$ in E^A and dividing by $|A| = |E^A|$, we have

$$\frac{1}{\left|E^A\right|}\sum_{\bar{\beta}\in E^A}\sum_{\varphi=\beta\varphi} W(\phi) = \frac{1}{|A|}\sum_{\beta\in A}\prod_{k=1}^{s} c\left(x^k, y^k\right)^{t_k(\beta)} \qquad \dots (10)$$

One can observe that here R.H.S. of (10) is $Z(A, c(x,y))$. Applying Burnside's Lemma given in theorem 8.5. We see that it is $C(x, y)$.

The sum of the weights of the orbit for the power group E^A is given by

$$\sum_{i=1}^{n} w(\sigma_i) = \sum C_{mn} x^m y^n = C(x, y) \qquad \dots (11)$$

But we can see from (1) that L.H.S. of (10) and (11) are equal, so that

$$Z(A, c(x, y)) = C(x, y)$$

Hence the theorem.

Let us go back to the necklace problem with four beads, we see that the cycle index of the dihedral group D_4 is

$$Z(D_4) = \frac{1}{8}\left(a_1^4 + 2a_1^2 a_2 + 3a_2^2 + 2a_n\right) \qquad \dots (12)$$

and the figure counting series is

$$c(x, y) = x^1 y^0 + x^0 y^1 = x + y.$$

Putting $x + y$ into (12) in accordance with (6), we obtain

$$Z(D_4, x+y) = \frac{1}{8}\left\{(x+y)^4 + 2(x+y)^2\left(x^2+y^2\right) + 3\left(x^2+y^2\right)^2 + 2\left(x^4+y^4\right)\right\}$$

$$= x^4 + x^3 y + 2x^3 y^2 + xy^3 + y^4 \qquad \dots (13)$$

The coefficient of $x^m y^n$ in (13) is the number of different necklaces with four beads, m yellow and n black.

The 6 different necklaces are shown in Fig. 8.6. It is to be noted here that necklaces can also be counted by using $1+x$ as the figure counting series instead of $x + y$. For this a yellow bead has weight 1 and a black bead weight 0. Then in $Z(D_4, 1 + x) = x^4 + x^3 + 2x^2 + x + 1$, the coefficient of x^m is the number of necklaces with m yellow beads and hence 4 - m black ones; comparing (13).

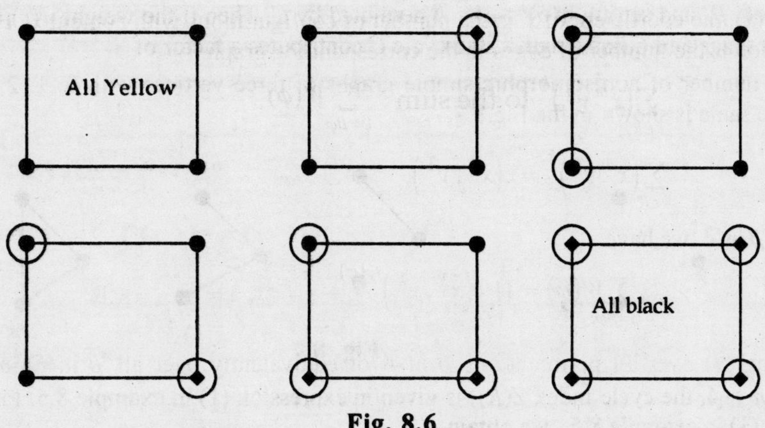

Fig. 8.6

NOTE

1. Figure counting series $1+x$ plays an important role in enumeration problems, since x^0 indicates absence of a figure and x^1 presence.
2. An n-subset of a set X is a subset with exactly n elements.

COROLLARY

If A is a permutation group acting on X, then the number of orbits of n-subsets of X induced by A is the coefficient of x^n in $Z(A, 1+x)$.

8.12 GRAPHS ENUMERATION WITH POLYA'S THEOREM

8.12-1 ENUMERATION OF SIMPLE GRAPHS

Let us take the problem of counting of all unlabelled, simple graphs of n vertices. We can regard such graph G as a mapping of the domain set D of all $n(n-1)/2$ unordered pair of vertices (for digraphs $n(n-1)$ pair of vertices). Range R consists of two elements s and t, with contents x^1 and x^0 respectively. Let us join a vertex pair by an edge in G, the vertex pair maps into s, an element with content x^1; otherwise, into t, an element with content $x^0 = 1$. Thus the figure counting series is

$$c(x) = \sum C_m x^m = 1 + x.$$

R_n, the relevant permutation group in this case is the group of permutations on the pair of vertices induced by the full symmetric group on the n vertices of the graph S_n.

∴ We will obtain configuration counting series by putting $1+x$ for y_1, $1+x^2$ for y_2, $1+x^3$ for y_3, and so on in $Z(R_n)$. Let us take some particular case

(a) For $n = 3$

$$Z(R_3) = \frac{1}{6}\left(y_1^3 + 3y_1 y_2 + 2y_3\right)$$

∴ the configuration counting series is

$$C(x) = \frac{1}{6}\left[(1+x)^3 + 3(1+x)(1+x^2) + 2(1+x^3)\right]$$

$$= 1 + x + x^2 + x^3.$$

In $C(x)$, the coefficient of x^j is the number of configurations with content x^j. Here, the content of configuration is the number of edges in the corresponding graph.

$\Rightarrow$ The number of nonisomorphic simple graphs of three vertices with 0, 1, 2, and 3 edges is each one. The same is shown in the Fig.8.7

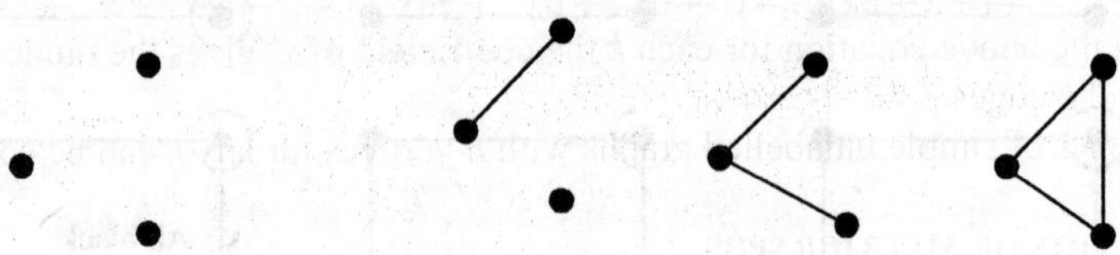

Fig. 8.7

(b) For $n = 4$, the cycle index $Z(R_4)$ is given in expression (1) in example 8.5. Putting $1+x^j$ for y_j in expression (1) in example 8.5 , we obtain

$$C(x) = \frac{1}{24}\left[(1+x)^6 + 9(1+x)^2(1+x^2)^2 + 8(1+x^3)^2 + 6(1+x^2)(1+x^4)\right]$$
$$= 1 + x + 2x^2 + 3x^3 + 2x^4 + x^5 + x^6$$

In the above equation coefficient of x^k gives the number of simple graphs with four vertices and k edges. The validity of above equation can be verified from the Fig. 8.8

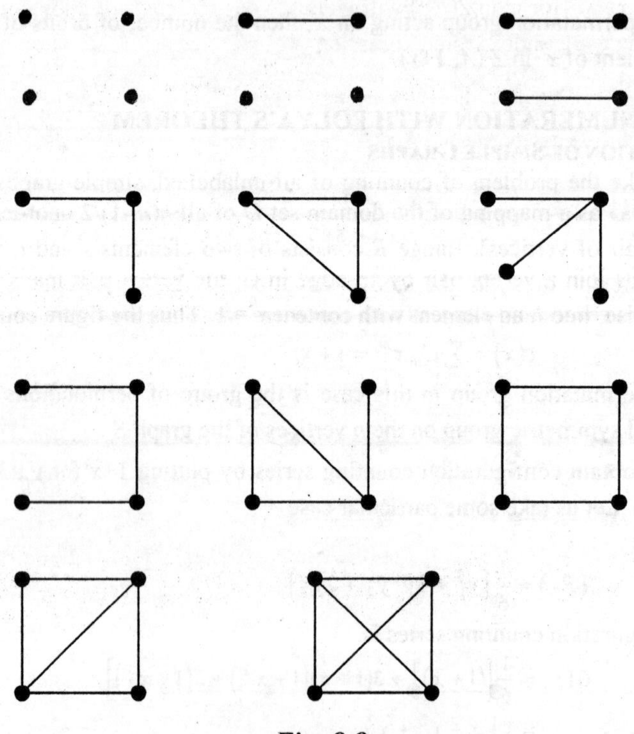

Fig. 8.8

(c) Similarly we can find for $n = 5$, the cycle index $Z(R_5)$. By putting $1+x^j$ for y_j in $Z(R_5)$.

$$C(x) = \frac{1}{120}\Big[(1+x)^{10} + 10(1+x)^4(1+x^2)^3 + 20(1+x)(1+x^3)^3 + 15(1+x)^2 (1+x^2)^4 + 30(1+x^2)(1+x^4)^2 + 20(1+x)(1+x^3)(1+x^6) + 24(1+x^5)^2\Big]$$

$$= 1 + x + 2x^2 + 4x^3 + 6x^4 + 6x^5 + 6x^6 + 4x^7 + 2x^8 + x^9 + x^{10}.$$

Again in the above equation for each k the coefficient of x^k gives the number of simple graph of five vertices and k edges.

The number of simple unlabelled graphs with n vertices for any n can be counted similarly.

8.12-2 ENUMERATION OF MULTIGRAPHS

Let us enumerate the multigraph of n-vertices in which at most two edges allowed between a pair of vertices. One can observe here that the domain and the permutation group in this case is similar to that of the simple graphs, but the range is different A pair of vertices may be joined by

(i) no edge
(ii) one edge, or
(iii) two edges.

The range R contains three elements, say p, q, r with contents x^0, x^1 and x^2 respectively; or in other words we can say x^j denotes the presence of j edges between a pair of vertex, for $j = 0, 1, 2$.

$\therefore$ the figure-counting series becomes $1+x+x^2$ ... (1)

Putting $1+x^k+x^{2k}$ for y_k in $Z(R_n)$ will yield the desired configuration counting series.

Now for $n = 4$, using cycle index from expression (1) in example 8.5, we get

$$\frac{1}{24}\Big[(1+x+x^2)^6 + 9(1+x+x^2)(1+x^2+x^4)^2 + 8(1+x^3+x^6)^2 + 6(1+x^2+x^4)(1+x^4+x^8)\Big]$$

$$= 1 + x + 3x^2 + 5x^3 + 8x^4 + 9x^5 + 12x^6 + 9x^7 + 8x^8 + 5x^9 + 3x^{10} + x^{11} + x^{12}. \quad \text{...(2)}$$

The coefficient of x^j in the above equation gives the number of distinct, unlabelled, multigraphs of four vertices and j edges (such that there are four edges between any pair of vertices).

Example 8.6

The coefficient of x^3 expression (1) of 8.9-2 is 5, hence there exist 5 unlabelled multigraph of four vertices, three edges, and at most two parallel edges, which is shown in the Fig. 8.9.

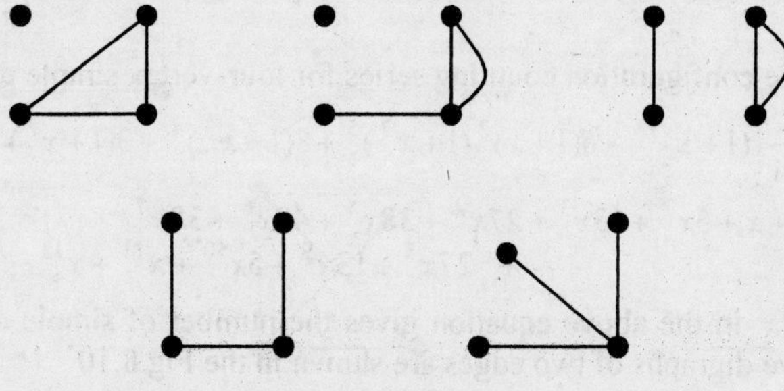

Fig. 8.9

Example 8.7

In the above enumeration of multigraphs we have allowed at most two parallel edges. Instead of allowing at the most two parallel edges between a pair of vertices had we allowed any number of parallel edges the figure-counting series would be the infinite series

$$C(x) = 1 + x + x^2 + x^3 + ... = \frac{1}{1-x}.$$

8.12-3 Enumeration of Digraphs

If we are interested in counting digraphs, then we have to consider all $n(n - 1)$ ordered pair of vertices as constituting the domain. The relevant permutation group in this case is permutations induced on all ordered pair of vertices by S_n. To obtain the cycle index of this group M_n, we will proceed on same pattern as we have done in the case of R_n.

Example 8.8

Let us take a particular case $n = 4$ in S_n. The following table gives the term in $Z(M_n)$ induced by each term in $Z(S_n)$.

Term in $Z(S_4)$	Induced term in $Z(M_4)$
y_1^4	y_1^{12}
$y_1^2 y_2$	$y_1^2 y_2^5$
$y_1 y_3$	y_3^4
y_2^2	y_2^6
y_4	y_4^3

∴ the cycle index is

$$Z(M_4) = \frac{1}{24}\left(y_1^{12} + 6y_1^2 y_2^5 + 8y_3^4 + 3y_2^6 + 6y_4^3\right).$$

8.12-4 Enumeration of Simple Digraph

Here in simple digraph for a given ordered pair of vertices (a, b) either does or does not have an edge (directed) from a to b. Therefore, in this case the figure counting series $c(x) = 1 + x$ is applicable. The configuration counting series is obtained by putting $1 + x^i$ for y_i in expression (1) in example 8.8

The following is the configuration counting series for four-vertex simple digraphs.

$$C(x) = \frac{1}{24}\left[(1+x)^{12} + 6(1+x)^2(1+x^2)^5 + 8(1+x^3)^4 + 3(1+x^2)^6 + 6(1+x^4)^3\right]$$

$$= 1 + x + 5x^2 + 13x^3 + 27x^4 + 38x^5 + 48x^6 + 38x^7 + 27x^8 + 13x^9 + 5x^{10} + x^{11} + x^{12}.$$

The coefficient of x^i in the above equation gives the number of simple digraphs with four-vertices and i edges. The five digraphs of two edges are shown in the Fig.8.10

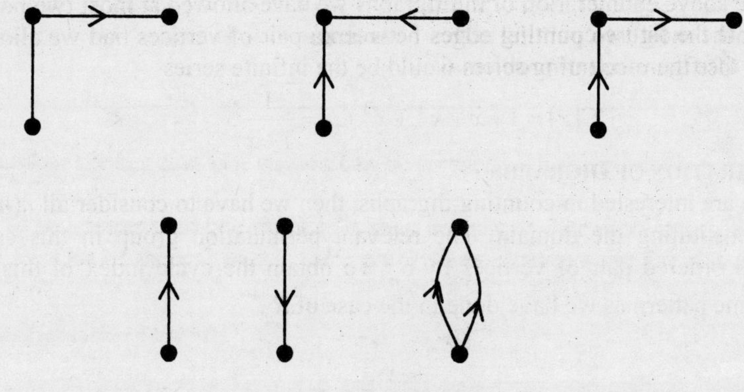

Fig. 8.10

EXERCISE 8.1

1. Show that the cycle index of a group consisting of the identity permutations only y_1^m, m being the number of elements in the object set.
2. Show that order of D_n, the group of symmetric regular n-sided polygon is $2n$. Find the cycle index of D_n.
3. Show that the cycle index of induced group R_3 is the same as that of S_3 i.e.

$$Z(R_3)=\frac{1}{6}\left(y_1^3+3y_1y_2+2y_3\right).$$

4. Find the counting series for unlabelled, simple, connected graphs with exactly one circuit.
5. Find the different ways of painting the six vertices of an octahedron with three colours. Two octahedron are coloured distinctly if they cannot be made to coincide by rotation.
6. Show that the cycle index of S_5, the full symmetric group of degree 5, is

$$Z(S_5)=\frac{1}{5!}\left(y_1^5+10y_1^3y_2+20y_1^2y_3\right)+15y_1y_2^2+30y_1y_4+20y_2y_3+24y_5$$

7. Find the number of different ways of painting the four faces of a pyramid with two colours.
8. Show that the cycle index of the unordered pair group R_5 (on the set of 10 unordered pairs induced by S_5) is

$$Z(R_5)=\frac{1}{5!}\left(y_1^{10}+10y_1^4y_3^2+20y_1y_3^3+15y_1^2\ y_2^4+30y_2y_4^2+20y_1y_3y_6+24y_5^2\right)$$

ANSWERES

4. **Hint** Use $Z(D_n)$ and consider the graph as consisting of a single circuit with one or more trees attached to its vertices.
5. **Hint** First show that the cycle index of the permutation group is

$$\frac{1}{24}\left(y_1^6 + 6y_1^2 y_4 + 3y_1^2 y_2^2 + 6y_3^2 + 8y_3^2\right)$$

then put the figure-counting series $1 + x + x'$.

8. **Hint** Use the result of problem 6.

9 APPLICATIONS OF GRAPHS IN COMPUTER SCIENCE

9.0 INTRODUCTION

Graphs have found applications in subjects as diverse as Sociology, Chemistry, Geography, Engineering Sciences and many more fields. They are also widely used in solving games and puzzles. In computer science, graphs are used in many areas such as in computer designing; scheduling problems in operating system; file management in data base management system; data-flow control between networks (routing algorithms); network of interconnected networks (internet), etc. In day-to-day applications, graphs find their importance as representations of many kinds of physical structure.

The structure of digital computers is shown in Fig. 9.1(a) and represented in the form of graph in Fig. 9.1(b).

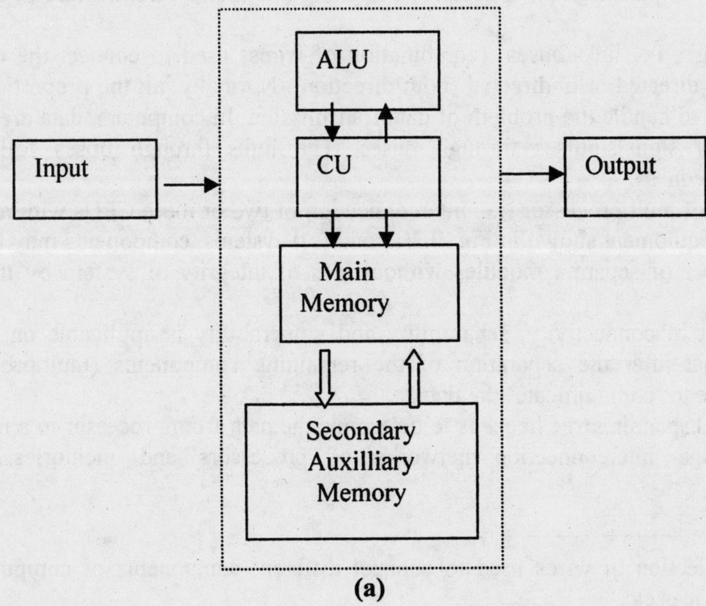

(a)

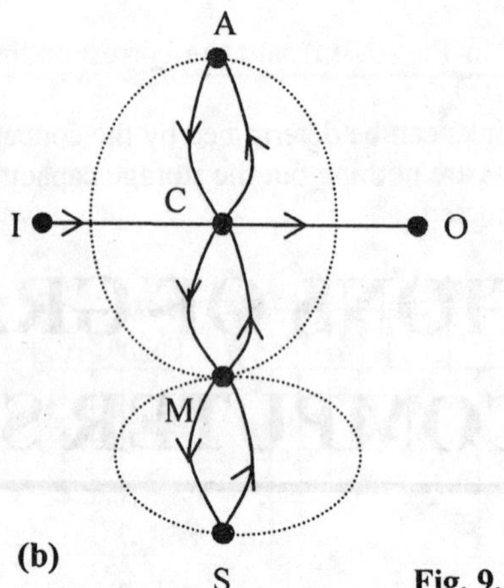

Fig. 9.1

We may use the cut set to show the different main components of computer such as input device, output device, central processing unit (CPU) and auxiliary/ secondary memory.

Computers are made up of diodes, triodes, transistors, TTL, decoders, registers, encoder etc. These components are connected with the help of wires. Graphs theoretical formulation is used to represent the circuits where all the components termed as nodes and wires or links between them known as edges. The theoretical or mathematical concepts of graphs such as connectedness, completeness, biconnected, bipartite, planarity, duality etc. are used for designing the circuits and simulate the effects of implementation before actual implementation. Some of the applications of graphs are stated below:

* Concept of planar graph is used for designing the internal architecture of computers on chip (motherboard).
* The edges i.e. links/buses* (combination of wires) used to connect the components/nodes may be directed or undirected (both direction). Normally, all the properties of digraphs can be used to handle the problem of data transmission. In computers, data are transferred in the form of signals/pulses through buses. The links through buses follow the rules of connectedness.
* In case of multiprocessor (i.e. interconnection of two or more CPUs with memory and input-output equipment shown in Fig. 9.2(a) and (c)) systems, components may be partitioned into a number of separate modules without loss of integrity of system by the help of cut-set theory.
* Concept of connectivity, separability and vulnerability is applicable on connected graphs such that after the separation of the remaining components (multiprocessors) can still continue to 'communicate' the data.
* Minimal spanning tree helps us to determine the path from processor to a memory module in multistage interconnection networks of processors and memories. The multistage

---

* Buses, are collection of wires used to connect different components of computer such as CPU, memory, I/O devices etc.

interconnected memories shown in Fig. 9.3(a) and the corresponding graph shown in Fig. 9.3(b).

* Maximum flow of data through links can be determined by the concept of network flow with the help of weights, where weights are nothing but the storage capacity of each bus.
* The graph (circuits) must be connected.

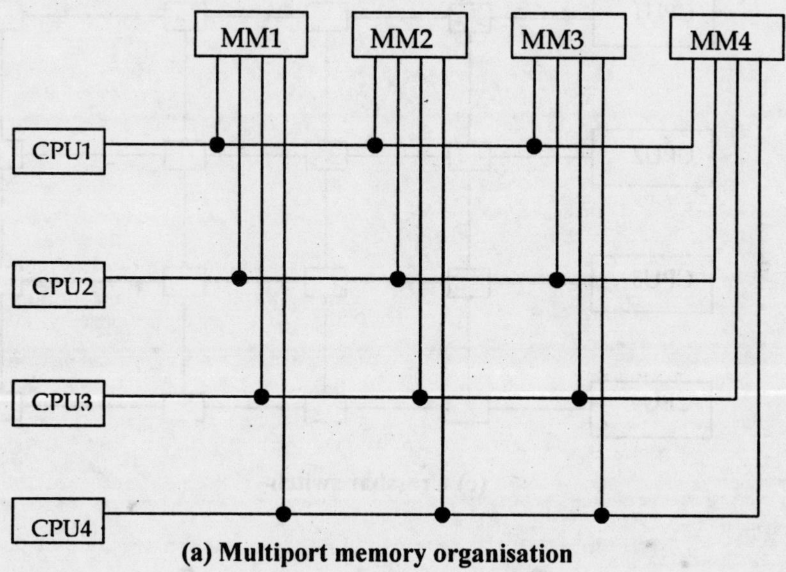

(a) Multiport memory organisation

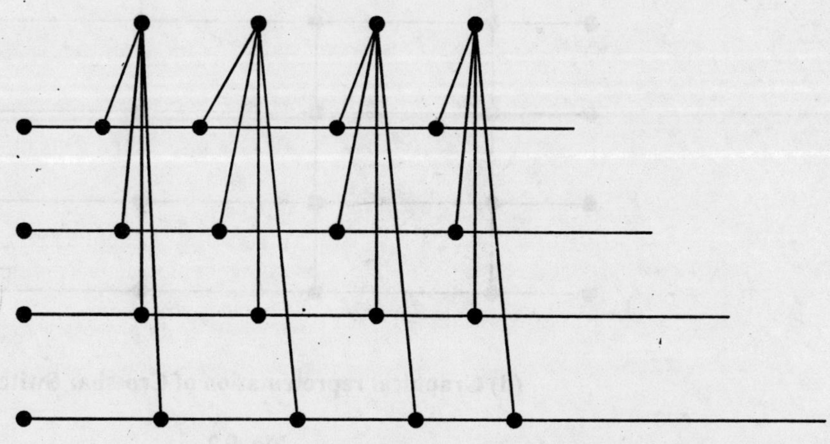

(b) Graphical representation of multiport memory organisation

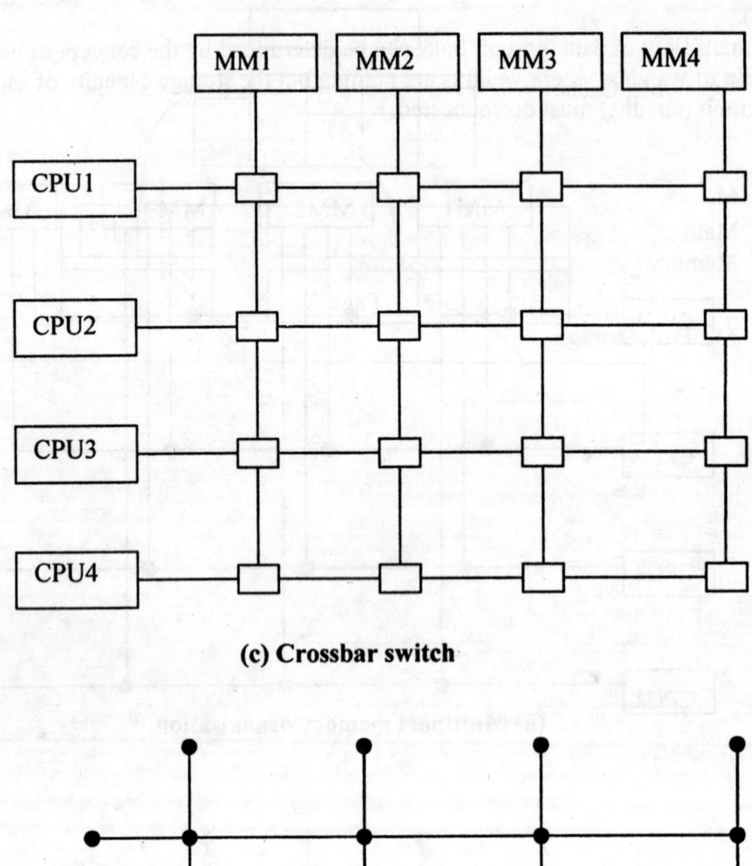

(c) Crossbar switch

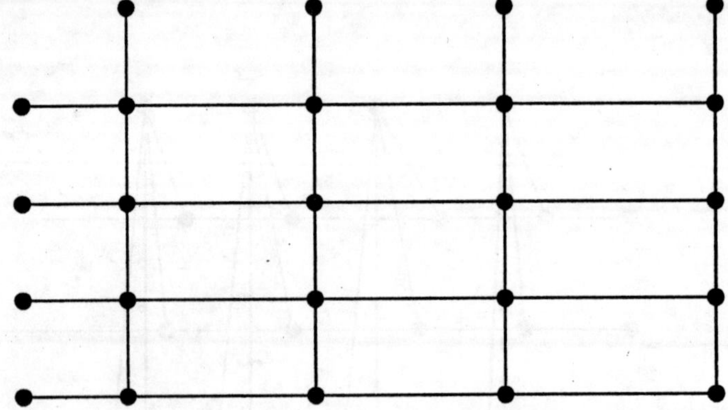

(d) Graphical representation of Crossbar Switch

Fig. 9.2

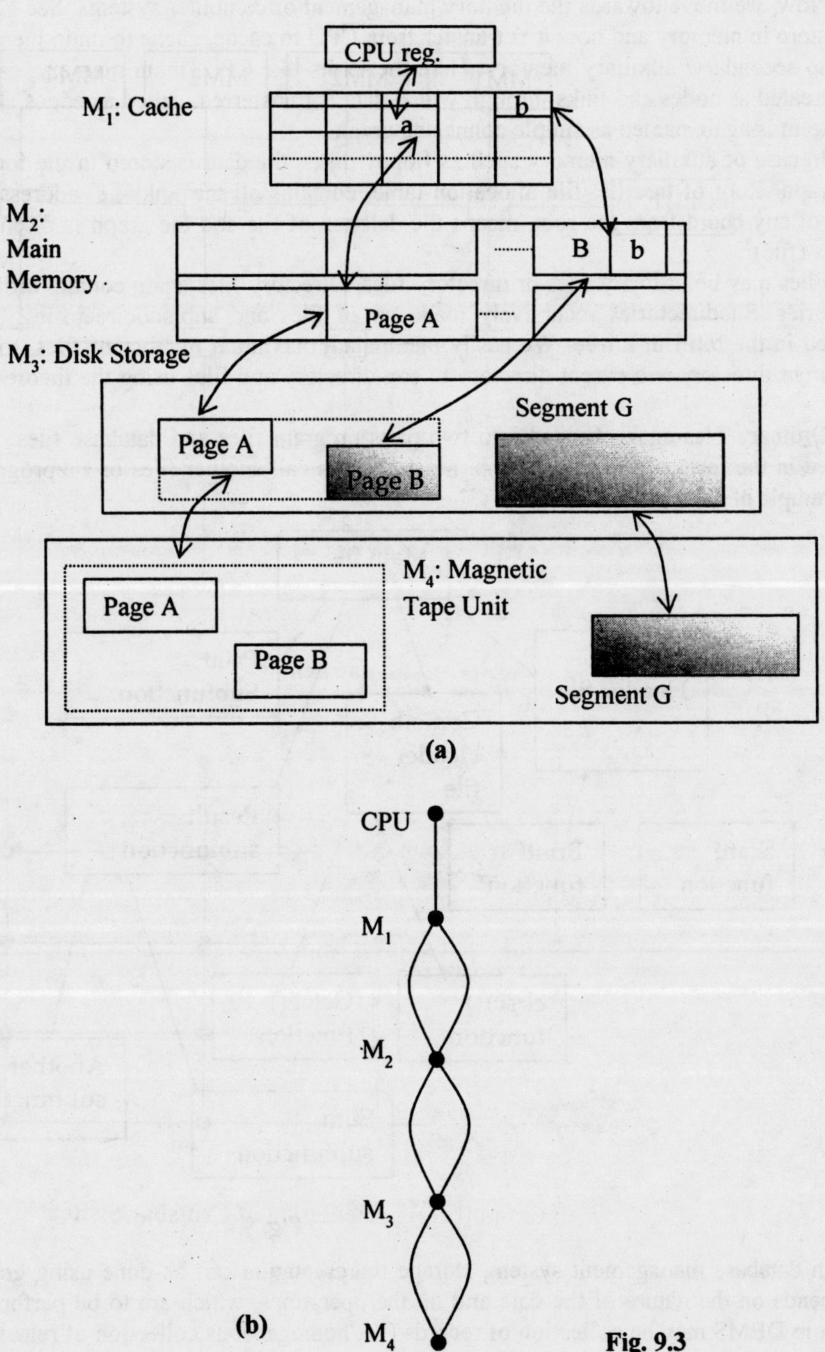
CPU reg.
M1: Cache
a
b
M2:
Main
Memory
B
b
Page A
M3: Disk Storage
Segment G
Page A
Page B
M4: Magnetic
Tape Unit
Page A
Page B
Segment G
(a)
CPU
M1
M2
M3
M4
(b)

Fig. 9.3

Now, we move towards the memory management of computer systems. See Fig.9.3 how the data can store in memory and how it is transfer from CPU to cache, cache to main memory and main memory to secondary/ auxiliary memory. Here the terms like CPU, main memory, cache, auxiliary memory treated as nodes and links through which data is transferred, called as edges. That's why the whole system is again named as simple connected graph.

In case of auxiliary memory such as floppy disks, the data is stored in the form of a tree, a special graph. Root of tree i.e. file allocation table, contains all the links i.e. address, of each file. Removal of any chord from the root, means the deletion of file and the graph is disconnected along that vertex (file).

Files may be ordinary files or directory files. Directory files again contain ordinary files and subdirectories. Subdirectories recursively made up of files and sub-subdirectories. This whole is represented in the form of a tree. We easily manipulate maximum number of files, total links from root to parent directory and parent directory to son-directory and files using the theorems in 'chapter 3'.

Ordinary files again divided into two parts program files and database files. Both files are represented in the form of trees and graph. Main file can call another files or subprograms. See Fig. 9.4, an example of C program.

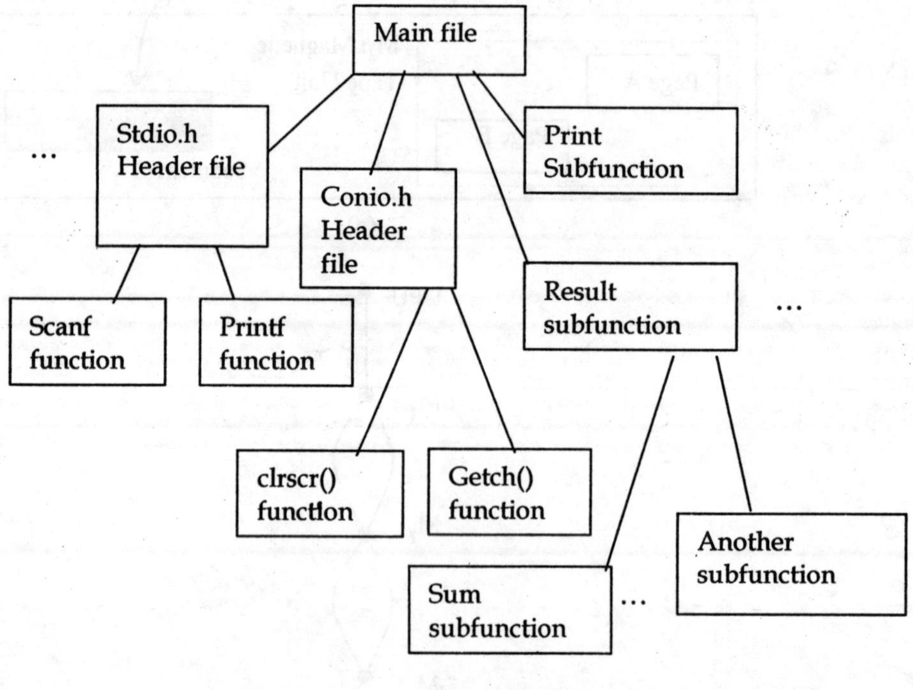

Fig. 9.4

In database management system, storage representation can be done using graphs. General graph depends on the nature of the data and on the operations, which are to be performed on these data. Data in DBMS may be collection of records (i.e. homogeneous collection of related fields). The choice of a suitable representation is affected by other factors, such as number of nodes, the average number of edges leaving a node, whether a graph is directed or not, the frequency of insertion and/or deletions to be performed etc. The technique used to represent and store the records on a file is called the file organisation. The file organisation determines the files record sequencing, which is the physical ordering of the records in storage and determines the set of operation necessary to find particular records. The most famous representations in file manipulation are binary tree, *m*-way tree, B tree, B^+ tree etc..

In these trees all the nodes except leaves contains indexes and leaves are used to store the data. Properties of trees are used to manipulate the problems of DBMS such as breadth first search (BFS) and depth first search (DFS) techniques which is used to find the shortest distance between some starting node and the remaining nodes of the graph. And the path may be represented using minimal spanning trees.

These techniques applicable in DBMS to search the specific record in minimum time. Another popular search technique binary-search method can be easily explained in terms of binary tree.

In order to prevent a file (data) unbalancing, the properties of height balanced trees are applicable that associate a balance indicator with each node in the tree-structure of file (data).

To arrange the records (data) in a specific manner we use the term 'sort'. Sorting is a technique to arrange the data in lexicographic order. Sorting are of two types internal sort and external sort. Internal sorting is done on data that are internally stored in the form of files but external sorting include the list of items (file) being stored in high-speed main memory and magnetic tapes. The index refers to a unique key associated with each record. B tree (balanced-sort tree) is optimum technique of organising a file and its index in database system. Another sorting techniques such as heap sort, merge sort also follow the rules of heap tree, binary trees respectively.

As, components of computers, inter-linked files, subprograms, memory devices forms the structure of trees and graphs and follow all properties of trees and graphs; the network of computers or networks, known as internet, also represented in the form of connected graph or digraph.

Computer network is a collection of interconnected autonomous computers. Autonomous computers termed as node and links between computers and networks at distant cities or countries not a simple copper wire, but use fibre optics, twisted pair, coaxial cable, fibre cables, microwaves and communication satellites. These links are unidirectional or bi-directional. Some examples of networks are shown in Fig. 9.5.

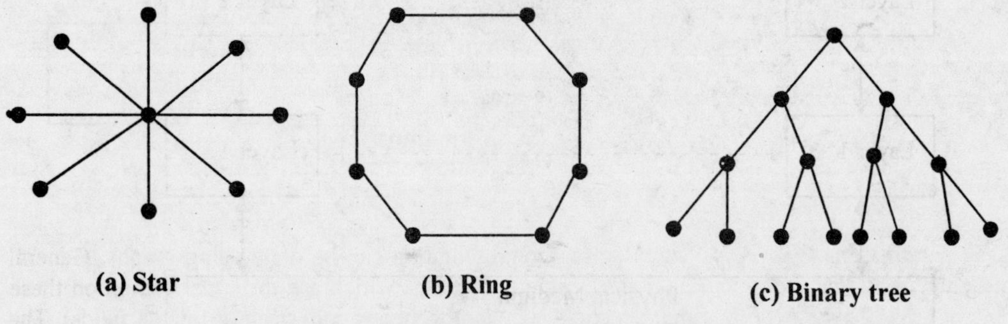

(a) Star **(b) Ring** **(c) Binary tree**

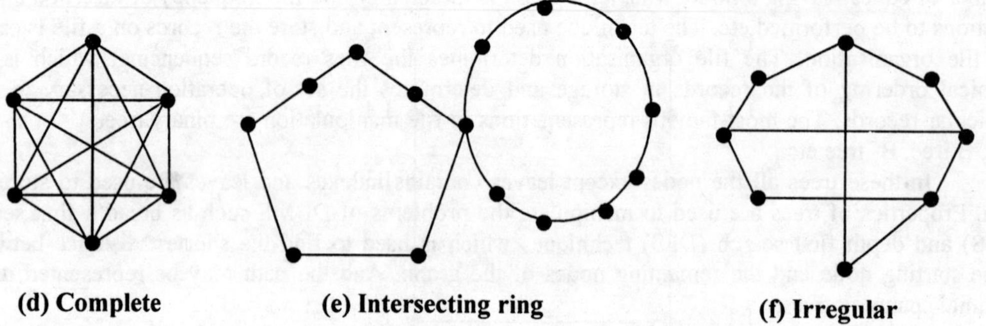

(d) Complete **(e) Intersecting ring** **(f) Irregular**

Fig. 9.5 Some possible topologies for a point-to-point subnet

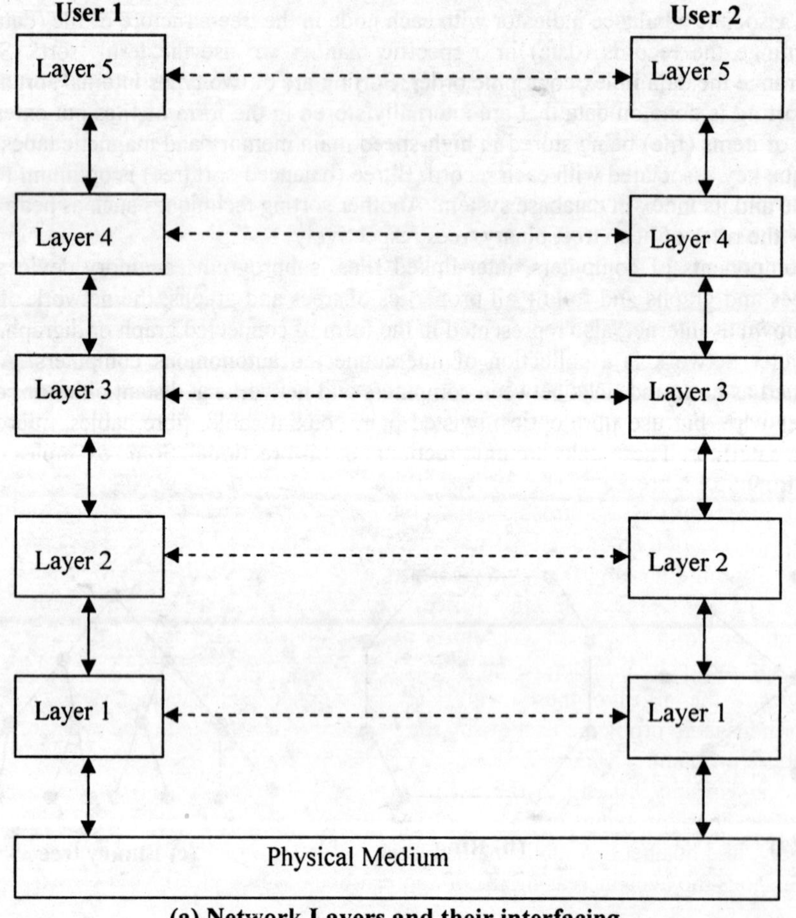

(a) Network Layers and their interfacing

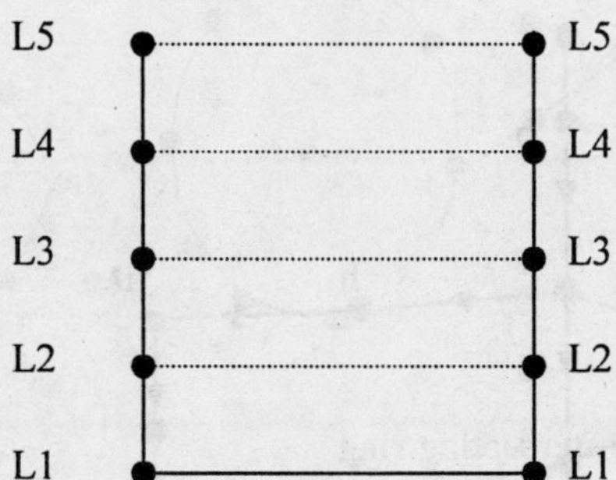

(b) Graphical representation of Network Layers

Fig. 9.6

Networks are organised as a series of layers or levels (see in Fig. 9.6). These layers are strongly connected on one machine and communicate with n^{th} layer of another machine. Using theorems and properties of digraph, computer engineers establish the connection between nodes, gateways, routes etc. and provide the facility of communication to users.

Due to mathematical concepts of graph theory, it is easier to choose the data structures and various algorithms of routing packets from source machines to estimation machines. To choose a route between a given pair of routers, the shortest path algorithm finds the shortest path between them on the graph. If capacity and average flow of data is considered, weighted graphs are used to analyse the flows mathematically to optimise the routing. Digraphs under the name Internet have been used to represent interconnection of individual computer networks. Connectedness, separability, complete sub-digraphs, size of fragments and so forth, in a network can be given immediate significance.

In analysis of any problem we may use the data flow diagram and flowchart before implementation on computers. The purpose of analysis could be to estimate the running time or storage requirement of a program. The vertices are the flowchart boxes and the connecting arrows are the edges. The program must be directed-connected graph and consists of start and end vertices. Cut set theory is applied to find an optimal partitioning of the program digraph into sub-digraphs.

If the program digraph has no loop, partitioning problem is solved easily. But in another case which there is a loop in the program, the difficulty in segmentation arises. We never want to cut the graph at articulation point. In this case, identify all maximal strongly connected sub-digraphs in the program diagram and then compute the memory requirement of each sub-digraph by adding weights of all its vertices. If the largest of these memory requirements does not exceed the size of the working memory, then there is no problem. That's why the problem of segmentation of program can be solved as graph-theoretic problem.

There is virtually no end to the list of problems that can be solved with graph theory in computer science. These problems originate not just in connection with computers, but throughout the science, industry, and business. A mathematical model of any problem has been attempted with graph theory and thus goes the ever-increasing list of applications of graph theory.

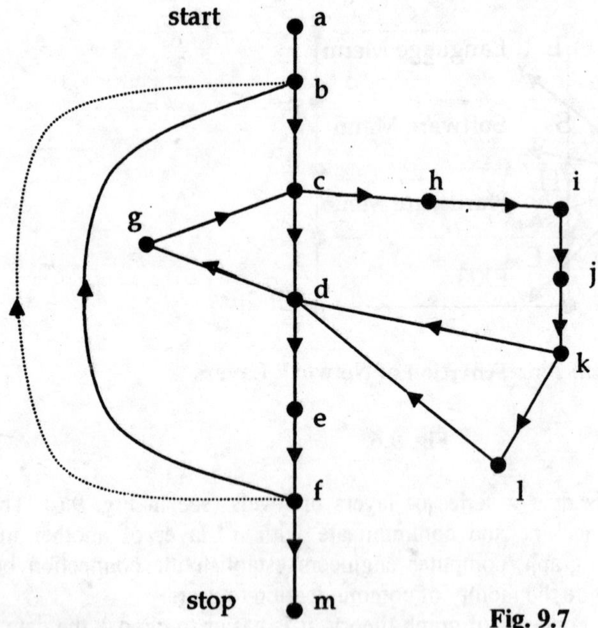

Fig. 9.7

9.1 DECISION TREES

A decision tree is made of nodes and branches. Each branch leading away from a node indicates one of the several possible courses of action available to the decision-maker. Any branch that makes the end of the decision tree is called a terminal branch. A terminal node indicates either a course of action or an outcome. A decision tree helps to show the paths that are possible in a design following an action or decision by user. Fig.9.8 illustrates this type of tree. As indicated, if the user selects 1, followed by M and A, the algebra menu would be displayed.

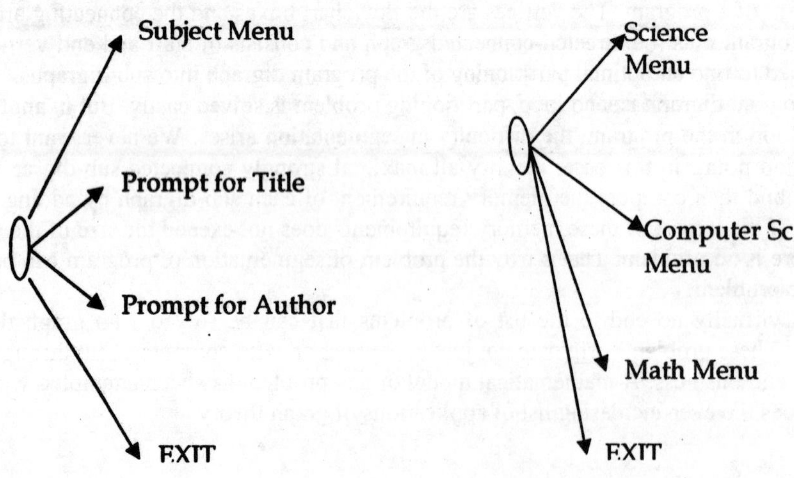

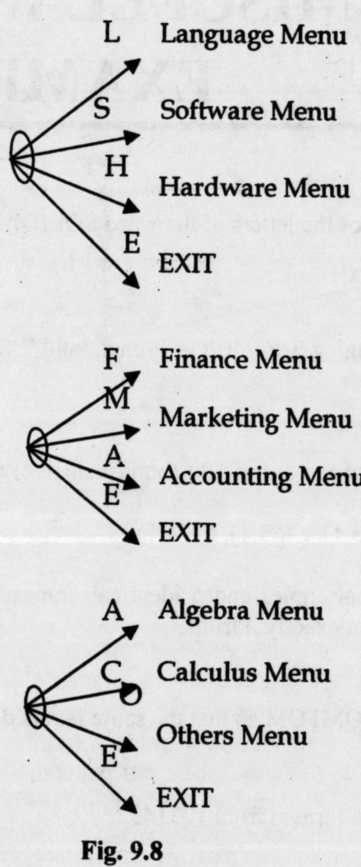

Fig. 9.8

What is the value of a tree such as this? It helps the designer visualise how the user will move through the design to reach a desired location. Thus, a decision tree provides an overview of the flow of control to be built into computer programs.

EXERCISE 9.1

1. Write any four applications of graphs in computer sciences.
2. Explain decision tree.

10 MISCELLANEOUS EXAMPLES

1. Find the number of arrangements of the letters of the word SOLIDARITY when vowels must occupy only even positions.
   Ans. $60 \times 6!$

2. Find the number of ways of arranging 3 English, 4 French, and 7 German books, so that all book of given language are together.
   Ans. 3!. 3!. 4!. 7!

3. How many numbers are there between 1 and 100 (both inclusive) which are such that their sum is divisible by 3?
   Ans. $2\,C(33, 3) + C(34, 3) + C(33,1).\ C(33,1).\ C(34,1)$

4. In how many ways can 5 identical apples and 5 identical oranges be distributed among 5 persons such that every persons get exactly 2 fruits.
   Ans. $1 + 30 + 20 = 5$

5. Find the permutations of letters TOMTOM so that the same letters do not appear together.
   Ans. $(6!/(2!)^3) - 3!$

6. How many 4-digit numbers can be formed from 1231432.
   Ans. $24 + 96 + 18 = 150$

7. Solve $y_{n+1} + 2y_n = 0$, given $y_0 = 1$
   Ans. $y_n = (-2)^n$

8. Solve $Q(k) - Q(k-1) = 2$ with $Q(0) = 9$.

9. Find the homogeneous solution of

$$T(k) - 7\,T(k-1) + 10T(k-2) = 6 + 8k$$

   Ans. $T^{(h)}(k) = b\,2^k + b_2 5^k]$

10. Find the closed form expression for generating function for sequence $< 3^0, 3^1, 3^2, 3^3, \ldots >$
    Ans.

$$G(s,z) = \sum_{n=0} 3^n z^n$$

11. Find the simple expression for the generating function of the following numeric function:

(a) $0 \times 5^0, 1 \times 5^0, 2 \times 5^0, \ldots, r \times 5^r, \ldots$.

Solution

We know $\frac{1}{(1-z)} = 1 + z + z^2 + \ldots + z^r + \ldots$

Also $\frac{1}{(1-z)^2} = 1 + 2z + 3z^2 + \ldots + (r+1)z^r + \ldots$

On multiplying both sides by z, we obtain

$$\frac{z}{(1-z)^2} = 0 + z + 2z^2 + 3z^3 + \ldots + rz^r + \ldots$$

$\Rightarrow \frac{z}{(1-z)^2}$ is the generating function for the numeric function (0, 1, 2, 3, …r, …).

Now applying the property (8) of generating function, i.e.

If $S(r) = \alpha^r T(r)$ then $G(S, Z) = G(T, \alpha Z)$.

Where α is any constant

Hence, the generating function of the given numeric function is $\frac{5Z}{(1-5z)^2}$.

(b) 1, -2, 3, -4, 5, -6, …, $(-1)^r (r+1)$.

Solution

$$G(S, z) = 1 - 2z + 3z^2 - 4z^3 + 5z^4 - 6z^6 + \ldots + (-1)^r (r+1) z^r + \ldots$$

$$= \frac{1}{(1+z)^2}.$$

12. For $A(r) = 3^{r+2}$, find the generating function.

Solution

We can write $A(r) = 3^r \cdot 3^2$

$= 9 \cdot 3^r$

Also, we know that generating function of $3^r = \frac{1}{1-3z}$

By property (3) of generating function $CG(S, Z) = \sum_{n=0}^{\infty} (CS_n) Z^n$

$$\Rightarrow \quad G(A, Z) = \frac{9}{1-3Z}.$$

Aliter

Using property (7b) of generating function.

Let $i = 2$ then $A(r) = 3^r$.

And. $G(A, Z) = Z^{-2}\left[\frac{1}{1-3Z} - 13z\right]$

$$=\frac{9}{1-3Z}.$$

13. For $G(A,Z)=\dfrac{Z^4}{(1-2Z)}$, determine the discrete numeric function.

Solution

$$G(A,Z)=z^4\left[\frac{1}{(1-2Z)}\right]$$

Using property [7a] of generating, let $i=4$, then

$$A(r)=\begin{cases}0 & 0\le r\le 3\\ 2^{r-4} & r\ge 4\end{cases}.$$

14. Find $C(r)$; where $C(r)=A(r)*B(r)^*$, $A(r)=3^r$ and $B(r)=2^r$, $r\ge 0$.

Solution

Here $A(r)=3^r$ and $B(r)=2^r$,

we know that

$$G(A,Z)=\frac{1}{1-3Z} \quad \text{and} \quad G(B,Z)=\frac{1}{1-2Z}$$

Therefore $G(C,Z)=G(A,Z)\cdot G(B,Z)$ [By Property (4)]

$$=\frac{1}{(1-3Z)}\cdot\frac{1}{(1-2Z)}=\frac{3}{1-3Z}-\frac{2}{1-2Z}$$

$$C(r)=3\cdot 3^r-2\cdot 2^r=3^{r+1}-2^{r+1}.$$

15. Let $A(r)$ be an arbitrary numeric function and $B(r)$ be the numeric function (1, 1, 1, ..., 1, ...)

Let $C(r)=A(r)*B(r)$

So that $C(r)=\sum_{i=0}^{r}A_i\ B_{r-i}=\sum_{i=0}^{r}A_i$

Thus $C(r)$ is the accumulated sum of $A(r)$, and the generating function

$$G(C,Z)=\frac{1}{1-Z}G(A,Z)$$

REMARK

If $G(A,Z)$ is generating function of numeric function $A(r)$ then $\dfrac{1}{1-Z}G(A,Z)$ is generating function of accumulated sum of $A(r)$.

---

* $C(r)=A(r)*B(r)$ is known as convolution of numeric function and its value $C(r)=\sum_{i=0}^{\infty}A_rB_{r-i}$.

16. Evaluate the sum $1^2+2^2+3^2+...+r^2$ using generating function.

Solution

Firstly, we determine the generating function of the numeric function $(0^2+1^2+2^2...,r^2)$. As we know that

$$\frac{1}{1-Z}=1+Z+Z^2+Z^3+...+Z^r+...$$

Differentiating w.r.t. Z both sides, we get

$$\frac{1}{(1-Z)^2}=1+2Z+3Z^2+...+rZ^{r-1}+...$$

Multiplying both sides by Z, we obtain

$$\frac{Z}{(1-Z)^2}=Z+2Z^2+3Z^3+...+rZ^r+...$$

Now,

$$\frac{d}{dz}\left[\frac{Z}{(1-Z)^2}\right]=1+2^2Z+3^2Z^2+4^2Z^3+...+r^2Z^{r-1}+...$$

Again, multiplying both sides by z, we get

$$Z\frac{d}{dz}\left[\frac{z}{(1-Z)^2}\right]=0^2Z^2+2^2Z^2+3^2Z^3+...+r^2Z^r+...$$

So, $Z\frac{d}{dz}\left[\frac{z}{(1-Z)^2}\right]=\frac{z(1+z)}{(1-Z)^3}$ is generating function of $\left(0^2,1^2,\ 2^2,...,r^2,...\right)$ by remark of example 15.

Thus $0^2+1^2+2^2+...+r^2$ is coefficient of z^r in $\frac{z(1+z)}{(1-Z)^4}$.

From binomial theorem, we know that the coefficient of z^r in $\frac{1}{(1-Z)^4}$

$$=\frac{(-4)(-4-1)(-4-2)...(-4-r+1)}{r}(-1)^r$$

$$=\frac{(r+1)(r+2)(r+3)}{1\cdot 2\cdot 3}$$

$\Rightarrow$ Coeff. of z^r in $\frac{z(1+z)}{(1-Z)^4}=\frac{z}{1-Z^4}+\frac{z^2}{(1-Z)^4}$

$$=\frac{r(r+1)(r+2)}{1\cdot 2\cdot 3}+\frac{r(r-1)(r+2)}{1\cdot 2\cdot 3}$$

i.e. $1^2+2^2+3^2+...+r^2=\frac{r(r+1)\ (2r+1)}{6}$.

17. Using generating function, evaluate the sum $3\cdot 2\cdot 1+4\cdot 3\cdot 2+5\cdot 4\cdot 3+....+(r+1)\,r(r-1)$

Solution

Let $A(r)=(r+1)\,r(r-1)$

We have

$$\frac{1}{1-z}=1+z+z^2+z^3+...+z^r+...$$

Also $$\frac{1}{(1-z)^2}=1+2z+3z^2...+rz^{r-1}+(r+1)\,z^r+...$$

And $$\frac{2}{(1-z)^3}=2+6z+12z^2+...+r(r-1)z^{r-2}+(r+1)rz^{r-1}+(r+2\,(r+1)z^r+...$$

Differentiating again w.r.t. z, we obtain

$$\frac{6}{(1-z)^4}=6+24z+...+(r-3)(r+2)\,(r-1)z^r+...$$

$$=3\cdot 2\cdot 1+4\cdot 3\cdot 2z+...+(r+1)\,r(r-1)\,z^{r-2}$$
$$+(r+2)\,(r+1)rz^{r-1}+(r+3)(r+2)\,(r+1)z^{r+}...$$

Now, multiplying z^2 both sides, we get

$$\frac{6}{(1-z)^4}=3\cdot 2\cdot 1z^2+4\cdot 3\cdot 2z^2+...+(r+1)\,r(r-1)z^r+...$$

From remark of example (15) the required sum is coefficient of z^r in $\dfrac{6z^2}{(1-z)^5}$.

The coefficient of z^r in $\dfrac{6z^2}{(1-z)^5}$ = Coeff. of z^{r-2} in $\dfrac{6}{(1-z)^5}$

$$=(-1)^5\,6\times\frac{(-5)\,(-5-1)\,(-5-2)...(-5-(r-2)+1)}{(r-2)!}$$

$$=6\times\frac{5\cdot 6\cdot 7\cdot ...(r+2)}{(r-2)!}=\frac{6(r+2)!}{4!(r-2)!}$$

$\Rightarrow$ $$3\cdot 2\cdot 1+4\cdot 3\cdot 2+...+(r+1)\,r(r-1)=\frac{6\,(r+2)!}{4!\,(r-2)!}z^r$$

18. Determine the discrete numeric function corresponding to each of the following generating functions:

(i) $G(A,Z)=\dfrac{7z^2}{(1-2z)\,(1+3z)}$ (ii) $G(A,Z)=\dfrac{1}{5-6z+z^2}$

(iii) $G(A,Z)=\dfrac{z^5}{5-6z+z^2}$ (iv) $G(A,Z)=\dfrac{1+z^2}{4-4z-z^2}$

(v) $G(A,Z)=\dfrac{(1+z)^2}{(1-z)^4}$ **Ans.** $a_r=\dfrac{(r+1(2r^2+4r+3)}{3}$

19. Define (i) undirected graph (ii) directed graph.

20. Define (a) simple graph (b) Multiple graph and give an example for each.

21. What is the complement of a graph G of n vertices?
Ans. If G is a graph of n vertices, the complement $\bar{G}$ of G is such that it has all the vertices of G; if there exists an edge e_{ij} between any two vertices i and j in G, an edge does not exist in $\bar{G}$ between the vertices i and j and conversely if an edge does not exist between i and j in G, there exists an edge e_{ij} between i and j in $\bar{G}$.

22. The sum of the degree of all vertices of a graph is equal … to edges.
Ans. Twice the number of edges

23. Suppose G is a non-directed graph of order 9 such that each vertex has degree 5 or 6. Prove that G has at least 5 vertices each of degree 6, or 6 vertices each of degree 5.
Ans. Let m be the number of vertices each of degree 5 and n the number of vertices each of degree 6. We know that m is even. Let us form a table for this combination.

Sr.	M	Sum of degrees of m	n	Sum of degrees of n	Sum
(a)	2	2×5 = 10	7	7×6 = 42	52
(b)	4	4×5 = 20	5	5×6 = 30	50
(c)	6	6×5 = 30	3	3×6 = 18	48
(d)	8	8×5 = 40	1	1×6 = 6	46

(a) and (b) have a minimum number of 5 vertices of degree 6 while (c) and (d) have a minimum number of 6 vertices of degree 6.

24. Find the in degrees and out degrees of the vertices v_4 and v_5 of the following graph.

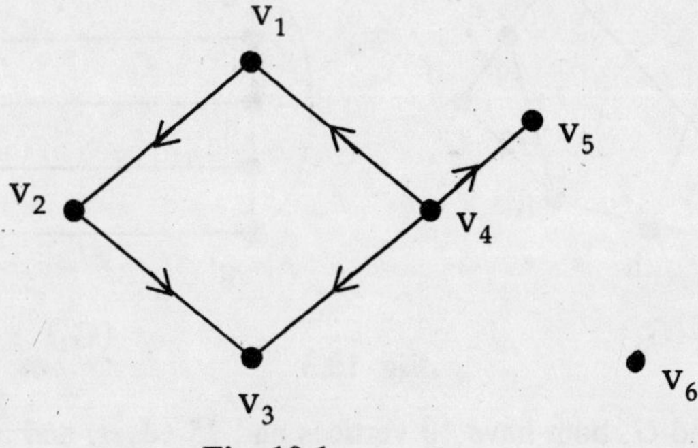

Fig. 10.1

Ans.	Vertex	Indegree	Outdegree
	v_4	0	2
	v_5	1	0

25. What is the largest number of vertices in a graph with 35 edges if all vertices are of degree at least 3.

Ans. Let m be the number of vertices, then $(2\times35)/m \geq 3$ or $3m \leq 70$. The largest integer value satisfying this inequality is 23.

26. Show that the following graphs are isomorphic.

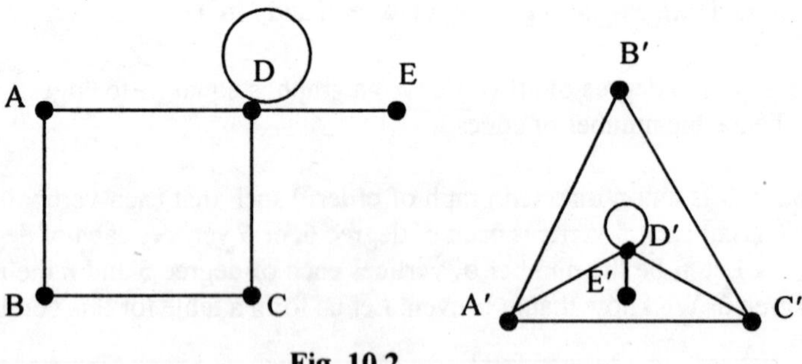

Fig. 10.2

Ans. Both graphs contain 5 vertices, 6 edges and 1 circuit; $d(A) = d(A') = 3$, $d(B) = d(B') = 2$, $d(C) = d(C') = 3$, $d(D) = d(D') = 4$, $d(E) = d(E') = 1$. Hence the correspondence is: $A - A'$, ... $E - E'$. Therefore the graph is isomorphic.

27. Find out whether the following graphs are isomorphic or not.

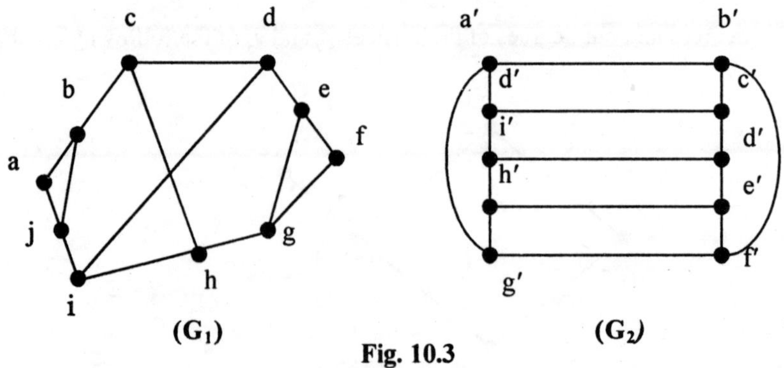

Fig. 10.3

Ans. Graphs G_1 and G_2 both have 10 vertices and 15 edges, and all vertices except a and f (these two are of degree 2) are of degree 3 in G_1. But G_2 has all vertices of degree 3. Also G_1

contains cycles (a, b, j, a) and (e, g, f, e) of length 3 while G_2 has no cycle of length 3. Hence, the two graphs are not isomorphic.

28. State whether the following graphs are isomorphic or not. Explain.

(i)

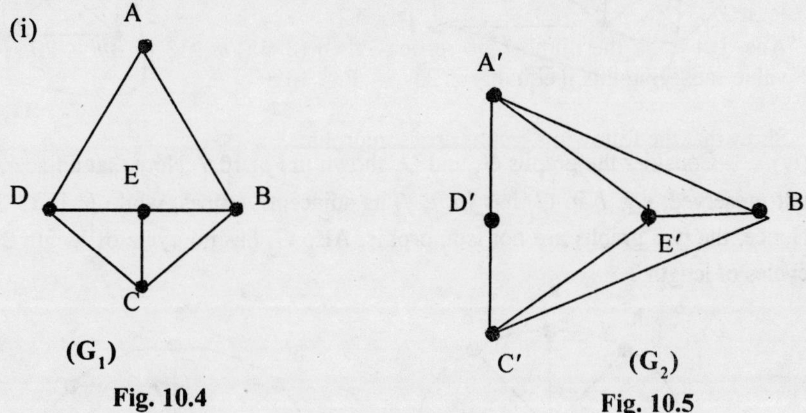

Fig. 10.4 Fig. 10.5

Ans. (i) Graph G_1 and G_2 are isomorphic; and the correspondence is A→D′, B→B′ D→A′, E→C′, C→E′. (See Fig.4)

(ii) The two given graphs are not isomorphic, since the vertices in G_2 are more than those in G_1.

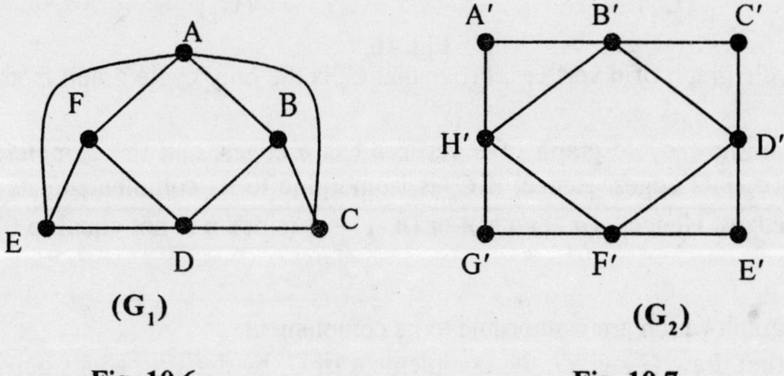

Fig. 10.6 Fig. 10.7

(iii) Consider the squares of the graphs G_1 and G_2. From the squares *ABEG* and *A′B′G′F′* in Fig.10.8 suppose we match *A-A′*, *B-B′*, *E-G*, *G-F*, and *D-D′*. Further match *H′*-*H′* being neighbours of A and *A′* respectively; similarly, since *F* is adjacent to E and *C* match *F-E′* and *C-C′*. Now, consider the vertices adjacent to *G*: they are (*C, A, D, E, H*) while the neighbours of *G′* are *F′*, *D′*, B′, E′, H′. Of these *H* matches with *H′* and *D* with *D′* while the remaining do not match. Hence, the two graphs are not isomorphic.

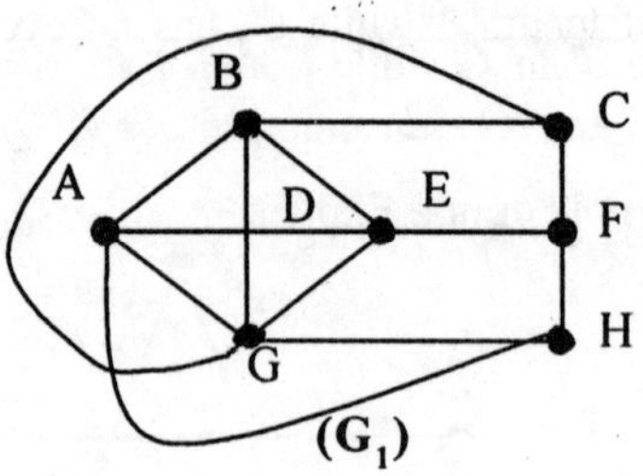

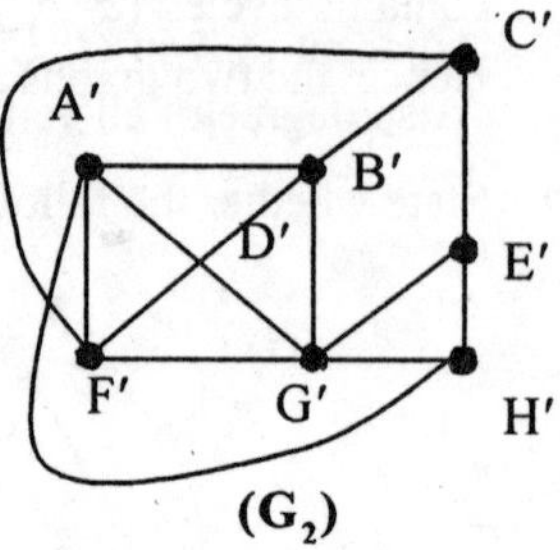

Fig. 10.8

(iv) Consider the graphs G_1 and G_2 shown in Fig.10.9. Note that adjacency of vertices is not preserved, e.g. *b* in G_1 has (*a, c, f*) as adjacent vertices, while *b′* in G_2 has (*a′, c′, h′,*). Hence, the two graphs are not isomorphic. Also G_1 has no cycle of length 3 while G_2 has 2 cycles of length 3.

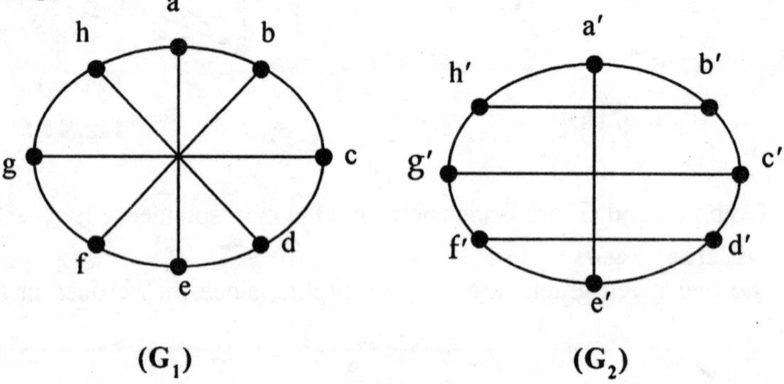

Fig. 10.9

29. Let C_n be a cycle graph of *n* vertices. Prove that C_5 is the only cycle graph isomorphic to its complement.

Ans. As we know cycle graph of *n* vertices has *n* edges, and the complete graph of *n* vertices has *n*(*n* -1)/2 edges. Hence, if C_n is isomorphic to its complement, its complement has *n*(*n* -1)/4 edges. Hence, *n*(*n* -1)/4 = *n* or (*n* -1) = 4 (since *n* is not equal to 0); hence, *n* = 5.

30. Draw a cycle graph which are isomorphic to its complement.

Ans. Let us first draw G and $\bar{G}$, the complement of G, by drawing edges between vertices which are non-adjacent in G. The vertices in $\bar{G}$ are labelled so as to correspond to those of G as follows: See Fig.10.10.

Vertices in G		Vertices in $\bar{G}$
A	→	A′
B	→	B′
C	→	C′
D	→	D′
E	→	E′

From the labelling we can observe that E' and B' are adjacent to A' in $\bar{G}$ while E and B are adjacent to A in G; C' and A' are adjacent to B' in $\bar{G}$ while C and A are adjacent to B in G. Also degree of all vertices are same. Hence G and $\bar{G}$ are isomorphic.

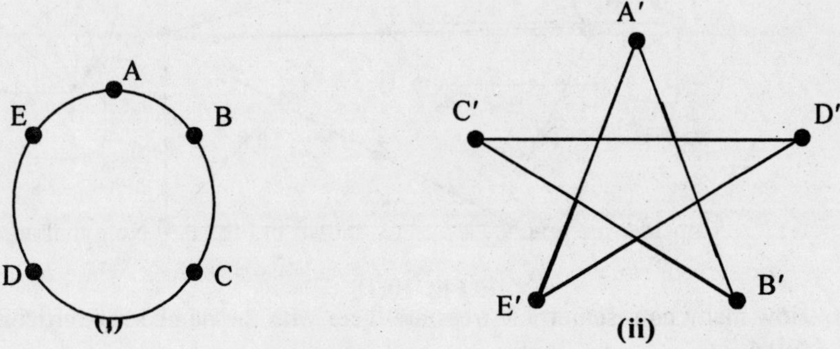

Fig. 10.10

31. Give 2 non-isomorphic graphs of order 6, size 6 and with degree sequence (2, 2, 2, 2, 2, 2).
Ans. The two graph G_1 and G_2 shown in Fig.10.11 are graphs of order 6, size 6 and degree sequence (2, 2, 2, 2, 2, 2). We can observe that G_1 is not isomorphic to G_2. Since G_1 has a cycle (a, b, c, d, e, f, a) while G_2 has no such cycle of length 6. Further neighbours of a in G_1 are b and f with no edge between f and b; in G_2 the neighbours of a' are no doubt d' and f' but there is an edge between them. In G_1, b and d are neighbours of c and there is no edge between them. In G_2 the neighbours of d' are e' and c' and there is an edge between them. Hence, there exists no one to one correspondence in the vertices, edges, degree sequences of the two graphs. Therefore, they are not isomorphic.

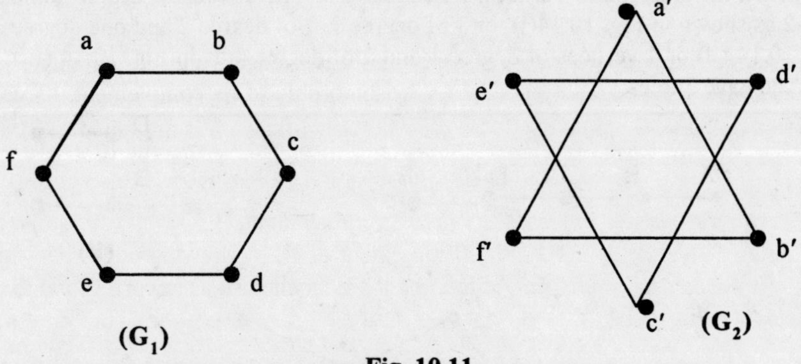

Fig. 10.11

32. (i) Define (*a*) walk (*b*) path for an undirected graph.
(ii) Define (*a*) simple path (*b*) elementary path for a digraph.

33. How many simple paths are there from v_1 to v_3 for the following digraph?
Ans. 4.

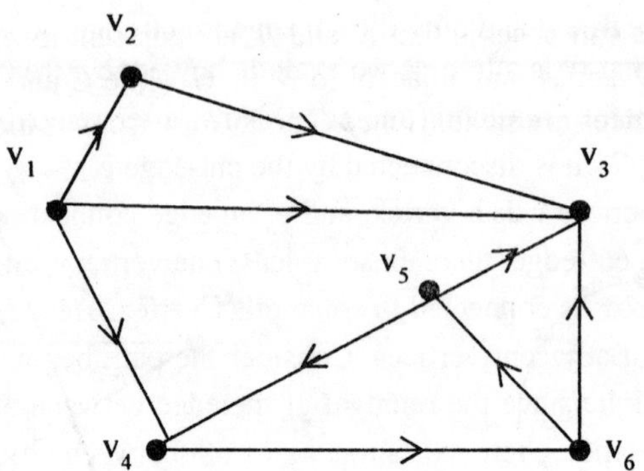

Fig. 10.18

34. How many non-isomorphic trees are there with the number of vertices equals to (i) 2 (ii) 3 (iii) 4 (iv) 5.

Ans. (i) 1.

(ii) 1.

(iii) When the number of vertices graph is 4, we can have either 2 vertices each of degree 1 and 2 of degree 2 as in Fig.10.13 (A) or 3 vertices each of degree 1 and one of degree 3 as in Fig.10.13(B). Hence, we have 2 such trees.

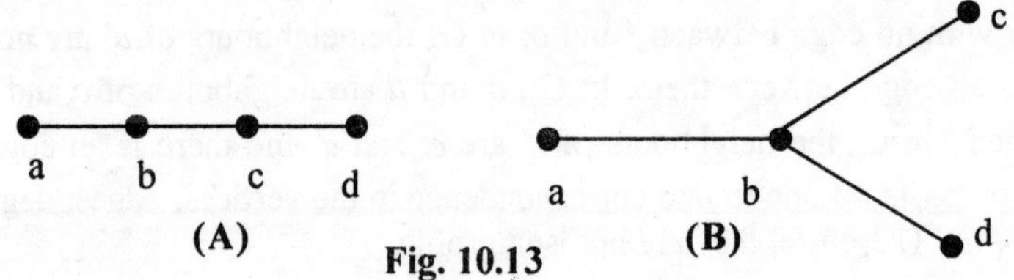

Fig. 10.13

(iv) When the graph has 5 vertices, we can have 2 vertices each of degree 1 and 3 of degree 2 as shown in Fig. 10.14(i), or 3 of degree 1, 1 of degree 2 and one degree 3 as in Fig. 10.14 (ii) or 4 vertices of degree 1 and 1 of degree 4 as in Fig. 10.14 (iii). Hence, we can have 3 trees.

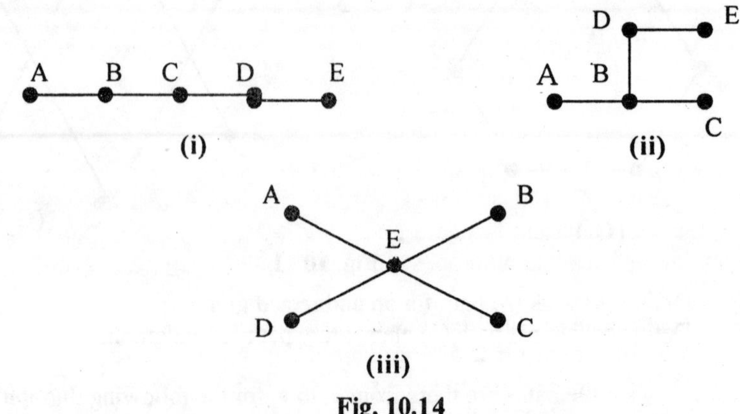

Fig. 10.14

35. Prove that a graph G is a tree if and only if each edge is cut-edge.
Proof. Let us assume that G is a tree. If we exclude any edge e from G, where v_1 and v_2 are adjacent vertices of e before removing the edge. Now, after removing the edge e, v_1 is not connected to v_2. Hence, $G - e$ is disconnected by the cut-edge e.

Conversely, suppose v_i is a vertex, and e_{ij} an edge connecting it with another vertex v_j. Since every edge is a cut-edge, there exists at least one vertex v_j and an edge e_{ij} connecting them. Thus, every vertex v_i is connected to some other vertex. Hence, G is connected. Let v_1, v_2 ... v_k is a sequence of adjacent vertices. Consider the path beginning with v_1 and ending with v_k ; this path is unique since the removal of the edge between any consecutive vertices disconnects the graph. Hence, there exists a unique path between any two vertices. Hence, G is a tree.

36. In any graph the number of vertices which have odd degree is an even number. True or False.
Ans. True

37. Suppose that a tree has m vertices of degree 1; 2, vertices of degree 2; 4 vertices of degree 3 and 3 vertices of degree 4. Find m.
Ans. Suppose the graph has e number of edges and m number of vertices. Then $m + (2 \times 2) + (4 \times 3) + (3 \times 4))$ gives sum of degrees $= 2e = 2(m-1)$ or $2(m-1) = 2(m + 9 - 1) = 2(m + 8)$, i.e. $m + 28 = 2m + 16)$, giving $m = 12$.

38. Suppose a tree has n_1 vertices of degree 1, n_2 vertices of degree 2, ... n_k vertices of degree k, find n_1 in terms of $n_2, n_3, \ldots n_k$.
Ans. $n_1 + 2n_2 + 3n_3 + \ldots k\,n_k = 2(n_1 + n_2 + n_k - 1)$, giving
$n_1 = (3 - 2)\,n_3 + (4 - 2)\,n_4 \ldots (k - 2)\,n_k + 2$.

39. What is a self-complimentary graph?
Ans. A graph is said to be **self-complementary** if it is isomorphic to its complement.

40. Define graphical.
Ans. An ordered n-tuple $(x_1, x_2, \ldots, x_n)$ of non-negative integers is said to be **graphical** if there exists a linear graph with no self-loops that has n vertices with the degree of vertices being $x_1, x_2, \ldots, x_n$.

41. Define Dominating and Minimal dominating set.
Ans. A **dominating set** is a set of vertices in an undirected graph in which every vertex not in the set is adjacent to one or more vertices in the set. A minimal dominating set is a dominating set such that no proper subset of it is also dominating.

42. Define Independent set and Maximal Independent Set.
Ans. An **independent set** is a set of vertices in an undirect graph such that no two vertices are adjacent. A **maximal independent set** is an independent set, which will no longer be one when any vertex is added to the set.

43. Define Outclassed Group.
Ans. A nonempty subset of the vertices of a complete directed graph G is said to be an **outclassed group** if any edge joining a vertex in the subset and a vertex not in the subset is always directed from the latter to the former.

44. Define spanning tree of a graph *G*.

45. If *G* is a connected graph, prove that any circuit and the complement of any spanning tree have a common edge.
Ans. Suppose *a, b, c* are three vertices on the spanning tree *T* with a circuit in *G*. Let *abca* is a circuit in *G*. Then, *c a* is a component of $\bar{T}$ as well as the circuit. In fact, $\bar{T}$ contains all edges which together with those in *T* make a circuit.

46. Define the terms: center of a tree, diameter and radius.

47. Define the terms: binary tree, edison tree.

48. Define proper ancestor and proper descendent in a digraph.
Ans. If there is a directed path from *u* to *v*, *u* is called a **proper ancestor** and *v* a **proper descendent.**

49. How many different non-isomorphic trees are possible for a graph of order 4? Draw all of them.
Ans. The sum of the degrees of the 4 vertices equals 2 (*e*) = 2(*n*-1) = 2*n* – 2 = 8 – 2 = 6. Hence, the degrees of the four vertices are (2, 2, 1, 1) or (3, 1, 1, 1). They are drawn as shown in Fig. 10.15.

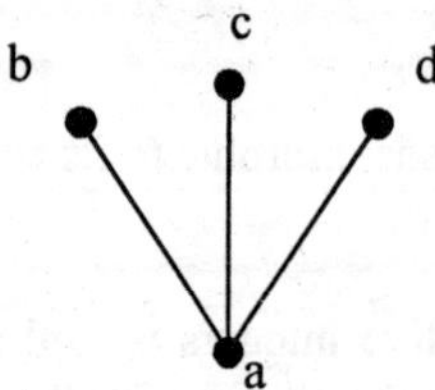

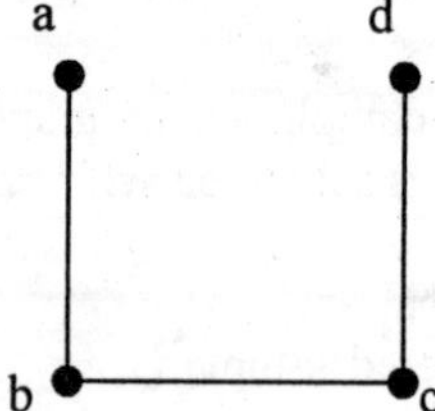

Fig. 10.15

50. Define the minimal spanning tree?

51. Prove that if a connected graph has only one spanning tree, then *G* itself is a tree.
Ans. Let *G* has a cycle c_1, c_2, ... c_{k-2}, c_{k-1}, c_k, where c_1, c_2, ... c_{k-2}, c_{k-1}, c_k is a path with c_1, c_2, c_3, ..., c_{k-1}, c_k as branches of the given tree. Then, by removing the edge c_{k-1} c_k from the given tree, we will have a second tree with branches c_1 ,c_k and c_1, c_2, ... c_{k-2}, c_{k-1}. But this contradicts the hypothesis that *G* has only one spanning tree. Hence, *G* has no cycles and being connected, *G* itself is a tree.

52. The number of internal vertices in a binary tree with n vertices is ..... .

Ans. $\left(\dfrac{n-2}{2}\right)$.

53. (i) Define (*a*) rooted tree (*b*) ordered tree.
(ii) Define (*a*) binary tree (*b*) full binary tree.

54. (i) Define the terms directed forest, parent, child and proper descendent of a directed tree.
(ii) Define a sub-tree in a directed forest.
(iii) What is a leaf ?

55. Define a B-tree of order k and show that a B-tree of order k and height h has at least $\lceil (k/2) \rceil^{h-1}$ leaves for $h \geq 1$.

Ans. A B-tree of order k is a directed tree s.t.

(i) All the leaves are at the same level;
(ii) Every internal vertex except possibly the root has at least $\lceil (k/2) \rceil$ children where $\lceil (x) \rceil$ means the least integer greater than or equal to x;
(iii) The root is a leaf or has at least two children
(iv) No vertex has more than k children.

An example is given in Fig.10.16.

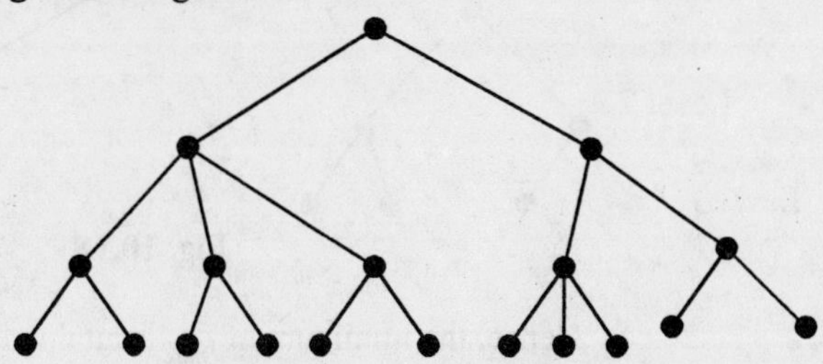

Fig. 10.16 B-Tree with order 3 or 4

Let us assume that a B-tree of order k and height (h-1) has at least 2 vertices. Then, by axiom (ii) above each of these leaves give rise to at least $\lceil (k/2) \rceil$ children at height h. Hence, the number of vertices at height h is at least $2\left\lceil \dfrac{k}{2} \right\rceil^{h-1} \cdot \left\lceil \dfrac{k}{2} \right\rceil^{h-1} = 2.\lceil (k/2) \rceil^{h-1}$. Also, for height, the B-tree has at least $2\lceil k/2 \rceil^{0} = 2$ children. Hence, the result follows by induction.

56. Prove that a graph G is a tree iff G is connected and $e = n - 1$, where e is the number of edges and n is the number of vertices.

57. Define a binary tree.

Ans. A tree in which there is exactly one vertex of degree 2 and each of the remaining vertices is of degree 1 or 3 is called a binary tree. See Fig.10.17.

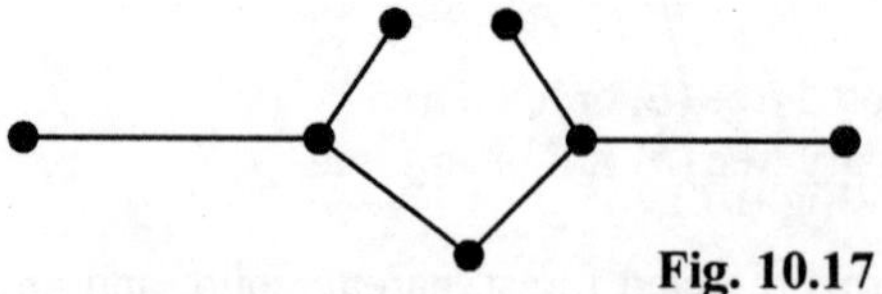

Fig. 10.17

58. Show that the number of vertices of degree 1 in a binary tree is $(n + 1)/2$.

59. Define the height or order of a binary tree and derive expressions for its maximum and minimum values for a n-vertex binary tree.
Ans. The maximum level of a binary tree G from its root is called the height or order of G. It is denoted by l_{max}. At level 0 we have 1 vertex or 2^0 vertices, at level 1, we have 2 vertices or 2^1 vertices; at level 3, we have 4 vertices or 2^2 vertices; at level k, we have 2^k vertices, where k is the maximum level.

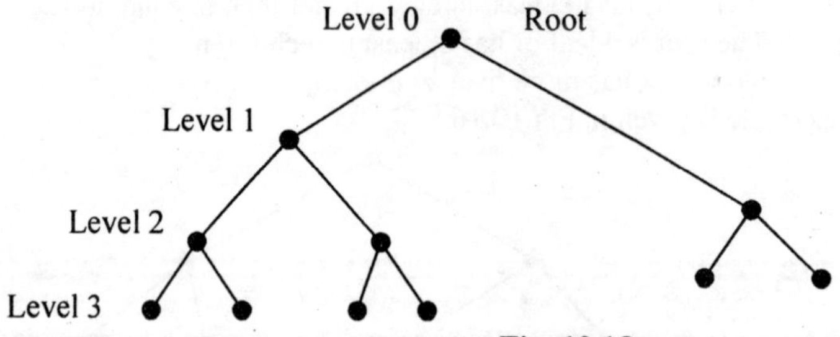

Fig. 10.18

Hence, $1 + 2^1 + 2^2 + \dots 2^k \geq n$. Hence, the minimum of l_{max} is attained when $1 + 2^1 + 2^2 + \dots 2^k = n$ or when $2^{k+1} - 1 = n$ or $k = [\log_2 (n + 1) - 1]$indicates the smallest integer greater than or equal. Max of l_{max} satisfies following relation, since all the vertices except the root have exactly 2 edges.

$$2 \max . l_{\text{max}} = n - 1 \quad \text{or} \quad \max . l_{\text{max}} = (n - 1)/2 .$$

60. Define a cut-set.

61. If a minimal set of edges S contains at least one branch of every spanning tree of a connected graph G, then it will be a cut-set of G.
Ans. Let $U = S \cap ST$, where ST is the set of all spanning trees of G. Then, $S - U$ disconnects G. Since S is the minimal set such that $n[S - U] = 0$ and since $U \leq S$, it follows that $U = S$. Thus, S is the minimal set of edges disconnecting G and hence is a cut-set of G.

62. Give an example of a binary tree which is not a full binary tree.

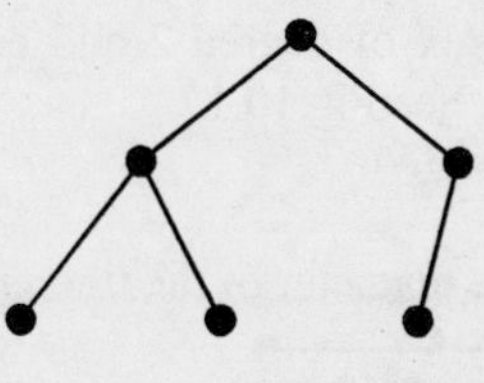

Fig. 10.19

63. Find the number of ways in which the four faces of a pyramid can be painted with 2 colours?
Ans. Here, each face can be painted with two colours in 2-ways. As there is no other restriction, the 4 faces can be painted in 2^4 ways.

64. Show that every circuit has an even number of edges in common with a cut-set.
Ans. Let S be the cut-set of *G*. If we remove *S* from *G* it partitions the vertices of *G* into two sets G_1 and G_2 as shown in Fig. 10.20.

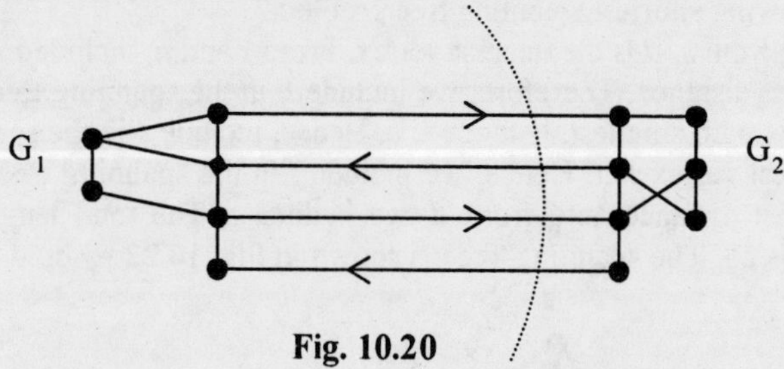

Fig. 10.20

Consider a circuit *C*. If all the vertices of *C* fall in either G_1 or G_2, then the number of edges it has common with *S* is zero. On the other hand, if *C* has some vertices in G_1 and some in G_2, then for each directed edge connecting a vertex in G_1 to G_2, there is a directed edge connecting a vertex in G_2 to a vertex in G_1. Hence, the number of edges common to *S* and *C* must be even.

65. Define the center of a tree.

66. What do you mean by the height of a tree?

67. (i) Find the centre (*s*) of the following tree.

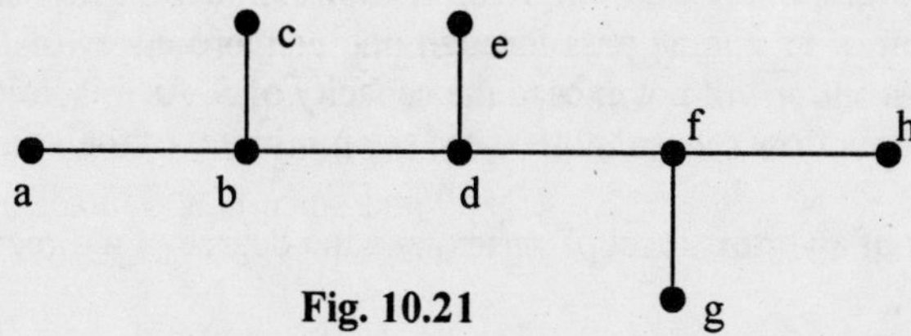

Fig. 10.21

Ans. *d*

(ii) What is the height of the tree given in (i)?

Ans. 4

(iii) Define diameters of a tree. What is the diameter of the tree given in (i) .

Ans. 4

68. (i) Define the edge connectivity of a connected graph.

(ii) Define the vertex connectivity of a graph.

69. A complete non-directed graph of *n* vertices is such that an edge exists between any pair of vertices. Prove that there are $n(n-1)/2$ edges in a complete graph.

Ans. Any two vertices out of *n* vertices can be chosen in ${}^nC_2$ ways, and each such choice gives one edge. Hence, there are $n(n-1)/2$ edges in a complete graph.

70. Solve the following shortest spanning tree problem?

Ans. Starting from *s, a* is the nearest vertex. From *s* and *a*, included in the spanning tree, *ab* has the smallest distance. Therefore, we include *b* in the spanning tree. From the vertices (*s, a, b*), the vertex with smallest distance is *c*. Hence, include *c* in the spanning tree. From (*s, a, b, c*), the nearest vertex is *f*. Hence, we include *f* in the spanning tree. From (*s, a, b, c, f*), *d* has the smallest distance, and from *d*, we include *t*. The total length of the edges in the spanning tree is 26. The spanning Tree is shown in Fig. 10.22 by bold lines.

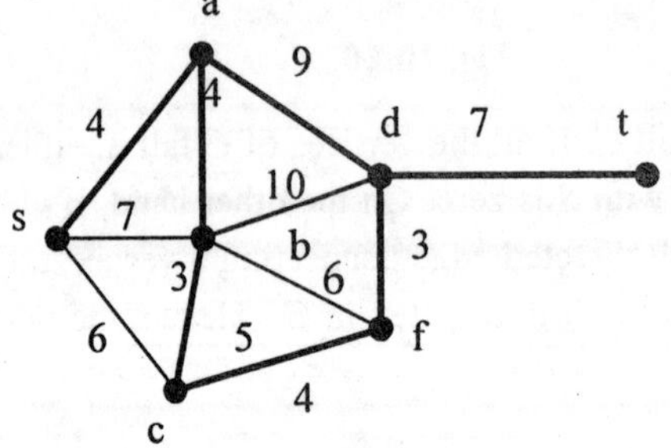

Fig.10.22

71. The maximum flow between any two vertices is equal to the minimum of the capacities of all the cut-sets with these two vertices.

Proof. Let us consider any cut-set *C* with respect to the vertices *u* and *v* in *G*. In the sub-graph *G* - *C* (i.e. the sub-graph left after removing *C* from *G*), there is no path between *u* and *v*. Thus, every flow from *u* to *v* must pass through one or more edges of *C*. Therefore, the total flow rate between *u* and *v* will not exceed the capacity of *S*. As this holds for all cut-sets with respect to *u* and *v*, this flow rate cannot exceed the minimum of their capacities.

72. Define the terms height of a vertex, level of vertex, and the degree of a directed tree.

73. Define a complete binary tree.

74. Show that the maximum number of edges in a complete bipartite graph is $n^2/4$.
Ans. As we know in a complete bipartite graph, every block has $n/2$ vertices, and each vertex in a block has $n/2$ edges connecting it to the $n/2$ vertices of the other block. Therefore, there are $n^2/4$ edges. Now to show that the maximum number of vertices in a complete graph is $n/2$. Let n_l be the vertices in one block and $(n - n_l)$ be the number of vertices in the other block. Then, the total number of edges is given by $f(n_l) = n_l\ (n - n_l) = n_l n - n_l^2$. $f'(n_l) = n - 2\,n_l = 0$ gives $n_l = n/2$, and this is obviously the maximum.

75. Show that in a complete binary tree the vertices at any level m are given by the label numbers ranging from $(2^{m+1} - 1)$ or from 2^m to n if n is less than $(2^{m+1} - 1)$.
Ans. Let us use principle of induction to prove this. Suppose that at level $(m - 1)$, we have (2^{m+1}) to (2^{m-1}) vertices. As we know that the vertices at level m are the children of vertices at level $m - 1$, vertex with index at level $(m - 1)$ produce at level m, the following vertices if $n = 2^{m+1} - 1$.

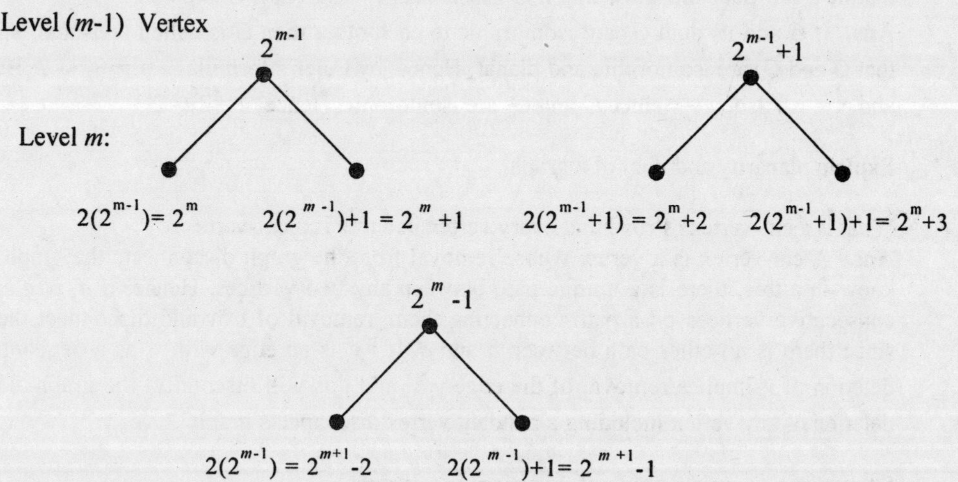

If n is less than $(2^{m+1} - 1)$, the index number at level m will terminate with n = one of the index number less than $(2^{m+1} - 1)$.

76. Draw a binary tree with level order indices: 1, 2, 3, 4, 5, 6, 7, 8, 10, 11, 15, 20, 30.
Ans. See Fig. 10.23.

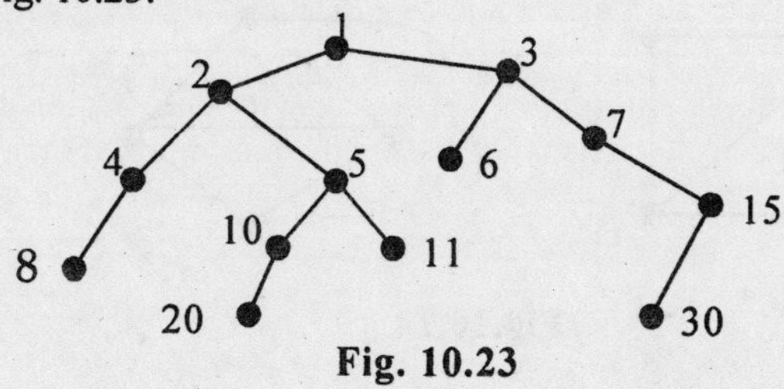

Fig. 10.23

77. Show that $K_{3,3}$ is non-planar.

78. Show that if G is planar with number of vertices $v \geq 11$ then the complement of G is non-planar.

Ans. Let us assume that both G and its complement are planar. Then G must satisfy $e \leq 3v - 6$, where e is the number of edges of G. Let e' be the number of edges of its complement $\bar{G}$. If $\bar{G}$ is also planar, then $e' \leq 3v - 6$. Hence $e + e' \leq 6v - 12$ or $v(v - 1)/2 \leq 6v - 12$ giving $v^2 - 13v + 12 \leq 0$ or $v \leq 12$ since v is not less than 1.

This together with $v \geq 11$ shows that v is equal to either 11 or 12. With $v = 11$ $v(v - 1)/2 = 55$ and $e' \leq 3 \cdot 11 - 6 = 27$ so that $e' \geq 28$, but if $\bar{G}$ is planar $e' \leq 3 \cdot 11 - 6 = 27$, but this leads to a contradiction. If $v = 12$, then $v(v\text{-}2)/2 = 66$, and if $\bar{G}$ is planar $e' \leq 3 \cdot 11 - 6 = 30$ while G being planar e is also less than or equal to 30 forcing $e' \geq 36$, thus leading to a contradiction. Hence, $\bar{G}$ is non-planar.

79. Define a self-dual and show that if G is self-dual $e = 2n - 2$, if G is planar.

Ans. If G and its dual G^* are isomorphic to each other then G is called self-dual. We know that G and G^* are isomorphic and planar. Hence, by Euler's formula. $e^* = v + r^* - 2$. But $e = e^*$ and $r^* = v$. Also, $v^* = v$ since G and G^* are isomorphic. Hence, $e = 2v - 2$.

80. Explain planarity and dual of a graph.

81. What is a cut-vertex? Prove that every vertex in a tree is a cut-vertex.

Ans. A cut-vertex is a vertex whose removal from the graph disconnects the graph. As we know in a tree, there is a unique path between any two vertices. Hence, if a, b, c are three consecutive vertices on a path connecting them, removal of b would disconnect the graph, since there is no other path between a and c. If v_iv_j is an edge with v_i as a pendant vertex, deletion of v_i implies removal of the edge $v_i\ v_j$ and this will disconnect the graph. Thus, the deletion of any vertex including a pendant vertex disconnects graph.

82. Show that a complete graph of 4 vertices is self-dual.

Ans. One can observe from the graphs G and G^* shown in Fig.10.24 that they are isomorphic, the correspondence being $v^*_i \leftrightarrow v_i$ for $i = 1$ to 4. Also $d(v_i) = d(v^*_i) = 3$ for $i = 1$ to 4, and adjacency of the vertices is preserved in both the graphs. Hence, G is isomorphic to G^* and hence G is self-dual.

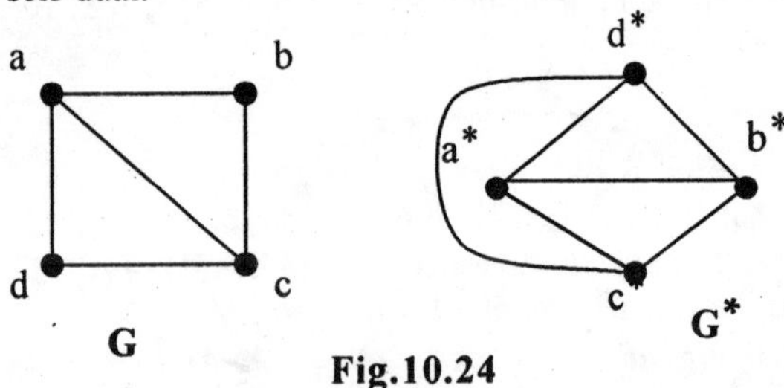

Fig.10.24

83. Draw a graph with edge connectivity 4, vertices connectivity 3 and vertex degree equal to or greater than 5.
Ans. See Fig. 10.25.

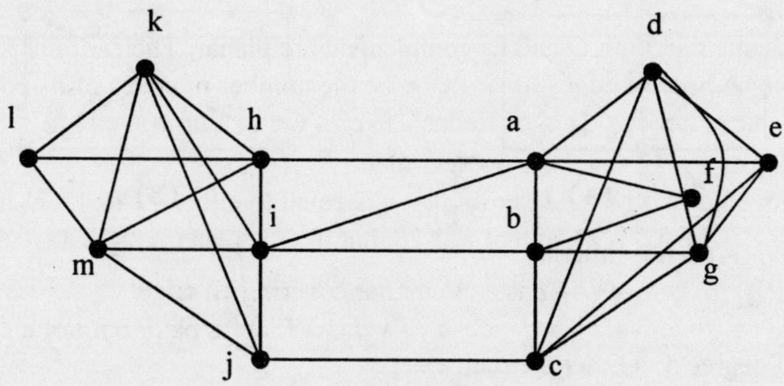

Fig.10.25

The cut-set for edge connectivity consists of: (*ha, ia, ib, jc*). The vertex-set for vertex connectivity is either (*a, b, c*) or (*h, i, j*). All vertices have degree equal to or greater than 5.

84. Is $K_{n,3}$ critical planar.

85. Prove that every vertex in a directed tree different from the root has a unique parent.
Ans. Suppose *u* be a vertex different from root in a directed tree. As *u* is not the root, there must be a vertex *v* from which there is a directed edge to *u*. If *v* is not unique, let *v′* be another vertex such the *v′u* is another edge. Then, the removal of *v′u* will not disconnect the graph. Since this contradicts the definition of a tree as well as a directed tree, there can not exist another vertex *v′* such that *v′u* is a directed edge. Hence, *v* is the unique parent of *u*.

86. Is there a bipartite graph with degree sequence (3, 3, 3, 5, 6, 6, 6, 6, 6)?
Ans. In a bipartite graph, if we have *m* vertices in one block and *n* in the other block, then there would be $m \cdot n$ edges. Hence, the degree sum is $2\,m\,n$, while $m + n = p$. In this problem, the degree sum is 50, while $p = 10$. Hence $2m(10 - n) = 50$ or $10\,m - n = 25$, or $n^2 - 10\,m + 25 = 0$. This gives $m = n = 5$ and the degree of each of the 10 vertices is 5. Hence, a bipartite graph with the given degree sequence is not possible.

87. Let $L(G)$ of a graph *G* be another graph which has a vertex for each edge in *G* and two of these vertices are adjacent if corresponding edges in *G* have common vertices. Show that $L(G)$ has a Euler's circuit if *G* has a Euler's circuit.

88. (i) Define a multi-graph.
(ii) Show that a non-directed multi-graph has a Euler's path if and only if it is connected and has 0 or exactly 2 vertices of odd degree, and in the latter case the vertices are the end-points of the Euler's path in the multi-graph.

89. Which of the following have Euler's path or circuits? See Fig.10.26(A) and (B).

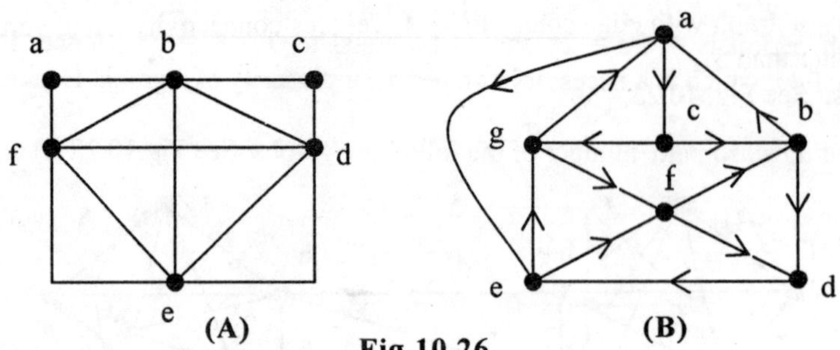

Fig.10.26

Ans. (i) Fig.10.26 (A) is not Eulerian.

(ii) In Fig.10.26(B) $b \to a \to c \to b$ is a circuit not covering all edges.

$c \to g \to a \to e \to g \to f \to b \to d \to e \to f \to d$ is a Euler's path, but not a circuit. Since all vertices are of degree 4, a Euler's circuit exists.

90. Define a Hamiltonian graph.

91. Show that the graph shown in Fig. 10.27 has neither a *H*-cycle nor a *H*-path.

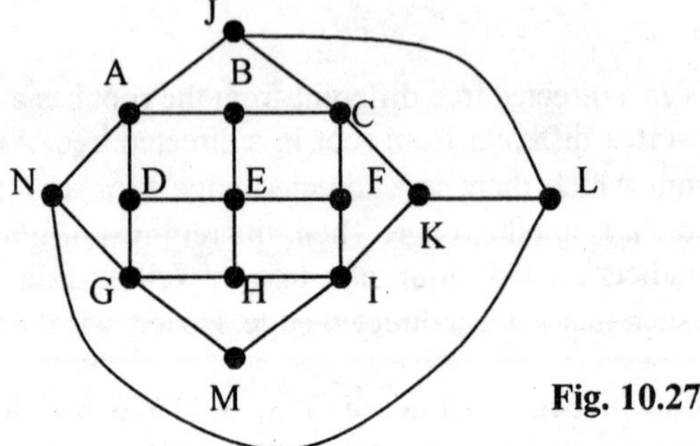

Fig. 10.27

Ans. Let at *E* we choose (*DE*), (*EF*) and remove (*EB*) and (*EH*). At *B* now its degree is 2. Hence, we must choose (*AB*), (*BC*) and similarly (*GH*) and (*HI*). At *A* suppose we choose (*AD*), then we have to delete (*AN*) and (*AJ*). Similarly, if we choose (*DG*), we have to remove (*GN*), (*GM*). Now, at *C* we can not choose (*CF*) since (*CFEDABC*) will be a cycle. Hence, we have to choose (*FI*) and remove (*IK*) and (*IM*). Now both at *J* and *K* their degree is 2. Hence, we have to include (*JC*) and (*CK*), which together with (*BC*) make the degree of $C = 3$. Hence, there exists no *H*-cycle, or a *H*-path.

92. What is the chromatic number of a graph?

93. State-five-colour theorem of a planar graph and prove it.

94. Show that a simple graph with 7 vertices each of degree 4 is non-planar.
Ans. Since $2e = 28,\ e = 14$.

$r = e - v + 2 = 14 - 7 + 2 = 9$. Therefore, we have 9 regions in G and, since the degree of each region is 4, $2e = 28$ should be equal to $9 \times 4 = 36$. Hence, Euler's formula is not satisfied, which is a necessary condition for planarity of a graph. Hence, G is non-planar.

95. Find the chromatic number of the following graphs. See Fig.10.28 (i) and (ii).

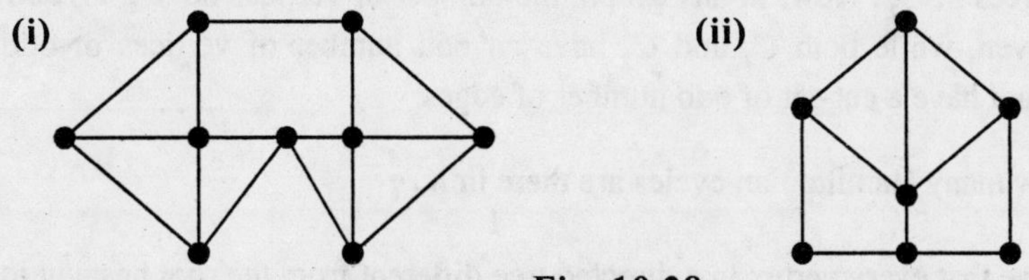

Fig. 10.28

96. Define the 4-colour problem. Give an example of a graph that needs 4 colours.

Ans. In a planar graph, the regions are said to be properly coloured if no two adjacent regions have the same colour. Two regions are said to be adjacent if there is a common edge between them.

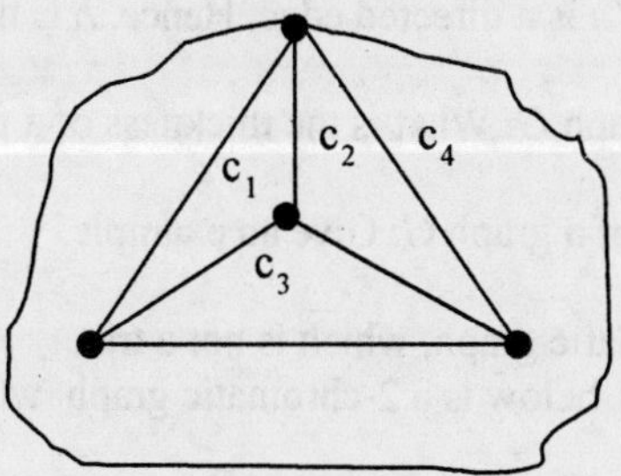

Fig. 10.29

The graph in Fig.10.29 is an example of a graph that needs 4 colours.

97. What is proper coloring of a graph? What is chromatic number?

98. Prove that every tree with n (number of vertices)equal to or greater than 2 is 2-chromatic.

Ans. Begin with any vertex and colour it 1, assuming that the tree is rooted at v. After this paint all vertices adjacent to it as 2. Then, colour all vertices adjacent as 1. This results in vertices at odd distances with colour 2 and vertices at even distances with colour 1. See Fig.10.30

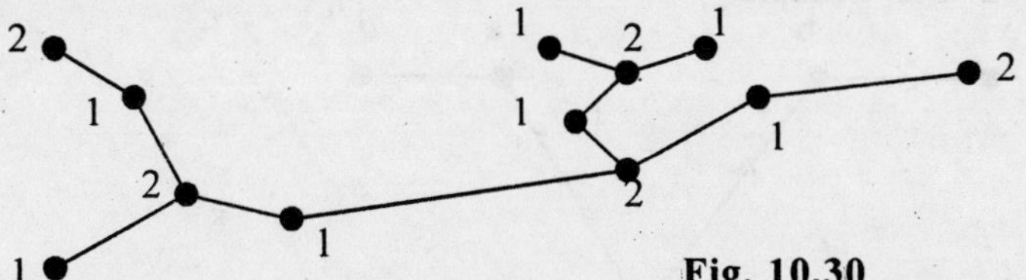

Fig. 10.30

99. Show that a Euler's graph can not have a cut-set of odd number of edges.

Ans. Let us assume that C is cut-set having an odd number of edges, say 3. After removing these edges, let the components be C_1 and C_2. Consider the 3 edges have end vertices u_1v_1, u_2v_2 and u_3v_3 then u_1 is in C_1 and v_1 is in C_2. Now, every vertex in G is of even degree since G is Eulerian. Therefore, after removing the 3 edges, each v_i would have an odd number of degrees in C_2. Now, in any graph, the number of vertices having an odd number of degrees is even, while both C_1 and C_2 have an odd number of vertices of odd degree. Hence, G cannot have a cut-set of odd number of edges.

100. How many Hamiltonian cycles are there in K_n ?

101. Prove that every vertex in a directed tree different from the root has a unique parent.
Ans. Let a be a vertex in a directed tree T. As a is not the root there must be a vertex b from which there is a directed edge to a. Let us suppose that b is not unique, let b' be another vertex such that $b'a$ is another edge. Then, the removal of $b'a$ will not disconnect the graph. Since this contradicts the definition of a tree as well as a directed tree, there can not exist another vertex b' such that $b'a$ is a directed edge. Hence, b is the unique parent of a.

102. Define the thickness of a Graph G. What is the thickness of a planar graph?

103. Define an incidence matrix of a graph G. Give an example.

104. Give an example of 2-chromatic graph, which is not a tree.
Ans. The graph in Fig.10.31 below is a 2-chromatic graph which is not a tree.

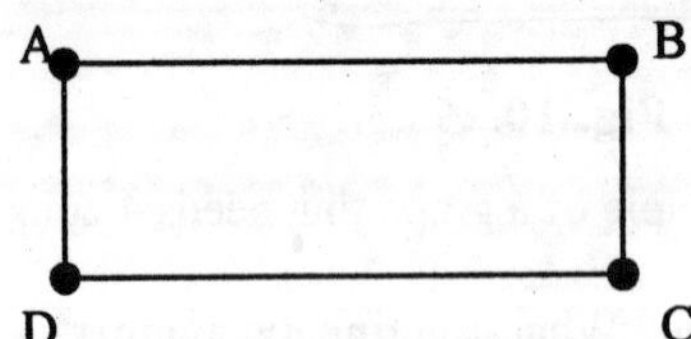

Fig. 10.31

105. What is an arborescence? Give an example.
Ans. A digraph is said to be an arborescence if
(a) it contains no directed circuits or semi-circuits
(b) there is one vertex r of precisely 0 in-degree (this vertex is called the root). Fig.10.32 is an example.

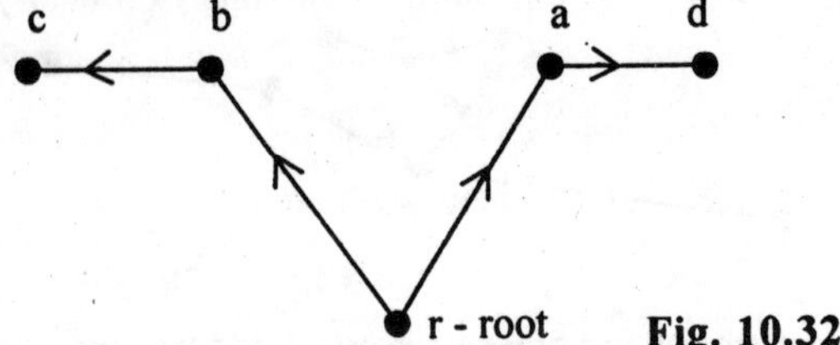

Fig. 10.32

106. Explain vertex connectivity and edge connectivity of a graph *G*. What is the relation between these connectivities?

107. How will you detect planarity? Explain.

108. The following digraph is strongly connected True or False?

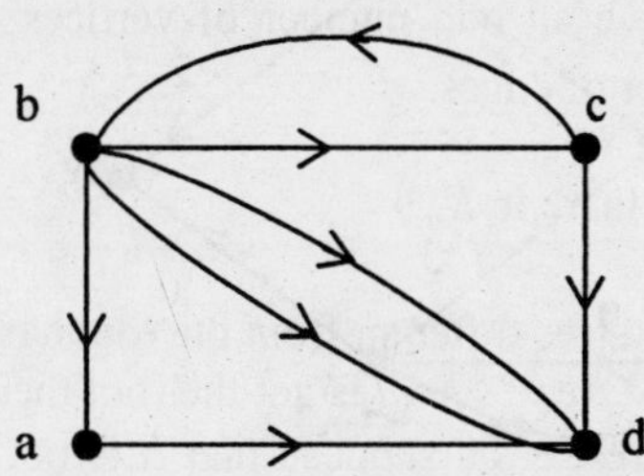

Ans. True

Fig. 10.33

109. Is the following graph (a) unilaterally connected (b) strongly connected?

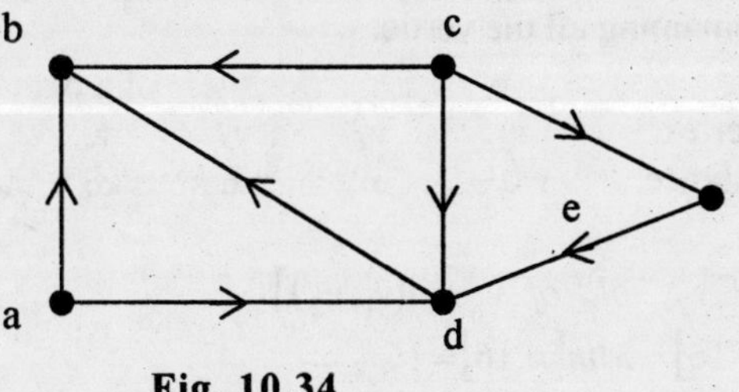

Fig. 10.34

Ans. (a) Yes (b) No.

110. Define chromatic polynomial of a graph *G*. Given an example.

111. Explain any one method of counting of labelled trees.

112. Prove that there are n^{n-2} labelled trees with *n* vertices ($n \geq 2$).

113. Prove that every tree with two or more vertices is 2- chromatic.

114. Define the distance between two spanning trees and show that it is a metric.

115. Discuss the relationship between the fundamental circuits and the fundamental cut-sets of a graph with reference to a spanning trees.

116. Define adjacency matrix of a graph G. Give an example.

117. *n* parallel lines in a plane are intersected by a family of *m* parallel lines. How many parallelograms are formed in the network thus formed.

Ans. $\cap\, C(m,2).\; C(n,2)$

118. Find the shortest path between vertex v_1 and v_7 with the help of Dijkstra's algorithm for the Fig. 10.35.

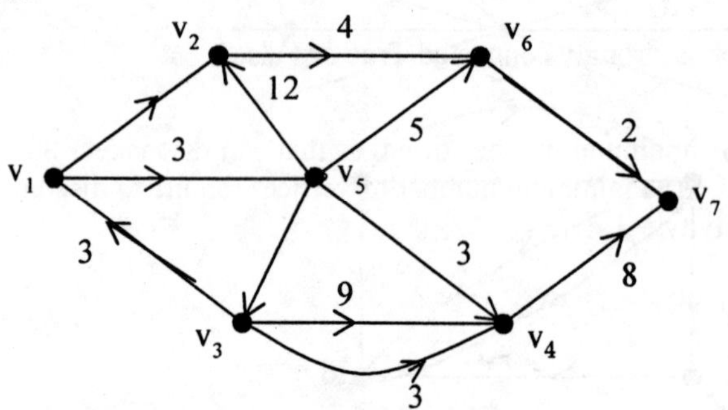

Fig.10.35

Ans. By applying Step 1 of Dijkstra algorithm assign permanent label zero to the starting vertex v_1 and ∞ to remaining all the vertices.

Vertex	v_1	v_2	v_3	v_4	v_5	v_6
Distance	0	∞	∞	∞	∞	∞

For vertex v_2

$$Min[old\ of\ v_2,\ \ old\ of\ v_1 + d(v_1, v_2)]$$
$$Min[\infty,\ 0+16] = Min[\infty, 16] = 16$$

Similarly for all the vertices we can calculate and finally we got following table

Vertex	v_1	v_2	v_3	v_4	v_5	v_6	v_7
Distance	0	∞	∞	∞	∞	∞	∞
	0	16	∞	∞	3	∞	∞
	0	15	5	6	3	8	∞
	0	15	5	6	3	8	∞
	0	15	5	6	3	8	14
	0	15	5	6	3	8	10

Thus shortest path or shortest spanning trees between vertex v_1 and v_7 is 10.

119. Find the dual of the given graph G.

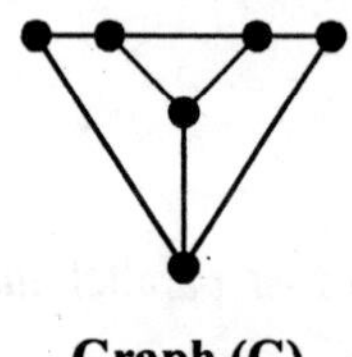

Graph (G)

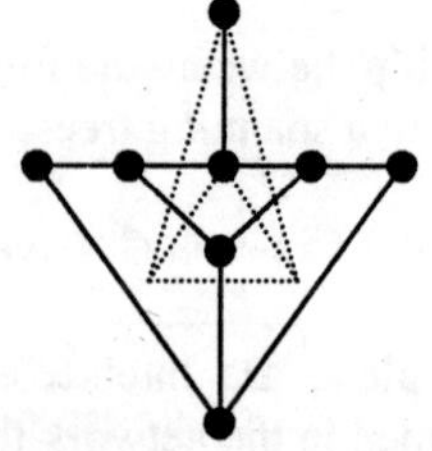

Dual of Graph (G)

120. Find the edge connectivity and the vertex connectivity of the following graph.

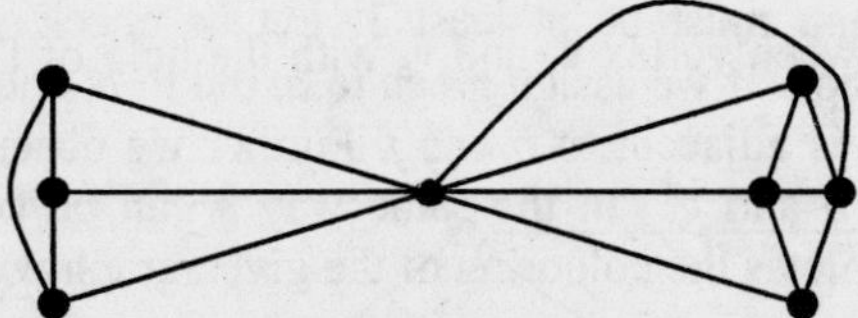

Ans. The removal of minimum number of edges that can disconnect the graph is 3. Hence edge connectivity = 3. The minimum number of vertices require to disconnect the graph is 1, hence vertex connectivity = 1.

121. What is chromatic number of K_5?

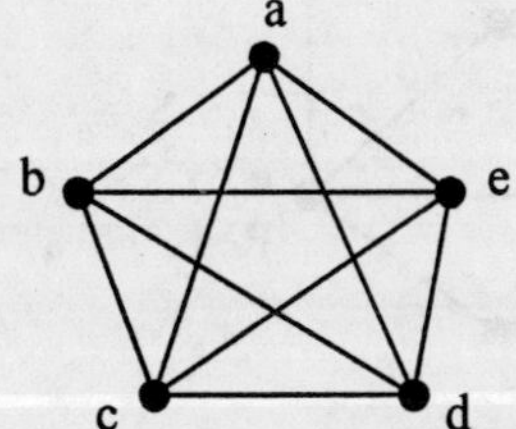

Ans. As in the complete graph, each vertex is adjacent to other vertices. Since in K_5 we need five colours for coluring of K_5. Hence chromatic number K_5 is 5. In general the chromatic number of K_n is n.

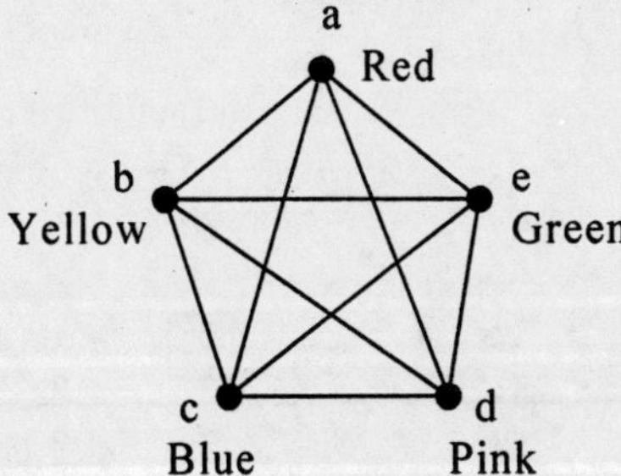

122. What is the chromatic number of the following graph.

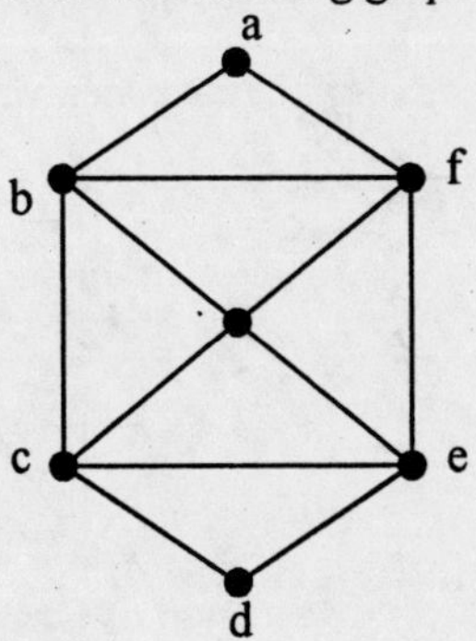

Ans. We can observe vertices *a*, *b* and *f* must be assigned different colours. Thus chromatic number of the given graph must be at least 3. Let us check whether three colours are sufficient to colour the graph. If we assign green to *a*, red to *b*, and yellow to *f*, then *g* can be coloured with green as it is adjacent to *b* and *f*. Further we observe that *c* can be coloured yellow, as it adjacent to *b* and *g*. In the same way *e* can be coloured red and *d* can be coloured green. This completes the colouring of the given graph with exactly 3 colours.

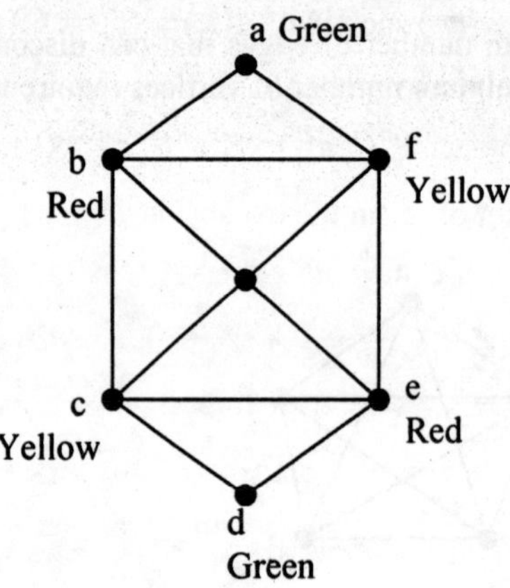

APPENDIX

SOLUTION OF COMBINATORIAL PROBLEMS USING GENERATING FUNCTIONS

Generating functions can also solve combinatorial problems. As we know combination of *r* distinct objects taken *n* at a time is given by $\frac{n!}{r!(n-r)!}$ denoted by ${}^nC_r$.

Now, consider the numeric functions $A(r)$ such that $A(r)={}^nC_r = C(n,r)$.

The generating function of $A(r)$ is given by

$$G(A,z)={}^nC_0+{}^nC_1z+{}^nC_2z^2+......+{}^nC_rz^r+......+{}^nC_nz^n=(1+z)^n \quad ...(1)$$

Let us consider the coefficient of z^r in the expansion $(1+z)^n$. In calculating the product of 'n' factors of $(1+z)$, each factor will contribute either a '1' on a '*z*'. Particularly, to make up the term z^r, r of the factors contribute a 'z' and $(n-r)$ of the factors contribute a '1' each. Hence, the coefficient of z^r is the number of ways of selecting *r* of the *n* $(1+z)$ factors to make up the term z^r that is equal to $C(n,r)$.

Now, if an object may be selected 0,1,2,...,*k* times or if there are *k* objects of a given kind, then (1) can be replaced by

$$G(A,z)=1+z+z^2+.......+z^k.$$

Now, we illustrate this with some examples to build generating functions for combinatorial problems.

Example 1

Using generating function, find the number of ways to select *r* objects from 10 objects $x_1, x_2 --- x_{10}$ among which x_1 can be selected at most twice, x_2 can be selected at most thrice and others can be selected only once.

Solution

We can choose the objects x_1 zero times, once or twice, then the contribution in generating function can be 1, *z* or z^2 respectively, So, the generating function for x_1 object will contain the facts ($1+z+z^2$). Similarly object x_2 can be selected zero times, once, twice or thrice. Hence generating function will also have the factor ($1+z+z^2+z^3$). The contribution of each of the eight factors ($1+z$) can 1 or z corresponding to selecting each of the remaining objects zero or once time. Hence generating function is $(1+z+z^2)(1+z+z^2+z^3)(1+z)^8$.

Corollary

(i) Setting $z=1$ in (1) we obtain ${}^nC_0+{}^nC_1+{}^nC_2+......+{}^nC_n=2^n$

(ii) Setting $z=-1$ in (1) we obtain

$${}^nC_0-{}^nC_1+{}^nC_2-{}^nC_3+......+(-1)^r\,{}^nC_r+......+(-1)^n\,{}^nC_n=0$$

or $${}^nC_0+{}^nC_2+......={}^nC_1+{}^nC_3+{}^nC_5+......$$

Example 2

Using generating function, find the number of ways of selecting six objects from three types of objects with repetition of upto four objects of each type.

Solution

The number of ways to select r objects from three types of objects with repetition of upto four objects of each type is $G(A,z) = (1+z+z^2+z^3+z^4)^3$.

Now we are required to find the number of ways of selecting six objects i.e. the coefficient of z^6 in $G(A,z)$.

We can write $$G(A,z) = \left(\frac{1-z^5}{1-z}\right)^3 = (1-z^5)^3(1-z)^{-3}$$

$$= (1-3z^5+3z^{10}-z^{15})(1-z)^{-3}$$

Coefficient of z^6 in $G(A,z)$ = Coefficient of z^6 in $(1-z)^{-3} - 3\times$ coeff. of z in $(1-z)^{-3}$•

Example 3

Suppose $A(r)$ denote the number of ways of selecting r objects from n objects with unlimited repetitions.

Solution

Since each object can be selected as many as we wish, $A(r)$ is equal to the coefficient of z^r in $\left(1+z+z^2+---\right)^n = \left(\frac{1}{1-z}\right)^n = (1-z)^{-n}$.

We get

$$A(r) = (-1)^r \frac{(-n)(-n-1.......(-n-r+1)}{r!}$$

$$= \frac{(n+r-1)}{r!(n-1)!} = (n+r-1, r).$$

Example 4

Using generating function, find the number of ways to select 10 balls from a large pile of red, white and blue balls if

(i) the selection has at least two balls of each colour,

(ii) the selection has at most two red balls,

(iii) the selection has an even number of blue balls.

• 3 is coefficient of z^5 in previous breaking × coefficient of z in $(1-z)^{-3}$ i.e. coefficient in z^6

$$= \frac{3.4.----8}{6!} - 3 = 25$$

Solution

Let n_1, n_2 and n_3 represent number of red, white and blue balls chosen respectively.

(a) If selection has at least two balls of each colour then

$$n_1 + n_2 + n_3 = 10, \qquad n_i \geq 2.$$

The generating functions is given by

$$G(A,z) = (z^2 + z^3 + ...)^3$$
$$= z^6\left(1 + z + z^2 + ...\right)^3 = z^6(1-z)^{-3}$$

$\therefore$ Coefficient of z^{10} in $G(A,z)$ = Coefficient of z^4 in $(1-z)^{-3}$

$$= {}^6C_4 = 15$$

(b) To select at most two red balls, the generating functions is given by

$$G(A,z) = \left(1 + z + z^2\right)\left(1 + z + z^2 + ...\right)^2$$

{For at most two red balls} {General solution: ways to select r objects}

$$= (1 + z + z^2)(1-z)^{-2}$$

$\therefore$ Coefficient of $z^{10} = {}^{11}C_{10} + {}^{10}C_9 + {}^9C_8 = 30$

(c) To select an even number of blue balls, the generating function is given by

$$G(A,z) = \left(1 + z + z^2 + ...\right)^2\left(1 + z^2 + z^4 + z^6 ......\right)$$

{Corresponds to the even number of blue balls}

$$= \left(1 + z^2 + z^4 + ...\right)(1-z)^{-2}$$

$\therefore$ Coefficient of z^{10} in $G(A,z)$ = Sum of coefficients of $z^{10}, z^8, z^6, z^4, z^2$ and 1 in $(1-z)^{-2} = {}^{11}C_{10} + {}^9C_8 + {}^7C_6 + {}^5C_4 + {}^3C_2 + 1 = 36$

Example 4

Use generating function to show that $\sum_{k-0}^{r} {}^mC_k . {}^nC_{r-k} = {}^{m+r}C_r$

Solution

Let $G(A,z) = (1+z)^n$ and $G(B,z) = (1+z)^n$

Then $A(r) = {}^nC_r$ and $B(r) = {}^mC_r$

Now $A(r)B(r) = (1+z)^{n+m}$

Coefficient of z^r in $G(A,z).G(B,z) = {}^{m+n}C_r$

$$\sum_{k=0}^{r} a_k b_{r-k} = \sum_{k=0}^{r} {}^nC_k . {}^mC_{r-k}$$

$$\therefore \quad \sum_{k=0}^{r} {}^{n}C_{k} \cdot {}^{m}C_{r-k} = {}^{m+n}C_{r}$$

Exercise

1. Using generating function, find the number of distributions of 12 identical objects into:
(i) Five different boxes with at most four objects in each box.
(ii) Seven different boxes with at least one object in each box.
(iii) Three different boxes with at most five objects in the first box.

Ans. (i) ${}^{16}C_{12} - 5\times{}^{11}C_{7} + 10\times{}^{6}C_{2}$

(ii) ${}^{11}C_{5}$

(iii) 63

2. Find a generating function for the number of ways to select five non-consecutive integers from 1,2,----,*n*. What is your answer for $n = 10$?

3. A set of *r* balls is selected from an infinite supply of red, blue, white & yellow balls. A selection must satisfy the condition that either the number of red balls is even and the number of blue balls is odd or the number of white balls is even and the number of yellow balls is odd. Let $A(r)$ denote the total number of such selection.
(a) Determine the generation function $G(A, z)$.
(b) Find a closed – form expression for $A(r)$. Evaluate the expression for $r = 23$.

4. Let $A(r)$ denote the number of ways to seat 10 students in a row of *r* chairs so that no two students will occupy adjacent seats. Determine the generating function of the discrete numeric function $(a_0, a_1, ---, a_r, ---)$.

5. Prove that
${}^{m}C_0 \times {}^{n}C_0 + {}^{m}C_1 \times {}^{n}C_1 + ----+ {}^{m}C_n \times {}^{n}C_n = {}^{m+n}C_n$ where $n \le m$.

6. Find the number of integer solutions to the equation $n_1 + n_2 + n_3 + n_4 = 7$ with $0 \le n_i \le 3$.

[**Hint:** Generating furction of the problem is $(1 + z + z^2 + z^3)^4$ and the solution is the coefficient of z^7].

7. Prove the relation ${}^{n}C_r = {}^{n-1}C_r + {}^{n-1}C_{r-1}$

INDEX